ENZYME INACTIVATION IN FOOD PROCESSING

Technologies, Materials, and Applications

Innovations in Agricultural and Biological Engineering

ENZYME INACTIVATION IN FOOD PROCESSING

Technologies, Materials, and Applications

Edited by

Megh R. Goyal, PhD
Junaid Ahmad Malik, PhD
Ravi Pandiselvam, PhD

APPLE ACADEMIC PRESS

First edition published 2023

Apple Academic Press Inc.
1265 Goldenrod Circle, NE,
Palm Bay, FL 32905 USA
760 Laurentian Drive, Unit 19,
Burlington, ON L7N 0A4, CANADA

CRC Press
6000 Broken Sound Parkway NW,
Suite 300, Boca Raton, FL 33487-2742 USA
4 Park Square, Milton Park,
Abingdon, Oxon, OX14 4RN UK

Apple Academic Press exclusively co-publishes with CRC Press, an imprint of Taylor & Francis Group, LLC

Library and Archives Canada Cataloguing in Publication

Title: Enzyme inactivation in food processing : technologies, materials, and applications / edited by Megh R. Goyal, PhD, Junaid Ahmad Malik, PhD, Ravi Pandiselvam, PhD.
Names: Goyal, Megh R., editor. | Malik, Junaid Ahmad, 1987- editor. | Pandiselvam, Ravi, editor.
Series: Innovations in agricultural and biological engineering.
Description: First edition. | Series statement: Innovations in agricultural and biological engineering | Includes bibliographical references and index.
Identifiers: Canadiana (print) 20230134920 | Canadiana (ebook) 20230135072 | ISBN 9781774911600 (hardcover) | ISBN 9781774911617 (softcover) | ISBN 9781003331797 (ebook)
Subjects: LCSH: Food industry and trade—Technological innovations. | LCSH: Food—Preservation. | LCSH: Enzymes.
Classification: LCC TP370 .E59 2023 | DDC 664—dc23

Library of Congress Cataloging-in-Publication Data

CIP data on file with US Library of Congress

ISBN: 978-1-77491-160-0 (hbk)
ISBN: 978-1-77491-161-7 (pbk)
ISBN: 978-1-00333-179-7 (ebk)

ABOUT THE BOOK SERIES: INNOVATIONS IN AGRICULTURAL AND BIOLOGICAL ENGINEERING

Under this book series, Apple Academic Press Inc. is publishing book volumes over a span of 8–10 years in the specialty areas defined by the American Society of Agricultural and Biological Engineers (<asabe.org>). Apple Academic Press Inc. aims to be a principal source of books in agricultural and biological engineering. We welcome book proposals from readers in areas of their expertise.

The mission of this series is to provide knowledge and techniques for agricultural and biological engineers (ABEs). The book series offers high-quality reference and academic content on agricultural and biological engineering (ABE) that is accessible to academicians, researchers, scientists, university faculty and university-level students, and professionals around the world.

Agricultural and biological engineers ensure that the world has the necessities of life, including safe and plentiful food, clean air and water, renewable fuel and energy, safe working conditions, and a healthy environment by employing knowledge and expertise of the sciences, both pure and applied, and engineering principles. Biological engineering applies engineering practices to problems and opportunities presented by living things and the natural environment in agriculture.

ABE embraces a variety of the following specialty areas (<asabe.org>): aquaculture engineering, biological engineering, energy, farm machinery and power engineering, food, and process engineering, forest engineering, information, and electrical technologies, soil, and water conservation engineering, natural resources engineering, nursery, and greenhouse engineering, safety, and health, and structures and environment.

For this book series, we welcome chapters on the following specialty areas (but not limited to):

- Academia to industry to end-user loop in agricultural engineering;
- Agricultural mechanization;
- Aquaculture engineering;
- Biological engineering in agriculture;

- Biotechnology applications in agricultural engineering;
- Energy source engineering;
- Farm to fork technologies in agriculture;
- Food and bioprocess engineering;
- Forest engineering;
- GPS and remote sensing potential in agricultural engineering;
- Hill land agriculture;
- Human factors in engineering;
- Impact of global warming and climatic change on agriculture economy;
- Information and electrical technologies;
- Irrigation and drainage engineering;
- Nanotechnology applications in agricultural engineering;
- Natural resources engineering;
- Nursery and greenhouse engineering;
- Potential of phytochemicals from agricultural and wild plants for human health;
- Power systems and machinery design;
- Robot engineering and drones in agriculture;
- Rural electrification;
- Sanitary engineering;
- Simulation and computer modeling;
- Smart engineering applications in agriculture;
- Soil and water engineering;
- Micro-irrigation engineering;
- Structures and environment engineering;
- Waste management and recycling;
- Any other focus areas.

For more information on this series, readers may contact:

Megh R. Goyal, PhD, PE
Book Series Senior Editor-in-Chief: Innovations in Agricultural and Biological Engineering
E-mail: goyalmegh@gmail.com

OTHER BOOKS ON AGRICULTURAL AND BIOLOGICAL ENGINEERING BY APPLE ACADEMIC PRESS, INC.

Management of Drip/Trickle or Micro Irrigation
Megh R. Goyal, PhD, PE, Senior Editor-in-Chief
Evapotranspiration: Principles and Applications for Water Management
Megh R. Goyal, PhD, PE and Eric W. Harmsen, PhD Editors

Book Series: Research Advances in Sustainable Micro Irrigation
Senior Editor-in-Chief: Megh R. Goyal, PhD, PE

- **Volume 1:** Sustainable Micro Irrigation: Principles and Practices
- **Volume 2:** Sustainable Practices in Surface and Subsurface Micro Irrigation
- **Volume 3:** Sustainable Micro Irrigation Management for Trees and Vines
- **Volume 4:** Management, Performance, and Applications of Micro Irrigation Systems
- **Volume 5:** Applications of Furrow and Micro Irrigation in Arid and Semi-Arid Regions
- **Volume 6:** Best Management Practices for Drip Irrigated Crops
- **Volume 7:** Closed Circuit Micro Irrigation Design: Theory and Applications
- **Volume 8:** Wastewater Management for Irrigation: Principles and Practices
- **Volume 9:** Water and Fertigation Management in Micro Irrigation
- **Volume 10:** Innovation in Micro Irrigation Technology

Book Series: Innovations and Challenges in Micro Irrigation
Senior Editor-in-Chief: Megh R. Goyal, PhD, PE

- Engineering Interventions in Sustainable Trickle Irrigation: Water Requirements, Uniformity, Fertigation, and Crop Performance
- Management Strategies for Water Use Efficiency and Micro Irrigated Crops: Principles, Practices, and Performance

- Micro-Irrigation Engineering for Horticultural Crops: Policy Options, Scheduling, and Design
- Micro-Irrigation Management: Technological Advances and Their Applications
- Micro-Irrigation Scheduling and Practices
- Performance Evaluation of Micro-Irrigation Management: Principles and Practices
- Potential of Solar Energy and Emerging Technologies in Sustainable Micro-Irrigation
- Principles and Management of Clogging in Micro-Irrigation
- Sustainable Micro-Irrigation Design Systems for Agricultural Crops: Methods and Practices

Book Series: Innovations in Agricultural and Biological Engineering
Senior Editor-in-Chief: Megh R. Goyal, PhD, PE

- Advanced Research Methods in Food Processing Technologies
- Advances in Food Process Engineering: Novel Processing, Preservation and Decontamination of Foods
- Advances in Green and Sustainable Nanomaterials: Applications in Energy, Biomedicine, Agriculture, and Environmental Science
- Advances in Sustainable Food Packaging Technology
- Analytical Methods for Milk and Milk Products, 2-volume set:
    - Volume 1: Sampling Methods, Chemical and Compositional Analysis
    - Volume 2: Physicochemical Analysis of Concentrated, Coagulated and Fermented Products
- Biological and Chemical Hazards in Food and Food Products: Prevention, Practices, and Management
- Bioremediation and Phytoremediation Technologies in Sustainable Soil Management, 4-volume set:
    - Volume 1: Fundamental Aspects and Contaminated Sites
    - Volume 2: Microbial Approaches and Recent Trends
    - Volume 3: Inventive Techniques, Research Methods, and Case Studies
    - Volume 4: Degradation of Pesticides and Polychlorinated Biphenyls
- Dairy Engineering: Advanced Technologies and Their Applications
- Developing Technologies in Food Science: Status, Applications, and Challenges

- Emerging Technologies in Agricultural Engineering
- Engineering Interventions in Agricultural Processing
- Engineering Interventions in Foods and Plants
- Engineering Practices for Agricultural Production and Water Conservation: An Interdisciplinary Approach
- Engineering Practices for Management of Soil Salinity: Agricultural, Physiological, and Adaptive Approaches
- Engineering Practices for Milk Products: Dairyceuticals, Novel Technologies, and Quality
- Enzyme Inactivation in Food Processing: Technologies, Materials, and Applications
- Field Practices for Wastewater Use in Agriculture: Future Trends and Use of Biological Systems
- Flood Assessment: Modeling and Parameterization
- Food Engineering: Emerging Issues, Modeling, and Applications
- Food Process Engineering: Emerging Trends in Research and Their Applications
- Food Processing and Preservation Technology: Advances, Methods, and Applications
- Food Technology: Applied Research and Production Techniques
- Functional Dairy Ingredients and Nutraceuticals: Physicochemical, Technological, and Therapeutic Aspects
- Handbook of Research on Food Processing and Preservation Technologies, 5-volume set:
    - Volume 1: Nonthermal and Innovative Food Processing Methods
    - Volume 2: Nonthermal Food Preservation and Novel Processing Strategies
    - Volume 3: Computer-Aided Food Processing and Quality Evaluation Techniques
    - Volume 4: Design and Development of Specific Foods, Packaging Systems, and Food Safety
    - Volume 5: Emerging Techniques for Food Processing, Quality, and Safety Assurance
- Modeling Methods and Practices in Soil and Water Engineering
- Nanotechnology and Nanomaterial Applications in Food, Health, and Biomedical Sciences
- Nanotechnology Applications in Agricultural and Bioprocess Engineering: Farm to Table

- Nanotechnology Applications in Dairy Science: Packaging, Processing, and Preservation
- Nanotechnology Horizons in Food Process Engineering, 3-volume set:
  - Volume 1: Food Preservation, Food Packaging and Sustainable Agriculture
  - Volume 2: Scope, Biomaterials, and Human Health
  - Volume 3: Trends, Nanomaterials, and Food Delivery
- Novel and Alternative Methods in Food Processing: Biotechnological, Physicochemical, and Mathematical Approaches
- Novel Dairy Processing Technologies: Techniques, Management, and Energy Conservation
- Novel Processing Methods for Plant-Based Health Foods: Extraction, Encapsulation and Health Benefits of Bioactive Compounds
- Novel Strategies to Improve Shelf-Life and Quality of Foods: Quality, Safety, and Health Aspects
- Phytochemicals and Medicinal Plants in Food Design: Strategies and Technologies for Improved Healthcare
- Processing of Fruits and Vegetables: From Farm to Fork
- Processing Technologies for Milk and Milk Products: Methods, Applications, and Energy Usage
- Quality Control in Fruit and Vegetable Processing: Methods and Strategies
- Scientific and Technical Terms in Bioengineering and Biological Engineering
- Soil and Water Engineering: Principles and Applications of Modeling
- Soil Salinity Management in Agriculture: Technological Advances and Applications
- State-of-the-Art Technologies in Food Science: Human Health, Emerging Issues and Specialty Topics
- Sustainable and Functional Foods from Plants
- Sustainable Biological Systems for Agriculture: Emerging Issues in Nanotechnology, Biofertilizers, Wastewater, and Farm Machines
- Sustainable Nanomaterials for Biomedical Engineering: Impacts, Challenges, and Future Prospects
- Sustainable Nanomaterials for Biosystems Engineering: Trends in Renewable Energy, Environment, and Agriculture
- Technological Interventions in Dairy Science: Innovative Approaches in Processing, Preservation, and Analysis of Milk Products

- Technological Interventions in Management of Irrigated Agriculture
- Technological Interventions in the Processing of Fruits and Vegetables
- Technological Processes for Marine Foods, From Water to Fork: Bioactive Compounds, Industrial Applications, and Genomics
- The Chemistry of Milk and Milk Products: Physicochemical Properties, Therapeutic Characteristics, and Processing Methods

ABOUT SENIOR-EDITOR-IN-CHIEF

Megh R. Goyal, PhD, PE

Megh R. Goyal, PhD, PE, is, currently a retired professor of agricultural and biomedical engineering from the General Engineering Department at the College of Engineering at the University of Puerto Rico–Mayaguez Campus (UPRM); and Senior Acquisitions Editor and Senior Technical Editor-in-Chief for Agricultural and Biomedical Engineering for Apple Academic Press Inc.

During his professional career, he has worked as a Soil Conservation Inspector; Research Assistant at Haryana Agricultural University and Ohio State University; Research Agricultural Engineer/Professor at the Department of Agricultural Engineering of UPRM; and Professor of Agricultural and Biomedical Engineering in the General Engineering Department of UPRM. He spent a one-year sabbatical leave in 2002–2003 at the Biomedical Engineering Department of Florida International University, Miami, USA.

Dr. Goyal was the first agricultural engineer to receive the professional license in agricultural engineering from the College of Engineers and Surveyors of Puerto Rico. In 2005, he was proclaimed the "Father of Irrigation Engineering in Puerto Rico for the Twentieth Century" by the American Society of Agricultural and Biological Engineers, Puerto Rico Section, for his pioneering work on micro irrigation, evapotranspiration, agroclimatology, and soil and water engineering.

During his professional career of 52 years, he has received many awards, including Scientist of the Year, Membership Grand Prize for the American Society of *Agricultural* Engineers Campaign, Felix Castro Rodriguez Academic Excellence Award, Man of Drip Irrigation by the Mayor of Municipalities of Mayaguez/Caguas/Ponce and Senate/Secretary of Agriculture of ELA, Puerto Rico, and many others. He has been recognized as one of the experts "who rendered meritorious service for the development of [the] irrigation sector in India" by the Water Technology Centre of Tamil Nadu Agricultural University in Coimbatore, India, and ASABE who bestowed on him the 2018 Netafim Microirrigation Award.

Dr. Goyal has authored more than 200 journal articles and edited more than 100 books. AAP has published many of his books, including *Management of Drip/Trickle or Micro Irrigation; Evapotranspiration: Principles and Applications for Water Management;* ten-volume set on *Research Advances in Sustainable Micro Irrigation.* He has also authored the textbooks *Elements of Agroclimatology* (Spanish) by UNISARC, Colombia, and two *Bibliographies on Drip Irrigation.* Dr. Goyal has also developed several book series with AAP, including Innovations in Agricultural & Biological Engineering, Innovations and Challenges in Micro Irrigation; and Innovations in Plant Science for Better Health: From Soil to Fork.

Dr. Goyal received his BSc degree in Engineering from Punjab Agricultural University, Ludhiana, India, and his MSc and PhD degrees from the Ohio State University, Columbus, Ohio, USA. He also earned a Master of Divinity degree from the Puerto Rico Evangelical Seminary, Hato Rey, Puerto Rico, USA.

Readers may contact him at goyalmegh@gmail.com.

ABOUT THE EDITORS

Junaid Ahmad Malik, PhD

Junaid Ahmad Malik, PhD, is a Lecturer with the Department of Zoology at Government Degree College, Bijbehara, Jammu and Kashmir, India, and is actively involved with teaching and research activities. He started his career as a Lecturer in the School Education Department, Govt. of Jammu and Kashmir for two years He now has more than eight years of research experience. His areas of interest are ecology, soil macrofauna, wildlife biology, conservation biology, etc. Dr. Malik has published over 20 research articles and technical papers in international peer-reviewed journals and has authored and edited books, book chapters, and more than 10 popular editorial articles. He also serves as an editor and reviewer of several journals. He has participated in several state, national, and international conferences, seminars, workshops, and symposia and has more than 20 conference papers to his credit. He is a life member of the Society for Bioinformatics and Biological Sciences.

Dr. Malik received his BSc (2008) in Science from the University of Kashmir, Srinagar, J&K; his MSc (2010) in Zoology from Barkatullah University, Bhopal, Madhya Pradesh; and PhD (2015) in Zoology from the same university. He completed his BEd program in 2017 from the University of Kashmir, Srinagar, Jammu and Kashmir, India.

Readers may contact him at: malik.junaidahmad@gmail.com.

Ravi Pandiselvam, PhD

Ravi Pandiselvam, PhD, is a scientist in the Physiology, Biochemistry and Post-Harvest Technology Division at the ICAR-Central Plantation Crops Research Institute, Kasaragod, India, where he contributes to developing value-added food products (viz., Kalpa Krunch and coconut-based frozen delicacies); to design and development of the tender coconut trimming machine, coconut testa removing machine, and tender coconut cutting machine; and to design preservation protocol for trimmed tender coconut, which are commercialized and sold to many entrepreneurs. In addition, he has authored or co-authored more than 60 international/national journals articles. He has been author or co-author of five books and over 30 book chapters. Recently he received a prestigious National Academy of Agricultural Sciences Young Scientist Award. He served as a guest editor and editorial board member for *Ozone Science and Engineering Journal* and *Journal of Food Process Engineering.* He is an associate editor of *Measurement: Food Journal* and an editorial board member of *Quality Assurance and Safety of Crops & Foods, Nutrition & Food Science*, and *Current Agriculture Research Journal.*

Readers may contact him at: anbupandi1989@yahoo.co.in.

CONTENTS

CONTRIBUTORS

S. Abdullah
PhD Research Scholar, Department of Food Process Engineering, National Institute of Technology, Rourkela 769008, Odisha, India; E-mail: abdullahfpe@gmail.com

Noshin Afshan
Assistant Professor, Department of Chemistry, Riphah International University, Faisalabad Campus, Faisalabad 38000, Pakistan; E-mail: noshinafshan@gcwuf.edu.pk

Roshan Ara
Research Assistant, Department of Biotechnology Engineering, Institute of Engineering and Technology, Bundelkhand University, Jhansi 284128, U.P., India; E-mail: roshanara_1998@gmail.com

G. N. Ashitha
PhD Research Scholar, Department of Processing and Food Engineering, Kerala Agricultural University (KAU), Thrissur 656680, Kerala, India; E-mail: ashithagn@gmail.com

Ardra C. Babu
M.Tech. Student, Department of processing and Food Engineering, Kelappaji College of Agricultural Engineering and Technology, Tavanur, Malappuram 679573, India; E-mail: ardracbabu@gmail.com

Sonia Bansal
Research Scholar, Institute of Molecular Biology, Academia Sinica, Taipei 11524, Taiwan; E-mail: sonya12323@gmail.com

Alina Bari
PhD Research Scholar, School of Chemistry and Chemical Engineering, Anhui Normal University, Wuhu, China; E-mail: alinabari19@gmail.com

Agrima Bhatt
Researcher, Department of Biotechnology, Biyani Girls College, Jaipur 302039, Rajasthan, India; E-mail: agrima2010@gmail.com

Manish Biyani
Professor, Department of Bioscience and Biotechnology, Japan Advanced Institute of Science and Technology, Ishikawa 9231292, Japan; E-mail: biyani@jaist.ac.jp

Jagadeesh Chandra Bose
Associate Professor, University Institute of Biotechnology, Chandigarh University, NH-95, Ludhiana 140413, Chandigarh, Punjab, India; E-mail: jagadeesh.e8748@cumail.in

Mariyam Chaudhary
PhD Research Scholar, University Institute of Biotechnology, Chandigarh University, NH-95, Ludhiana140413, Chandigarh, Punjab, India; E-mail: mariyam.a.chaudhary@gmail.com

Namrata Deka
PhD Research Scholar, Jacob's School of Medicine and Biomedical Sciences, State University of New York at Buffalo, New York 14203, USA; E-mail: deka.namrata24@gmail.com

B.L. Dinesha
Research Associate, Department of Agricultural Engineering, College of Horticulture, Kerala Agricultural University (KAU), Vallanikkara, Thrissur 680656, Kerala, India; E-mail: dinirbdgtc@gmail.com

Wasiya Farzana
PhD Research Scholar, Department of Agricultural Processing and Food Engineering, SV College of Engineering and Technology and Research Station, IGKV, Raipur, India; E-mail: wasee92@gmail.com

Rashi Garg
Researcher, Department of Biotechnology, Biyani Girls College, Jaipur 302039, Rajasthan, India; E-mail: rashigarg8599@gmail.com

Madhuri Girdhar
Assistant Professor, Lovely Professional University, Phagwara 144401, Punjab, India; E-mail: madhuri.23745@lpu.co.in

Megh R. Goyal
Senior Editor-in-Chief (Agriculture and Biomedical Engineering) for AAP; Retd. Professor in Agric. & Biomedical Eng., University of Puerto Rico-Mayaguez, USA; E-mail: goyalmegh@gmail.com

Annika Durve Gupta
Assistant Professor, Department of Biotechnology, B.K. Birla College (Autonomous), Kalyan 421304, Thane, Maharashtra, India; E-mail: annikadurve@yahoo.com

Saqib Hassan
Research Associate (ICMR), Department of Microbiology, School of Life Sciences, Pondicherry University, Puducherry 605014, India; E-mail: hassansaqib1@gmail.com

Habiba Ibrahim
Department of Biotechnology, Faculty of Science, Ain Shams University, Cairo 11566, Egypt; E-mail: habibaibrahim12310@yahoo.com

Rakesh R. Jadhav
Assistant Professor, Dr. B.S. Konkan Krishi Vidyapeeth (Agricultural University), Ratnagiri 415629, Maharashtra, India; E-mail: rakeshcof169@gmail.com

T. Jayasree Joshi
PhD Research Scholar, Department of Agriculture and Food Engineering, Indian Institute of Technology, Kharagpur 721302, India; E-mail: jayasreejoshi48@gmail.com

B. Kamalapreetha
Assistant Professor, Department of Agriculture Engineering, Mahendra Engineering College, Mallasamudrum, Namakkal 637503, Tamil Nadu, India; E-mail: kpreetha25@gmail.com

Anandu Chandra Khanashyam
Post Graduate Fellow, Department of Food Science and Technology, Kasetsart University, 50 Ngamwongwan Road, Ladyao, Chatuchak 10900, Bangkok, Thailand; E-mail: Khanashyamac@gmail.com

Roshna Khawas
M.Tech Student, Department of Agricultural Engineering, Assam University, Silchar 788011, Cachar District, Assam, India; E-mail: khawashroshna60@gmail.com

George Seghal Kiran
Assistant Professor, Department of Food Science and Technology, School of Life Sciences, Pondicherry University, Puducherry 605014, India; E-mail: seghalkiran@gmail.com

Anjineyulu Kothakota
Scientist, Agro-Processing & Technology Division, CSIR-National Institute for Interdisciplinary Science and Technology (NIIST), Trivandrum 695019, Kerala, India; E-mail: anjineyuluk@niist.res.in

Tarun Kumar Kumawat
Assistant Professor, Department of Biotechnology, Biyani Girls College, University of Rajasthan, Jaipur 302039, Rajasthan, India; E-mail: tarun.kumar1419@gmail.com

Varsha Kumawat
Director, Naturilk Organic and Dairy Foods Pvt. Ltd. Jaipur 302012, Rajasthan, India; E-mail: varsha29jun1993@gmail.com

Junaid Ahmad Malik
Department of Zoology, Govt. Degree College Bijbehara, Kashmir 192124, (J&K), India; E-mail: malik.junaidahmad@gmail.com

Aamir Afzal Magray
PhD Research Scholar, Department of Chemistry, School of Physical, Chemical and Applied Sciences, Pondicherry University, Puducherry 605014, India; E-mail: aamirmagray91@gmail.com

Sajid Mahmood
Post-doctoral Researcher, College of Chemistry and Life Sciences, Zhejiang Normal University, Jinhua, China; E-mail: sajidmahmood1987@zjnu.edu.cn

Mallika Mazumder
Research Assistant (Guest), Department of Botany, Dukhulal Nibaran Chandra College, Murshidabad, 742201, West Bengal, India; E-mail: mallika.flk@gmail.com

Ramu Meenatchi
Dr. D.S Kothari Post-Doctoral Fellow, SRM Institute of Science and Technology (SRMIST), Chennai 603203, Tamil Nadu, India; E-mail: meenatchiramu20@gmail.com

Mudasir Bashir Mir
Assistant Professor, Department of Botany, Government Degree College, Baramulla 193101, Jammu and Kashmir, India; E-mail: mirmudasir786@gmail.com

Sobia Noreen
M.Phil Research Scholar, Department of Chemistry, Government College Women University, Faisalabad 38000, Pakistan; E-mail: nosh.chem@yahoo.com

Kalpana Pachillu
PhD Research Scholar, Center for Development Research (ZEF), University of Bonn, Bonn, Germany; E-mail: pkalpana@uni-bonn.de

Anand Kumar Pandey
Assistant Professor, Department of Biotechnology Engineering, Institute of Engineering and Technology, Bundelkhand University, Jhansi U.P., India; E-mail: akpandey.bme@gmail.com; dr.akpandey@bujhansi.ac.in

Ravi Pandiselvam
Scientist, Physiology, Biochemistry and Postharvest Technology Division, ICAR – Central Plantation Crops Research Institute, Kasaragod 671124, Kerala, India; E-mail: anbupandi1989@yahoo.co.in

Anjali Pandit
Researcher, Department of Biotechnology, Biyani Girls College, Jaipur 302039, Rajasthan, India; E-mail: panditanjali04@gmail.com

Sonali Zankar Patil
Assistant Professor, Department of Bioanalytical Science, B.K. Birla College (Autonomous), Kalyan421304, Thane, Maharashtra, India; E-mail: sonali1386@gmail.com

Abhipriya Patra
M.Tech. Student, Department of Food Process Engineering, National Institute of technology, Rourkela Odisha; E-mail: abhipriyapatra1004@gmail.com

Anushara Prabhakaran
Junior Research Fellow, Department of Microbiology, School of Life Sciences, Pondicherry University, Puducherry 605014, India; E-mail: anushara0010@gmail.com

Rama Chandra Pradhan
Associate Professor, Department of Food Process Engineering, National Institute of technology, Rourkela 769008, Odisha; E-mail: pradhanrc@nitrkl.ac.in

Yashaswini Premjit
PhD Research Scholar, Department of Agricultural and Food Engineering, Indian Institute of Technology, Kharagpur 721302, West Bengal, India; E-mail: premjit.yashaswini@gmail.com

V. Prithviraj
M.Sc. Student, Food Processing Engineering and Management, National Institute of Food Technology Entrepreneurship and Management, Sonipat 131028, Haryana, India; E-mail: prithvi.rjv@gmail.com

Tamanna Ramesh
Senior Scientist, Innovation and Technology, Kraft Heinz Company; Address: 43 Fountainhead Dr, Apt 203, Westmont 60 559, IL, USA; E-mail: tamannaramesh@gmail.com

Niket Rana
BSc Student, University Institute of Biotechnology, Chandigarh University, NH-95, Ludhiana 140413, Chandigarh, Punjab, India; E-mail: rananiket5@gmail.com

Rahiel Rasool
Project Associate, Department of Applied Mechanics, Indian Institute of Technology (IIT-Madras), Chennai 600036, Tamil Nadu, India; E-mail: bzzy007@gmail.com

Shayanti Roy
BSc Student, University Institute of Biotechnology, Chandigarh University, NH-95, Ludhiana 140413, Chandigarh, Punjab, India; E-mail: royshayanti8@gmail.com

Jatindra K. Sahu
Associate Professor, Indian Institute of Technology Delhi, India; Address: Block III, Room 382, Indian Institute of Technology, New Delhi 110016, India; E-mail: jksahu@iitd.ac.in

K.B. Sankalpa
Research Associate, Department of Agricultural Engineering, College of Horticulture, Kerala Agricultural University (KAU), Thrissur 656680, Kerala, India;
E-mail: sankalpabgowda.27@gmail.com

Anup Kumar Sarkar
Assistant Professor, Department of Botany, Dukhulal Nibaran Chandra College, Murshidabad 742201, West Bengal, India; E-mail: anupsarkar.jpg@gmail.com

Jyoti Sarwan
PhD Research Scholar, University Institute of Biotechnology, Chandigarh University, NH-95, Ludhiana140413, Chandigarh, Punjab, India; E-mail: jyotisarwan710@gmail.com

Joseph Selvin
Professor and Head, Department of Microbiology, School of Life Sciences, Pondicherry University, Puducherry 605014, India; E-mail: josephselvinss@gmail.com

M. Anjaly Shanker
Post Graduate Fellow, Department of Agriculture and Environmental Science, National Institute of Food Technology and Entrepreneurship and Management, Kundli 31028, Sonepat, Haryana, India; E-mail: anjushnkr1992@gmail.com

Heenu Sharma
PhD Research Scholar, University Institute of Biotechnology, Chandigarh University, NH-95, Ludhiana 140413, Chandigarh, Punjab, India; E-mail: sharmaheenu507@gmail.com

Vishnu Sharma
Assistant Professor/Research Scientist, Department of Biotechnology, Biyani Girls College, University of Rajasthan, Sector 3, Vidyadhar Nagar, Jaipur 302039, Rajasthan, India; E-mail: vishnusharma666@yahoo.com

N. U. Sruthi
PhD Research Scholar, Agricultural and Food Engineering Department, Indian Institute of Technology, Kharagpur 721302, WB, India; E-mail: sruthisparks@gmail.com

Zeba Tabassum
PhD Research Scholar, Lovely Professional University, Phagwara 144401, Punjab, India; E-mail: zebatabassum326@gmail.com

Priyamvada Thorakkattu
PhD Research Scholar, Department of Animal Sciences and Industry, Food Science Institute, Kansas State University, Manhattan 66506, KS, USA; E-mail: priyamvadats11@ksu.edu

Nazim Uddin
PhD Research Scholar, University Institute of Biotechnology, Chandigarh University, NH-95, Ludhiana 140413, Chandigarh, Punjab, India; E-mail: Nazimuddinahmmed89@gmail.com

Atif Khurshid Wani
PhD Research Scholar, Lovely Professional University, Phagwara 144401, Punjab, India; E-mail: atifkhurshid61200216@gmail.com

ABBREVIATIONS

Δ E	color change
1°	primary
2°	secondary
3-D	three-dimensional
3D-QSAR	three-dimensional quantitative structure-activity relationships
3°	tertiary
4CL	4-coumarate CoA ligase
4HR	4-hexylresorcinol
4°	quaternary
A	adenine
AA	ascorbic acid/alpha-amylase
AAT	alcohol acyltransferase
ABTS	2,2′-azinobis(3-ethylbenzothiazoline-6-sulfonate)
ACE	angiotensin converting enzyme
AChE	acetylcholinesterase
ACO	ACC oxidase
ACP	atmospheric cold plamsa
ACS	ACC synthase
ACSOs	alk(en)yl cysteine sulphoxides
AD	Alzheimer's disease
ADC	antibody-drug conjugate
AFases	alpha L arabinofuranosidases
AFM	atomic force microscopy
Al_2O_3-NPs	aluminum oxide nano particles
AMP	adenosine monophosphate
AMPK	AMP-activated protein kinase
APA	*Aegopdium podagraria* agglutinin
APX	ascorbate peroxidase
AuNPs	gold nanoparticles
a_w	$_w$ater activity
BChE	butyrylcholinesterase
BCWL	black cotton wood lignin
b_E	regression coefficient

BG	β-glucosidase
BHET	bis(2-hydroxyethyl) terephthalate (BHET)
BI	Browning index
Bio-PE	bio-based polymers like bio-polyethylene
BIO-PET	bio-polyethylene terephthalate
Bio-PP	bio-polypropylene
BSA	bovine serum albumin
C	cytosine
c	speed of light in air
CA	citric acid
CT	controlled temperature
$CaCO_3$	calcium carbonate
CAD	cinnamyl alcohol dehydrogenase
CADD	computer aided drug designing
CANPA	computer-assisted natural products anticipation
CAT	catalase
CBD	convention on biodiversity
CC	calcium chloride
CCC	counter current chromatography
CD22	cluster of differentiation-22
CE	capillary electrophoresis
CEs	carbohydrate esterases
CHB	cellobiohydrolases
CLL	chronic lymphocytic leukemia
CMBs	carbohydrate-binding modules
CMCases	carboxymethyl cellulases
CNT	carbon nanotubes
CO_2	carbon dioxide
COVID-19	coronavirus disease 2019
COX	cyclooxygenase
COX-2	cyclooxygenase-2
CPCB	Central Pollution Control Board
CRISPR	clustered regularly interspaced short palindromic repeats
CS-NPs	chitosan nanoparticles (CS-NPs)
CTAB	hexadecyltrimethylammonium bromide
Cys	cysteine
DAMPs	damage-associated molecular patterns
DCS	differential scanning calorimetry
DDT	dichlorodiphenyltrichloroethane

DMF	N,N-dimethylformamide
DMSO	dimethylsufoxide
DNA	deoxyribonucleic acid
DOPA	dihydroxyphenylalanine
DOS	diversity-oriented synthesis
d_p	penetration depth
DPCD	dense phase carbon dioxide
DPPE	1,2-bis diphenyl-phosphino-ethylene
DTT	dithiothreitol
D-value	decimal reduction time
E	electric field strength
E_a	activation energy
EC	Enzyme Commission
E_c	critical value of electric field strength
EDAC	ethyl-3-(3-dimethyl aminopropyl) carbodiimide
EDTA	ethylene diamine tetra-acetic acid
EFSA	European Food Safety Authority
EG	endo-β-glucanase
EMA	European Medicines Agency
EOs	essential oils
EPR	electron paramagnetic resonance spectroscopy
EPs	exopolysaccharides
ESP	electrostatic precipitator
eV	electronvolt
EVOH	ethylene-vinyl alcohol
f	frequency
FA	fatty acid
FC	foaming capacity
FDA	U.S. Food and Drug Administration
FDC	Food Drug and Cosmetic Act
FFA	free fatty acids
FS	foaming stability
FSC	food supply chain
GC-MS	gas chromatography coupled with high resolution mass spectroscopy
GGT	gamma glutamyl transpeptidase
GHG	greenhouse gases
GHs	glycoside hydrolases
GIA	Global Industry Analysts Inc.

GIP	gastric inhibitory peptide
GLP	glucagon-like peptides
GLP-1	glucagon-like peptide-1
GPC	gel permeation chromatography
GPX	guaiacol peroxidase
GR	glutathione reductase
GRAS	generally recommended as safe
HAARF	hot air-assisted radio frequency heating
HBT	1-Hydroxybenzotriazol
HBV	hepatitis B Virus
HC	hemicellulose
HCoV-229E	human coronavirus 229E
HDPE	high-density polyethylene
HHP	high hydrostatic pressure
HIPEF	high-intensity pulsed electric field
His	histidine
HIV	human immunodeficiency virus
HMF	hydroxy-methyl-furfural
HMG-CoA	3-Hydroxy-3-methylglutaryl coenzyme A
HOBr	hypobromous acid
HOCl	hypochlorous acid
HPCD	high pressure carbon dioxide
HPL	hydroperoxide Lyase
HPLC	high-performance liquid chromatography
HPLC-MS	high performance liquid chromatography–mass spectrometry
HPLC-NMR	high performance liquid chromatography-nuclear magnetic resonance
HPM	high pressure mercury
HPP	high pressure processing
HRMS	high resolution mass spectrometry
HRP	horseradish peroxide
HSV	herpes simplex virus
HTS	high throughput screening
Hz	Hertz
IAA	indoacetate
IC_{50}	the half maximal inhibitory concentration
ICF	inertial confinement fusion
IL-1	Interleukin-1
ISB	infrared steam blanching

IUB	International Union of Biochemistry
j	imaginary unit
k	reaction constant
KKS	kallikrain-kinin systems
kV	kilovolt
kV/cm	kilovolts per centimeter
L/h	liter per hour
LA	lipase
LC	lignocellulose
LCEs	lignocellulolytic Enzymes
LC-MS	liquid chromatography–mass spectrometry
LDH-A	lactate dehydrogenase subtype-A
l-DOPA	l-3,4-dihydroxyphenylalanine
LDPE	low-density polyethylene
LED	light emitting diode
L-EPLP	lignin produced by ethanol-pretreated lodgepole pine
LiP	lignin peroxidase
LMS	laccase mediator system
LOX	lipoxygenase
LPM	low pressure mercury (lamps)
L-SELP	lignin derived by steam exploding lodgepole pine
M&Y	mold and yeast
MALDI-TOF	matrix-associated laser desorption or ionization-time of flight
MAP	modified atmospheric packaging
MAP	modified atmospheric packaging
MC	moisture content
MCC	microcrystalline cellulose
MDR	multidrug resistant
Met	methionine
METLIN	metabolite and chemical entity database
MH	mild heat
MHET	mono(2-hydroxyethyl) terephthalic acid
MHz	megahertz
MIP	molecular imprinted polymer
MMAE	monomethyl auristatin E
MnP	manganese peroxidase
MNPs	metal nanoparticles
MOF	metal–organic framework
MPM	medium pressure mercury (lamps)

MS/MS	tandem mass spectrometry
MTS	mano-thermosonication
n	reaction order
NaCl	sodium chloride
NAD^+	nicotinamide adenine dinucleotide
NBs	nanozymes-based biosensors
NBS/N-BS	N-bromosuccinimide
NFκB	nuclear factor kappa light chain enhancer of activated B cells
N-HL	non-Hodgkin lymphoma
NMR	nuclear magnetic resonance
NPD	natural product discovery
NPs	nanoparticles
NT	nucleotides
O_2	oxygen
OAC	oil absorption capacity
p	shape parameter
P	enzyme activity
PACA	poly alkylcyanoacrylate
PAL	phenylalanine ammonia lyase
PAMAM	polyamidoamine
PBH	poly-b-hydroxyl butryate
PCL	polycaprolactone
PCR	polymerase chain reaction
PE	pectinesterase
PEDV	porcine epidemic diarrhea virus
PEF	pulsed electric field
PEG	polyethylene glycol
PET	polyethylene terephthalate
PG	polygalacturanose
PGE	platinum group elements
pH	hydrogen ion concentration
PHA	polyhydroxyalkanoate
PHB	polyhydroxybutyrate
PHBV	poly-3-hydroxybutyrate-co-3-hydroxyvalerate
Phe	phenylalanine
PHV	polyhydroxyvalerate
PLA	polylactic acid
PME	pectin methyl esterase
PNP-GAase	p-nitrophenyl α-D-glucuronide-hydrolyzing enzyme

POD	peroxidase
PP	polypropylene
PPM	parts per million
PPO	polyphenolase/polyphenol oxidase
PS	polystyrene/protein solubility
PTFE	polytetrafluroethylene
PUFA	polyunsaturated fatty acids
PUV	pulsed ultraviolet
PV	peroxide value
PVC	polyvinyl chloride
PVDF	poly(vinylidene fluoride)
Q	total electric energy density
QCM	quartz Crystal Microbalance
QSAR	quantitative Structure Activity Relationship
R	ideal gas constant
R&D	Research and Development
RA	residual activity
RAS	renin-angiotensin system
RF	radiofrequency
RNA	ribonucleic Acid
ROS	reactive oxygen species
RR	*Rosa roxburghii*
RSM	response surface methodology
RT	reverse transcriptase
S	survival fraction at treatment time
SAR	structure–activity relationship
SARS-CoV-2	severe acute respiratory syndrome coronavirus 2
SAzymes	single-atom nanozymes
SB	steam blanching
SCF	supercritical fluid
SDS	sodium dodecyl sulfate
SELEX	systematic evolution of ligands by exponential enrichment
SEM	scanning clectron microscopy
SFC	supercritical fluid chromatography
SGLT2	sodium-glucose co-transporter-2
SiO_2-NPs	silicon dioxide nanoparticles
SOD	superoxide dismutase
SP	spruce lignin
t	reaction time

TAG	triacylglycerols
TAXI	*Triticum aestivum* xylanase inhibitor
TBA	thiobarbituric acid
t_c	extrapolated critical value of t for 100% survival
TEMED	tetramethylethylenediamine
TfH	*Thermobifida* spp.
TGEV	transmissible gastroenteritis coronavirus
TiO_2	titanium oxide
TiO_2-NPs	titanium dioxide nanoparticles
TLXI	taumatin like xylanase inhibitor
TPA	terephthalic acid
TPC	total plate count
Trp	tryptophan
TSS	total soluble solids
Tyr	tyrosinase
Tyr	tyrosine
U	uracil
UDA	*Urtica dioica* agglutinin
UHPLC	ultra-high performance liquid chromatography
US	ultasonication/ultrasound
US FDA	United States Food and Drug Administration
UV	ultraviolet
V	volume
VP	versatile peroxidase
w/w	weight by weight
WAC	water absorption capacity
WHO	World Health Organization
WPI	whey proteins isolates
WRK	Woodward's reagent K
x	treatment time or specific energy density
XEGIP	xyloglucans-specific endoglucanase inhibitor protein
XIP	xylanase protein inhibitor
XT-I	xylosyltransferase-I
ZnO-NPs	zinc oxide nanoparticles
α-L-ara	α-L-arabinopyranoside
βCD	β-cyclodextrin
β-D- gal	β-D-galactopyranoside
δ	scale parameter
ε'	dielectric constant

ε”	dielectric loss factor
ε_r	relative permittivity
μF	micro-Farad
μs	microseconds
Ω	Ohm

PREFACE

Everything we do in the field of food science and technology revolves around food preservation. There are several advanced technologies for food characteristics, and redesigning is yet another possible method for preserving and processing food. Food enzymes, which are proteinaceous compounds, are extremely essential. The preservation of food is done by activating or deactivating the enzymes. This action can be done by thermal treatment, cooking, drying, cooling, and freezing. Several better inactivation technologies have shown potential applications have also been investigated by scientists and researchers

This book is a collection of novel emerging technologies dealing with inactivation of enzymes for the design of food and preservation through advanced technologies. This book serves as an excellent resource on the inactivation of food enzymes through basic concepts and chemical methods. It presents novel processing technologies for inactivation of food enzymes. It looks at the inactivation of food enzymes through ozone technology, pulsed electric fields, microwave processing and ohmic heating technology, dense-phase carbon dioxide and high-pressure processing, radiofrequency technology, ultraviolet and light emitted diode technologies, cold plasma technology, and ultrasonication of foods and food products. It discusses the extraction of phytonutrients and biocompounds from cereals and considers the inactivation of enzymes in the design of nutraceutical-based functional foods, in the design of specific foods for gut-microbiodata, in conjunction with omega-3 fatty acids, and for the fortification in food products. The volume also investigates the characteristics of dairy-based dry powders and millet starches. The book also explores the role of the bioactive compounds and metal ions, catalases secreted by medicinal plants and mushrooms for enzyme inactivation and biosensing. The role of bionanomaterials in nanoencapsulation and catalysis is also discussed.

This book, *Enzyme Inactivation in Food Processing: Technologies, Materials, and Applications*, will serve many professionals working in the area of food science, technology and engineering around the world. The book will also serve as a reference book for researchers, students, scholars, industries, universities, and research centers.

This book has taken its present shape due to the excellent contribution by all contributing authors who have been this soul of this compendium. We have mentioned their names in each chapter and also in the list of contributors. We are indeed indebted to them for their knowledge, dedication, and enthusiasm. We expect this book to prove a helpful resource for all the food processing and engineering academicians, food processors, and students.

We also extend our sincere thanks to the editorial office of Apple Academic Press, Inc. We hereby appeal to our readers for their productive suggestions that may form the basis for improving future editions.

We take this opportunity to thank: (1) our families for their motivation, moral support, and blessings in counteracting every obstacle coming our way, (2) our spouses for their understanding, patience, and encouragement throughout this project, (3) the contributors for contributing their research in the form of chapters.

— *Editors*

PART I

NANOMATERIALS FOR MEDICAL APPLICATIONS

CHAPTER 1

CONTROL MEASURES FOR FOOD ENZYMES: BASIC CONCEPTS AND CHEMICAL METHODS

RAVI PANDISELVAM, V. PRITHVIRAJ, ARDRA C. BABU, and ANJINEYULU KOTHAKOTA

ABSTRACT

Enzymes are proteinaceous substances that act as catalysts for various biochemical reactions and their study is known as enzymology. As far as postharvest processing industry is concerned, they have a plethora of applications in the area of dairy, fruits and vegetables, confectionery, food preservation etc. Moreover, depending on the scenario, enzymes need to be activated or inactivated for obtaining favorable processing conditions. This is usually carried out through control of pH, temperature, concentration of species, etc. that affect the enzyme–substrate chemical reaction. The fundamentals and primary working mechanism regarding the enzymes are documented here along with their applications in post-harvest technology. This chapter also deals with chemical enzyme inactivation methods and their recent advancements.

1.1 INTRODUCTION

Enzymology is a scientific branch dealing with the study of enzymes, their mechanism of action, kinetics, structure, etc. Enzymes are complex molecular catalysts synthesized inside the living cells. Furthermore, like

Enzyme Inactivation in Food Processing: Technologies, Materials, and Applications. Megh R. Goyal, PhD, Junaid Ahmad Malik, PhD & Ravi Pandiselvam, PhD (Eds.)

all other catalysts they gear up the rate of various biochemical reactions. Most of them are naturally occurring proteins and are unique in their action. Various sectors related to our day-to-day life require the applications of these biocatalysts. Different domains like food processing, medicine, textiles, paper, fuels, and feeds employ diverse enzymes. In most of occasions the biocatalysts are able to successfully replace chemical treatments, thereby providing a greener alternative. Moreover, food industry employs numerous enzymes to improve texture and flavor of products, to achieve desired quality foods of high nutritional and sensorial aspects.

Recently, the need for functional foods has triggered the use of biocatalysts. The rise in demand of nutritional as well as safe food products is expected to propel the usage of enzymes and related processes. With the recent developments of biotechnology catalysts from bacteria, fungi, yeast, and other micro-organisms are gaining popularity than those from plant or animal origin. Many of its applications in different fields are yet to be explored and developed. Novel methodologies like recombinant DNA, immobilization techniques etc. may put forward environmentally viable and cost-effective methods for enzyme applications.[10] Similarly, the conversion of harmful pollutants and other synthetic chemicals could be done in the presence of biocatalysts. This process of bioremediation employs the catalytic functionalities to reduce the energy requirement as well as to work in conditions unfavorable to micro-organisms.[1]

This chapter deals with the basics of enzymes and specifically on its retardation using chemical agents. The catalytic activity needs to be reduced when it imparts any type of deteriorating effects on food products. Chemical inhibitors enjoy a huge part in this process by preventing off-flavor development, surface discoloration, rancidity, and so on. In fresh-cut apples, browning was retarded synergistically by ethanol and ascorbic acid treatment.[100] Likewise, similar results were obtained by using acid agents in banana samples.[102] Moreover, mixtures of certain organic acids successfully inhibited lipase-induced rancidity in wheat germs.[2] Certain reducing agents too come up with the property of limiting enzymatic browning in wine.[95] Deviating from the normal approaches we have made an effort to incorporate various aspects of enzymology. Mostly, the action mechanism is described with the aid of two models, namely, the lock and key and the induced fit. Here, the other models describing enzyme action have been put together. Besides the phenomena of browning, there are various other drawbacks associated with the biological catalysts. All those developments till date accompanied by their solution in the form of chemical agents have been elucidated here.

This chapter focuses on various aspects of enzymes and their usage. The applications of these biological catalysts in different fields of the food processing sector, structural models defining its action mechanism, and more specifically the chemical inhibition methods to reduce undesirable enzyme action are discussed here.

1.2 FUNDAMENTALS OF FOOD ENZYMES

Enzymes are complex proteinaceous substances synthesized by living cells that catalyze various biochemical reactions. Specificity is their main characteristic that is clearly visible in the reactions they catalyze as well as in the selection of substrates.[9] Food enzymes are synthesized inside living organisms, despite which they can actively work in vitro also. When compared with chemical catalysts these biocatalysts are highly effective, specific and own better catalytic properties. There are four following types of specificity of enzymes[46]:

- Absolute (catalyze single reaction)
- Group (acts upon molecules having specific functional groups)
- Linkage (acts only on certain chemical bonds)
- Stereochemical (catalyze reactions having steric or optical isomer)

1.3 CHARACTERISTICS OF ENZYMES

Enzymes are catalysts to activate different biochemical reactions without themselves being consumed.[8] Due to specificity of action, they catalyze only a single or a set of closely related chemical reactions. Some of these catalysts require a helper molecule to assist their activities. This helper molecule is known as cofactor. Cofactors can be organic molecules which bind to the inactive part of enzymes called apoenzyme. Cofactors together with apoenzyme are known as the holoenzyme. Coenzymes and prosthetic groups are the two types of cofactors.[46]

Nomenclature or naming of the enzymes is perplexing, as most of them have systematic names, trivial names, and also an Enzyme Commission number—EC number. Trivial names are mostly historically derived that may not represent the enzyme action.[92] According to an International Agreement in 1971, enzyme nomenclature has been standardized. According to the International Union of Biochemistry (IUB) two parts make up the name of an enzyme: name of substrate (first part) and the type of reaction they catalyze (second part). The second part of the name ends with a suffix "ase". Each enzyme is allocated with a unique EC number having four digits:

- First number indicates class of enzyme.
- Second number shows the subclass.
- Third number denotes sub-subclass.
- Last number indicates the serial number of enzymes in sub-subclass.

Enzymes are classified into six main divisions based on the reactions they catalyze, such as: Oxidoreductase (EC 1), Transferases (EC 2), Hydrolases (EC 3), Lyases (EC 4), Isomerases (EC 5), and Ligases (EC 6).

1.3.1 STRUCTURE OF ENZYMES

Generally, catalysts are of two types: homogeneous and heterogeneous. Homogeneous group include soluble catalysts which performs the reaction in solution, whereas in heterogeneous category an insoluble surface is required. Most of the enzymes have the properties of homogeneous catalyst; however, some of them are insoluble and hence belong to heterogeneous catalysts.[9] Moreover, in both the categories to gear up the rate of reaction catalysts are needed in minimal quantities compared to the amount of substrate. Protein molecules in them are made of amino acids that are bonded by peptide linkages.

The structural configurations of enzymes include primary, secondary, tertiary, and quaternary. The linear chains of covalent bonded amino acids and cofactors constitute the primary structures. Hydrogen bonds are formed within the protein molecule due to intermolecular reactions within them, which further leads to the formation of a new configuration. This makes up the secondary structures. In other words, the localized folding of primary structure can be referred to as secondary while complete folding is termed tertiary structure. The quaternary structure could be called as, how individual molecules are held in an oligomeric enzyme (an enzyme that has two or more polypeptide chain nor linked covalently neither by peptide chains).[9,22] Commercially, these are isolated from three main sources, namely, plant and animal tissue and microbes.[40]

1.3.2 ENZYME UNIT

It is important to determine the amount of enzyme in any biochemical reaction. It can be found either qualitatively by the reaction catalyzed or quantitatively by the rate of that reaction. Thus, enzyme units are used to measure enzyme concentration and express the same.[18] Enzyme Commission of IUB recommends the enzyme activity to be expressed as in International units (IU).[46]

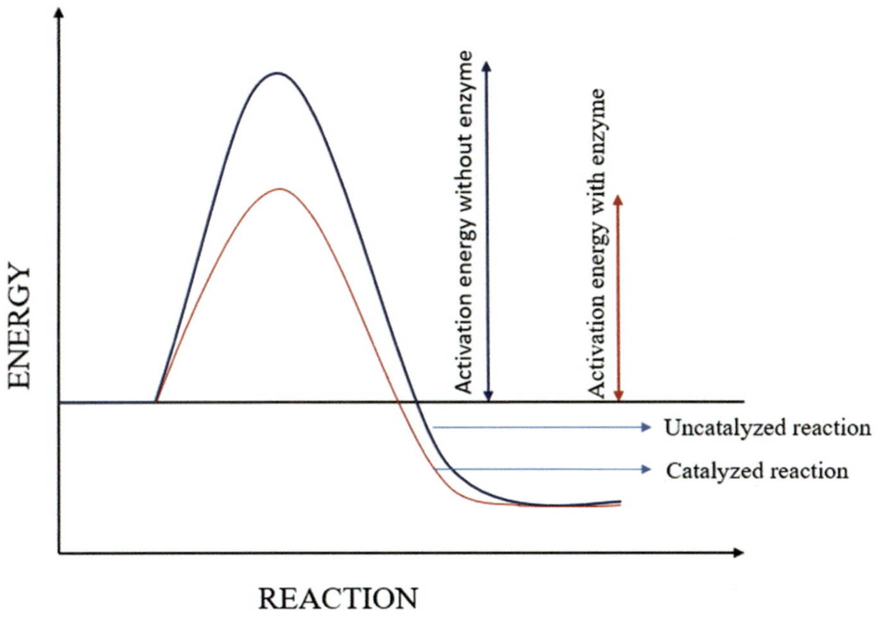

FIGURE 1.1 Use of enzymes will reduce the energy requirement of the reaction.

1.3.3 MECHANISM OF ENZYMATIC ACTION

For a reaction to take place the substrates need a minimum amount of energy, called activation energy (Fig. 1.1). Once this threshold energy is crossed the products are formed.[46] Addition of catalysts reduces this energy barrier, thus increasing reaction rates in the range of 10^6–10^{24}.[35] A Swedish chemist Arrhenius in 1888 proposed that enzymes and substrates react together to form an intermediate complex molecule (Enzyme-Substrate complex). Once the reaction is completed the complex gets decomposed into active enzyme and product. Two different models have been put forward to understand the mechanism of enzyme action.

1.3.3.1 LOCK AND KEY OR FISHER TEMPLATE MODEL

Emil Fischer in 1894 suggested that an active site of the enzyme has a unique structure that is complementary to the conformation of substrate (Fig. 1.2). This allows both the molecules to fit into each other as a key fitting into a lock. Thus, it is referred to as the lock and key mechanism.[49] This tells us about the rigidity of an enzyme. Enzyme flexibility as well as its structural changes in the presence of allosteric modulators are not accounted here.

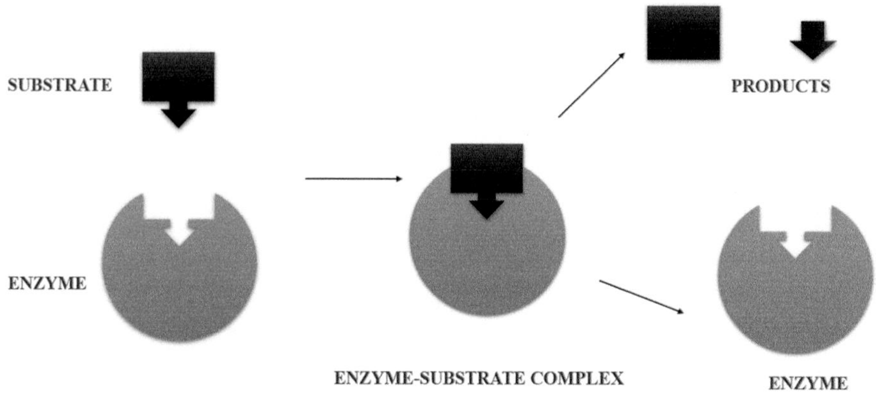

FIGURE 1.2 Lock and key or fisher template model, substrate exactly fits into the enzyme.

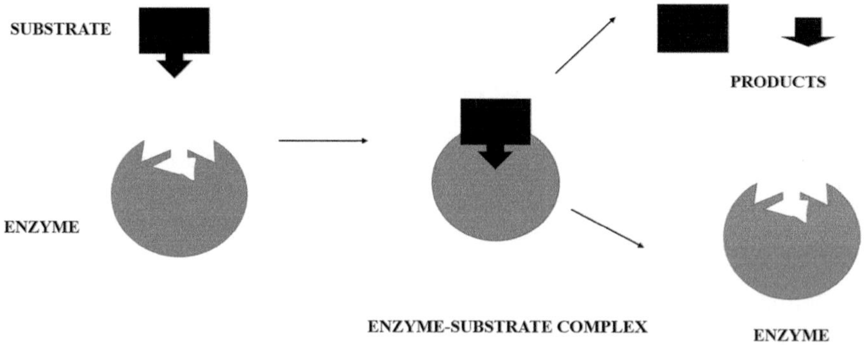

FIGURE 1.3 Induced-fit model, where the substrate will cause changes in catalytic groups of the active site of enzyme.

1.3.3.2 INDUCED-FIT MODEL

The shortcomings of the lock and key model lead Daniel Koshland to another hypothesis called the induced-fit model (Fig. 1.3). In this, he accounted for the flexibility of active site of an enzyme. Here, the reactant (substrate) is said to alter structure of active site (conformational change) so that it can fit into the enzyme.[42] This model advocates competitive inhibition, enzyme inactivation, and allosteric modulation. The model was proposed as follows[43]:

- The catalytic groups should be precisely oriented for enzyme action
- The three-dimensional relation of amino acids will be changed by the substrate
- Conformational alterations in the protein by substrate will put the catalytic groups in perfect alignment.

1.3.3.3 CONFORMATIONAL AND EXTENDED CONFORMATIONAL SELECTION

The existences of proteins as a unit of different structural states lead to the model conformational selection. This model unlike induced-fit suggested that structural changes are not always induced by substrate binding.[90,91] It demonstrated that proteins are a cluster of certain related structures with varying actions and binding interests.[14] The binding of different substrates to enzyme active site has been well explained by this model.[61,93] Complexity of mutual selection and adjustment of protein dynamics process create so called extended conformational model that includes the induced-fit, classic lock and key, and original conformational models.[15] All these proposed theories together help to understand various phenomena, namely, enzyme–substrate binding, enzyme inhibition, selection of substrates, structural changes, etc.

1.4 SAFETY CONCERNS

Most biocatalysts are marketed as preparations with the desired enzyme in it along with certain preservatives and stabilizers. Primarily, safety of a food enzyme is determined by the organism producing it. The safety of production strain, techniques used, and host organism will determine whether the biocatalyst used is secure or not.[78] In United States, for commercialization, it should be sanctioned by USFDA under Food, Drug, and Cosmetic (FDC)

Act. To approve the use of enzymes as food additives. the authorization of European Food Safety Authority (EFSA) is required. These governing bodies guarantee safe utilization of food enzymes.[46]

1.5 NECESSITY OF ENZYME INHIBITION

Enzymes are applied as beneficial processing aids in food processing sector. Fruit and vegetable processing, cheese making, meat tenderization etc. are some of their major applications. Apart from all these usages, they also exhibit many undesirable effects in food products. Among the various deteriorating effects of enzymes, browning or discoloration has the most pronounced effect. This factor is of paramount importance as it is a major contributor in quality losses of food products. The mechanical stresses experienced by food materials during processing cause enzymes and their substrates to get delocalized, which in turn leads to browning, change in texture, off-flavor development etc.[29]

Polyphenol oxidase (PPO) causes browning reaction, through oxidation reactions in phenolic compounds. Browning reaction is said to cause major financial loss in fruits and vegetables (around one-half).[34,97] Peroxidase (POD), an iron-containing enzyme, also adds to the total browning phenomena.[29] The activities of PPO and POD have reduced the shelf-life and nutritional qualities of various food commodities viz. fresh-cut fruits,[48] sweet potato,[101] wheat pastry dough,[77] cloudy apple juice,[98] sugarcane juice,[29] and also minimally processed foods like stored mung beans.[84] Along with browning by PPO and POD many other biocatalysts are also responsible for numerous quality degradations. Rancidity in rice bran by ligases and lipoxygenase,[52] greening of garlic by γ-glutamyl transpeptidase and alliinase,[32] loss of turbidity and clarification in cloudy apple juice,[98] pectin esterase and peroxidase induced texture and nutritional losses[17] are some of the disadvantages.

All the above- discussed quality losses point to the necessity of enzyme inactivation. Various types of inhibition methods are used in food industry to curtail the negative effects of enzymes. Naturally occurring substances, chemical agents, thermal and nonthermal processes, or combination of any these methods could be employed in the inactivation.[5,32,53,55,102] Use of natural agents as anti-browning agents is gaining popularity in processing sector since they are inexpensive, bio-degradable, health promoting, and safe to use.[98,101,103]

Chemical inhibitors consisting of acids like citric acid, ascorbic acid, phosphoric acid,[102] adsorbents like bentonite,[84] salts of calcium are able to successfully cease unwanted enzyme activities.[44,74] Conventional thermal treatments (blanching, microwave, and infrared heating and many more) as well as novel nonthermal methods (high pressure carbon dioxide treatment, pulsed electric field, cold plasma treatment etc.) play a vital role in the inhibition process.[32,37,44,88] It is noteworthy that certain natural agents or plant extracts from various species were effectively used to inhibit activities of lipase,[28] α-glucosidase, α-amylase,[19,39] thio-glucosidase,[58] and so on. These methods are yet to gain popularity in the food processing sector whose further developments will provide promising outcomes in the phenomena of inhibition. Table 1.1 shows chemicals methods of inactivation of enzymes in various products.

1.6 APPLICATIONS OF ENZYMES IN FOOD PROCESSING SECTOR

Enzymes are employed in food processing sector for numerous applications. Different characteristics, such as low temperature activity, pH adaptation, tolerance to high salinity and pressures, thermostability add to the uniqueness of enzymes, which are employed for numerous uses,[104] such as meat tenderization, cheese making, juice clarification, wine production, and starch processing.[13]

1.6.1 PROCESSING OF FRUITS AND VEGETABLES

From time immemorial an inevitable part of human diet are fruits and vegetables, which are highly perishable in nature which in turn decline their quality. Thus, conversion of fresh produce into juice, jam, jelly, pickles, fruit bar etc. extends their shelf-life as well as is economical to the producers.[80] Fruit and vegetable juices, nectars, and purees contain all major components of the original produce. Use of exogeneous enzymes before pulping of fruit purees and nectars gives higher yields and also results in cloud stability.[89]

Application of enzymes improves clarification and extraction of fruit and vegetable juices.[20] Moreover, immobilized pectinases successfully provided clearer orange juices[16] as well as the synergistic effect of cellulases and pectinases has given a massive yield of quality banana juice[26] which otherwise is tedious. Fruit firmness is an important characteristic that decides the quality and level of maturity of fruit. During mechanical as well as thermal

TABLE 1.1 Chemicals Methods of Inactivation in Various Products.

Product	Enzyme inactivated	Inactivation method	Specifications	Inference	References
Apple juice	Peroxidase (POD), Superoxide dismutase (SOD), Polyphenol oxidase (PPO)	Natural method	Oligochitosans used as anti-browning agent	Oligochitosan with one kDa molecular weight is the effective antioxidant, which is successfully retarded browning	[105]
Asian Pear	Polyphenol oxidase (PPO)	Chemical method	Ascorbic acid at 0.16, 0.20, 0.24%	Acid concentration of 0.20% (w/v) had better results of larger inactivation	[37]
Banana	Polyphenol oxidase (PPO)	Chemical method	Ascorbic acid Calcium chloride Hot water (65°C)	Banana slices treated with 1% ascorbic acid showed least enzyme activity after 2 week-storage at 0–4°C	[102]
Chrysanthemum indicum L. flower	Polyphenol oxidase (PPO), Peroxidase (POD)	Steam blanching (SB)	SB conducted in stainless steel steamer with 1 L water (100 g sample)	ISB treatments allowed stable surface temperature at 99°C and enhanced PPO, POD activity.	[51]
		Infrared blanching (IB) Combined infrared and steam blanching (ISB)	IB done with far infrared dryer at infrared intensity of 1 W/cm^2 (550 W IR bulb)		
Garlic	γ-glutamyl transpeptidase (GGT), Alliinase	Blanching	(70, 80, 90)°C at different time viz. (1, 2, 3, 4, 5 min) respectively	Blanching at 90°C for 4 min was found to be the best condition	[32]

TABLE 1.1 *(Continued)*

Product	Enzyme inactivated	Inactivation method	Specifications	Inference	References
Multi-grain dough	PPO, POD	Combination of microwave and chemical treatment	Citric acid (CA) Ascorbic acid (AA) L-cysteine (L-cys) Microwave (80 s)	0.3% CA + AA (1:2 w/w basis) along with microwave treatment retarded enzyme activities	[5]
Peach juice	POD, PPO	Ozone treatment	Ozone concentration 0.11 and 0.20 mg O_3/min/1 mL in bubble column up to 12 min	Inhibition increased with concentration of ozone and time of operation. Most of the reductions were seen after 12 min.	[55]
Potato	PPO, POD and Phenylalanine ammonia-lyase (PAL)	Natural methods	COD Peptides	A concentration of 0.1% (w/w) inactivated PAL, POD, PPO activities	[53]
Rice bran	Lipase (LA), Lipoxygenase (LOX) activities	Hot air-assisted radio frequency heating (HAARF)	6 kW, 27.12 MHz pilot-scale free running 105 oscillator RF system	Radiofrequency heating 100°C with an electrode gap of 10 cm at a holding time of 15 min in hot air retarded enzyme activity	[52]
Sugarcane juice	Peroxidase (POD), Polyphenol oxidase (PPO)	Chemical treatment	Adsorbent (Bentonite) in combination with acidulants (ascorbic acid, Citric acid)	0.4% bentonite along with ascorbic acid completely inhibited catalytic action at pH 3.9	[29]
Tomato	Polygalacturanose (PG), Pectin methyl esterase (PME)	UV-C irradiation and calcium chloride	Experiment I-$CaCl_2$ at 0, 1, 1.5, 2% Experiment II-1.5, 3, 4.5 kJ m^{-2} UV-C radiation for 5, 10, 15 min respectively	3, 4.5 kJ m^{-2} UV-C and 2% $CaCl_2$ conditions had good effects on product quality	[57]

TABLE 1.1 *(Continued)*

Product	Enzyme inactivated	Inactivation method	Specifications	Inference	References
Tomato juice	Pectinmethylesterase (PME)	High Pressure Carbon Dioxide (HPCD)	Pressures-8.5–20 MPa	PME completely inactivated at 55°C and 20 MPa however PG was more resistant to HPCD	[37]
	Polygalacturonase (PG)		Temperature-35–55°C		
Wheat germs	Lipase, Lipoxygenase	Chemical treatment	Ascorbic acid Succinic acid Fumaric acid	Mixture of three acids in a concentration of 3–5% by weight of product decline enzyme action.	[2]
Wheat germs	Lipase, Lipoxygenase	Atmospheric cold plasma treatment (ACP)	20 and 24 kV of 50 HZ power supply	Low plasma temperature short time (25 min) ACP improved shelf-life	[88]
			Different holding time (5–35 min)		

treatments, the texture of most of soft fruits gets damaged. Immobilization of various enzymes, namely, pectinase and xylanse, pectinase and naringinase, and pectinase alone could be successfully used in processing of fruit juices like apple juice, grape juice, and jamun for clarification operation (Indian Black Berry) juice respectively.[24,27,47]

Addition of pure pectin methyl esterase to fruits leads to increased firmness, which allows fruit to keep up its shape and consistency.[21] Most of the vegetable oils are extracted using hexane, an organic solvent that is a potent carcinogen. Thus, to make cooking oils free from harmful residues and to improve its quality, cell-wall-degrading pectinolytic enzymes could be used which increase extraction of phenolic compounds having high antioxidant activity.[25]

1.6.2 DAIRY INDUSTRY

The dairy industry uses biocatalysts for the production of various milk products. Enzymes could be used to enhance flavor, color, aroma, as well as to increase yield of milk products in dairy industry.[46] There are four main categories of milk-coagulating enzymes: vegetable and animal rennet, microbial milk coagulants, and genetically engineered chymosin.[86] In the dairy processing sector, hydrolysates of casein are used in foods based on milk proteins for new born babies, lipases used to improve flavor in cheese, as well as proteases to reduce allergic properties of bovine milk.[76] Microbial transglutaminase is a hopeful tool to improve functional properties of milk and milk products. Amylase, proteinase, and lipases are significant in degrading milk into monomeric molecules that are responsible for the chief flavors of cheese.[41,72]

1.6.3 PRODUCTION OF NUTRIENT-DENSE FOODS

Numerous varieties of biocatalysts can be employed in the production of nutrient-enriched products (foods having high nutrient content and low calories). This can be accomplished by breaking rigid plant-based foods into different enzymes to release nutrients.[104] The enzymes like sortase, butelase, peptiligase, or omniligase can be used in enzyme-mediated peptide ligation which is an important technique to prepare complex peptides as food additives.[75] The enzymatic conversion of date fiber concentrate is an effective way to enrich it with soluble dietary fiber. This could be used as a prebiotic and an antioxidant food ingredient.[65]

1.6.4 MEAT PROCESSING

Meat obtained from each individual is unique. Its quality is based on various religious, cultural, and functional factors. Enzyme proline iminopeptidase in softening of chicken muscles and recombinant metalloprotease in tenderization could be used successfully in meat processing.[12,62] Tenderness and flavor of meat are important when it comes to the value of cooked meat.[13] Due to different enzyme activities and loosening of connective tissue (particularly collagen) meat tenderization takes place. Enzymes such as papain, bromelain from plants are mainly utilized for this process.[13] Cohesiveness, shelf-life extension, as well improved texture of meat could be obtained by the enzymatic action of transglutaminases.[20]

1.6.5 BAKERY AND CONFECTIONARY PRODUCTS

Food products rich in sugar and carbohydrates are termed confectionery. Their shelf-life is limited due to the crystallization of sucrose. The enzyme invertase prevents sucrose crystallization by converting it into simple sugars which otherwise shorten the shelf stability of such products.[40] Furthermore, in bread making also exogeneous enzymes are used for shelf-life extension, improving bread features, compensating flour availability, and so on.[20]

Proteases are employed to hydrolyze gluten for easy preparation of baking mass. Similarly, lipases is used in bakery products for shelf-life extension and flavor enhancement.[46] During the preparation of fondant, invertases soften it and also improve sensory characters.[69] Likewise, food enzymes are said to improve quality attributes and shelf-life of gluten-free bread made out of chickpea.[63] Bread staling is a source of huge economic loss in baking industry. Enzymes can be added during bread making to prevent staling.[86]

1.6.6 PRODUCTION OF BEVERAGES

The production of different types of beverages is closely related to catalytic activity. Among all those commonly used drinkables, green tea is an immensely popular and widely consumed product, primarily due its physiological benefits. At various stages of its production enzymes have been successfully utilized for improving its beverage quality as well as other characteristics.[66] Tannase-catalyzed hydrolysis improved antioxidant activities in inferior oolong tea.[50] Furthermore, its treatment lowered cream formation and also increased stability of storage. Similarly in wine-making industries different types of enzymes have been used for clarification, enhancing the quality and stability, color extraction, aroma and protein stabilization of various commodities.[56,60]

Proteolytic biocatalyst from papaya (Papain) is used in post-fermentation stages of beer production to restrict the presence of protein and tannin-containing chill haze.[13] Enzymatic methods of production of beer[94] and instant coffee[7] have been proved as a great alternative to the conventional process of manufacturing. Findings demonstrated the ability of used up coffee to produce the instant type through enzymatic hydrolysis. Furthermore, this technique is ecofriendly with less energy expenditure as well as it has been substantiated as a potential alternative that lowers the deterioration of carbohydrates in coffee. Likewise, similar results were obtained in enzyme-assisted beer production process where it reduced energy requirement and inferior impacts on environment.

1.7 ENZYME INHIBITION BY CHEMICAL METHODS

Natural, chemical, thermal, and nonthermal technologies have been used for this purpose. This section is intended to provide information about the chemical methods of enzyme inhibition.

1.7.1 ENZYME INHIBITORS

Catalytic inhibitors could be any type of organic or inorganic substances employed to retard the deleterious effects caused by enzymes in food products. Furthermore, numerous thermal and nonthermal methods may also be employed. Inactivation by certain low-molecular weight chemicals, metal ions, or other inhibitors involves their interaction with the active sites of catalyst causing a reduction in the rate of reaction or may prevent the enzyme from acting in its usual manner.[46,81] Such organic or inorganic chemical substances are known as enzyme inhibitors. They are of two types: specific and nonspecific inhibitors. Specific inhibitors target specific components of the enzyme while nonspecific include all physical and chemical factors denaturing proteins.[81]

1.7.2 POLYPHENOL OXIDASES AND PEROXIDASES (PPO, POD)

Most of the horticultural crops are subjected to various quality losses and a diminished market value due to enzymatic browning. Loss of physical, nutritional, and organoleptic properties is the adverse effects caused by browning.[85] During the processing of food products they undergo different unit operations viz. peeling, cutting, bruising that cause many physical and mechanical damages that induce browning reactions.[33] Polyphenol oxidases (PPO) category of enzymes causes browning reaction, which creates discoloration of fresh-cut fruits and vegetables. PPO uses oxygen and some phenols as its substrates which are initially hydrolyzed, later on oxidized to quinones and then finally to the brown colored pigments called melanin that causes the surface discoloration.[96]

The rate of enzymatic browning is governed by several factors that include active content of PPO, pH and phenolic content of tissue as well as temperature and oxygen availability.[33] In addition to PPO, the enzyme peroxidases (POD) also play a significant role in browning, but their part in the reaction is a bit convoluted and requires more clarity and simplification.[85] POD is able to oxidize phenols to quinones in the presence of hydrogen peroxide.[29] Both the enzymes together or alone are the prime cause for browning in fruits and vegetables. However, the same process is a boon for certain products like raisins, cakes. Nevertheless, the hindrances caused by these biocatalysts in processing of food materials need to be destructed. The retardation of catalytic activity may be achieved by utilizing diverse inactivation methods, namely, natural agents, chemical agents, heat treatments, nonthermal techniques, and so on. The discoloration of fruits and vegetables are controlled using chemicals that act as anti-browning agents. Based on the primary mechanism of action, anti-browning agents are classified as follows[33]:

- Reducing agents
- Complexing agents
- Acidulants
- Enzyme inhibitors
- Chelating agents
- Enzyme treatments

1.7.2.1 REDUCING AGENTS

Chemical substances that cause reduction in pigment precursors are called reducing agents.[38] They decrease the *o*-quinones by the irreversible inhibition of PPO or by converting them back to phenols.[85] Initially, sulfites were used to control browning, but it has been restricted due to increased health concerns.[33] Increased consumer awareness paved the way for other alternatives like ascorbic acid along with its derivatives, glutathione, cysteine, etc. as reducing group of anti-browning agents.[33,38] Moreover, this category of chemical inhibitors finds their applications in fresh-cut products, liquid foods, etc. For instance, fresh-cut banana samples treated with several chemical agents showed a reduced enzymatic browning.[102] Among all chemical agents treated, ascorbic acid was found to be more fruitful in ceasing browning even after a storage of 2 weeks. Furthermore, this reducing agent prevented any type of quality degradation in the banana slices.

Ascorbic acid treatment provided promising results for inhibiting browning in sprouts of mung bean.[84] Postharvest losses in sprouts due to browning could be controlled via anti-browning agents viz. L-cysteine, citric acid, and ascorbic acid with the later having the most pronounced effect. The inhibition of PPO and POD was done without damaging the nutritional and antioxidant capacity of the mung bean sprouts. Thus, both the studies indicate the ability of ascorbic acid as a chemical inhibitor of the catalytic activity.

The synergistic effect of chemical agents and other treatments has provided favorable outcomes in a number of fruits and vegetables. Chitosan coating having calcium chloride (CC) or ascorbic acid (AA) in apples,[54] ascorbic acid and plant proteases in apricot,[70] ethanol-ascorbic acid treatment in apples[100] etc. present such cases of combined effects. In fresh-cut apples, chitosan coating with containing calcium chloride is a finest choice to retard browning. Ascorbic acid is a good anti-browning agent which is least toxic and also it prevents quality degradation of the processed product. However, it does not have a prolonged effect since it oxidizes faster. Nonetheless, it comes up with superior results when used in combination with other agents like ethanol and calcium chloride. The combined effort of these agents provides promising alternatives to enzymatic browning, quality losses, and microbial activity.

Compounds like glutathione and cysteine also belong to this category of anti-browning agents. They are widely used in products like fruit juices, wine, pears, apples, litchi to cease the discoloration by enzymes. Addition of reduced glutathione to sparkling wine after its disgorging as well as to apple wine during storage exhibited lesser browning.[95,99] Earlier sulfur compounds were used to decline the browning reaction. But both the experiments present glutathione as a better anti-browning compound. Reduced glutathione improves the quality and flavor of wine. Use of cysteine to alleviate browning has been encouraged in pears,[107] litchi,[3] plums,[6] and so on. In three of the fruit's cysteine emerged to be a potent chemical agent which not only reduced browning but also enhanced their antioxidant capacities without causing any hazards. Pear treated with L-cysteine, ascorbic, citric, and kojic acids among which L-cysteine turned out to be the best chemical inhibitor.

Therefore, we can elucidate that L-cysteine can be employed in food industry as an anti-browning substance for soft fruits (apples, plums, pears). Furthermore, ascorbic acid also successfully inactivated enzymes, but is more efficient when used in combination with other chemical agents. The reducing agent glutathione is well suited in inhibition of PPO, POD in liquid foods like wine, juice. It modifies the flavor as well as preserves quality. Hence, all these antioxidants or reducing agents can be claimed as healthy and safe alternatives to sulphites in controlling enzymatic browning.

1.7.2.2 CHELATING AGENTS

The PPO have copper present in the active sites which react with chelating substances and form complexes that in turn reduce enzyme. activity.[38,64] Ethylene diamine tetra acetic acid (EDTA), phosphate-based compounds, different types of carboxylic acids etc. are the chelating compounds that prevent browning reaction and thus used widely in food industry. The lowering of pH or metal-chelating activities by this class of compounds inhibit browning.[64] Oxalic acid, a GRAS (Generally Recognized as Safe) compound of this category, simultaneously retards browning and extends shelf-life in bamboo shoots[105] and slices of lotus roots.[3]

In addition to POD and PPO inactivation in bamboo shoots, oxalic acid inhibited activities of enzymes like 4-coumarate CoA ligase (4CL), cinnamyl alcohol dehydrogenase (CAD), and phenylalanine ammonia lyase (PAL) which cause lignification. Hence, it can be spotted as an excellent approach toward the prevention of discoloration in fruits, vegetables, and edible portions of plants. Moreover, citric acid is also one of the mostly used inhibitors of this group. Citric acid (CA) treatments in various food commodities like fresh-cut drumstick,[83] sugarcane juice,[29] pear puree[106] significantly reduced enzymatic browning. Color retention, preservation of antioxidants, and enzyme inhibition were the results obtained when pear puree was treated separately with CA. When used in conjunction with thermal and nonthermal preservation methodologies, it enhanced the inactivation in the same experiment. Similarly, in sugarcane juice CA in combination with adsorbents like bentonite showed promising results.

Citric and oxalic acids can be employed in various processing sectors of food industry to reduce the deteriorating effects of enzymes. Both the chemicals restrict the catalytic activities either by altering pH (lower than optimal range) and also their ability to form chelating complexes. They can be exclusively used in the fresh-cut industry to prevent the discoloration on ready-to-eat products and at the same time to enhance their visual quality. Liquid products could also be benefited from the enzyme-inhibiting activity of these chelating agents and that too without any deleterious change in the quality attributes. EDTA, as an anti-browning compound, provides better results when used in combination with other agents.[68] All these chelating agents provide better outcomes in reducing catalytic activities. To achieve complete alleviation of enzyme activity these anti-browning agents may be used in combination with any thermal or nonthermal effects.

1.7.2.3 COMPLEXING AGENTS

Cyclodextrins and chitosan are commonly used complexing agents. Cyclodextrins cause entrapment of PPO forming inclusion complex products and substrates to eliminate browning.[68] These compounds are nontoxic cyclic oligosaccharides of α-D-glucopyranose from the biotransformation of starch.[11] The α-cyclodextrins and β-cyclodextrins showed good anti-browning effects in avocadoes[11] and pears.[59] In avocadoes browning was prevented using chemical agents: β-cyclodextrin (βCD), paeonol, and 4-hydroxybenzaldehyde. The β-cyclodextrin was the least efficient chemical, while paenol had the most inhibitory impact on the fruit. However, the inclusion complex of paeonol and β-cyclodextrin provided better results and could be employed as a PPO inhibitor. Similar results were obtained in pears where the inclusion complex of βCD and 4-tert-butylcatechol retarded the browning phenomena.

Chitosan, a natural polymer, is also seen to retard browning.[38,68] Chitosan-coated fresh-cut slices of mushroom containing ascorbic acid had a good inhibition potential.[67] Anti-browning capacity of ascorbic acid has been enhanced by chitosan. Hence, chitosan coatings holding acidifying agents or other chemicals are a possible alternative to browning. Citric acid as well as calcium chloride may also be employed along with chitosan to attain superior results in declining browning. Moreover, other than coatings in solid foods, juice products can also use chitosan as an inhibitor of browning.[103] Since chitosan is a natural substance as it did not impart any sort of toxicity to juice, besides it enhanced antioxidant capacity, nutritional, as well as organoleptic properties of the product.

1.7.2.4 ACIDULANTS

Mostly, PPO is active in the pH range 6–7, and is inactive below a pH 3. Acidifying agents or acidulants functions are used for creating suitable conditions for catalyst which can maintain pH below optimal range. Ascorbic acids, citric acids, phosphoric acids are the most common in this class. However, tartaric acids, erythorbic acids, acetic, and malonic acids are also acidulants but are not commonly used alone.[38,64,68] PPO and POD inhibition activities of citric and ascorbic acids have already been discussed in the previous sections. Two or more acidifying agents could be used together to provide their synergistic effect on browning of food products. They do not leave behind any chemical residue in the course of the reaction. Moreover, they prevent the development of off-flavors and also help in retaining the visual quality of solid as well as liquid food products. Acidulants can even be combined with heat treatment methods or nonthermal methods to enhance the efficacy of those preservation methods.

1.7.2.5 PPO INHIBITORS

Substituted resorcinol, peptides, aromatic carboxylic acids, aliphatic alcohols, anions are some compounds that belong to this category of chemical anti-browning agents. Many 4-substituted resorcinol are used in food industry to retard discoloration as they are non-carcinogenic and nontoxic additives.[38] The inhibition of PPO in pears was studied using certain chemical agents like citric acid, ascorbic acid, 4-hexylresorcinol (4HR), sodium isoascorbate.[59] In this treatment 4-hexylresorcinol emerged out to be the best inhibitor. Nevertheless, fresh-cut apples treated with the same reagent presented contrary outcomes.[71] Ineffectiveness of 4HR is attributable to its inability to control catechol oxidase as well as changes in structure of tissues and permeability of cell membrane.

Cinnamic acid and its analogues, benzoic acid etc. are certain aromatic carboxylic compounds that inhibit the browning. Besides PPO another biocatalyst called tyrosinase can also be held responsible for the phenomena of browning. Tyrosinase is a copper containing oxidase that causes surface discoloration in plant-derived foods.[23] Moreover, it is noteworthy that some ester derivatives of cinnamic acid can be employed as promising inhibitor of mushroom tyrosinase.[82] Most of the derivatives presented better retardation capacities than their parent compound. Furthermore, it opens up to a new possibility of freshly produced derivative compounds as effective inhibitors of the catalytic action of tyrosinase. Browning and softening in fresh-cut apple slices limits the shelf-stability thereby reducing its market demand. The combined action of ascorbic acid along with ethanol decreased browning to lower levels in apple slices.[100]

Furthermore, it reduced microbial count which in turn preserved its sensorial properties as well as increased the life of product. Additionally, identical results were obtained during ethanol treatment on fresh-cut sugarcane.[30] Besides PPO inactivation, phenylalanine ammonia-lyase was also controlled during the treatment. Ethanol treatment could efficiently maintain the quality of the fresh-cut fruits. Hence, acidulants along with ethanol is a promising alternative to preserve the sensory and physicochemical attributes of food products in the fresh-cut industry. Kojic acid, a metabolite from species like *Aspergillus* and *Penicillium*, also finds its place in this class of chemical agents in inhibiting catalytic activity.[79] Kojic acid applications before storing of litchi fruit controlled the pericarp browning of the fruit, and simultaneously maintained its quality aspects.

Therefore, anti-browning agents provide promising outcomes to the food processing sector to remove its most important menace of discoloration. The harmful effects of enzymatic browning could successfully be declined with usage of most of these agents either as separate treatments or in combination with each other as well as other thermal or nonthermal methods. The choice of anti-browning agent can be done on the basis of its availability, cost, efficiency of operation, safety of treatments involved, and other health concerns related to it, the type of food product to be treated.

1.7.3 LIPASES

This group of catalysts perform reactions including esterification, transesterification, and inter esterification. In food industry usage of lipases is there in milk fat hydrolysis, low bitterness, cheese flavor, etc.[46] However, these classes of enzymes cause rancidity in many commodities. One of the main by-products of rice milling is rice bran whose use is limited due to rancidity because of the contact between aleurone lipase and oil in rice bran.[31] Coarse, fine, and superfine cultivars of rice bran were treated with acid and alkali.[87] Alkali (sodium hydroxide) produced soap during the treatment while hydrochloric acid at a level of 5% has been the best option to stabilize lipase.

Furthermore, another study conducted in rice bran[63] gave similar results where lipase inactivation by ultrasound was enhanced due to the presence of hydrochloric acid. Thus, it is evident that hydrochloric acid can effectively retard the catalytic action by lowering the pH. Unlike other agents it does not hinder the oil extraction from rice bran as well as its nutrient content. Activity of lipase and lipoxygenase causes unpleasant odor and rancidity in wheat germ. The composite mixtures of organic acids (fumaric, ascorbic, and succinic acids) reduced the catalytic activity in a noncompetitive inhibition.[2] The three organic acids when used as a mixture provided much better results than when solely utilized. Henceforth, all these acid agents could be beneficial in the inhibition of lipase and lipoxygenase without causing any quality losses. They can be used in conjunction with other nonthermal treatments to enhance the inactivation process.

1.7.4 PECTINASES

The enzyme, pectin methyl esterase (PME), polygalacturonase (PG) belongs to the family of pectinases. The former performs the de-esterification of pectin by hydrolyzing methyl ester groups.[97] While PG degrades pectin and will give rise to softening of fruit tissue. PME activity causes loss of stability of juices, reduction in viscosity and texture in tomato purees etc.[70] The ill-effects due to its action in tomato have been discontinued successfully using the synergistic effect of two treatments (calcium chloride and UV-light treatment).[57]

It is reported that calcium chloride along with UV-treatment has inactivated pectin methyl esterase as well polygalacturonase (PG) which significantly caused decay of tomato. Furthermore, it is seen that calcium chloride solely cannot prolong the shelf-life to a much greater extent. Thus, any kinds of heat treatments or chemicals combined with calcium salts may be employed to retard enzyme actions, prevent viscosity reductions, and retain fruit firmness which forms the main negative impacts of PME and PG catalytic activity. Ascorbic acid or citric acid in combination with calcium salts could be encouraged as a significant method having greater inhibiting impact on pectinases in fruits and vegetables, their purees, and also in fruit juices.

1.8 SUMMARY

Enzymes are promising aids in the food industry which ensure high-quality products without any residual chemicals. Moreover, these biocatalysts are highly preferred as they are eco-friendly, low cost, highly efficient, and are perfect greener alternative to chemicals. When it comes to the mechanism of enzyme action, various models have been suggested. However, this part needs to be explored more since their possibilities in the processing sector may provide an aid in facing the foreseen challenges. As already discussed in this chapter the catalytic activity of enzymes is advantageous in improving texture, increasing nutritional attributes, and preserving the sensorial aspects of the products. Most of the processing industries owe to biocatalysts for their immense applications in each field which includes fruit and vegetable processing, dairy products, meat products, bakery items, and so on. Nonetheless, it can also impart certain degrading impacts on food commodities like surface discoloration, rancidity, softening, and so on, and various methodologies have been put forward to remediate this problem. Furthermore, the chemical agents so far discussed proved to be a success in this as most of them synergistically as well as individually inhibit these negative impacts of biocatalysts. Advanced investigations related to the inhibition process as well as their mode of action may help in studying the effects of chemicals in inactivation process in a wider way. The incorporation of the new models, conformational selection, and its extended version in the process could dig out huge possibilities in activation and inhibition phenomena.

Emerging technologies like encapsulation as well as immobilization of enzymes open up a vast scope in their usage. In addition, various novel techniques could be used to produce and regenerate the biocatalysts with new characteristic features. Thus, it could assist certain reactions conducted in extreme conditions, namely, high temperature, pressure, salinity. Hence, enzymes provide us with a pool of opportunities that on its further developments and investigations can upgrade the food processing sector to a higher standard in terms of food quality and quantity.

KEYWORDS

- **blanching**
- **citric acid**
- **enzymes**
- **peroxidase**
- **polyphenol oxidase**

REFERENCES

1. Ain, Q. U.; Basit, I. A Step Forward Towards Uses of Bio-Engineering Enzymes in Bioremediation. *Int. J. Sci. Eng. Res.* **2020,** *3*, 1119–1144.
2. Alekseeva, T.; Tokareva, D.; Eds. Study of the Regularities of Inhibition by Organic Acids of the Enzyme Complex Wheat Germ. *Conf. Proc. Int. Symp. Innov. Life Sci.* **2019,** *7*, 6–10. DOI: http://dx.doi.org/10.2991/isils-19.2019.2.
3. Ali, S.; Khan, A. S.; Anjum, M. A.; Nawaz, A.; Naz, S.; Ejaz, S.; Hussain, S. Effect of Postharvest Oxalic Acid Application on Enzymatic Browning and Quality of Lotus Root Slices. *Food Chem.* **2020,** *312*, 110–116.
4. Ali, S.; Khan, A. S.; Malik, A. U. Postharvest L-Cysteine Application Delayed Pericarp Browning, Suppressed Lipid Peroxidation and Maintained Antioxidative Activities of Litchi Fruit. *Postharvest Biol. Technol.* **2016,** *121*, 135–142.
5. Banerji, A.; Ananthanarayan, L. Dough Browning Inhibition of Multigrain Indian Flatbread Using a Combination of Chemical and Microwave Treatment. *J. Food Meas. Charact.* **2018,** *10*, 23–31.
6. BaninSogvar, O.; Razavi, F.; Rabiei, V.; Gohari, G. Postharvest Application of L-Cysteine to Prevent Enzymatic Browning of "Stanley" Plum Fruit During Cold Storage. *J. Food Process. Preserv.* **2020,** *44* (10), 1–12.
7. Baraldi, I. J.; Giordano, R. L. C.; Zangirolami, T. C. Enzymatic Hydrolysis as an Environmentally Friendly Process Compared to Thermal Hydrolysis for Instant Coffee Production. *Braz. J. Chem. Eng.* **2016,** *33* (4), 763–771.
8. Berg, J. M.; Tymoczko, J. L.; Stryer, L. *Biochemistry 5th Edition-Textbook*; 2002; pp 0–794.
9. Bhatia, S; Bhatia, S. Introduction to Enzymes and Their Applications. In *Introduction to Pharmaceutical Biotechnology*; Vol. 2; IOP Publishing Ltd.: Bristol, UK, 2018; 571pp.
10. Bilal, M.; Iqbal, H. M. N. State-of-the-Art Strategies and Applied Perspectives of Enzyme Biocatalysis in Food Sector—Current Status and Future Trends. *Crit. Rev. Food Sci. Nutr.* **2020,** *60* (12), 2052–2066.
11. Campo, A. F.; Sancho, M. I.; Melo, G.; Dávila, Y. A.; Gasull, E. In Vitro and in Vivo Inhibition of Hass Avocado Polyphenol Oxidase Enzymatic Browning by Paeonol, b-Cyclodextrin, and Paeonol: b-Cyclodextrin Inclusion Complex. *J. Biosci. Bioeng.* **2018,** *127* (6), 124–542.

12. Chanalia, P.; Gandhi, D.; Attri, P.; Dhanda, S. Extraction, Purification and Characterization of Low Molecular Weight Proline Iminopeptidase from Probiotic L. Plantarum for Meat Tenderization. *Int. J. Biol. Macromol.* **2018,** *109*, 651–663.
13. Chaudhary, S.; Sagar, S. The Use of Enzymes in Food Processing: A Review. *South Asian J. Food Technol. Environ.* **2015,** *1* (3 and 4), 190–210.
14. Cross, T. A. M2 Proton Channel from Influenza A: Example of Structural Sensitivity. *Environment* **2013,** 1075–1084.
15. Csermely, P.; Palotai, R.; Nussinov, R. Induced Fit, Conformational Selection and Independent Dynamic Segments : An Extended View of Binding Events. *Trends Biochem. Sci.* **2010,** *35* (10), 539–546.
16. Dal, L.; Pedro, J.; Pessoa, S.; Klein, M. P.; Fernandez-lafuente, R.; Rodrigues, R. C. Enzymatic Clarification of Orange Juice in Continuous Bed Reactors: Fluidized-Bed Versus Packed-Bed Reactor. *Catal. Today* **2020,** *2020* (Feb), 184–190.
17. Deepika, S.; Sutar, P. P. Combining Osmotic—Steam Blanching with Infrared—Microwave—Hot Air Drying : Production of Dried Lemon (Citrus limon L.) Slices and Enzyme Inactivation. *Dry. Technol.* **2018,** *2018*, 1–19.
18. Dewdney, P.A. Enzymes in Food Processing. *Nutr. Food Sci.* **2003,** *73* (4), 20–22.
19. Esatbeyoglu, T.; Rodríguez-werner, M.; Schlösser, A.; Rimbach, G. Fractionation, Enzyme Inhibitory and Cellular Antioxidant Activity of Bioactives from Purple Sweet Potato (*Ipomoea batatas*). *Food Chem.* **2016,** *127* (*6*), 26–30.
20. Fernandes, P.; Carvalho, F. Enzymes in Food Processing. In *Agro-Industrial Wastes as Feedstock for Enzyme Production*; Tucker, G. A., Woods, L. F. J.; Eds.; Elsevier Inc.: New York, 2016; pp 703–709.
21. Fleuri, L. F.; Delgado, C. H. O.; Novelli, P. K.; Pivetta, M. R.; Do Prado, D. Z.; Simon, J. W. Enzymes in Fruit Juice and Vegetable Processing. *Enzym. Food Beverage Process.* **2015,** *2015*, 255–280.
22. Foldes, F. F.; Ed. *Enzymes in Anesthesiology*; Springer: New York, 1978; 176pp.
23. Garcia-Jimenez, A.; García-Molina, F.; Teruel-Puche, J. A.; Saura-Sanmartin, A.; Garcia-Ruiz, P. A.; Ortiz-Lopez, A.; Rodríguez-López, J. N.; Garcia-Canovas, F.; & Munoz-Munoz, J. Catalysis and Inhibition of Tyrosinase in the Presence of Cinnamic Acid and Some of Its Derivatives. *Int. J. Biol. Macromol.* **2018,** *119*, 548–554.
24. Ghosh, P.; Pradhan, R. C.; Mishra, S. Low-Temperature Extraction of Indian Black Berry and Optimization of Enzymatic Clarification Using Box-Behnken Design. *J. Food Process Eng.* **2017,** *40* (2), 110–115.
25. Gupta, R.; Bhardwaj, K. N.; Choudhary, C.; Chandna, P.; Jain, K. K.; Kaur, A.; Kumar, S.; Shrivastava, B.; Ninawe, S.; Singh, A.; Kuhad, R. C. Biofuels: The Environment-Friendly Energy Carriers. *Biotechnol. Environ. Manag. Resour. Recovery* **2013,** *2013*, 125–148.
26. Handique, J.; Jyoti, S.; Nandan, B. Optimization of Banana Juice Extraction Using Combination of Enzymes. *J. Food Sci. Technol.* **2019,** *56* (8), 3732–3743.
27. Hassan, S. S.; Williams, G. A.; Jaiswal, A. K. Computational Modelling Approach for the Optimization of Apple Juice Clarification Using Immobilized Pectinase and Xylanase Enzymes. *Curr. Res. Food Sci.* **2020,** *1* (3), 243–255.
28. Hernandez, C. Lipase Inhibitory Activity of *Carica papaya, Chrysophyllumcainito, Corcorus*. *Food Res.* **2018,** *2* (1), 51–55.

29. Hithamani, G.; Medappa, H.; Chakkaravarthi, A.; Ramalakshmi, K.; Raghavarao, K. S. M. S. Effect of Adsorbent and Acidulants on Enzymatic Browning of Sugarcane Juice. *J. Food Sci. Technol.* **2018,** *55* (10), 4356–4362.
30. Homaida, M. A.; Yan, S.; Yang, H. Effects of Ethanol Treatment on Inhibiting Fresh-Cut Sugarcane Enzymatic Browning and Microbial Growth. *LWT—Food Sci. Technol.* **2017,** *77*, 8–14.
31. Huang, S.; Benchamas, G.; & Huang, G.Whole Processing and Use of Rice Polishings. *Innov. Food Sci. Emerg. Technol.* **2020,** *63* (May), 102373.
32. Huang, Z. Thermal Kinetics of Enzyme Inactivation, Color Changes, and Allicin Degradation of Garlic under Blanching Treatments. *J. Food Process Eng.* **2019,** *42* (3), 1–9.
33. Hui, Y. H.; Editors, A.; Nip, W.; Nollet, L. M. L.; Paliyath, G.; Simpson, B. K. *Food Biochemistry and Food Processing*; Simpson, B. K., Ed.; Wiley-Blackwell, 2012.
34. Ichihara, A. Recent Advances in Chemistry of Phytotoxins. *J. Synth. Org. Chem. (Jpn.)* **1987,** *45* (4), 357–368.
35. Illanes, A. Stability of Biocatalysts. *Electron. J. Biotechnol.* **1999,** *2* (1), 1–9.
36. Illera, A. E.; Sanz, M. T.; Beltrán, S.; Melgosa, R.; Solaesa, A. G.; Ruiz, M. O. Evaluation of HPCD Batch Treatments on Enzyme Inactivation Kinetics and Selected Quality Characteristics of Cloudy Juice from Golden Delicious Apples. *J. Food Eng.* **2018**, *221*, 141–150.
37. Illera, A. E.; Sanz, M. T.; Trigueros, E.; Beltrán, S.; Melgosa, R. Effect of High Pressure Carbon Dioxide on Tomato Juice: Inactivation Kinetics of Pectin Methylesterase and Polygalacturonase and Determination of Other Quality Parameters. *J. Food Eng.* **2018,** *239*, 64–71.
38. Iyengar, R.; Mcevily, A. J. Anti-Browning Agents: Alternatives to the Use of Sulfites in Foods. *Trends Food Sci. Technol.* **1992,** *31* (March), 60–64.
39. Kalita, D.; Holm, D. G.; LaBarbera, D. V.; Petrash, J. M.; Jayanty, S. S. Inhibition of α-Glucosidase, α-Amylase, and Aldose Reductase by Potato Polyphenolic Compounds. *PLoS One* **2018,** *13* (1), e0191025, 8.
40. Kannaujia, P. K.; Prasad, H. Enzymes and Its Importance in Food Processing: A Review Enzymes and Its Importance in Food Processing: A Review. *J. Prospect. Agric.* **2019,** *6* (August), 71–74.
41. Konkit, M.; Kim, W. Activities of Amylase, Proteinase, and Lipase Enzymes from Lactococcus Chungangensis and Its Application in Dairy Products. *J. Dairy Sci.* **2016,** *99* (7), 4999–5007.
42. Koshland, D. E. Application of a Theory of Enzyme Specificity to Protein Synthesis. *Proc. Natl. Acad. Sci.* **1958,** *44* (2), 98–104.
43. Koshland, Daniel E. The Key-Lock Theory and the Induced Fit Theory Introduction of the Induced Fit Theory. *Angew. Chem. Int. Ed.* **1994,** *33* (510), 2375–2378.
44. Kubo, M. T. K.; Curet, S.; Augusto, P. E. D.; Boillereaux, L. Multiphysics Modeling of Microwave Processing for Enzyme Inactivation in Fruit Juices. *J. Food Eng.* **2019,** *263* (July), 366–379.
45. Kubo, M. T. K.; Rojas, M. L.; Curet, S.; Boillereaux, L.; Augusto, P. E. D. Peroxidase Inactivation Kinetics Is Affected by the Addition of Calcium Chloride in Fruit Beverages. *LWT* **2018,** *89*, 610–616.
46. Kuddus, M. Introduction to Food Enzymes. In *Enzymes in Food Biotechnology*; Elsevier: New York, 2019; pp 1–18.

47. Ladole, M. R.; Pokale, P. B.; Varude, V.; Belokar, P. G.; Pandit, A. B. One Pot Clarification and Debittering of Grapefruit Juice Using Co-Immobilized Enzymes @ Chitosan MNPs. *Int. J. Biol. Macromol.* **2020,** *167*, 1297–1307.
48. Lante, A.; Tinello, F.; Nicoletto, M. UV-Light Treatment for Controlling Enzymatic Browning of Fresh-Cut Fruits. *Innov. Food Sci. Emerg. Technol.* **2016,** *34*, 141–147.
49. Lemieux, R. U.; Spohr, U. How Emil Fischer Was Led To The Lock and Key Concept for Enzyme Specificity. *Adv. Carbohydr. Chem. Biochem.* **1994,** *50* (C), 1–20.
50. Li, J., Xiao, Q.; Huang, Y.; Ni, H.; Wu, C.; Xiao, A. Tannase Application in Secondary Enzymatic Processing of Inferior Tieguanyin Oolong Tea. *Electron. J. Biotechnol.* **2017,** *28*, 87–94.
51. Li, Y.; Wu, Z.; Wang, X.; Yu, F.; Yang, M. Effects of Combined Infrared and Steam Blanching on Enzyme Inactivation and Product Quality of Chrysanthemum Indicum L. Flower. *J. Food Process. Preserv.* **2019,** *43* (10), 1–8.
52. Ling, B.; Lyng, J. G.; Wang, S. Effects of Hot Air-Assisted Radio Frequency Heating on Enzyme Inactivation, Lipid Stability and Product Quality of Rice Bran. *LWT—Food Sci. Technol.* **2018,** *91*, 453–459.
53. Liu, X.; Lu, Y.; Yang, Q.; Yang, H.; Li, Y.; Zhou, B.; Li, T.; Gao, Y.; Qiao, L. Cod Peptides Inhibit Browning in Fresh-Cut Potato Slices: A Potential Anti-Browning Agent of Random Peptides for Regulating Food Properties. *Postharvest Biol. Technol.* **2018,** *146* (May), 36–42.
54. Liu, X.; Ren, J.; Zhu, Y.; Han, W.; Xuan, H.; Ge, L. Physicochemical and Engineering Aspects the Preservation Effect of Ascorbic Acid and Calcium Chloride Modified Chitosan Coating on Fresh-Cut Apples at Room Temperature. *Coll. Surf. A: Physicochem. Eng. Asp.* **2016,** *502*, 102–106.
55. Loredo, A. B. G.; Contigiani, E. V.; Paula, L. G.; Alzamora, S. M. Inactivation Kinetics of Peroxidase and Polyphenol Oxidase in Peach Juice Treated with Gaseous Ozone. *Int. J. Food Sci. Technol.* **2017,** *53* (2), 347–355.
56. Lu, Y.; Zhu, J.; Shi, J.; Liu, Y.; Shao, D.; Jiang, C. Immobilized Enzymes from Geotrichum spp. Improve Wine Quality. *Appl. Microbiol. Biotechnol.* **2017,** *101* (17), 6637–6649.
57. Mansourbahmani, S.; Ghareyazie, B.; Kalatejari, S.; Mohammadi, R. S. Effect of Post-Harvest UV-C Irradiation and Calcium Chloride on Enzymatic Activity and Decay of Tomato (*Lycopersicon esculentum* L.) Fruit During Storage. *J. Integr. Agricult.* **2017,** *16* (9), 2093–2100.
58. Marcinkowska, M. Inactivation of Thioglucosidase from *Sinapis alba* (White Mustard) Seed by Metal Salts. *Molecules* **2020,** *25* (19), 8. E-article 4363, online.
59. Melo, G. M.; Filippa, M. A.; Ragazzo Sanchez, J. A.; Gasull, E. I. PPO Activity of Two Varieties of Pears: Control of Enzymatic Browning for Temperature Effect, Presence of Inhibitors and Complexation with b-Cyclodextrin. *Revista Mexicana de Ingeniería Química (Mex. J. Chem. Eng.)* **2019,** *19* (2), 877–887.
60. Merín, M. G.; Morata de Ambrosini, V. I. Application of a Grape Surface Majority Pectinolytic species, Aureobasidium pullulans, to Low-Temperature Red Winemaking: Development and Stability of Wine Color. *J. Wine Res.* **2020,** *31* (3), 218–239.
61. Michel, D. Conformational Selection or Induced Fit? New Insights from Old Principles. *Biochimie* **2016,** *128–129*, 48–54.
62. Minaev, M. Y.; Makhova, A. A. Recombinant Metalloprotease as a Perspective Enzyme for Meat Tenderization. *Potravin. Slovak J. Food Sci.* **2019,** *13* (1), 628–633.

63. Mohammadi, F.; Marti, A.; Nayebzadeh, K.; Marzieh, S. Effect of Washing, Soaking and pH in Combination with Ultrasound on Enzymatic Rancidity, Phytic Acid, Heavy Metals and Coliforms of Rice Bran. *Food Chem.* **2020,** *334* (1), E-article 127583.
64. Moon, K. M.; Kwon, E. Bin.; Lee, B.; Kim, C. Y. Recent Trends in Controlling the Enzymatic Browning of Fruit and Vegetable Products. *Molecules* **2020,** *25* (12), 98–102.
65. Mrabet, A.; Rodríguez-Gutiérrez, G.; Rubio-Senent, F.; Hamza, H.; Rodríguez-Arcos, R.; Guillén-Bejarano, R.; Sindic, M.; Jiménez-Araujo, A. Enzymatic Conversion of Date Fruit Fiber Concentrates Into a New Product Enriched in Antioxidant Soluble Fiber. *LWT* **2017,** *75*, 727–734.
66. Murugesh, C. S.; Subramanian, R. Applications of Enzymes in Processing Green Tea Beverages: Impact on Antioxidants. *Process. Impact Antioxid. Beverages*, **2014,** *2014,* 99–108.
67. Ojeda, G. A.; Sgroppo, S. C.; Martín-Belloso, O.; Soliva-Fortuny, R. Chitosan/Tripolyphosphate Nanoaggregates Enhance the Antibrowning Effect of Ascorbic Acid on Mushroom Slices. *Postharvest Biol. Technol.* **2019,** *156* (6), E-article 110934.
68. Otwell, W. S.; Iyengar, R. Inhibition of Enzymatic Browning in Foods and Beverages. *Crit. Rev. Food Sci. Nutr.* **1992,** *32* (3), 253–273.
69. Ozcan, O.; Yildirim, R. M.; Toker, O. S.; Akbas, N.; Ozulku, G.; Yaman, M. The Effect of Invertase Concentration on Quality Parameters of Fondant. *J. Food Sci. Technol.* **2019,** *56* (9), 4242–4250.
70. Pretzler, M.; Kampatsikas, I.; Barkat, M.; Rompel, A. Food Chemistry : X Inhibition of Apricot Polyphenol Oxidase by Combinations of Plant Proteases and Ascorbic Acid. *Food Chem.: X* **2019,** *4* (September), E-article 100053.
71. Ribeiro, J. A.; Cantillano, R. F. F.; Nora, F. R. Effect of 4-Hexylresorcinol on Post-Cut Browning and Quality of Minimally Processed Fuji Apple Fruits. *J. Food Measurement Charact.* **2020,** *14* (5), 2461–2471.
72. Romeih, E.; & Walker, G. Recent Advances on Microbial Transglutaminase and Dairy Application. *Trends Food Sci. Technol.* **2017,** *62*, 133–140.
73. Santos, F. G.; Aguiar, E. V.; Centeno, A. C. L. S.; Rosell, C. M.; Capriles, V. D. Effect of Added Psyllium and Food Enzymes on Quality Attributes and Shelf-Life of Chickpea-Based Gluten-Free Bread. *LWT* **2020,** *134* (May), 98–102.
74. Sarpong, F.; Yu, X.; Zhou, C.; Hongpeng, Y.; Bernard, B.; Junwen, U. Influence of Anti-browning Agent Pretreatment on Drying Kinetics, Enzymes Inactivation and Other Qualities of Dried Banana Under Relative Humidity-Convective Air Dryer. *J. Food Measurement Charact.* **2018,** *12* (2), 1229–1241.
75. Schmidt, M.; Toplak, A.; Quaedflieg, P. J.; Nuijens, T. Enzyme-Mediated Ligation Technologies for Peptides and Proteins. *Curr. Opin. Chem. Biol.* **2017,** *38*, 1–7.
76. Selamoglu, Z. Use of Enzymes in Dairy Industry. *Rev. Curr. Prog.* **2020,** *75* (1), 131–136.
77. Serrano, S.; Sta, C.; Windhab, E. J.; Bru, L.; Rugiero, S.; Fischer, P.; Kuster, S. Targeted Inhibition of Enzymatic Browning in Wheat Pastry Dough. *J. Agric. Food Chem.* **2018,** *66* (46), 12535–12360.
78. Sewalt, V.; Shanahan, D.; Gregg, L.; La Marta, J.; Carillo, R. The Generally Recognized as Safe Process for Industrial Microbial Enzymes. *Ind. Biotechnol.* **2016,** *12* (5), 295–302.
79. Shah, H. M. S.; Khan, A. S.; Ali, S. Pre-Storage Kojic Acid Application Delays Pericarp Browning and Maintains Antioxidant Activities of Litchi Fruit. *Postharvest Biol. Technol.* **2017,** *132* (3), 154–161.

80. Sharma, H. P.; Patel, H.; Sugandha. Enzymatic Added Extraction and Clarification of Fruit Juices. *Crit. Rev. Food Sci. Nutr.* **2017,** *57* (6), 1215–1227.
81. Sharma, R. R.; Ed. Enzyme Inhibition: Mechanisms and Scope. In *Enzyme Inhibition and Bioapplications*; Sharma, R. R., Ed.; IntechOpen, 2012 May; DOI: 10.5772/39273. https://www.intechopen.com/chapters/36550 (accessed Aug 2, 2021).
82. Sheng, Z.; Ge, S.; Xu, X.; Zhang, Y.; Wu, P.; Zhang, K.; Xu, X.; Li, C.; Zhao, D.; Tang, X. Design, Synthesis and Evaluation of Cinnamic Acid Ester Derivatives as Mushroom Tyrosinase Inhibitors. *Med. Chem. Commun.* **2018,** *9* (5), 853–861.
83. Siddiqui, M. W.; Homa, F.; Nayyer, M. A.; Ghatak, A.; Yousuf, B.; Aftab, M. A.; Ali, A. Browning Metabolism and Quality of Fresh-Cut Drumstick as Influenced by Acidulant Treatments. *J. Food Sci. Technol.* **2020,** *57* (6), 2010–2016.
84. Sikora, M.; Michał, S. Effect of Ascorbic Acid Postharvest Treatment on Enzymatic Browning, Phenolics and Antioxidant Capacity of Stored Mung Bean Sprouts. *Food Chem.* **2018,** *239*, 1160–1166.
85. Singh, B.; Suri, K.; Shevkani, K.; Kaur, A.; Kaur, A.; Singh, N. Enzymatic Browning of Fruit and Vegetables. In *Enzymes in Food Technology*; Springer: Singapore, 2018; pp 63–78.
86. Singh, R.; Singh, A.; Sachan, S. Enzymes Used in the Food Industry: Friends or Foes? In *Enzymes in Food Biotechnology: Production, Applications, and Future Prospects*; Elsevier Inc.: New York, 2018; pp 827–843.
87. Singh, T. P.; Sogi, D. S. Inhibition of Lipase Activity in Commercial Rice Bran of Coarse, Fine, and Superfine Cultivars. *Cogent Food Agric.* **2016,** *2016*, 1–9.
88. Tolouie, H.; Mohammadifar, M. A.; Ghomi, H.; Yaghoubi, A. S.; Hashemi, M. The Impact of Atmospheric Cold Plasma Treatment on Inactivation of Lipase and Lipoxygenase of Wheat Germs. *Innov. Food Sci. Emerg. Technol.* **2018,** *47*, 346–352.
89. Toushik, S. H.; Lee, K. T.; Lee, J. S.; Kim, K. S. Functional Applications of Lignocellulolytic Enzymes in the Fruit and Vegetable Processing Industries. *J. Food Sci.* **2017,** *82* (3), 585–593.
90. Tsai, C.; Kumar, S.; Ma, B.; Nussinov, R. Folding Funnels, Binding Funnels, and Protein Function. *Protein Sci.* **1999,** *8 (6)*, 1181–1190.
91. Tsai, C.; Ma, B.; Nussinov, R. Commentary Folding and Binding Cascades: Shifts in Energy Landscapes. *Proc. Natl. Acad. Sci. U. S. A.* **1999,** *96*(8), 9970–9972.
92. Tucker, G. A.; Woods, L. F. J.; Eds. *Enzymes in Food Processing.* Springer: v, 1995; pp 26–40.
93. Ugur, I.; Schatte, M.; Marion, A.; Glaser, M.; Boenitz-dulat, M.; Id, I. A. Ca^{2+} Binding Induced Sequential Allosteric Activation of Sortase A: An Example for Ion-Triggered Conformational Selection. *PLoS One* **2018,** *2018* (October), 1–19.
94. van Donkelaar, L. H. G.; Mostert, J.; Zisopoulos, F. K.; Boom, R. M.; van der Goot, A. J. The Use of Enzymes for Beer Brewing: Thermodynamic Comparison on Resource Use. *Energy* **2016,** *115*, 519–527.
95. Webber, V.; Dutra, S. V.; Spinelli, F. R.; Carnieli, G. J.; Cardozo, A.; Vanderlinde, R. Effect of Glutathione During Bottle Storage of Sparkling Wine. *Food Chem.* **2017,** *216*, 254–259.
96. Whitaker, J. R.; Lee, C. Y. Recent Advances in Chemistry of Enzymatic Browning. *J. Synth. Org. Chem.—Jpn.* **1995,** *15* (1), 2–7.
97. Wong, D. W. S. Food Enzymes. *J. Chem. Inform. Model.* **1995,** *53* (9), 53–74.

98. Xu, J.; Zhou, L.; Miao, J.; Yu, W.; Zou, L.; Zhou, W.; Liu, C.; Liu, W. Effect of Cinnamon Essential Oil Nanoemulsion Combined with Ascorbic Acid on Enzymatic Browning of Cloudy Apple Juice. *Food Bioprocess Technol.* **2020,** 13 (5), 860–870.
99. Xu, J.; Qi, Y.; Zhang, J.; Liu, M.; Wei, X.; Fan, M. Effect of Reduced Glutathione on the Quality Characteristics of Apple Wine During Alcoholic Fermentation. *Food Chem.* **2019,** *300* (6), 125–130.
100. Yan, S.; Luo, Y.; Zhou, B.; Ingram, D. T. Dual Effectiveness of Ascorbic Acid and Ethanol Combined Treatment to Inhibit Browning and Inactivate Pathogens on Fresh-Cut Apples. *LWT—Food Sci. Technol.* **2017,** *80*, 311–320.
101. Yee, W.; Cing, L.; Cheun, F.; Wai, C. Inhibition of Enzymatic Browning in Sweet Potato with Chemical and Natural Anti-Browning Agents. *J. Food Process. Preserv.* **2019,** *43* (11), 1–8.
102. Yildiz, G. The Effect of Different Chemical Agents on the Prevention of Enzymatic Browning in Banana. *J. Food Sci. Eng.* **2018,** *7*, 91–96.
103. Zhang, J.; Zhao, P.; Liu, B.; Meng, X. Use of Oligochitosan as an Inhibiting Agent of Apple Juice Enzymatic Browning. *J. Food Process. Preserv.* **2017,** *41* (4), 1–9.
104. Zhang, Y.; He, S.; Simpson, B. K. Enzymes in Food Bioprocessing—Novel Food Enzymes, Applications, and Related Techniques. *Curr. Opin. Food Sci.* **2018,** *19*, 30–35.
105. Zheng, J.; Li, S.; Xu, Y.; Zheng, X. Effect of Oxalic Acid on Edible Quality of Bamboo Shoots (*Phyllostachys prominens*) Without Sheaths During Cold Storage. *LWT* **2019,** *2019*, 194–200.
106. Zhou, L.; Liu, W.; Stockmann, R.; Terefe, N. S. Effect of Citric Acid and High Pressure Thermal Processing on Enzyme Activity and Related Quality Attributes of Pear Puree. *Innov. Food Sci. Emerg. Technol.* **2018,** *45*, 196–207.
107. Zhou, X.; Xiao, Y.; Meng, X.; Liu, B. Full Inhibition of Whangkeumbae Pear Polyphenol Oxidase Enzymatic Browning Reaction by L-Cysteine. *Food Chem.* **2018,** *266*, 1–8.

CHAPTER 2

NOVEL FOOD PROCESSING TECHNOLOGIES FOR INACTIVATION OF FOOD ENZYMES

B. KAMALAPREETHA, RAVI PANDISELVAM,
ANJINEYULU KOTHAKOTA, WASIYA FARZANA,
B. L. DINESHA, and V. PRITHVIRAJ

ABSTRACT

The trend of consuming fresh fruits and vegetables has been continuously increasing all over the globe. Enzymes cause an endogenous deteriorative reaction along with other microbial growth which decreases the shelf-life of horticultural produce. Industries have encountered these challenges using the conventional heat treatment processes to inactivate enzymes; however, these thermal processing techniques cause undesirable changes in sensory parameters and a reduction of food nutritional profile. Nearly 20 years the industries and academia were in quest of alternative techniques to solve these issues; thus, the nonthermal techniques emerged with minimal use of heat and preservatives. Nonthermal technologies have helped in quality retention and shelf-life enhancement of the produces by eliminating the enzymatic effects. This chapter emphasizes on effectiveness of cold plasma and hybrid processing technologies for inactivation of food enzymes that can degrade the food quality.

2.1 INTRODUCTION

Fresh horticultural produces are subjected to processing to enhance their edibility and storability. These are major source of sugars, vitamins,

Enzyme Inactivation in Food Processing: Technologies, Materials, and Applications. Megh R. Goyal, PhD, Junaid Ahmad Malik, PhD & Ravi Pandiselvam, PhD (Eds.)

minerals, fiber, antioxidants, phytochemicals, and many more. The main aim of fruit and vegetable processing is not only to increase the shelf-life but also to retain the nutritional quality, texture, and color by constraining the activities of enzymes and microorganisms. During the processing of fresh horticultural produce, exposure of cut pieces to atmospheric condition enhances the metabolic process and results in a significant variation in their physical and biochemical properties. Quality degradation (such as discolorations and rancidity) in fresh and processed fruits and vegetables (FAV) is due to peroxidase (POD), pectinesterase (PE), pectin methylesterase (PME), polyphenol oxidase (PPO), β-glucosidase, lipase, polygalacturonase (PG), and lipoxygenase enzymes.[78] For example, PPO and POD cause browning in minimally processed fruits and fruit juices[3] and also affect the subjective characteristics of wine.[24] PE causes undesired cloudy nature in orange juice.[3]

Enzymes are specialized proteins that catalyze the biological reaction, which initiates from active sites bonded by the three-dimensional molecular configuration. Enzymes lose their activity, even at a small change in active sites. Hence, protein degradation is linked with changes in the functionality of the enzyme.[15] In fresh fruits and vegetables, enzymes and the substrates are positioned in different compartments. During the food processing operation, this compartment gets damaged, which results in enzyme–substrate interaction leading to low-quality characteristics.[15]

Oxidative enzymes are responsible for the quality deterioration of processed food in terms of color, flavor, and nutritional profile.[30,64] Structural properties of fruits and vegetables depend on the composition, cell wall, and middle lamella. Enzymes including PME and PG are rupturing the pectin network in the cell wall.[14] Thus, enzyme inactivation would increase the shelf-life of food products.

Thermal or blanching treatment is widely used for enzyme inactivation in horticultural produce. Thermal processing causes denaturation of proteins at a higher temperature.[38] However, blanching (≥65°C) causes an objectionable effect on the quality of horticultural produce, especially heat-sensitive food produces.[7,32,63] Tanimoto[68] and Buzrul[6] observed that pasteurization (at 65°C) leads to loss in its flavor and taste.

To eradicate the limitations of blanching, developments of novel nonthermal methods are essential for enzyme inactivation in solid and liquid foods.[63] Advances in nonthermal technologies in recent decades have proved to enzymes inactivation either solely or in hybrid treatment. Technologies like Ozone, Cold Plasma, Ultraviolet, Pulsed Light, Ultrasound, and Irradiation have proved the efficiency against enzymes in food matrices prolonging their shelf-life. These advanced processing techniques eliminate high-temperature treatment.[49,50] The use of these treatments has become a more attractive in food industries to inactivate the enzyme because these are non-flammable, nontoxic, and leave no toxic components and/or residue.[7,66]

The focus of this paper is on analyzing the effectiveness of nonthermal technologies on enzyme inactivation, mechanism of enzyme inactivation, and the combination of various food processing technologies (hybrid technologies) for inactivating the quality degrading enzymes.

2.2 COLD PLASMA (CP)

CP application in food industry extended due to its dry nature and production at atmospheric temperature and pressure.[44] The CP is potential decontamination technique to control the enzyme and microbial population of various food products including dairy, meat, oils, fats, as well as fruits and vegetables, processes for decomposition of mycotoxins, treatment of wastewater from food industries, as well as complex dyes that take complex procedures can be simplified, effective tool for food surface modification, modification of functional properties of edible materials can be effectively carried out to utilize low cost by products, surface modulation of packaging materials for expanding their use, traditional use of chemicals on food can be eliminated for stabilization, and an amateur technology for minimally processed and fresh foods. Its delayed introduction in the food sector owes to its safety on complex food matrices. However, a combination of applications in the food industry alone has led to its increasing popularity and acceptability as well as inciting the scientists and researchers to exploit it at its fullest. Not only has it been successful in the destruction of microorganisms and beetles, *Tribolium castaneum*[54] but also as a pretreatment technology for binding of films in the packaging industry.[17]

The efficacy of the use of cold plasma can generate a number of primary and secondary products. Formation of secondary products has been reported only in extended treatment time with CP rendering it safe on certain food products.[62] Surface Dielectric Barrier Discharge (SDBD) had initially been used in the aerospace industry and then extended its application in food industry.[22] Hertwig et al.[16] suggested the use of CP as an alternate technology for almond pasteurization. Timmons et al.[72] revealed the SDBD effect on the inactivation of different subspecies of *Salmonella enterica* and *Listeria monocytogenes* and showed its potential use over the existing CP methods to be adaptable, easily scalable, and economic for use in large-scale food production applications. Though CP has become more popular in recent decades, yet current findings have further widened their scope and product versatility. This has led the researchers, scientists, and industrialists to foresee its benefits and thereby give a different direction to the research.

2.2.1 ENZYME INACTIVATION BY COLD PLASMA

Cold plasma (CP) being a relatively newer technology has shown to inactivate bacteria, spores, and fungi, thereby giving a scope for enzyme inactivation that is an important aspect for quality maintenance being a marker for efficiency of any applied food process.[18,19,20,25,65] Plasma application can enhance seed germination during drought conditions.[60] A recent study on fresh-cut melons showed an increased stability with slight reduction of enzymatic activity endemically evident and minimal effect on the quality. Few observations were defined, the dry matter content was increased, and the appearance was translucent during storage, along with a reduction of metabolic heat.[69] Pankaj et al.[51] highlighted the recent application of cold plasma technology on in-packed fresh fruits for stabilization which could possibly replace chemical sanitization used traditionally. Yin et al.[76] depicted that the plasma treatment there was an increase in the protease enzymes activity due to the increase of soluble protein (almost 8 mg/g increase). Plasma-treated samples showed better digestibility and increase in soluble protein content which contributes to an increase in protease activity.[27,31,40,74]

2.2.2 MECHANISM OF INACTIVATION

The decontamination in the substance or components occurs due to reactive species generation during plasma discharge. The reactive species in the Dielectric Barrier Discharge (DBD) includes nitrogen (N_2^+, NO^-) and oxygen (OH^-).[53,55] The inactivation mechanism is due to the partial or total destruction of the plasma membrane by membrane lipids oxidation with consequent leakage of intracellular components such as proteins and amino acids. The oxidation may lead to cell injury and death.[10]

Microorganisms could be effectively targeted utilizing atmospheric cold plasma (ACP) technique, wherein interactions of atoms, radical species, and charged molecules along with microorganisms inactivate the latter. Physical interaction of high energy ions, oxidized radicals, ROS (including ozone, peroxide, superoxide, and hydroxide molecules) with the enzymes and cell walls of bacterium are effective in inactivating the undesirable activities of the enzymes.[4,35,77]

2.2.3 EFFECT OF CP ON BIOMOLECULES

Biomolecules such as starch, proteins, lipids, and nucleic acids are an integral part of any food industry. Processing them in their available form does not always serve the purpose due to the drawbacks like nonsolubility of starch in cold water, coagulation of proteins when exposed to thermal treatments. A considerable modification using physical, chemical, and enzymatic processes in the fundamental biomolecules can lead to better functionality broadening the areas of application in the food industry. Atmospheric cold plasma (ACP) as a nonthermal treatment has proved to modify the basic functional properties.[41,75] Natural resins and gums have been consistently used in food formulations. Xanthan gum powders, when treated with ACP, have shown elevated viscosity and emulsion stabilizing properties that are beneficial characteristics for salad dressings and instant dry soup formulations minimally affecting the molecular structure of the compound.[41] The physicochemical properties of the protein powdered can be effectively altered using ACP. Zhang et al.[78] tested for the feasibility of cold plasma treatment as a pretreatment on chili peppers.

Pertaining to the quality attributes, the retention of the red pigment was shown to have improved with 30 s exposure with an increase in the antioxidant property of the chili peppers. The probable reason explained was that the cold plasma can oxalate or damage the cell wall having the pigment, thereby enhancing the extraction rate of the pigment contents. Longer exposure times lead to a loss in the color pigment. Thirumdas et al.[71] investigated the nutritional composition, textural and cooking characteristics of brown rice treated by plasma. Results showed a reduction in chewiness and contact angle with no change in proximate composition. Cooking time and hardness showed a significant decrease from 29.1 to 21.1 min and from 40.47N to 30.09N, respectively. The cooking time of rice grains was reduced because of increase in surface energy causing the surface to become hydrophilic thereby absorbing more water.[25]

2.3 HYBRID TECHNIQUES FOR INACTIVATION OF ENZYMES

Combining elevated temperature with pressure[34] and pulsed electric fields[13] have proven synergistic effect on the inactivation of enzyme in terms of energy saving. Chemat et al.[8] have addressed that in terms of ultrasonication, inactivation of enzymes is more effective only with the combination of heat and pressure compared to treatment used individually. Mano thermosonication (MTS) is the process of combining ultrasound with pressure and heat.

Liao et al.[29] and Lerasle et al.[26] have reported that combination of HPP and pressurized CO_2 at 50 MPa has the ability in affecting the microorganisms and enzymes by molecular effect. In Coconut water,[37,42] microfiltration can remove the enzymes, but it causes loss in some of the essential nutrients, change of flavor, and decrease in the product acceptability.[57] Combination of thermal treatment along with the preservative is the most common form of process on coconut water preservation. But, it results in the negative impact on the nutritional and the sensory properties so an alternative method is more advisable.

Generally, the inactivation of oxidative enzyme and keeping quality of foods by phenolic activity gave better results when combining pressure treatment along with physical or chemical methods.

2.3.1 HYBRID TECHNIQUES ON INACTIVATION OF ENZYME IN HORTICULTURAL PRODUCE

2.3.1.1 INACTIVATION OF ENZYMES IN FRUITS

During storage and processing of horticultural produces, the most important enzymes are classified as hydrolases and oxidoreductases. During post-harvest storage, increased POD activity will result in close association with anthocyanin degradation and discoloration (browning) was reported in a number of studies.[73,79] A study shows that in fresh-cut fruits, browning (limiting factor) is influenced by peroxidase and polyphenol oxidase enzymes.[21] Browning of fruits and vegetables are neglected by the consumers. In this regard, inactivation of enzymes is required. Enzymes responsible for enzymatic browning have been inactivated by combining ultrasonication with ascorbic acid treatment.

A recent study was done on the inactivation kinetics and thermal stability of blueberry peroxidase using the combination of thermal treatment and high pressure.[70] This shows the complete inactivation of enzyme was investigated at an optimized combination of 690 MPa (HPP) and 90°C (thermal treatment) for 15 min. It has also been concluded that practically considerable inactivation of Blueberry POD occurred under high pressure only when the treatment temperature reaches 80°C.

Rapeanu et al.[56] reported the combined effect of HPP at 600 MPa with mild heat treatment (20–55°C) on the Polyphenol oxidase inactivation rate constant in white grape. This study concluded that PPO inactivation was achieved with increasing temperature. In Mango, combination of HPP and thermal treatment achieved the enzyme inactivation.[28] MacDonald et al.[36] investigated the combined effect of moderate pressure (50–139 MPa) and mild thermal treatment on Peroxidase enzyme in banana. In all combinations, inactivation was achieved. Maximum inactivation of 98.6% was attained in cell-free extract in a combination of 110 MPa for 5 min at 70°C. Representatives of different studies on enzyme inactivation in different fruits are shown in Table 2.1.

TABLE 2.1 Different Studies on Enzyme Inactivation in Fruits.

Fruits	Enzyme inactivated	Combination of technologies	Optimized treatment condition	Reference
Banana	Peroxidase	HPP and thermal	110 MPa and 70°C for 5 min	[36]
Blueberry	Peroxidase	HPP and thermal	690 MPa and 90°C	[70]
Fresh cut apple	Polyphenol oxidase and peroxidase	Sonication and ascorbic acid	40 kHz	[48]
Mango	Soluble acid Invertase	HPP and thermal treatment	50–60 MPa, 40–50°C for 1–60 min	[28]
White grapes	Polyphenol oxidase	HPP and mild heat treatment	600 MPa and 20–55°C	[56]

HPP, high pressure processing; US, ultra sonication; DPCD, dense phase carbon dioxide; PEF, pulsed electric field; LED, light emitting diode; UV, ultraviolet

2.3.1.2 *INACTIVATION OF ENZYMES IN VEGETABLES*

Lopez et al.[33] observed the combined ultrasound at 20 kHz and thermal treatment in the inactivation of soybean lipoxygenase buffer. Taste and odor of the food was not altered during the treatment. Duong et al.[12] studied the combined effect of HPP and dense phase CO_2 on poly phenol oxidase inactivity in mushroom. Synergistic effect of HHP and CO_2 treatment could decrease the REA (residual enzyme activity) up to 30% at 1 min and 12% at 9 min of treatment time. It was concluded that addition of CO_2 to HPP would increase the enzyme inactivation by altering the secondary protein conformations.

In feijoa puree, a combined synergistic effect of DPCD and HPP was reported on the enzyme inactivation at the pressures level of below 600 MPa for ≥13 min.[11,46] Park et al.[52] have experimented the combination effect of HPP and HPCD on PPO inactivation in carrot juice. This study has compared the residual activity of PPO in combination effect (HPP & HPCD), HPP and HPCD attained 19%, 39%, and 61%, respectively. The same study also reported on the Lipoxygenase inactivation. Synergetic effect of HPP and HPCD is reported to attain residual activity of less than 25% than the individual treatment having the residual activity of 30%. It is concluded that combination of treatment had increased enzyme inactivation. Kim et al.[23] have reported against the 95% enzyme (Lipoxygenase, Poly Methyl Esterase and Poly Phenol Oxidase) inactivity in carrot juice when applying optimized condition of combined processing at 400 MPa and at 70°C for 10 min and it also retained the stability of the α-Carotene and β-Carotene (related to Lipoxygenase).

Combined effect of thermal blanching (85°C, and up to 3 min) along with 11 J/m² UV-C radiation pretreatment was investigated[43] on the kinetics of POD inactivation in zucchini slices represented in first-order kinetics. The study showed a highest inactivation of approximately 35% using combined processing than thermal blanching alone. Representatives of different studies on enzyme inactivation of different fresh vegetables are shown in Table 2.2.

TABLE 2.2 Different Studies on Enzyme Inactivation in Vegetables.

Vegetable	Enzyme inactivated	Combination of technologies	Optimized treatment condition	Reference
Feijoa puree	Pectin methyl esterase, Polyphenol oxidase Peroxidase	HHP and DPCD	600 MPa, 20–25°C	[11, 46]
Mushroom	Polyphenol oxidase	HPP and DPCD	600 MPa, 3.6% w/w of CO_2 for 1–9 min	[12]
Soyabean	Soybean lipoxygenase	US and Heat	20 kHz	[33]

TABLE 2.2 *(Continued)*

Vegetable	Enzyme inactivated	Combination of technologies	Optimized treatment condition	Reference
Zucchini	Peroxidase	UV-C radiation and thermal blanching	11 J/m^2 and 85°C for 3 min	[43]
Carrot Juice	Lipoxygenase	HPP and thermal	100–600 MPa & 25–70°C	[23]
	Poly phenol oxidase	HPP and HPCD	400 MPa, 25°C, 5 min & 4.9 MPa,5°C, 5 min	[52]
	Lipoxygenase		200 ~ 400 and 2.9 ~ 4.9 MPa (HPCD)	

2.3.1.3 INACTIVATION OF ENZYMES IN FRUIT JUICES

The enzyme (polyphenolase, pectin methyl esterase, and peroxidase) inactivation of apple juice was evaluated using the combined effect of ultrasonication (US) and HPP.[2] Residual activity of POD, polyphenolase, and PME at US-HHP at 450 MPa are 32.67 ± 0.06%, 21.13 ± 0.05%, and 23.93 ± 0.05% respectively. Even though all the combined US-HHP treatment achieved the inactivation, the highest reductions of all three enzymes were achieved for US-HHP 450 MPa.

Ortuño et al.[47] highlighted the effect of using DPCD and high-power ultrasound on PME inactivation in a batch system for orange juice. Increased inactivation is achieved at increased pressure and temperature on PME. The combining UV and thermal processing could enhance the enzyme inactivation rate. Enzymes were inactivated permanently without regaining its activity even after its storage.[61] Rojas et al.[59] emphasized that ultrasonication (US) is a promising technology in inactivating the enzyme (Peroxidase) before thermal processing for green coconut water. The enzymes (Peroxidase) get inactivated to heat (at 85°C) after US pretreatment (286 W/L at 25°C, 20 kHz) required 46% less time compared to untreated product to achieve 99% inactivity. Seyderhelm et al.[67] highlighted that combining thermal treatment between 45°C and 55°C along with pressures treatment between 600 and 900 MPa tends to inactivate pectin esterase and phosphatase in different extensions.

A study emphasized the use of ultrasound-(19 kHz) assisted ozone (1.50 ppm) treatment on the enzyme inactivation in acai juice.[45] For this combined process, increasing the ozone dose and the energy density of ultrasound resulted in influencing the decrease of enzyme activity (Peroxidase). In contrast, polyphenol oxidase activity was reported to be higher while using the combined treatment than the individual process. Merve and Sevcan[39] reported that UV-LED irradiation in combination of UV-C rays was found to have a significant reduction in PPO activity to 32.58% in apple juice. Aadil et al.[1] quoted synergistic effect of ultrasonication and thermal effect on the fresh grape fruit juice samples at different processing variables for the destruction of enzymes. Maximum reduction of PME, PPO, and POD activity was achieved to be 91%, 90%, and 89%, respectively, at 60°C for 60 min of exposure.

TABLE 2.3 Different Studies on Enzyme Inactivation of Fruit Juices.

Fruit juice	Enzyme inactivated	Combination of technologies	Optimized treatment condition	Reference
Acai pulp	Peroxidase Polyphenol oxidase	US and Ozone	19 kHz at 350 and 700 J/mL 1.50 ppm for 5 and 10 min	[45]
Apple juice	Polyphenolase, Peroxidase Pectin methyl esterase	US and HPP	25 kHz and 70% amplitude 250, 350, or 450 MPa for 10 min	[2]
Cloudy apple juice	Poly phenol oxidase	UV A and UV C LED irradiation	254/280/365/405 nm for 20 and 40 min	[39]
Fresh Grape juice	Pectin methyl esterase, Peroxidase Poly phenolase	US and thermal	28 kHz at power 70% for 420 W. 60°C for 60 min	[1]
Green Coconut water	Peroxidase	US and thermal	286 W/L at 25°C, 20 kHz for 30 min 80°C, 85°C and 90°C	[59]
Orange juice	Pectin methyl esterase	DPCD and US	–	[47]
	Poly Methyl Esterase	HPP and CO_2	CO_2 (1 MPa) and HHP (800 MPa, 25°C, 1 min)	[9]
			CO_2 (0.8 MPa) and HHP (600 MPa, 25°C, 130 s)	[5]

TABLE 2.3 *(Continued)*

Fruit juice	Enzyme inactivated	Combination of technologies	Optimized treatment condition	Reference
Enzyme dissolved in orange juice	Peroxidase (POD), Lipoxygenase (LOX) and Poly phenol oxidase (PPO)	Thermal and UV-C	25–65°C for 1–10 min	[61]
Blended (orange—80% and carrot—20%) juice	Pectin methyl esterase	PEF and thermal	25 kV/cm and 68°C and 25 kV/cm and 71°C	[58]

Experiment was conducted in blended (orange—80% and carrot—20%) juice to emphasize the effect of different intensities of PEF and conventional heat treatment (98°C for 21 s) on inactivation of Pectin methyl esterase. Inactivation was achieved to be 75.6% and 81% for P1 and P2, but a maximum of 98% reduction was found in the pasteurized samples (HTST). Increase in intensity also shows a less increase in the inactivation percentage.[58] Representatives of different studies on enzyme inactivation of fruit juices are shown in Table 2.3.

2.4 SUMMARY

Quality of the food gets reduced due to the enzyme activity which is not desirable. In this regard, novel food preservation is getting increased interest for satisfying the demand of food safety and to provide nutritious and healthy food to eventual consumer. Various hybrid technologies are discussed in this review for the enzyme inactivation or destruction of various fruits, vegetable, and fruit juices. Resistance of certain enzymes to thermal treatment paves the way for alternative technologies. The combination of different technologies has been shown potential for inactivation of enzyme. There is a demand for finding the deteriorative enzyme inactivation through a combined approach. To make it suitable for the practical application keen attention is required in terms of considering the different combination of technologies to achieve the satisfactory inactivation of deteriorative enzymes.

KEYWORDS

- **cold plasma**
- **food preservation**
- **food safety**
- **high pressure processing**
- **nonthermal technology**
- **sonication**

REFERENCES

1. Aadil, R. M.; Zeng, X.; Zhang, Z. H.; Wang, M. S.; Han, Z.; Jing, H.; Jabbar, S. Thermo-Sonication That Influences Quality of Grapefruit Juice. *Int. J. Food Sci. Technol.* **2015,** *50*, 1275–1282.
2. Abid. M.; Saqib, J.; Hu, B.; Hashim, M. M.; Wu, T.; Wu, Z.; Khan, M. A.; Zeng, X. Synergistic Impact of Sonication and High Hydrostatic Pressure on Microbial and Enzymatic Inactivation of Apple Juice. *LWT—Food Sci. Technol.* **2014,** *2014*, 1–7.
3. Bayindirli, A.; Alpas, H.; Bozoglu, F.; Hizal, M. Efficiency of High-Pressure Treatment on Inactivation of Pathogenic Microorganisms and Enzymes in Apple, Orange, Apricot and Sour Cherry Juices. *Food Control* **2006,** *17* (1), 52–58.
4. Birmingham, J. G. Mechanisms of Bacterial Spore Deactivation using Ambient Pressure Non-thermal Discharge. *IEEE Trans. Plasma Sci.* **2004,** *32*, 1526–1531.
5. Boff, J. M.; Truong, T. T.; Min, D. B.; Shellhammer, T. H. Effect of Thermal Processing and Carbon Dioxide-Assisted High-Pressure Processing on Pectin Methylesterase and Chemical Changes in Orange Juice. *J. Food Sci.* **2003,** *68*, 1179–1184.
6. Buzrul, S. High Hydrostatic Pressure Treatment of Beer and Wine: A Review. *Innov. Food Sci. Emerg. Technol.* **2012,** *13*, 1–12.
7. Cappelletti, M.; Ferrentino, G.; Endrizzi, I.; Aprea, E.; Betta, E.; Corollaro, M. L.; Charles, M.; Gasperi, F.; Spilimbergo, S. High Pressure Carbon Dioxide Pasteurization of Coconut Water—A Sport Drink with High Nutritional and Sensory Quality. *J. Food Eng.* **2015,** *145,* 73–81.
8. Chemat. F.; Huma, Z.; Khan, M. K. Applications of Ultrasound in Food Technology: Processing, Preservation and Extraction. *Ultrason. Sonochem.* **2011,** *18*, 813–835.
9. Corwin, H.; Shellhammer, T. H. Combined Carbon Dioxide and High-Pressure Inactivation of Pectin Methylesterase, Polyphenol Oxidase, Lactobacillus Plantarum and *E. coli*. *J. Food Sci.* **2002,** *67*, 697–701.
10. Critzer, F.; Kellywintenberg, K.; South, S.; Golden, D. A. Atmospheric Plasma Inactivation of Foodborne Pathogens on Fresh Produce Surfaces. *J. Food Prot.* **2007,** *70* (10), 2290–2296.
11. Duong. T.; Balaban, M. Optimisation of the Process Parameters of Combined High Hydrostatic Pressure and Dense Phase Carbon Dioxide on Enzyme Inactivation in

Feijoa (*Accasellowiana*) Puree Using Response Surface Methodology. *Innov. Food Sci. Emerg. Technol.* **2014,** *26*, 93–101.
12. Duong. T.; Balaban, M.; Perera, C. Effects of Combined High Hydrostatic Pressure and Dense Phase Carbon Dioxide on the Activity, Structure and Size of Polyphenoloxidase. *J. Food Sci.* **2015,** *80* (11), 2486–2494.
13. Espachs-Barroso, A. G. V.; Barbosa-Cánovas, O.; Martín-Belloso. Microbial and Enzymatic Changes in Fruit Juice Induced by High-Intensity Pulsed Electric Fields. *Food Rev. Int.* **2003,** *19* (3), 253–273.
14. Giovane, A.; Servillo, L.; Balestrieri, C.; Raiola, A.; D'avino, R.; Tamburrini, M.; Ciardiello, M.; Camardella, L. Pectin Methylesterase Inhibitor. *Biochim. Biophys. Acta-Prot. Proteom.* **2004,** *1696* (2), 245–252.
15. Hendrickx, M.; Ludikhuyze, L.; Broeck, I.; Van Den, Weemaes, C. Effects of High Pressure on Enzymes Related to Food Quality. *Trends Food Sci. Technol.* **1998,** *9* (5), 197–203.
16. Hertwig, C.; Leslie, A.; Meneses, N.; Reineke, K.; Rauh, C.; Schlüter, O. Inactivation of *Salmonella Enteritidis* PT30 on the Surface of Unpeeled Almonds by Cold Plasma. *Innov. Food Sci. Emerg. Technol.* **2017,** *44*, 242–248.
17. Honarvar, Z.; Farhoodi, M.; Khani, M. R.; Mohammadi, A., Shokri, B.; Ferdowsi, R.; Shojaee, S. Application of Cold Plasma to Develop Carboxymethyl Cellulose Coated Polypropylene Films Containing Essential Oil. *Carbohydr. Polym.* **2017,** *176*, 1–10.
18. Hosseini, S. I.; Farrokhi, N.; Shokric, K.; Khanid, M. R.; Shokri, B. Cold Low-Pressure O_2 Plasma Treatment of *Crocus Sativus*: An Efficient Way to Eliminate Toxicogenic Fungi with Minor Effect on Molecular and Cellular Properties of Saffron. *Food Chem.* **2018,** *257*, 310–315.
19. Huang, K.; Tian, H.; Gai, L.; Wang, J. Kinetic Models for Inactivating Microorganisms and Enzymes by Pulsed Electric Field Processing. *J. Food Eng.* **2012,** *111* (2), 191–207.
20. Innocente, N.; Segat, A.; Manzocco, L.; Marino, M.; Maifreni, M.; Bortolomeoli, I.; Ignat, A.; Nicoli, M. C. Effect of Pulsed Light on Total Microbial Count and Alkaline Phosphatase Activity of Raw Milk. *Int. Dairy J.* **2014,** *39* (1), 108–112.
21. Jang. J. H.; Moon, K. D. Inhibition of *Polyphenol Oxidase* and *Peroxidase* Activities on Fresh-cut Apple by Simultaneous Treatment of Ultrasound and Ascorbic acid. *Food Chem.* **2011,** *124*, 444–449.
22. Jia, C.; Chen, P.; Liu, W.; Li, B.; Wang, Q. Surface Treatment of Aramid Fiber by Air Dielectric Barrier Discharge Plasma at Atmospheric Pressure. *Appl. Surf. Sci.* **2011,** *257* (9), 4165–4170.
23. Kim. Y. S.; Park, S. J. Cho, Y. H.; J. Park. Effects of Combined Treatment of High Hydrostatic Pressure and Mild Heat on the Quality of Carrot Juice. *J. Food Sci.* **2001,** *66*, 1355–1360.
24. Kobayashi, F.; Ikeura, H.; Odake, S.; Hayata, Y. Inactivation Kinetics of Polyphenol Oxidase using a Two-Stage Method with Low Pressurized Carbon Dioxide Microbubbles. *J. Food Eng.* **2013,** *114,* 215–220.
25. Lee, K. H.; Kim, H. J.; Woo, K. S.; Jo, C.; Kim, J. K.; Kim, S. H.; Park, H. Y.; Oh, S. K.; Kim, W. H. Evaluation of Cold Plasma Treatments for Improved Microbial and Physicochemical Qualities of Brown Rice. *LWT—Food Sci. Technol.* **2016,** *73*, 442–447.
26. Lerasle. M.; Federighi, M.; Simonin, H.; Anthoine, V.; Rezé, S.; Cheretand, R.; Guillou, S. Combined use of Modified Atmosphere Packaging and High Pressure to Extend the Shelf-Life of Raw Poultry Sausage. *Innov. Food Sci. Emerg. Technol.* **2014,** *23*, 54–60.

27. Li, H. P.; Wang, L. Y.; Li, G.; Jin, L. H.; Le, P. S.; Zhao, H. X.; Xing, X. H.; Bao, C. Y. Manipulation of Lipase Activity by the Helium Radio-Frequency, Atmospheric-Pressure Glow Discharge Plasma Jet. *Plasma Process. Polym.* **2011,** *8*, 224–229.
28. Li. R.; Hou, Z.; Zou, H.; Wang, Y.; Liao, X. Inactivation Kinetics, Structural, and Morphological Modification of Mango Soluble Acid Invertase by High Pressure Processing Combined with Mild Temperatures. *Food Res. Int.* **2018**, *105*, 845–852.
29. Liao. H. M.; Zhang, F. S.; Liao, X. J.; Hu, X. S.; Chen, Y.; Deng, L. Analysis of Escherichia coli Cell Damage Induced by HPCD Using Microscopies and Fluorescent Staining. *Int. J. Food Microbiol.* **2010,** *144* (1), 169–176.
30. Liavoga, A.; Matella, N. J. Enzymes in Quality and Processing of Tropical and Subtropical Fruits. In *Tropical and Subtropical Fruits: Postharvest Physiology, Processing and Packaging*; Siddiq, M., Ed.; Wiley-Blackwell: Oxford, UK, 2012; pp 35–51.
31. Ling, L.; Jiafeng, J.; Jiangang, L.; Minchong, S.; Xin, H.; Hanliang, S.; Yuanhua, D. Effects of Cold Plasma Treatment on Seed Germination and Seedling Growth of Soybean. *Sci. Rep.* **2014,** *4*, 8. E-article 5859.
32. Liu, X.; Gao, Y.; Xu, H.; Hao, Q.; Liu, G.; Wang, Q. Inactivation of Peroxidase and Polyphenol Oxidase in Red Beet (*Beta vulgaris* L.) Extract with Continuous High Pressure Carbon Dioxide. *Food Chem.* **2010,** *119* (1), 108–113.
33. Lopez, P.; Sala, F. J.; De La Fuente, J. L.; Condon, S.; Raso, J.; Burgos, J. Inactivation of Peroxidase, Lipoxygenase, and Polyphenol Oxidase by Manothermosonication. *J. Agric. Food Chem.* **1994,** *42*, 252–256.
34. Ludikhuyze. L.; Van Loey, A.; Indrawati, C.; Smout, M. Hendrick. Effects of Combined Pressure and Temperature on Enzymes Related to Quality of Fruits and Vegetables: From Kinetic Information to Process Engineering Aspects. *Crit. Rev. Food Sci. Nutr.* **2003,** *43* (5), 527–586.
35. Ma, Y.; Zhang, G. J.; Shi, X. M.; Xu, G. M.; Yang, Y. Chemical Mechanisms of Bacterial Inactivation Using Dielectric Barrier Discharge Plasma in Atmospheric Air. *IEEE Trans. Plasma Sci.* **2008,** *36*, 1615–1620.
36. MacDonald. L.; Schaschke, C. J. Combined Effect of High Pressure, Temperature and Holding Time on Polyphenoloxidase and Peroxidase Activity in Banana (*Musa acuminata*). *J. Sci. Food Agric.* **2000,** *80*, 719–724.
37. Magalhães. M. P.; Gomes, F. D. S.; Modesta, R. C. D.; Matta, V. M. D.; Cabral, L. M. C. *Conservação De Agua De Coco Verde Por Filtração Com Membrana* (Conservation of Coconut Water with Membrane Filter). *Food Sci. Technol. (Camp.)* **2005,** *25*, 72–77.
38. Martinez, M. V.; Whitaker, J. R. Biochemistry and Control of Enzymatic Browning. *Trends Food Sci. Technol.* **1995,** *6* (6), 95–200.
39. Merve, P. A.; Sevcan, U. Effects of Ultraviolet Light Emitting Diodes (LEDs) on Microbial and Enzyme Inactivation of Apple Juice. *Int. J. Food Microbiol.* **2017,** *260*, 65–74.
40. Misra, N. N.; Kaur, S.; Tiwari, B. K.; Kaur, A.; Singh, N.; Cullen, P. J. Atmospheric Pressure Cold Plasma (ACP) Treatment of Wheat Flour. *Food Hydrocoll.* **2015,** *44*, 115–121.
41. Misra, N. N.; Yong, H. I.; Phalak, R.; Jo, C. Atmospheric Pressure Cold Plasma Improves Viscosifying and Emulsion Stabilizing Properties of Xanthan Gum. *Food Hydrocoll.* **2018,** *82*, 29–33.
42. Nakano. L. A.; Leal, Jr. W. F.; Freitas, D. G. C.; Cabral, L. M. C.; Penha, E. M.; Penteado, A. L.; Matta, V. M. Coconut Water Processing Using Ultrafiltration and Pasteurization.

In *Proceeding of International Congress on Engineering and Food*; National Technical University of Athens: Athens—Greece, 2011; pp 1–5.
43. Neves, F. I.G.; Vieira, M. C.; Silva, C. L.M. Inactivation Kinetics of Peroxidase in Zucchini (*Cucurbita pepo* L.) by Heat and UV-C Radiation. *Innov. Food Sci. Emerg. Technol.* **2012,** *13*, 158–162.
44. Niemira, B. A. Cold Plasma Decontamination of Foods. Annual Reviews in Food of Cold Plasma Treatment on Seed Germination and Seedling Growth of Soybean. *Sci. Rep.* **2012,** *4*, 6. E-article 5859.
45. Oliveira, A. F.A.; Mar, J. M.; Santos, S. F.; da Silva Júnior, J. L.; Kluczkovski, A. M.; Bakry, A. M.; Bezerra, J.; Nunomura, R.; Sanches, E. A.; Campelo, P. H. Non-Thermal Combined Treatments in the Processing of Açai (*Euterpe oleracea*) Juice. *Food Chem.* **2018,** *265*, 57–63.
46. Ortuno, C.; Duong, T.; Balaban, M.; Benedito, J. Combined High Hydrostatic Pressure and Carbon Dioxide Inactivation of Pectin Methylesterase, Polyphenol Oxidase and Peroxidase Infeijoa Puree. *J. Supercrit. Fluid.* **2013,** *82*, 56–62.
47. Ortuño, C.; Balaban, M.; Benedito, J. Modeling the Inactivation Kinetics of Escherichia coli, Saccharomyces Cerevisiae and Pectin Methylesterase in Orange Juice Treated with Ultrasonic-Assisted Supercritical Carbon Dioxide. *J. Supercrit. Fluid.* **2014,** *90*, 8–26.
48. Pan. X.; Niu, G.; Liu, H. Microwave-Assisted Extraction of Tea Polyphenols, Tea Caffeine from Green Tea Leaves. *Chem. Eng. Process* **2003,** *42*, 129–134.
49. Pandiselvam, R.; Subhashini, S.; Banuu Priya, E. P.; Kothakota, A.; Ramesh, S. V.; Shahir, S. Ozone Based Food Preservation: A Promising Green Technology for Enhanced Food Safety. *Ozone: Science and Engineering*, **2019**, *41*, 17–34.
50. Pandiselvam, R.; Sunoj, S.; Manikantan, M. R.; Kothakota, A.; Hebbar, K. B. Application and Kinetics of Ozone in Food Preservation. *Ozone: Sci. Eng.* **2017,** *39* (2), 115–126.
51. Pankaj, S. K.; Bueno-Ferrer, C.; Misra, N. N.; Milosavljevic, V.; O'Donnell, C. P.; Patange, A.; Boehm, D.; Ferrer, C. B.; Cullen, P. J.; Bourke, P. Controlling *Brochothrix Thermosphacta* as a Spoilage Risk Using in Package Atmospheric Cold Plasma. *Food Microbiol.* **2017,** *66*, 48–54.
52. Park S. J.; Lee, J. I.; Park, J. Effects of a Combined Process of High-Pressure Carbon Dioxide and High Hydrostatic Pressure on the Quality of Carrot Juice. *J. Food Sci.* **2002,** *67*, 1827–1834.
53. Ragni, L.; Berardinelli, A.; Vannini, L.; Montanari, C.; Sirri, F.; Guerzoni, M. E.; Guarnieri, A. Non-Thermal Atmospheric Gas Plasma Device for Surface Decontamination of Shell Eggs. *J. Food Eng.* **2010,** *100*, 125–132.
54. Ramanan, K. R.; Sarumathi, R.; Mahendran, R. Influence of Cold Plasma on Mortality Rate of Different Life Stages of *Tribolium castaneum* on Refined Wheat Flour. *J. Stored Prod. Res.* **2018,** *77*, 126–134.
55. Ramazzina, I.; Berardinelli, A.; Rizzi, F.; Tappi, S.; Ragni, L.; Sacchetti, G.; Rocculi, P. Effect of Cold Plasma Treatment on Physicochemical Parameters and Antioxidant Activity of Minimally Processed Kiwi Fruit. *Postharvest Biol. Technol.* **2015,** *107*, 55–65.
56. Rapeanu. G.; Van Loey, A.; Smout, C.; Hendrickx, M. Thermal and High-Pressure Inactivation Kinetics of Polyphenol Oxidase in Victoria Grapes. *J. Agric. Food Chem.* **2005,** *53* (8), 2988–2994.
57. Reddy. K. V.; Das, M.; Das, S. K. Nonthermal Sterilization of Green Conutwater for Packaging. *J. Food Qual.* **2007,** *30*, 466–480.

58. Rivas. A.; Rodrigo, D.; Martínez, A.; Barbosa-Cánovas, G. V.; Rodrigo, M. Effect of PEF and Heat Pasteurization on the Physical–Chemical Characteristics of Blended Orange and Carrot Juice. *LWT* **2006,** *39*, 1163–1170.
59. Rojas. M. L.; Trevilin, J. H.; Dos, S. E.; Funcia, J. A.; Gut, W.; Augusto, P. E. D. Using Ultrasound Technology for the Inactivation and Thermal Sensitization of Peroxidase in Green Coconut Water. *Ultrason. Sonochem.* **2017,** *36*, 173–181.
60. Sadhu, S.; Thirumdas, R.; Deshmukh, R. R.; Annapure, U. S. Influence of Cold Plasma on the Enzymatic Activity in Germinating Mung Beans (*Vigna radiate*). *LWT—Food Sci. Technol.* **2017,** *78*, 97–104.
61. Sampedro, F.; Fan, X. Inactivation Kinetics and Photoreactivation of Vegetable Oxidative Enzymes after Combined UV-C and Thermal Processing. *Innov. Food Sci. Emerg. Technol.* **2014,** *23*, 107–113.
62. Sarangapani, C.; Keogh, D. R.; Dunne, J.; Bourke, P.; Cullen, P. J. Characterization of Cold Plasma Treated Beef and Dairy Lipids Using Spectroscopic and Chromatographic Methods. *Food Chem.* **2017,** *235*, 324–333.
63. Savadkoohi, S.; Bannikova, A.; Hao, T. T.; Kasapis, S. Inactivation of Bacterial Protease and Food Borne Pathogens in Condensed Globular Proteins following Application of High Pressure. *Food Hydrocoll.* **2014,** *42*, 244–250.
64. Scanlon, M. G.; Henrich, A. W.; Whitaker, J. R. Factors Affecting Enzyme Activity in Food Processing. *IN Proteins Food Process* **2018,** 337–365.
65. Segat, A.; Misra, N. N.; Cullen, P. J.; Innocente, N. Effect of Atmospheric Pressure Cold Plasma (ACP) on Activity and Structure of Alkaline Phosphatase. *Food Bioprod. Process.* **2016,** *98*, 181–188.
66. Seiquer, L.; Delgado-Andrade, C.; Haro, A.; Navarro, M. P. Assessing the Effects of Severe Heat Treatment of Milk on Calcium Bioavailability: In Vitro and In Vivo Studies. *J. Dairy Sci.* **2010,** *92*, 5635–5643.
67. Seyderhelm, I.; Boguslawski, S.; Michaelis, G.; Knorr, D. Pressure-Induced Inactivation of Selected Food Enzymes. *J. Food Sci.* **1996,** *61,* 308–310.
68. Tanimoto, S. Thermal Inactivation Behavior of Enzymes in Fresh Sake During Pasteurization. *J. Brew. Soc. Jpn.* **2004,** *99*, 208–214.
69. Tappi, S.; Gozzi, G.; Vannini, L.; Berardinelli, A.; Romani, S.; Ragni, L.; Rocculi, P. Cold Plasma Treatment for Fresh-Cut Melon Stabilization. *Innov. Food Sci. Emerg. Technol.* **2016,** *33*, 225–233.
70. Terefe, N. S.; Delon, A.; Versteeg, C. Thermal and High Pressure Inactivation Kinetics of Blueberry Peroxidase. *Food Chem.* **2009,** *232*, 820–826.
71. Thirumdas, R.; Trimukhe, A.; Deshmukh, R. R.; Annapurea, U. S. Functional and Rheological Properties of Cold Plasma Treated Rice Starch. *Carbohydr. Polym.* **2017,** *157*, 1723–1731.
72. Timmons, C.; Pai, K.; Jacob, J.; Zhang, G.; Ma, L. M. Inactivation of *Salmonella enterica*, Shiga Toxin-Producing *Escherichia coli*, and *Listeria monocytogenes* by a Novel Surface Discharge Cold Plasma Design. *Food Control* **2018,** *84*, 455–462.
73. Underhill; S. J. R.; Critchley, C. Cellular Localisation of Polyphenol Oxidase and Peroxidase Activity in *Litchi chinensis* Sonn, Pericarp. *Funct. Plant Biol.* **1995,** *22* (4), 627–632.
74. Wu, Z. H.; Chi, L. H.; Bian, S. F.; Xu, K. Z. Effects of Plasma Treatment on Maize Seeding Resistance. *J. Maize Sci.* **2007,** *15*, 111–113.

75. Yang, R.; Liu, Y.; Meng, D.; Wang, D.; Blanchard, C. L.; Zhou, Z. Effect of Atmospheric Cold Plasma on Structure, Activity, and Reversible Assembly of the Phytoferritin. *Food Chem.* **2018,** *264*, 41–48.
76. Yin, M. Q.; Huang, M. J.; Ma, B. Z.; Ma, T. C. Stimulating Effects of Seed Treatment by Magnetized Plasma on Tomato Growth and Yield. *Plasma Sci. Technol.* **2005,** 7, 3143–3147.
77. Yun, H.; Kim, B.; Jung, S.; Kruk, Z. A.; Kim, D. B.; Choe, W. Inactivation of Listeria Monocytogenes Inoculated on Disposable Plastic Tray, Aluminum Foil, and Paper Cup by Atmospheric Pressure Plasma. *Food Control*, **2010,** *21*, 1182–1190.
78. Zhang, Y.; He, S.; Simpson, B. K. Enzymes in Food Bioprocessing—Novel Food Enzymes, Applications, and Related Techniques. *Curr. Opin. Food Sci.* **2018,** *19*, 30–35.
79. Zhang, Z.; X. Pang; D. Xuewu; Z. Ji; Y. Jiang. Role of Peroxidase in Anthocyanin Degradation in Litchi Fruit Pericarp. *Food Chem.* **2005,** *90*, 47–52.
80. Zhao, W.; Yang, R. Effect of High-Intensity Pulsed Electric Fields on the Activity, Conformation and Self-Aggregation of Pepsin. *Food Chem.* **2009,** *114* (3), 777–781.

CHAPTER 3

SCOPE OF OZONE TECHNOLOGY FOR INACTIVATION OF FOOD ENZYMES

ANANDU CHANDRA KHANASHYAM, M. ANJALY SHANKER, and RAVI PANDISELVAM

ABSTRACT

Enzymes are a specialized class of proteins; the presence of many endogenous enzymes in food samples causes quality and nutritional losses in raw and processed food products. Today, thermal and chemical treatments are conspicuous methods involved in the enzyme inactivation to sustain the quality of food systems. But these treatments can decrement the color and flavor characteristics of food, also with the nutritional losses. Disproportionate chemical usage can have negative implications in health. These limitations catalyze the usage of emerging nonthermal technologies as an alternative to traditional treatments for enzyme inactivation. The interest in ozone, as a greener, safer, and cost-effective nonthermal technology, is currently gaining momentum. The affirmation of ozone as GRAS led researchers to focus on its application and study. The inactivation efficacy on food matrix can be affected by internal and external factors such as type of ozone (gaseous or aqueous), ozone level, exposure time, flow rate, temperature, pH of the medium, and organic matter. Recent reports show that ozone treatment can alter the activity of various enzymes, such as amylase, polyphenol oxidase, and catalase. The modifications of this enzyme activity in the system can be by direct or indirect reaction of enzymes with ozone or its decomposed by-products.

Enzyme Inactivation in Food Processing: Technologies, Materials, and Applications. Megh R. Goyal, PhD, Junaid Ahmad Malik, PhD & Ravi Pandiselvam, PhD (Eds.)

3.1 INTRODUCTION

In 1886, Europeans recognized the potential of ozone as a disinfectant and in 1906, set up the first commercial water treatment plant in France.[8] Ozone was listed as Generally Recognized as Safe (GRAS) chemical in 1997 and was classified into a food additive by US FDA in 2001. Ozone is a powerful disinfectant and a strong sanitizer.[31] As the final decomposition product of ozone is oxygen, it leaves no toxic residues on food or processing equipment like other chlorine-based chemicals making it a greener environment-friendly technology.

Today, ozone has found its application in food industry as a disinfectant and an antimicrobial agent with proven efficacy against bacteria, virus, protozoa, fungi, and spores.[25] Ozone can also successfully inactivate mycotoxins,[51] pesticides,[45] and storage insects.[28] Compared to other oxidizing agents ozone is easy to use, has higher efficiency, and can be generated on-site with a comparatively lower cost. The on-site production of ozone also eliminates the need for transportation, storage, and other inventory issues.

Ozone can be applied to food either directly or indirectly in gaseous or aqueous forms. As ozone solubility increases with decrease in temperature and pH, it also allows us to use ozone in meat and poultry processing facilities, where cool, damp, and refrigerated conditions are to be maintained. As there is no heat generated during processing, ozone is also highly suitable for heat-sensitive foods. Furthermore, it is also more energy efficient technology than irradiation, microwave, or traditional thermal treatments, which in turn reduce processing cost.[57] As consumers today are moving to high quality, conveniently packed fresher like ready-to-eat products, ensuring product safety and quality remains a major challenge. As traditional processing methods can cause quality and nutritional losses, one way to dramatically extend the shelf-life is to change the way it has been processed. Due to its high oxidizing potential (+2.07 V) ozone can significantly affect the microbial load, phenolic content, total flavonoids, enzyme kinetics, texture, color, respiration rate, and nutritional contents in food.[12,22,25,45–47] Enzymes are one of the key players in determining the final quality of food produce.

This chapter focuses on the effect of ozone treatment on food enzymes, ozone-enzyme inactivation kinetics, and influence of ozone treatment on food quality parameters (such as texture, color, flavor) and safety aspects of ozone application.

3.2 MECHANICS AND KINETICS OF ENZYME INACTIVATION

Enzymes are a particular class of proteins accountable for catalyzing chemical reactions within the cells and thus are ultimate drug targets. Structurally they are compact proteins of relatively spherical in shape. The three-dimensional specific conformation of these enzymes, which are stabilized by noncovalent forces, is called native state (Fig. 3.1). For a biological activity to take place, this stable state of enzyme is necessary.

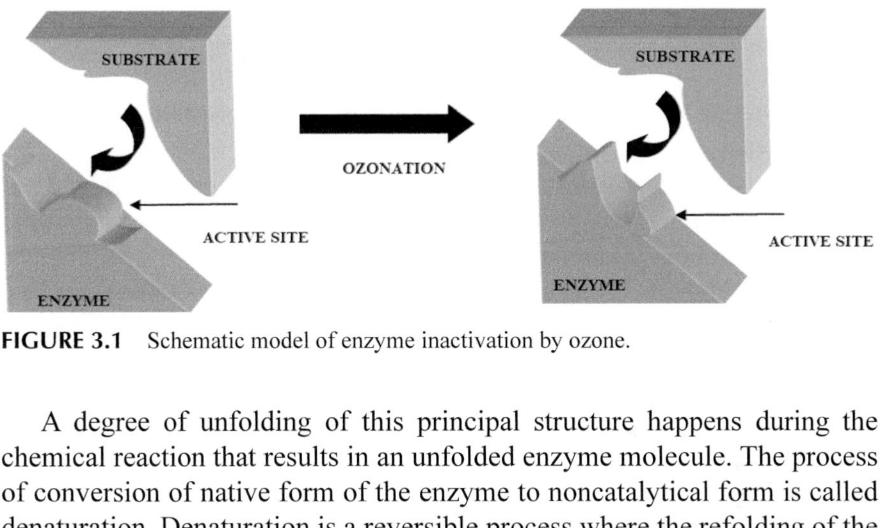

FIGURE 3.1 Schematic model of enzyme inactivation by ozone.

A degree of unfolding of this principal structure happens during the chemical reaction that results in an unfolded enzyme molecule. The process of conversion of native form of the enzyme to noncatalytical form is called denaturation. Denaturation is a reversible process where the refolding of the enzymes occurs after the elimination of factors that leads to the breakage of the principal structure. But the refolding is not always complete because of the series of reactions that took place during the unfolding process.

Let N_E be the native form of the enzyme before the reaction, M_E be a mixture of partially unfolded and folded mixture of enzyme, and U_E be the irreversible inactivated enzyme, then the process can be represented as

$$N_E \underset{K_M}{\leftrightarrow} M_E \underset{K_U}{\rightarrow} U_E \tag{3.1}$$

where K_M is the equilibrium constant between native and partially conversion state of enzyme and K_U is the rate constant of the reaction between native and unfolded states of enzyme.

Thorough study of these states is necessary to get a perception on the changes that happen in an enzyme with respect to the change in different conditions like temperature, pH, presence of catalysts, etc. Chemical kinetics gives an insight about the rate and mechanism of chemical reactions. It explains the mechanism by which reactants are converted and the dependency of the mechanism on possible factors like concentration, temperature, and pH. The rate of reaction can be correlated with the change in moles of a constituent, per unit of volume of total mixture. The rate is associated with the stoichiometric variables of a chemical reaction. If a reaction is expressed in the given format,

$$A + B \rightarrow C + D \tag{3.2}$$

The rate of this reaction can be expressed as

$$V = k\,x[A]^a\,[B]^b \tag{3.3}$$

where k = rate constant, which is dependent on factors like presence of catalysts, temperature etc.

The order of the given reaction which explains the empirical relation of rate on concentration is a with respect to A and b with respect to B and has an overall reaction order of $a + b$. Like any other kinetic study, the enzyme inactivation kinetics gives an insight about the mechanism of inactivation under hostile conditions. It entails the detailed determination of molecular process involved in the inactivation procedure. As explained earlier, the basic inactivation process involves the change of the native state of enzyme to unfolded state. The rate of this enzyme inactivation can be expressed as

$$V = K_U\,[M_E] \tag{3.4}$$

Residual activity of an enzyme is dependent on the native and partially unfolded form of enzyme remaining at the end of the reaction. Relating the activity of the enzyme with this, we can explain observed inactivity rate of enzyme as

$$V = K_{ob}\left([M_E] + [N_E]\right) \tag{3.5}$$

$$K_U\,[M_E] = K_{ob}\left([M_E] + [N_E]\right) \tag{3.6}$$

For particular instance when the enzyme inactivation is correlated with the temperature change, which is a principal method of inactivation in food system, the rate constant can be expressed as $K_T = U_E/N_E$. Combining this with the previous equation, we will get a clear insight about the observed inactivity rate of enzyme as

$$K_{ob} = \frac{K_U K_T}{1 + K_T} \tag{3.7}$$

At higher temperatures the observed inactivity rate does not depend on the reversible denaturation and K_{ob} is found to be equal to K_U. At thermal denaturation temperature, the value is found to be a combination of rate constants K_U and K_T. Alternative to this thermal inactivation technique is to reduce the negative effects like color modifications, nutrient losses etc. that is associated with high temperatures. Similar to the thermal inactivation conditions, the model given can be applied in many cases. According to the complexity of the system and the reactions involved there will be change in the pattern and kinetics of inactivation.

Reaction of ozone with organic compounds in a food system is an intricate process involving a series of different chemical reactions such as reaction of desired compound with ozone and its intermediates and the indirect reaction among the secondary oxidation products. Presence of ozone not only facilitates the inactivation process, it also leads to the development of highly reactive species which aids in the process. The inactivation process involving ozone and enzyme in fluid form follows first- or second-order kinetics. The reaction between enzyme and ozone can be expressed as

$$a[O+E] \rightarrow P \tag{3.8}$$

where a is the stoichiometric ratio of enzyme–ozone reaction, resulting in the production of different products. Depending upon the ozone and enzyme concentration, the reaction follows first-order kinetics at the equilibrium state. The rate can be expressed as

$$V = k[O]^a [E]^b \tag{3.9}$$

where k represents the rate constant of the reaction, [O] indicates the ozone concentration during the treatment, and [E] represents the enzyme concentration.

As the reaction involves the degradation of enzyme involved, the equation can be reordered as follows:

$$\frac{dE_t}{dt} = k[O]^a [E]^b \tag{3.10}$$

Maintaining the ozone concentration as constant, the equation can be rearranged as

$$-\frac{dE_t}{dt} = kE_O \tag{3.11}$$

Rearranging the equation and integrating, the equation can be expressed as

$$\ln \frac{E_t}{E_o} = -kt \tag{3.12}$$

This general eq (3.12) articulates the complete breakdown of any organic compound after prolonged ozone treatment.[61] Type of enzyme involved in the process, temperature change, and the reaction between the secondary products increases the complexity of the process which in turn results in changes in the given equation.

3.3 APPLICATIONS OF OZONE ON FOOD ENZYME INACTIVATION

Rapid deterioration in quality occurs if proper enzyme inactivation is not achieved during processing. Based on the relationship between enzyme and its effect on food quality, enzymes can be grouped into: flavor-related enzymes, color-related enzymes, and texture-related enzymes.

3.3.1 FLAVOR-RELATED ENZYMES

The main group of flavor-related enzymes are those that hydrolyze fats, protein, and low molecular weight precursors (mainly organo-sulfur compounds).[43] The major flavor-related enzymes include lipase/phospholipase, lipoxygenase, super oxidase dismutase, thio-glucosidase etc. out of which lipid hydrolyzing enzymes are most significant, as lipids can be easily degraded. The lipase enzymes (triacylglycerol hydrolases) are those that hydrolyze triacylglycerols (TAG) into glycerol and free fatty acids. The release of these enzyme by-products into food matrix can cause soapy and rancid flavor development. As the substrate for lipase enzyme (TAG) is insoluble in water, the enzyme can act only on lipid–water interface of the dispersed substrate in an emulsified state. This enzyme is widely studied in cereal and dairy-based products. The changes caused by ozonation have been studied on lipids present in cereal grains, peas, pulse, and beans.[42,45] The study showed that ozone treatment under low ozone concentrations (0.05–5.0 ppm) had no effect on lipid composition, whereas treatment at higher concentrations (50 ppm) leads to considerable lipid oxidation in the products.

The lipase activity for oil extracted from ozonated wheat flour (5 L/min) was studied by Mohammed et al.[39] According to the study, the initial lipase activity of the untreated oil was 12.60 U/g, on subsequent ozone exposure, there was a gradual, but an insignificant decrease in the enzyme activity with respect to the treatment time upto 35 min of exposure. But on prolonged exposure above 35 min, the lipase activity reduced to 7.73 U/g as ozone was able to affect the enzyme activity by altering the enzyme structure on its active side. Today, lipase also finds its application in detergent formulas, as incorporating lipase enzymes into detergent can increase its cleaning efficiency on fatty soils.[27] But care should be taken as lipase residues on the surface can later contaminate the food samples.

The inactivation efficacy of ozone on detergent lipase was studied by Encarnacion et al.[19] This study assessed the effect of ozone treatment on lipase activity in acidic, neutral, and alkaline conditions. Due to the high inactivation potential of O_3, the enzyme activity was affected immediately after contact. The inactivation was higher at acidic pH, as ozone decomposes faster in neutral and alkaline pH. The study also proved that O_3 concentration of 0.3 ppm was sufficient to inactivate 1 g/L of commercial residual lipase enzyme after the wash cycle.

Lipoxygenase (LOX) is another oxidative enzyme that is naturally present in plants and animals. LOX oxidizes the Polyunsaturated Fatty acids (PUFA) into peroxide derivatives. The peroxides by-product of the LOX catalyses will later degrade into secondary products and give characteristic flavors to many fruits and vegetables. However, improper processing conditions can alter the LOX activity causing bitter or rancid off-flavors. Excessive oxidation of PUFA by LOX enzyme is also reported to cause nutrient deterioration in some foods.[63] A decrease in LOX activity after ozonation was reported in grapes during postharvest storage.[38]

The LOX activity of ozone-treated strawberry fruits was assessed by Perez et al.[50] The study found no significant effect on LOX activity immediately after ozone treatment. But, during the simulated transport of fruits (2°C) after ozonation, a decrease in the enzyme activity was observed; this was followed by storage study (20°C) where there was a reported increase in enzyme activity. Contradictory data with an increase in LOX activity with respect to ozone treatment was reported in wheat flour[51] and lentil seeds.[35] LOX activity of ozonated wheat flour showed an increasing trend with respect to time and ozone concentration, and a fourfold increase in enzyme activity was reported after ozonation of flour (30 ppm for 30 min).

Superoxide dismutase (SOD) is a ubiquitous enzyme present in plants, animals, fungi, and bacteria. It is a primary antioxidant enzyme that catalyzes the dismutation reaction of superoxide anion (O_2^-) that is formed as a by-product of oxygen metabolism. SOD enzyme combines O_2^- ions with hydrogen ions to produce hydrogen peroxide (H_2O_2) and oxygen molecule. The effect of ozonation on SOD activity, O_2^- radicals, and H_2O_2 production on pericarp and pulp of cantaloupe fruits was investigated by Zhang et al.[68] For ozone-treated samples, a rapid increase in enzyme activity was observed for the initial 10 days of storage followed by a slow decline. The SOD activity of control sample remained unaffected throughout the study duration and was 50% lower when compared to ozone-treated samples. The higher enzyme activity was found on the pericarp than on the pulp of the cantaloupe fruits. In the case of O_2^- anion, an opposite trend of SOD activity was observed for both pulp and pericarp. As H_2O_2 is the by-product of SOD on superoxide anion, the presence of H_2O_2 was high in the control samples than the ozone-treated samples.

A similar increase in SOD activity was also reported in tangerine fruit peels.[9] Ozone treatment rapidly increased the SOD activity in tangerine peel irrespective of the concentration used. As SOD can reduce lipid peroxidation of fruits and vegetables, its activities can be associated with senescence. The enzyme activity of fresh-cut green pepper treated with ozone and stored under modified atmospheric packaging (MAP) was studied by Chen et al.[11] The SOD activity of the green pepper was assessed for 21 days of storage. The study showed that the time required for reaching the maximum enzyme activity varied according to the treatment. The control was the first to hit the peak enzyme activity on day 6, so as to protect its tissues from oxidative stress. But, for the ozone and combined ozone and MAP-treated samples, the maximum enzyme activity was reached on the 15th day of storage indicating prolonged sample quality. When compared with the control samples, all ozonated samples showed a higher enzyme activity. A similar increased SOD activity was also reported for ozone-treated grapes.[38] The effect of ozone in more flavor-related enzyme is discussed in Table 3.1.

TABLE 3.1 Effect of Ozone Treatment in Flavor-Related Enzyme.

Enzyme	Product	Treatment conditions	Remarks	Reference
Lipase	Whole wheat flour	Flow rate: 5 L/min Time: 5, 15, 25, 35, and 45 min	The lipase activity remains unaffected until 35 min of exposure.	[39]

TABLE 3.1 *(Continued)*

Enzyme	Product	Treatment conditions	Remarks	Reference
Lipase	Wheat flour	Ozone concentration: 5, 15, and 30 ppm Time: 5, 15, and 30 min	Enzyme activity decreased with increasing exposure time and concentration. The ozone concentration showed a linear dependence and the time had a quadratic dependence.	[51]
Lipoxygenase (LOX)	Wheat flour	Ozone concentration: 5, 15, and 30 ppm Time: 5, 15, and 30 min	LOX activity increased with increasing ozone concentration	[51]
Lipoxygenase (LOX)	Grapes	Concentration: 20 g/L Flow rate: 150 L/h Time: 12 h Temp: 10°C	LOX activity was found to decrease during storage and dehydration of fruits	[38]
Lipoxygenase (LOX)	Strawberry (*Fragaria ananassa* Duch.)	Temp: 2°C Concentration: 0 and 0.35 ppm	Ozone had no significant effect on LOX	[50]
Superoxide dismutase (SOD)	Cantaloupes (xizhoumi 25)	Concentration: 0. 15 mg/m^3 Time: 2 h	The enzyme activity was found to increase with ozone treatment	[68]
Superoxide dismutase (SOD)	tangerine (*Citrus reticulate*)	Concentration: 200 μL/L Time: 0, 2, 4, and 6 h Temp. 25°C	The enzyme activity increased irrespective of treatment time	[9]
Superoxide dismutase (SOD)	Green pepper	Concentration: 6.42 mg/cm^3 Time: 15 min	The SOD activity was enhanced by ozone treatment	[11]

TABLE 3.1 *(Continued)*

Enzyme	Product	Treatment conditions	Remarks	Reference
Superoxide dismutase (SOD)	Grapes	Concentration: 20 g/L Flow rate: 150 L/h Time: 12 h Temp: 10°C	The SOD activity was enhanced by ozone treatment	[38]
Superoxide dismutase (SOD)	Mushroom (*Lentinusedodes*)	Concentration: 3.21 mg/m^3 Time: 30 min	There was an initial increase in the enzyme activity during first few days followed by a steep decline. The activity was significantly higher than the control sample throughout the storage time	[32]

3.3.2 COLOR-RELATED ENZYMES

Pigments, such as carotenoids, chlorophylls, anthocyanins, betalains, and myoglobin, are the major components in food products responsible for the perceived colors. Degradation of any of these components in food can lead to significant loss in color quality. Polyphenol oxidase (PPO) is one of the main enzymes responsible for browning reaction in fruits and vegetable. It is an oxidoreductase enzyme with four Cuionin its active site. It has dual catalytic function and catalyzes the oxidation of monohydroxy phenols to *o*-dihydroxy phenols and subsequently, dehydrogenates the *o*-dihydroxy phenols into corresponding *o*-quinones. The quinones thus formed can later react with amino compounds to result in brown colored pigments. Therefore, inhibition and control of PPO activity is important.

Modesti et al.[38] investigated the effect of ozonation on the postharvest storage and dehydration of grapes. The PPO activity does not show any significant difference during the first 3 days of storage after the ozone treatment, but a significant decrease in PPO activity was reported for the treated sample during the dehydration process of the grapes. Ozone was also successful in inactivating the PPO activity in fresh-cut green peppers, and the time required for the treated samples to reach the maximum enzyme activity was significantly higher than the control samples.[11] A combined treatment of ozone with nisin for extending the shelf-life of tender coconut water showed a compete inactivation of PPO enzyme with a reduced browning index.[57] A similar result on ozone efficacy in inactivation of PPO is reported in fresh-cut lettuce,[58] celery[69], mushrooms,[32] and kiwi.[7]

Peroxidase (POD) is oxidoreductase enzyme that reduces hydrogen and other hydro-peroxides to water. Since they have multiple isoforms, plant PODs act on several substrates. Depending upon the catalytic center, peroxidase can be classified as heme and non-heme peroxidases. The heam peroxidase has a heme moiety in the reactive site while a selenol or thiol group is seen for the later. PODs are responsible for a variety of physiological functions such as abiotic and biotic stress, suberization, lignification, senescence, and wound healing.[49] POD is also responsible for the post-harvest degradation of flavonoids and other aromatic compounds in fruits and vegetables. POD is a heat stable enzyme and on heating, POD can form new isoenzymes and haemo proteins that can oxidize lipids. Thus, POD can be a limiting factor in thermal treatments.[1] As POD is one of the main enzymes controlling oxidative stress in plants, ozone treatment at lower concentrations has reported to increase the POD activity. When shiitake mushroom was treated with ozone at low concentration range (ranging from 1 to 10 mg/m^3), a gradual increase in POD activity with respect to O_3 concentration was observed until 3.21 mg/m^3.

The enzyme activity was stabilized (till 7.46 mg/m^3) before falling down.[32] POD activity of mushrooms was also higher for intermittent ozone treatment than one time treatment. A 77.8% increase in POD activity was also reported in kiwi fruits when ozone was used for decontamination of molds (*Botrytis cinerea* and *Penicillium expansum*),[7] whereas a complete inactivation of POD enzyme on tender coconut water was reported by Rajashri et al.[56] Application of 1 mg/L of ozone to fresh-cut lettuce reduced the POD activity from approximately 1000 to 300 units/g.[58] The effect of ozone in more color-related enzyme is presented in Table 3.2.

TABLE 3.2 Effect of Ozone Treatment in Color-Related Enzyme.

Enzyme	Product	Treatment conditions	Remarks	Reference
Peroxidase (POD)	Tender coconut water	Concentration: 20 mg/L Flow rate: 1 L/min Temp: 25°C Time: 10 min	Ozone completely inactivated POD in tender coconut water.	[39]

TABLE 3.2 *(Continued)*

Enzyme	Product	Treatment conditions	Remarks	Reference
Peroxidase (POD)	Lettuce	Concentration: 1 mg/L Temp: 18–20°C	A significant reduction in POD were observed for the first 7 days of storage.	[57]
Peroxidase (POD)	Mushroom (*Lentinusedodes*)	Concentration: 3.21 mg/m^3 Time: 30 min	As the ozone concentration increased, there was a gradual increase in the POD activity until a certain level from which it then stabilised	[32]
Peroxidase (POD)	Kiwi fruits	Concentration: 79.44 ppm Temp: 0°C Time: 1 h	The enzyme activity peaked rapidly and then declined during the storage	[7]
Peroxidase (POD)	Peach juice	Concentration: 0.11 and 0.20 mg/min Time: 12 min Temp: 20°C	A 99.5% and 99.8% reduction in enzyme activity was reported depending on ozone concentration.	[24]
Peroxidase (POD)	Pear (*Pyruspyrifolia*)	Concentration: 2.14, 6.24, and 21.4 mg/m^3 Time: 1 h	A significant increase in POD activity with respect to ozone concentration was reported in the study.	[69]
Polyphenol oxidase (PPO)	Grapes	Concentration: 20 g/L Flow rate: 150 L/h Time: 12 h Temp: 10°C	There was a insignificant decrease in enzyme activity during storage, but a significant decrease was observed during dehydration period	[38]
Polyphenol oxidase (PPO)	Green pepper	Concentration: 6.42 mg/cm^3 Time: 15 min	Ozone was successful in inactivation PPO activity and delayed the time required for the enzyme to reach its peak activity.	[11]
Polyphenol oxidase (PPO)	Tender coconut water	Concentration: 20 mg/L Flow rate: 1 L/min Temp: 25°C Time: 10 min	PPO was completely inactivated by Ozone.	[56]

TABLE 3.2 *(Continued)*

Enzyme	Product	Treatment conditions	Remarks	Reference
Polyphenol oxidase (PPO)	Lettuce	Concentration: 1 mg/L Temp: 18–20°C	PPO activity was significantly reduced by ozone	[57]
Polyphenol oxidase (PPO)	Celery	Concentration: 0.03, 0.08, 0.18 ppm Temp: 17°C Time: 2, 6, 10 min	The efficacy of enzyme inhibition increased with concentration	[68]
Polyphenol oxidase (PPO)	Mushroom (*Lentinusedodes*)	Concentration: 3.21 mg/m³ Time: 30 min	Ozonation significantly reduced the PPO activity throughout the storage period.	[32]
Polyphenol oxidase (PPO)	Kiwi fruits	Concentration: 79.44 ppm Temp: 0°C Time: 1 h	Ozone inhibited the enzyme activity and there was an initial increase followed by a steep decline on enzyme activity.	[7]
Polyphenol oxidase (PPO)	Mulberry	Concentration: 2 ppm Temp: 5°C	The PPO activity of the ozone-treaded fruits was higher during the initial 3 days of storage.	[5]
Polyphenol oxidase (PPO)	Peach juice	Concentration: 0.11 and 0.20 mg/min Time: 12 min Temp: 20°C	A 93.9% and 97.3% reduction in enzyme activity was reported depending on ozone concentration.	[24]
Polyphenol oxidase (PPO)	Sugar cane juice	Concentration: 150, 300, and 450 mg/L	The enzyme activity decreased with treatment time.	[4]

3.3.3 TEXTURE-RELATED ENZYME

The major texture-related plant enzyme is amylase in cereal grains and pectin/cellulose degrading enzymes in fruits and vegetables. Plant amylases can be classified into endo- and exo-amylases. The endo amylases cleave the 1,4-glycosidic bonds whereas the exo-amylase cleave the 1,4- and 1,6-glycisidic bonds in starch molecule. Amylase enzyme is largely used as additives in baking industry.[36] Addition of α-amylase into bread dough can increase the fermentable compounds, improve the softness of the products, and act as anti-stalling agents. But excessive addition of amylase can yield sticky dough making it difficult to handle. Ozone treatment has reported to reduce the amylose activity in wheat flour.[17,20,26,51]

The falling number apparatus records the time in seconds required for a plunger to drop down through a gelatinized flour sample and is used to measure the α-amylase activity. A higher falling number indicates a lower enzyme activity. Treatment of hard wheat flour with ozone concentration of 5 mg/L showed a significant decrease in amylase activity, as evidenced by Hagberg falling number. The falling number of the ozone-treated flour increased from 250 to 400 after 2 h of exposure. Similar findings where falling number was increased with augmentation in time and ozone concentration were reported by Felipe Trombete et al.[20] Interestingly, this influence of ozone concentration in amylase activity was only reported at lower quantities of grains, at higher masses this correlation was not observed. Ozonation significantly increased the peak viscosity of waxy rice flour due to the inactivation of the α-amylase.[17] A significant nonlinear decrease in amylase activity with respect to ozone concentration was also reported by Piechowiak et al.[51] Table 3.3 shows effect of ozone treatment in texture-related enzymes.

TABLE 3.3 Effect of Ozone Treatment in Texture-Related Enzyme.

Enzyme	Product	Treatment conditions	Remarks	Reference
Amylase	Wheat flour	Ozone concentration: 5, 15, and 30 ppm Time: 5, 15, and 30 min	There was a nonlinear decline in enzyme activity with increasing time and concentration	[51]
Amylase	Wheat flour	Concentration: 5 g/L Flow rate: 3.3 L/min Time: 0–2.0 h with an interval of 30 min	The enzyme activity decreased with increase in time.	[26]

TABLE 3.3 *(Continued)*

Enzyme	Product	Treatment conditions	Remarks	Reference
Amylase	Wheat flour	Concentration: 60 mg/L Flow rate: 0.5 L/min Time: 2–5 h Temp: 25°C	The enzyme inactivation depends on time and concentration of ozone.	[20]
Amylase	Wheat flour	Flow rate: 0.06 L/min Time: 0.5, 1, and 2 h	The peak viscosity increased from 21%, 33%, and 45% after 0.5, 1, and 2 h of ozone treatment.	[17]
Protease	Wheat flour	Ozone concentration: 5, 15, and 30 ppm Time: 5, 15, and 30 min	There was a linear decline in enzyme activity with increasing time and concentration	[51]
α-L-arabinopyranoside (α-L-ara)	Cantaloupe melon	Concentration: 0.15 and 0.3 ppm Time: 13 days Temp: 6°C	A significant decrease in enzyme activity was observed from 6th day of storage	[67]
α-L-arabinopyranoside (α-L-ara)	Kiwi	Concentration: 0.3 µL/L	Ozone had no effect on enzyme activity	[23]
β-D-galactopyranoside (β-D-gal)	Cantaloupe melon	Concentration: 0.15 and 0.3 ppm Time: 13 days Temp: 6°C	A significant decrease in enzyme activity was observed from 6th day of storage	[67]
β D galactopyranoside (β-D-gal)	Tomato	Concentration: 10 µL/L Time: 10 min	The enzyme activity decreased during storage, but the decrease was not significant	[33]
β-D-galactopyranoside (β-D-gal)	Kiwi	Concentration: 0.3 µL/L	Ozone had no effect on enzyme activity	[23]

TABLE 3.3 *(Continued)*

Enzyme	Product	Treatment conditions	Remarks	Reference
Pectin methyl esterase (PME)	Cantaloupe melon	Concentration: 0.15 and 0.3 ppm Time: 13 days Temp: 6°C	Ozone has no effect on PME	[67]
Pectin methyl esterase (PME)	Lettuce	Concentration: 1 mg/L Temp: 18–20°C	A reduction in PME activity was observed after ozone treatment.	[58]
Pectin methyl esterase (PME)	Tomato	Concentration: 10 μL/L Time: 10 min	No difference in PME activity was found immediate to ozone treatment, but a 50% reduction in activity as observed at 9th day of storage.	[33]
Polygalactorunase (PG)	Cantaloupe melon	Concentration: 0.15 and 0.3 ppm Time: 13 days Temp: 6°C	A steep decrease in enzyme activity was observed till 3rd day followed by further stabilization in activity	[67]
Polygalactorunase (PG)	Tomato	Concentration: 10 μL/L Time: 10 min	Ozone does not show any effect on PG activity	[33]
Polygalactorunase (PG)	Kiwi	Concentration: 0.3 μL/L	Ozone treatment inhibited the PG activity in kiwi fruits	[23]
ACC oxidase (ACO)	Kiwi	Concentration: 0.3 μL/L	Even though a slight ripening depended increase was reported while storage, ACO activity of control was always higher than ozonated samples	[23]
ACC synthase (ACS)	Kiwi	Concentration: 0.3 μL/L	A strong enzyme inhibition was observed for the ozone-treated samples.	[23]

Pectic substances are complex polysaccharides of galactouronic acid with α-1–4 linkages. Pectinases are enzymes that hydrolyze pectin and are widely seen in bacteria, certain fungi, and plants. The major pectin lytic enzymes are pectinesteases, protopectinase, polygalacturonase, and polygalacturonaselyase.[13] Toti et al.[67] studied the activity of cell wall enzymes such as α-L-arabinopyranoside (α-L-ara), β-D-galactopyranoside (β-D-gal), Pectin methyl esterase (PME), and polygalactorunase (PG) on cantaloupe melon stored in an ozone atmosphere for 13 days. The ozone treatment inhibited both the glycosidase enzymes (α-L-ara and β-D-gal) and a significant reduction in activity was reported from the 6th day of storage. The PG activity declined steeply after the ozone treatment until the 3rd day and then stabilized further. Interestingly, ozone does not have any significant effect on the PME activity. In contradiction, a decrease in PME activity with respect to ozone treatment was reported in fresh-cut lettuce[58] and tomato.[33] A similar study on kiwi fruit stored in ozone atmosphere also showed a significant inhibition on PG and 1,4-β-glucosidase enzymes.[23] Ozone also inhibited major ethylene biosynthesis enzymes such as ACC oxidase (ACO) and ACC synthase (ACS) activity in kiwi fruits.[23] The effect of ozone in texture-related enzyme and other food enzymes are discussed in Table 3.4.

TABLE 3.4 Effect of Ozone Treatment in Other Food Enzymes.

Enzyme	Product	Treatment conditions	Remarks	Reference
Hydroperoxide lyase (HPL)	Strawberry (*Fragaria ananassa* Duch.)	Temp: 2°C Concentration: 0 and 0.35 ppm	No correlation between ozone treatment and enzyme activity	[50]
Alcohol acyltrans-ferase (AAT)	Strawberry (*Fragaria ananassa* Duch.)	Temp: 2°C Concentration: 0 and 0.35 ppm	No correlation between ozone treatment and enzyme activity	[50]
Catalase	Tangerine (*Citrus reticulate*)	Concentration: 200 μL/L Time: 0, 2, 4 and 6 h Temp: 25°C	Catalayse activity barely changed on the 1st day of treatment, but a significant increased from 2nd day was observed	[9]

TABLE 3.4 *(Continued)*

Enzyme	Product	Treatment conditions	Remarks	Reference
Catalase	Grapes	Concentration: 20 g/L Flow rate: 150 L/h Time: 12 h Temp: 10°C	The enzyme activity was significantly higher than the control samples during storage	[38]
Catalase	Mushroom (*Lentinusedodes*)	Concentration: 3.21 mg/m^3 Time: 30 min	Ozonation increased the SOD activity of mushrooms in the initial days of storage	[32]
Ascorbate peroxidse (APX)	Tangerine (*Citrus reticulate*)	Concentration: 200 μL/L Time: 0, 2, 4, and 6 h Temp: 25°C	APX activity increased with increase in treatment time	[9]
Ascorbate peroxidse (APX)	Grapes	Concentration: 20 g/L Flow rate: 150 L/h Time: 12 h Temp: 10°C	Ozone treatment increased the APX activity	[38]
Phenylalanine ammonia lyase (PAL)	Tender coconut water	Concentration: 20 mg/L Flow rate: 1 L/min Temp: 25°C Time: 10 min	A significant reduction in enzyme activity was reported for the combined ozone+ nisin treatment.	[56]
Phenylalanine ammonia lyase (PAL)	Kiwi fruits	Concentration: 79.44 ppm Temp: 0°C Time: 1 h	Ozone inhibited PAL activity and the enzyme activity remained lower than control throughout the storage.	[7]

TABLE 3.4 *(Continued)*

Enzyme	Product	Treatment conditions	Remarks	Reference
Phenylalanine ammonia lyase (PAL)	Hot pepper (Capsicum annuum L)	Concentration: 2 mg/L Temp: 25°C Time: 1 and 3 h	There was a significant reduction in PAL activity for the treated samples when compared with control	[40]
Guaiacol peroxidase (GPX)	Grapes	Concentration: 20 g/L Flow rate: 150 L/h Time: 12 h Temp: 10°C	GPX activity was higher in treated samples than control	[38]

3.4 EFFECT OF PROCESSING FACTORS ON EFFICIENCY OF OZONE TREATMENT

Finest conditions of control parameters are important to ensure that the treatment is yielding an optimistic outcome. The parameters holding a role in the effectiveness of ozone treatment are divided into extrinsic and intrinsic factors.[53] The extrinsic and intrinsic factors under consideration in the treatment are concentration, flow rate, temperature etc. stability and reactivity of ozone in a system is dependent on these parameters.

3.4.1 CONCENTRATION

Amount of ozone inoculated to the system effects the efficacy of ozone treatment. Customarily there is a tendency of increasing the ozone concentration to increase the reactivity of the system. Contrarily, high levels of ozone introduced to the system may result in detrimental effects. Higher amount of ozone results in saturation of the system, which in turn hinders further reactions and reactivity of ozone. The ozone concentration is dependent on the type of enzyme present in the system and its complexity. The intricacy of the reaction between ozone and enzyme is also one of the parameters that help in determining the right concentration of ozone.

3.4.2 TEMPERATURE

Compared to oxygen, this allotropic form of O_2 is more than a dozen times soluble in water at 0-30°C. With a raise in temperature, the solubility and stability of ozone decreases; this in turn affects the decomposition rate.[29] Degree of inactivation competence is dependent on the decomposition rate. Increase in decomposition rate implies progress in inactivation process. So the inactivation efficacy of ozone is directly related to decline in temperature.

3.4.3 PH

Decomposition rate of ozone varies markedly with change in pH. Inactivation efficacy is high throughout lower pH values and efficiency decreases with higher pH. It is observed that the variation from lower to higher pH shows a little more than five times increase in treatment time. For higher inactivation efficiency, it is better to maintain a lower pH throughout the process.

3.4.4 FLOW RATE

Flow rate and bubble size is closely related, which controls the efficacy of the process. Different bubble sizes are produced depending on the gas flow rate. This bubble size has an effect on ozone solubility and thereby the efficiency of the system. Smaller bubble size is constructive as this increases the availability of interfacial area for mass transfer. A decrease of 10 times in bubble diameter will increase the contact area availability by more than dozen times. Controlled flow rate is essential to maintain considerably small bubble size which in turn controls the treatment rate.

3.4.5 OZONE-CONSUMING COMPOUNDS

Ozone-consuming compounds in the system control the effectiveness of the treatment. Dissolved organic matter decreases the activity by the introduction of by-products with less effectiveness, thereby reducing the amount of ozone that should be available for desired reaction. Reaction by products not only controls the efficacy of the process but also alters the organoleptic properties and the safety of the final product. Therefore, the set of undesired reactions that takes place and the resultant by-products are equally contagious to the ozone treatment.

3.5 EFFECT OF OZONE TREATMENT ON FOOD QUALITY

Ozone treatment of wheat kernels before milling has found to reduce the amount of coarse bran and damaged starch during processing.[15] Ozone treatment has found to significantly alter the pasting and thermal properties of flour. Starch from waxy rice, when ozonated for 0.5, 1, and 2 h, resulted in an increased gelatinization temperature and peak viscosity. The 0.5, 1, and 2 h treatment increased the peak viscosity by 27.4%, 32.8%, and 45.5%, respectively.[17] This is because the amylase activity of the flour was significantly reduced during ozone treatment. Ozonation of hard wheat flour at 5 mg/L showed no significant effect on protein or fat content of the flour until 1.5 h.[26] But, a prolonged exposure of wheat flour to ozone damaged the starch and gluten structure which led to a significant decrease in water-holding capacity and level of wet gluten in flour. The high oxidizing potential ozone can also increase the whiteness of the flour.[26,60] Jin Mei et al.[26] have shown that flours treated with 5 mg/L ozone for 1 h can improve the dough properties and can positively affect the sensory, volume, and structure of steamed bread.

Ozone also alters the cell wall enzyme activity in fruits and vegetables. Cantaloupe melons stored in ozone-rich atmosphere showed higher firmness and lower ethylene production during 13 days of storage.[67] This better firmness was due to reduced glycosidases enzyme activity (α-L-ara and β-D-gal), as these enzymes degrade the galactans in side chain of pectin back bone leading to an increased pectin solubility and thereby increasing its accessibility to pectin degrading enzymes. Ethylene production in climacteric fruits enhances its softening during ripening by activating cell wall hydrolyzing enzymes.

Ozone also prevents the softening of kiwi fruits by preventing ethylene biosynthesis and reduces ACS and ACO activity.[23] Ozonation of tomato with 10 μL/L for 10 min significantly reduced its softening until 9th day of storage without altering its color, acidity, or sugar content. In contrast, a higher concentration of ozone treatment correlated to a decrease in textural property, as 1 mg/L of ozone treatment not only inactivated the PME enzyme in fresh-cut lettuce and also thereby reduce its crispiness.[58]

As industrial standards today require for 5 log microbial reductions, application of ozone in concentrations high enough to achieve effective decontamination can in times lead to surface oxidation, development of off-odors and undesirable flavors. Strawberries treated with 0.35 ppm of ozone showed a significant decrease in anthocyanin content and aroma compounds.[5] Ozone (2 mg/L) also reduced the ascorbic acid, total flavonoids and total phenolic contents in tender coconut water by 85%, 30% and 36%, respectively.[56] A decrease in anthocyanin and ascorbic acid content by 98.2% and 85.5%was also reported in strawberry juice when treated with 7.8% w/w of ozone.[66] Although ozone treatment has no significant effect in pH, titratable acidity, brix content or color values of fresh orange juice, yet it decreased the ascorbic acid content and color values in a concentration and time depended manner.[10] Tiwari et al.[65] have demonstrated with the help of response surface methodology that, anthocyanin and color of blackberry juice degrades in a concentration and time depended manner during ozone processing.

3.6 SAFETY AND LEGAL ASPECTS OF OZONE APPLICATION

The principal objective of every novel technology is to render food with right quality and safety along with the innovations associated with the final product. This aspect of food leads to increase in apprehension about ingredients and technology associated with bringing the food to the hands of the consumer. The idea of consumer acceptance is at this time is correlated to the safety aspect of any technology associated with the product. Therefore, it is important to have valuable and ample idea about the applied technology and its distinctiveness for acceptance.

Interest in inclusion of this technology is in return to the customer demands for a greener alternative for potential food applications. The multi-functionality of ozone makes it a promising food processing agent with decomposition of surplus ozone into oxygen leaving no residues. Globally the status of ozone application is still in a developing state and in certain areas the apprehension about food applications is not even addressed. The limitation in widespread of this technology is due to the high initial cost associated and lack of education about the safety aspect of this treatment.

The application of ozone technology was initiated during late 1980s and early 1990s for drinking water treatment. Since then, the technology was associated with meat, seafood, water etc., with an application in preservation as well as purification. The application of treatment in food processing sector developed initially from water treatment applications, which later progressed to surface cleaning and direct food product washes and at last as a food additive. The application became widespread with the affirmation of ozone from US FDA as GRAS chemical[34] and an antimicrobial additive. This federal ruling cleared the usage of ozone in food industry in agreement with the present industry standards of manufacturing practices.

Along with the effects the treatment has on food, the safety of workers and surroundings associated with the treatment should be ensured. A safe environment should be ensured irrespective of form of ozone applied in food. The toxic effect of ozone varies from common issues like headache, irritation etc. to severe illness and even death. The safe levels of treatment exposures and contact time suitable for maintaining optimal environment should be ensured for the smooth functioning of the system and personnel involved.

Another aspect related to the safety of the treatment refers to the awareness about the interactions between the equipment and surfaces involved with ozone. The interactions associated can be favorable or unfavorable according to materials involved in the system. Therefore, it is important to opt for materials that are compatible with the treatment involved. The common materials used in the system that exhibit resistance to the adverse effects due to higher exposures involve PVC, PTFE (Teflon) etc. The exposure levels and period should be decided considering different factors that control the safety of the whole process.

3.7 CURRENT STATUS AND FUTURE TRENDS

Ozone has wide range of applications, such as: sterilization and disinfection of food contact surfaces, fumigation, plant sanitization, preservation of fruits and vegetables and its shelf-life extension, etc. The application of ozone as a potential fumigant in grain storage is well demonstrated and is proved to have minimal to no effect on grain quality.[28,47,48] Ozone is also used as an alternative in fruit and vegetable preservation. Ozone pre-treatment of fresh produce have shown to reduce the spoilage microbial count in a verity of fruits and vegetables such as apple, cantaloupe, carrot, blueberries, pepper, celery, spinach and strawberries.[22] Ozone reduces the amount of foodborne microorganism in food produce and decrease human health risk.

Ozone has been effective against many foodborne pathogens, like, *Escherichia coli*,[6] *Salmonella*,[37] *Shigella*,[62] and Listeria.[2] Ozone is also used in inactivating various mycotoxins such as aflatoxins,[57] Ochratoxin A,[18] Patulin,[16] Zearalenone,[31] and Fumonisins.[41] Ozone efficacy in removing pesticides such as captan and pyraclostrobinin apples,[59] difenoconazole and linuron in carrots,[14] malathion and profenofos in dried chilli[64] and chlorofenapyr in lettuce[3] is well studied.

Ozone is also used in treatment of contaminants in wastewater and drinking water. Recently, ozone is used in combination with other nonthermal technologies (such as UV radiation, pulse electric field, ultra sound treatment) and other chemical treatments (such as organic acids, and hydrogen peroxides). Although there are limited literatures in the combination of ozone with other hurdles, ozone in combination with other treatments show a synergistic effect in microbial inactivation.[21] To clearly establish the use of ozone as hurdle with other technologies, further studies have to be done to evaluate the mechanism behind the ozone synergy when used as a hurdle with other technologies. There are also limited studies on mechanism and inactivation kinetics of ozone on various food enzymes. Therefore, a detailed study is necessary to further elucidate the exact mechanism on how ozone influences the tertiary enzyme structure and its activity.

3.8 SUMMARY

Ozone, the allotropic form of oxygen, is widely used in food industry with an aim of increasing the safety and quality of food products. Starting from disinfecting the food contact surfaces to controlling the activity of particular components like enzymes, the uses of ozone vary significantly. The treatment with ozone is found to be effective in controlling the activity of flavor, color, and texture-related enzymes in the food system. The enzyme inactivation process follows first-order kinetics with complexity related to the complexity of enzyme, reaction between enzyme and ozone, and reaction by products. Parameters under consideration should be optimally adjusted in line to ensure the efficacy of process. Even though ozone treatment is considered to be an effective food processing additive, the usage of the treatment for enzyme inactivation needs strict policies and regulations considering the toxicity. This enzyme inactivation process is constructive over traditional methods like heat inactivation and chemical inactivation as the detrimental effects on food are fewer. Increasing awareness about the treatment, promoting the research activities based on enzyme inactivation, and considering the variation in parameters affecting the treatment ensures the growth and application of ozone in the enzyme inactivation process.

KEYWORDS

- **enzyme inactivation**
- **nonthermal technology**
- **ozone processing**
- **processing parameters**

REFERENCES

1. Adams, J. B. Review: Enzyme Inactivation during Heat Processing of Food-Stuffs. *Int. J. Food Sci. Technol.* **1991,** *26*, 1–20.
2. Adhikari, A.; Chhetri, V. S.; Bhattacharya, D.; Cason, C.; Luu, P.; Suazo, A. Effectiveness of Daily Rinsing of Alfalfa Sprouts with Aqueous Chlorine Dioxide and Ozonated Water on the Growth *of Listeria Monocytogenes* During Sprouting. *Lett. Appl. Microbiol.* **2019,** *69*, 252–257.
3. Al-Antary, T. M.; Shaderma, A. M.; Al-Dabbas, M. B. Effect of Ozonation Treatment on Spiked Chlorfenapyr Pesticide on Lettuce. *Fresenius Environ. Bull.* **2019,** *28* (1), 370–375.
4. Ana, C. B.; Flávio, L.; Lorena, L.; Angela, L.; Sharline, F.; Josivanda, P.; Figueroa, J. Enzymatic Polyphenoloxidase Inactivation with Temperature and Ozone in Sugarcane to Produce Lower Color Sugar. *Braz. J. Food Technol.* **2019,** *22*, 1–7.
5. Ana, G. P.; Carlos, S.; Rıos, J.; Olıas, R. Effects of Ozone Treatment on Postharvest Strawberry Quality. *J. Agric. Food Chem.* **1999,** *47*, 1652–1656.
6. Ana, M. A.; Yousef, A. Efficacy of Ozone Against *Escherichia Coli* O157: H7 on Apples. *J. Food Sci.* **2001,** *66*, 1380–1384.
7. Anwei, L.; Junqing, B.; Rui, L.; Yimeng, F.; Lin, L.; Dan, W.; Lu, Z.; Jin, L.; Tianzi, H.; Kou, L. Effects of Ozone Treatment on the Quality of Kiwifruit During Postharvest Storage Affected by Botrytis Cinerea and Penicillium Expansum. *J. Phytopathol.* **2018,** *167*, 470–478.
8. Colm O'Donnell, C.; Tiwari, B. K.; Cullen, P. J; Rice. R. G.; Eds. *Ozone in Food Processing*; Wiley-Blackwell: London, 2012; 312pp.
9. Boonkorn, P.; Gemma, H.; Sugaya, S.; Setha, S.; Uthaibutra, J.; Whangchai, K. Impact of High-Dose, Short Periods of Ozone Exposure on Green Mold and Antioxidant Enzyme Activity of Tangerine Fruit. *Postharvest Biol. Technol.* **2012,** *67*, 25–28.
10. Brijesh K. T.; Kasiviswanathan M.; Colm, P.; Cullen, P. J. Kinetics of Freshly Squeezed Orange Juice Quality Changes During Ozone Processing. *J. Agric. Food Chem.* **2008,** *56*, 6416–6422.
11. Chen, J.; Hu, Y.; Wang, J.; Hu, H.; Cui, H. Combined Effect of Ozone Treatment and Modified Atmosphere Packaging on Antioxidant Defense System of Fresh-Cut Green Peppers. *J. Food Process. Preserv.* **2015,** *2015*, 1–6.
12. Cullen, P. J., Valdramidis, V. P.; Tiwari, B. K.; Patil, S.; Bourke, P.; O'Donnell, C. P. Ozone Processing for Food Preservation: An Overview on Fruit Juice Treatments. *Ozone: Sci. Eng.* **2010,** *32*, 166–179.

13. De, Kashyap; Vohra, P; Chopra, S.; Tewari, R. Applications of Pectinases in the Commercial Sector: A Review. *Bioresour. Technol.* **2001,** *77*, 215–227.
14. De-Souza, L. P.; Faroni, L. R.; Heleno, F. F.; Pinto, F. G.; Queiroz, M. E.; Prates, L. H. F. Ozone Treatment for Pesticide Removal from Carrots: Optimization by Response Surface Methodology. *Food Chem.* **2018,** *243*, 435–441.
15. Desvignes, C.; Chaurand, M.; Dubois, M.; Sadoudi, A.; Abecassis, J.; Lullien, V. Changes in Common Wheat Grain Milling Behavior and Tissue Mechanical Properties Following Ozone Treatment. *J. Cereal Sci.* **2008,** *47*, 245–251.
16. Diao, E.; Wang, J.; Li, X.; Wang, X.; Song, H.; Gao, D. Effects of Ozone Processing on Patulin, Phenolic Compounds and Organic Acids in Apple Juice. *J. Food Sci. Technol.* **2019,** *56*, 957–965.
17. Ding, W.; Wang, Y.; Zhang, W.; Shi, Y.; Wang, D. Effect of Ozone Treatment on Physicochemical Properties of Waxy Rice Flour and Waxy Rice Starch. *Int. J. Food Sci. Technol.* **2015,** *50*, 744–749.
18. E, Torak Use of Gaseous Ozone for Reduction of Ochratoxin A and Fungal Populations on Sultanas. *Aust. J. Grape Wine Res.* **2019,** *25* (1), 25–29.
19. Encarnación, J. A.; Miguel, G. R.; Deisi, A. V.; Luis, J. P. Assessment of the Use of Ozone for Cleaning Fatty Soils in the Food Industry. *J. Food Eng.* **2012,** *110* (1), 44–52.
20. Felipe, T.; Adriana, M.; Yuri, P.; Otniel, F.; Daniela, F.; Sidinéa, F.; Carlos, C.; Tatiana, S.; Fraga, M. Chemical, Technological and Sensory Properties of Wheat Grains (*Triticum aestivum L*) as Affected by Gaseous Ozonation. *Int. J. Food Prop.* **2016,** *19* (12), 2739–2749.
21. Garud, S. R.; Negi, P. S.; Rastogi, N. K.; Eds. *Non-Thermal Processing of Foods*; 1st ed.; CRC Press: Boca Raton, FL, 2019; pp 21–30.
22. Glowacz, M.; Colgan, R.; Rees, D. The Use of Ozone to Extend the Shelf-Life and Maintain Quality of Fresh Produce. *J. Sci. Food Agric.* **2015,** *95*, 662–671.
23. Ioannis, S. M.; Ariel, R. V.; Arun, P. D.; George, A. M.; Vlasios, G.; Miltiadis, V.; Carlos, H. C.; Molassiotis, A. Ozone-Induced Kiwifruit Ripening Delay Is Mediated by Ethylene Biosynthesis Inhibition and Cell-Wall Dismantling Regulation. *Plant Sci.* **2014,** *229*, 76–85.
24. Jaramillo Sánchez, G. M.; Garcia Loredo, A. B.; Contigiani, E. V.; Gómez, P. L.; Alzamora, S. M. Inactivation Kinetics of Peroxidase and Polyphenol Oxidase in Peach Juice Treated with Gaseous Ozone. *Int. J. Food Sci. Technol.* **2018,** *53*, 347–355.
25. Jin-Gab Kim; Ahmed E. Yousef; Dave, S. Application of Ozone for Enhancing the Microbiological Safety and Quality of Foods: A Review. *J. Food Prot.* **1999,** *62*, 1071–1087.
26. Jin, Mei; Gang Liu; Xiongwei Huang; Ding, W. Effects of Ozone Treatment on Medium Hard Wheat (*Triticum Aestivum L.*) Flour Quality and Performance in Steamed Bread Making. *CyTA—J. Food* **2016,** *14* (3), 449–456.
27. Jurado, E.; Garcia-Roman, M.; Luzon, G.; Altmajer-Vaz, D.; Jiménez-Pérez, J. L. Optimization of Lipase Performance in Detergent Formulations for Hard Surfaces. *J. Ind. Eng. Chem. Res.* **2011,** *50*, 11502–11510.
28. Kells, S. A.; Mason, L. J.; Maier, D. E.; Woloshuk, C. P. Efficacy and Fumigation Characteristics of Ozone in Stored Maize. *J. Stored Prod. Res.* **2001,** *37*, 371–382.
29. Khadre, M.; Yousef, A.; Kim, J. G. Microbiological Aspects of Ozone Applications in Food: A Review. *J. Food Sci.* **2001,** *66*, 1242–1252.

30. Kim, J. G.; Yousef, A. E.; Khadre, M. A. Ozone and Its Current and Future Application in the Food Industry. *Adv. Food Nutr. Res.* **2003,** *45*, 167–218.
31. Krstović, S.; Krulj, J.; Jakšić, S.; Bočarov-Stančić, A.; Jajić, I. Ozone as Decontaminating Agent for Ground Corn Containing Deoxynivalenol, Zearalenone, and Ochratoxin. *J. Cereal Chem.* **2020,** *2020*, 1–9.
32. Liu, J.; Chang, M. C.; Meng, J. L.; Liu, J. Y.; Cheng, Y. F.; Feng, C. P. Effect of Ozone Treatment on the Quality and Enzyme Activity of *Lentinus Edodes* During Cold Storage. *J. Food Process. Preserv.* **2020,** *44* (8), 1–11.
33. Luis, R.; Natalia, C.; Concello, N.; Alicia, R.; Vicente, A. R. Effect of Short-Term Ozone Treatments on Tomato (*Solanum Lycopersicum L.*) Fruit Quality and Cell Wall Degradation. *J. Agric. Food Chem.* **2010,** *58*, 594–599.
34. Maccarrone, M.; Rossi, A.; Finazzi-Agro, A. Activity and EXpression of Diamine Oxidase in Lentil Seedling under Different Growth Conditions. *Plant Sci.* **1991,** *79* (1), 51–55.
35. Maccarrone, M.; Veldink, G. A.; Vliegenthart, J. F. G.; Finazzi-Agro, A. Ozone Stress Modulates Amine Oxidase and Lipoxygenase Expression in Lentil (*Lens Culinaris*) Seedlings. *FEBS Lett.* **1997,** *408* (2), 241–244.
36. Marc, J. E. C. Van der Maarel; Bart Van der Veen; Joost C. M. Uitdehaag; Hans Leemhuis; Dijkhuizen, L. Properties and Applications of Starch-Converting Enzymes of the A-Amylase Family. *J. Biotechnol.* **2002,** *94* (2), 137–155.
37. Mattioli, S.; Ortenzi, R.; Scuota, S.; Mancinelli, A. C.; Dal Bosco, A.; Cotozzolo, E.; Castellini, C. Impact of Ozone and UV Irradiation Sanitation Treatments on the Survival of *Salmonella* and the Physical–Chemical Characteristics of Hen Eggs. *J. Appl. Poult. Res.* **2020,** *29* (2), 409–419.
38. Modesti, M.; Petriccione, M.; Forniti, R.; Zampella, L.; Scortichini, M.; Mencarelli, F. Methyl Jasmonate and Ozone Affect the Antioxidant System and the Quality of Wine Grape During Postharvest Partial Dehydration. *Food Res. Int.* **2018,** *112*, 1–35.
39. Mohammed, O.; Ke-Xue, Z.; Wei, P.; Anwar, N.; Khalid, M.; Hui-Ming, Z. Characterization of Oil Extracted from Whole Grain Flour Treated with Ozone Gas. *J. Cereal Sci.* **2018,** *79*, 527–533.
40. Monika, S.; Małgorzata, M.; Barbara, C.; Monika, K.; Anna, J.; Irena, P.; Jackowska, I. Ozone-Induced Changes in the Content of Bioactive Compounds and Enzyme Activity During Storage of Pepper Fruits. *Food Chem.* **2016,** *211*, 59–67.
41. Mylona, K.; Kogkaki, E.; Sulyok, M.; Magan, N. Efficacy of Gaseous Ozone Treatment on Spore Germination, Growth and Fumonisin Production by *Fusarium verticillioides* In Vitro and In Situ in Maize. *J. Stored Prod. Res.* **2014,** *59*, 178–184.
42. Naito, S. The Influence of Ozone Treatment on Lipids Contained in Cereal Grains, Cereal Grain Powders, Peas, Beans and Pulse Products. *Nippon Shoku Hin Ko Gyo Gakkaishi* **1989,** *36* (11), 878–883.
43. Pabst, A.; Barron, D.; EtiEvant, P.; Schreier, P. Studies on the Enzymatic Hydrolysis of Bound Aroma Constituents from Raspbeny Hit Pulp. *J. Am. Food Chem.* **1991,** *39* (1), 173–175.
44. Pandiselvam, R.; Kaavya, R.; Jayanath, Y.; Veenuttranon, K.; Lueprasitsakul, P.; Divya, V.; Kothakota, A.; Ramesh, S. Ozone as a Novel Emerging Technology for the Dissipation of Pesticide Residues in Foods – a Review. *Trends Food Sci. Technol.* **2020,** *97*, 38–54.

45. Pandiselvam, R.; Kothakota, A.; Thirupathi, V.; Anandakumar, S.; Krishnakumar, P. Numerical Simulation and Validation of Ozone Concentration Profile in Green Gram (*Vigna Radiate*) Bulks. *Ozone: Sci. Eng.* **2016,** *39* (1), 54–60.
46. Pandiselvam, R.; Subhashini, S.; Banuu Priya, E. P.; Kothakota, A.; Ramesh, S. V.; Shahir, S. Ozone Based Food Preservation: A Promising Green Technology for Enhanced Food Safety. *Ozone: Sci. Eng.* **2018,** *41* (1), 17–34.
47. Pandiselvam, R.; Sunoj, S.; Manikantan, M. R.; Kothakota, A.; Hebbar, K. B. Application and Kinetics of Ozone in Food Preservation. *Ozone: Sci. Eng.* **2016,** *39* (2), 115–126.
48. Pandiselvam, R.; Thirupathi, V.; Chandrasekar, V.; Kothakota, A.; Anandakumar, S. Numerical Simulation and Validation of Mass Transfer Process of Ozone Gas in Rice Grain Bulks. *Ozone: Sci. Eng.* **2017,** *40* (3), 191–197.
49. Passardi, F.; Cosio, C.; Penel, C.; Dunand, C. Peroxidases Have More Functions than a Swiss Army Knife. *Plant Cell Rep.* **2005,** *24* (5), 255–265.
50. Perez, A. G.; Sanz, C.; Rı´os, J. J.; Olıas, R. Effects of Ozone Treatment on Postharvest Strawberry Quality. *J. Agiric. Food Chem.* **1999,** *47* (4), 1652–1656.
51. Piechowiak, T.; Józefczyk, R.; Balawejder, M. Impact of Ozonation Process of Wheat Flour on the Activity of Selected Enzymes. *J. Cereal Sci.* **2018,** *84*, 30–37.
52. Porto, Y. D.; Trombete, F. M.; Freitas-Silva, O.; De Castro, I. M.; Direito, G. M.; Ascheri, J. L. R. Gaseous Ozonation to Reduce Aflatoxins Levels and Microbial Contamination in Corn Grits. *J. Microorg.* **2019,** *7* (8), 220.
53. Prabha, V.; Barma, R. D.; Singh, R.; Madan, A. Ozone Technology in Food Processing: A Review. *Trends Biosci.* **2015,** *8* (16), 4031–4047.
54. Qi, L.; Li, Y.; Luo, X.; Wang, R.; Zheng, R.; Wang, L.; Li, Y.; Yang, D.; Fang, W.; Chen, Z. Detoxification of Zearalenone and Ochratoxin by Ozone and Quality Evaluation of Ozonized Corn. *Food Addit. Contam.: Part A* **2016,** *33* (11), 1700–1710.
55. Qiang Han; Haiyan Gao; Hangjun Chen; Xiangjun Fang; Wu, W. Precooling and Ozone Treatments Affects Postharvest Quality of Black Mulberry (*Morus Nigra*) Fruits. *Food Chem.* **2017,** *221*, 1947–1953.
56. Rajashri, K.; Roopa, B. S.; Negi, P. S.; Rastogi, N. K. Effect of Ozone and Ultrasound Treatments on Polyphenol Content, Browning Enzyme Activities, and Shelf-Life of Tender Coconut Water. *J. Food Process. Preserv.* **2020,** *44* (3), e14363.
57. Rice, R. G.; Graham, D. M.; Lowe, M. T.. Recent Ozone Applications in Food Processing and Sanitation. *Food Safety Mag.* **2002,** *8* (5), 10–17.
58. Rico, D.; Martín-Diana, A. B.; Frías, J. M.; Henehan, G. T. M.; Barry-Ryan, C. Effect of Ozone and Calcium Lactate Treatments on Browning and Texture Properties of Fresh-Cut Lettuce. *J. Sci. Food Agric.* **2006,** *86* (13), 2179–2188.
59. Sadło, S.; Szpyrka, E.; Piechowicz, B.; Antos, P.; Józefczyk, R.; Balawejder, M. Reduction of Captan, Boscalid and Pyraclostrobin Residues on Apples Using Water Only, Gaseous Ozone, and Ozone Aqueous Solution. *Ozone: Sci. Eng.* **2017,** *39* (2), 97–103.
60. Sandhu, H. P.; Manthey, F. A.; Simsek, S; Ohm, J. B. Comparison Between Potassium Bromate and Ozone as Flour Oxidants in Breadmaking. *Cereal Chem.* **2011,** *88* (1), 103–108.
61. Schokker, E. P.; and Van Boekel, A. J. S. Kinetic Modelling of Enzyme Inactivation: Kinetics of Heat Inactivation at 90−110°C of Extracellular Proteinase from *Pseudomonas Fluorescens* 22F. *Schokker. J. Agric. Food Chem.* **1997,** *45* (12), 4740–4747.

62. Selma, M. V.; Beltrán, D.; Allende, A.; Chacón-Vera, E.; Gil, M. I. Elimination by Ozone of Shigella Sonnei in Shredded Lettuce and Water. *J. Food Microbiol.* **2007,** *24* (5), 492–499.
63. Shi, Y.; Mandal, R.; Singh, A.; Singh, A. P. Legume Lipoxygenase: Strategies for Application in Food Industry. *Legum. Sci.* **2020,** *2* (3), 44–50.
64. Sintuya, P.; Narkprasom, K.; Jaturonglumlert, S.; Whangchai, N.; Peng-Ont, D.; Varith, J. Effect of Gaseous Ozone Fumigation on Organophosphate Pesticide Degradation of Dried Chilies. *Ozone: Sci. Eng.* **2018,** *40* (6), 473–481.
65. Tiwari, B. K.; O'Donnell, C. P.; Muthukumarappan, K.; Cullen, P. J. Anthocyanin and Color Degradation in Ozone Treated Blackberry Juice. *Innov. Food Sci. Emerg. Technol.* **2009,** *10* (1), 70–75.
66. Tiwari, B. K.; O'Donnell, C. P.; Patras, A.; Brunton, N.; Cullen, P. J. Effect of Ozone Processing on Anthocyanins and Ascorbic Acid Degradation of Strawberry Juice. *Food Chem.* **2009,** *113* (4), 1119–1126.
67. Toti, M.; Cristian Carboni; Botondi, R. Postharvest Gaseous Ozone Treatment Enhances Quality Parameters and Delays Softening in Cantaloupe Melon During Storage at 6°C. *J. Sci. Food Agric.* **2018,** *98* (2), 487–494.
68. Zhang, H.; Zhang, X.; Dong, C.; Zhang, N.; Ban, Z.; Li, L.; Yu, J.; Hu, Y.; Chen, C. Effects of Ozone Treatment on SOD Activity and Genes in Postharvest Cantaloupe. *RSC Adv.* **2020,** *10* (30), 17452–17460.
69. Zhang, L.; Lu, Z.; Yu, Z.; Gao, X. Preservation of Fresh-Cut Celery by Treatment of Ozonated Water. *Food Control* **2005,** *16* (3), 279–283.
70. Zheng, Z.; Gangming Xu; Zhinan Han; Qingqing Li; Yilun Chen; Li, D. Effect of Ozone on the Antioxidant Capacity of Pear During Postharvest Storage. *J. Food Qual.* **2012,** *36* (3), 190–197.

CHAPTER 4

POTENTIAL OF PULSED ELECTRIC FIELD FOR INACTIVATION OF FOOD ENZYMES

YASHASWINI PREMJIT and N. U. SRUTHI

ABSTRACT

Pulsed electric field (PEF) is both efficient for advancing food functionality by explicitly targeting and inactivating undesirable microbes and enzymes alike in foods. As an emerging nonthermal processing technique, it utilizes electrical currents for the induction of electroporation, thereby permitting noninvasive tissue structure alterations. It extends the shelf-life of foods by inactivating specific food enzymes. This chapter focuses on the application of PEF for enzyme inactivation. The role of enzymes in food systems and their effect on food quality have been briefly discussed. The mechanism of action via PEF for enzyme inactivation has been described, and the setup and components of PEF have been explained. The various kinetic models that have been fitted for the enzyme inactivation have been reviewed. The role of PEF in preventing the loss of food quality has been examined. The chapter presents various applications of PEF for the inactivation of food enzymes.

4.1 INTRODUCTION

The emergence of nonthermal processing techniques that involve the application of ambient temperatures is a suitable alternative to thermal processing techniques. Nonthermal techniques aid in improving food safety, enhancing

Enzyme Inactivation in Food Processing: Technologies, Materials, and Applications. Megh R. Goyal, PhD, Junaid Ahmad Malik, PhD & Ravi Pandiselvam, PhD (Eds.)

product quality, and maintaining economic feasibility. Pulsed electric field (PEF) relies on high-frequency electric pulses with set field strengths that can modulate membrane structures. The application of pulsed treatments was implemented primarily to prevent microbial proliferation and pasteurization.[15] However, with a more in-depth elucidation of the PEF mechanism along with the rapid rate of evolution of pilot-plant equipment with decreased treatment time and continuous processes, there have been extensive applications in the food processing industries.[34]

The PEF relies on the principle of high-voltage pulse application, ranging from 20 to 80 kV/cm in foods to induce permeabilizing effects on biological membranes in order to inactivate enzymes and microbial cells, modify textural properties, and enhance mass transport with treatments of mild or moderate intensity at ambient temperatures for short durations.[29,36,37] The most advantageous effect of PEF is the minimal loss of nutritional compounds and bioactive substances from foods, as well as a decreased loss of color, flavor, and organoleptic characteristics of foods (Fig. 4.1). Although the impact of employing PEF was known early in the 20th century, its application and research became widespread since the late 1990s.[37] Moreover, the burgeoning market demand for good quality food products with minimal processing has also paved the way for the increased interest and development of PEF since it presents a better alternative to traditional thermal processing, like pasteurization and commercial sterilization.[22,38,45]

Furthermore, since the treatment occurs at controlled temperatures for durations in microseconds, it has been shown to minimize the energy losses caused by heating.[6] Additionally, PEF ensures that the thermal effects are relatively lower, thereby differentiating it from techniques that employ thermal processing combined with electrical aspects, such as ohmic heating and moderate electrical field.[13] The application of PEF has predominantly increased because it is characteristically low with regard to energy consumption, continuous operations, and decreased treatment duration. It generates less waste, thereby permitting the establishment of advanced and sustainable technology for food processing, biotechnological, and pharmaceutical industry.[29]

When foods are subjected to PEF, they undergo a dielectric breakdown that leads to cell membrane disruption. This enhances the solute transportation rate across membranes.[6] PEF also enhances the extraction of specific intracellular compounds from plant-based materials, improves the secondary metabolite production rates by stress-reaction induction in plants, and inactivates vegetative microorganisms at ambient temperatures.[36,37] The advantages and disadvantages of PEF have been summarized in Figure 4.1.

The present chapter describes various effects of PEF on food enzymes during processing. A brief overview of the applications of PEF technology for enzymatic inactivation and inhibition has been provided. The role that enzymes play in food systems and their consequent impact on food quality has also been briefly discussed. The mechanism of action of PEF on enzymes is described, and the setup and components of PEF are detailed. The various kinetic models that have been fitted for the enzyme inactivation have been briefly reviewed. The chapter also enlists several applications of PEF for food enzymes and expands on the future scope of this technology.

4.2 SETUP AND COMPONENTS OF PEF PROCESSING

PEF unit comprises a high-voltage pulse generator system, control and monitoring system, and a treatment unit. The energy storage element, pulse shaping and activating circuit, power supply, and switches comprise the critical components of the PEF power generator (Fig. 4.2). The treatment unit consists of a chamber and a continuous circulation pump.[12]

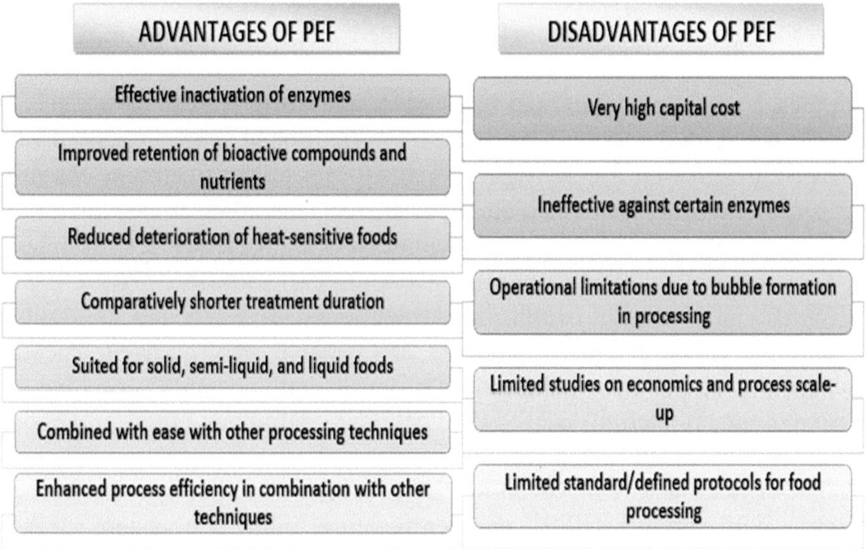

FIGURE 4.1 Advantages and disadvantages of PEF for enzyme inactivation in food processing.

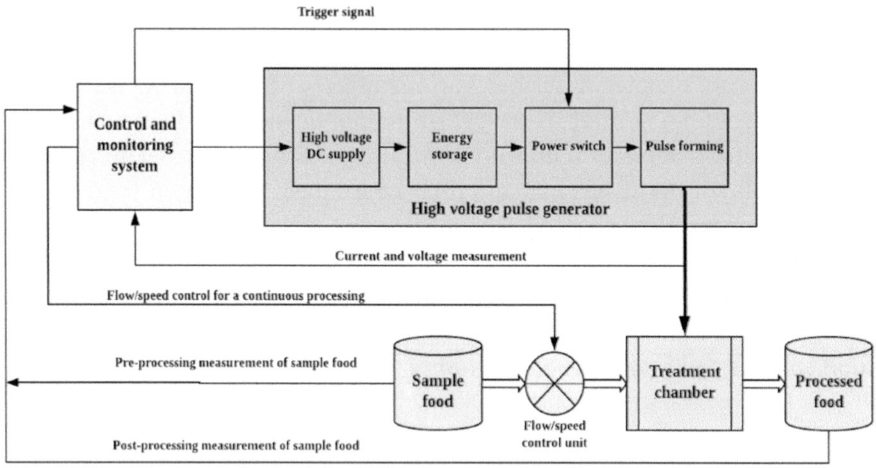

FIGURE 4.2 Depiction of the components of the PEF food processing system.
Source: Reprinted with permission from Ref. [3]. © 2020 Elsevier.

This treatment chamber has electrodes placed at specific positions by insulators, forming an enclosure to contain material to be treated. The distance between these electrodes is set based on the product type and processing intensity necessary. The electrode shape and size are designed to be suitable for specific products, either solid or liquid, and the treatment volume. Some electrode shapes like discs or round-edged electrodes aid in minimizing the electric field and decreasing the chances of the dielectric breakdown of samples. Based on an electrical stand-point, the treatment chamber signifies the electrical load and the food material to be treated. The success of PEF processing relies on the configurational aspects of the treatment chamber for a specific food. Several crucial parameters, including the electric field intensity, treatment uniformity, as well as process efficiency, are influenced by the treatment chamber design.[3]

Treatment chambers are broadly categorized as batch and continuous chambers. Although batch chambers are capable of processing specific volumes of liquids and solids, continuous or dynamic chambers can accomplish similar productivity as the conventional continuous liquid processing necessary for industrial applications.[3]

The critical factors involved in PEF processing include specific energy, electric field strength (EFS), and temperature, while the system's effectiveness relies on the characteristics of the pulse, including its waveform, number, adequate exposure time, frequency, and width. The other technical considerations, such as the system configuration and the treatment chamber volume or the flow rate, are also important. The electrical properties of the food material, including its resistivity and conductivity, as well as the pH and temperature conditions, play an essential role that is equally important.[12]

PEF creates an EFS, a critical parameter measured in kV/cm, that is, the ratio of the potential difference between the electrodes to their distance of separation.[21] The EFS and intensity are both terms that can be used interchangeably to express the peak voltage distribution between the electrodes. The uniformity of the electric fields generated is mainly dependent on the type of apparatus in place. The various configurations of electrodes in PEF processing include parallel, coaxial, and collinear types. The delivery of uniform electric fields is made possible by employing parallel plate electrodes separated by a distance suitably smaller than the surface of the electrode.[12] Although the intensity gradient in collinear and coaxial PEF setup is generated within the treatment chamber, in the parallel PEF setup, uniformity of treatment is observed along the food matrices.[34]

Another important consideration is the voltage applied in the chamber to provide an electric field exceeding a specified critical strength.[3] A higher applied voltage results in energy wastage since the food sample gets heated. This diminishes the energy efficiency of the process as well as increases the temperature of the treated food. In terms of addressing safety concerns, the system is designed to prevent excessive increments of EFS, as there is a risk of electric spark generation between the electrodes. Additionally, another safety aspect that is considered is Joule heating, caused by energy-dependent high-temperature increments.[15]

The frequently generated pulse waveforms for PEF processing and electroporation research are square and exponentially decaying pulses.[33,38] Others include monopolar or bipolar pulses employed as exponential, narrow, and wide pulses, rectangular and oscillatory waveforms.[31] The classification of pulses based on polarity includes mono-/unipolar or bipolar pulses. While monopolar pulses possess either a positive or a negative amplitude, bipolar pulses possess both.[33] Another difference arises wherein the monopolar pulses have the capacity of retaining positive charges with no polarity reversal in the electric field.[38] However, in bipolar pulses, there are instant charge reversals that lead to the rapid reversal of the electric field orientation.[21] Monitoring, observations, and recording of current, voltage signals, waveforms, and frequency are done through digital data acquisition systems.

The fluid volume passing through a control volume per unit of time is termed volumetric flowrate. This is considered a crucial factor in PEF and has a vital role in chamber design and its power supply.[3] The pulse width, measured in μs, is calculated as the time during which a pulse is held at an effective voltage. The pulse duration is identified by the shape of the applied pulse.[21] The treatment time, which is usually measured in seconds, is a function of the pulse number and width. Longer effective treatment times are obtained with an increased pulse width.[33]

The pulse frequency, measured in Hz, is calculated as the number of pulses applied per unit time. For most PEF applications, the treatment time is a function of pulse width and applied pulse frequency. When the EFS are fixed, the intensity increases with increments in the pulse width and/or frequency. This promotes electroporation (pore formation) based on the structural characteristics of the treated sample.[3] The specific energy in the PEF system is a function of treatment time, voltage, temperature, and electrical properties of the food materials treated (resistance and conductivity). It can be altered and modulated on the basis of prior experimentation.[34]

Though high-intensity PEF (HIPEF) treatments are considered to have minimized thermal effects, with temperatures not exceeding 40°C, the temperature of the treatment media tends to increase as the consequence of ohmic heating. This has been termed Joule heating. It occurs because an ohmic heating system intends the conversion of electrical to thermal energy, which rapidly elevates the temperature of the treated samples. Nevertheless, cooling systems are employed as an attachment to the standard equipment.[21]

4.3 PRINCIPLE OF PEF PROCESSING TECHNIQUE

Figure 4.3 shows the mechanisms of reversible and irreversible microbial cell breakdown upon exposure to PEF treatment. A reversible breakdown results in a damaged cell structure or wherein a cell would be capable of complete recovery. In the case of an irreversible breakdown, the microbial cell loses its membrane integrity, leading to cell death. This figure also illustrates that as the strength of the electric field attains values greater than critical threshold limits, the occurrence of cell rupture and the formation of large pores can be observed.[13]

A reliable mechanistic depiction of the influence and effect of PEF with regard to biological cells and membranes is provided by the "dielectric breakdown theory." Cell membranes are assumed to be capacitors packed with fluids of low dielectric constant. Upon exposure to strong electric fields, the migration of ions occurs along the fluid and toward the membrane. This results in the formation of free, oppositely charged ions that tend to accumulate on both sides of the membrane. The cell wall undergoes compression because of the attractive interaction of charges, which decreases the membrane thickness.[21] This continues till the EFS reaches a critical value, at which point the formation of micropores occurs on the surface of the membrane. This leads to heightened permeability, and the phenomenon is termed electroporation. Electroporation is defined as a physical mechanism for the induction of a permeabilizing effect on biological membranes by the application of intense external electric pulses.[13]

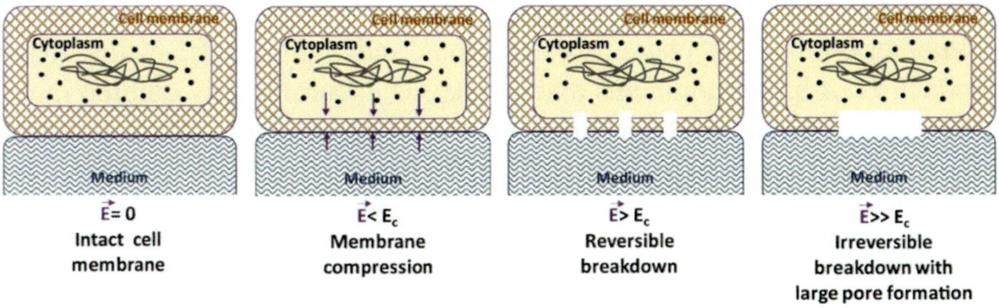

FIGURE 4.3 Schematic diagram depicting the breakdown of cells as a result of pore formation due to PEF, and E_c is the critical electric field.
Source: Reprinted with permission from Ref. [13]. © 2019 Elsevier.

Membranes that undergo electroporation could get damaged, attributable to its rupture or due to membrane surface overheating. In some instances, the formation of membrane pores facilitates transmembrane transport of substances, and this phenomenon contributes to the development of chemical imbalance in the cells. The PEF efficiency with regard to cell permeabilization depends on the following specific parameters:

- **Processing factors**: treatment time, temperature, EFS, pulse shape, frequency, width, specific energy, flow rate.
- **Properties of food materials**: pH, temperature, conductivity.
- **Properties of target cells** or enzymes: shape, membrane structure, size, the ionic composition of extracellular media, osmotic pressure.

The performance and efficiency of the process can be increased by adjusting the process parameters.[3,21]

4.4 NEED FOR THE INACTIVATION OF FOOD ENZYMES

In food processing, a crucial step is the control of enzymes and their activity since they have deteriorative effects on specific food quality attributes, cause the generation of off-flavors and odors, and bring about rheological property variations, among other undesirable effects. Therefore, the inactivation of these specific enzymes is of prime importance.[24]

Peroxidases (E.C. 1.11.1.7) severely affect the quality of processed foods.[45] With the presence of H_2O_2, it catalyzes the oxidation reaction of several substances to bring about color, flavor, and nutritional degradation.[35] Polyphenol oxidases (E.C. 1.14.18.1) are a class of enzymes that bring about monophenols to diphenol shydroxylation and further boost theortho-di-phenols oxidation. This generates o-quinones that subsequently polymerize with other o-quinones, phenolic compounds, and amino acids, leading to the production of brown pigments and condenses to form highly colored mela-nines.[35,23] For instance, cloudy apple juice is highly sensitive, in the presence of atmospheric oxygen, to enzymatic browning as it has substantial amounts of polyphenol oxidase. This causes deteriorative effects on the organoleptic quality of the juice.[14]

Lipoxygenase (E.C. 1.13.11.12) is an iron-containing class of enzymes. It catalyzes the molecular oxygen addition to fatty acids (FA) that contain pentadiene systems to generate unsaturated FA hydroperoxides, which further decompose to form acids, ketones, and aldehydes. Additionally, the hydroperoxides and free radicals resulting from this process have the ability to degrade vitamins and proteins present in food products.[23] This set of enzymes play a crucial role in the generation of rancid off-flavors in several vegetables.[22] The undesirable flavors produced in protein-rich products, the staleness of beer, the beany flavor of soybean (resulting from lipoxygenase-catalyzed oxidation that produces ketones and aldehydes), the co-oxidation of carotenoids that causes discoloration of foods, and nutrient decrease are all instances of lipoxygenase involvement. Therefore, the inactivation of lipoxygenases is essential for processing.[30]

Pectin methylesterase (E.C. 3.1.1.11) brings about the de-esterification reaction of pectin, leading to its conversion to low-ester pectin or pectic acids and methanol. Furthermore, these could be hydrolyzed by polyga-lacturonase, another pectic enzyme. They are vital in fruit and vegetable processing industries since they are known to have a significant impact on properties like viscosity, color, firmness, clarity, flavor, stability, and yield.[11] The inactivation of these pectic enzymes can inhibit undesirable textural modifications, cloudy appearance of juices, and gelation of concentrates and purees.[10]

4.5 MECHANISM OF DESTRUCTION OF ENZYMES BY PEF

PEF treatments are known to cause conformational changes to proteins ranging from microscopic modifications to macroscopic alterations. Various research studies have elucidated the structural modifications of proteins and enzymes due to PEF treatment.[44] Upon exposure to electric fields, the mechanism by which proteins are impacted involves the absorption of energy by the polar groups of proteins to produce free radicals or causing protein aggregation and unfolding. It has been suggested that free radicals have the ability to disrupt several interactions among protein molecules, including electrostatic, hydrogen bonding, van der Waals forces, hydrophobic interactions, salt bridges, and disulfide linkages. Thus, various types of protein structures and functional properties could be changed.

For enzyme inactivation to occur, the initial step involves the unfolding of the protein. It has been suggested that the protein molecules are polarized at a low PEF strength initially, leading to the exposure of the hydrophobic amino acids in the solvent. Gradually, as the EFS increases, protein unfolding occurs.[16] These structural alterations occur as a result of covalent bond changes that subsequently lead to the chemical modification of proteins (1° structure) or noncovalent changes that might cause conformational modifications in the protein spatial structure (from 2° structure to 4° structure).[44]

- **Primary structure:** There is a lack of concrete proof to suggest the influence of PEF with regard to the primary structure of proteins. It has been reported that since there are no peptide cleavages in the polypeptide chains after PEF, it does not cause the destruction of enzymes. The modifications in protein spatial structures could be the reason for the inactivation of enzymes. Additionally, it has also been established that PEF does not impact peptides and amino acids.[7] Several other studies have suggested that since the apparent molecular size of enzymes remains the same post-PEF treatment, this is an indication that enzyme modifications brought about by proteolysis were absent in PEF processing.[28]
- **Secondary structure:** To comprehend the influence of PEF onenzymes and proteins, it is essential to evaluate its role in the alteration of the secondary structure of proteins. It has been stipulated that

α-helix disruption in ceproteins have an intense dipole moment. This explains the effect of external electric fields on α-helix and its potential to be affected attributable to its nature and properties. From Figure 4.4, it can be observed that the α-helix in proteins, peptide dipoles are aligned in similar directions to create a macrodipole of considerable strength. The positive pole is pointed toward the N terminus of the peptide, while the negative pole of the dipole is pointed toward the C terminus.[44] It has been confirmed in several studies that there exists close parallelism between enzymatic inactivation and α-helical fraction loss in 2° structure due to PEF processing treatments.[41,42,46] In the case of those enzymes that are highly rich in β-sheet protein, for instance, pepsin, the inactivation has been correlated to β-sheet structural loss and the presence of randomly coiled regions and spaces in the secondary structure under PEF treatment. The link between residual enzymatic activity (RA) and residual β-sheet content for PEF treatments has also been reported.[43]

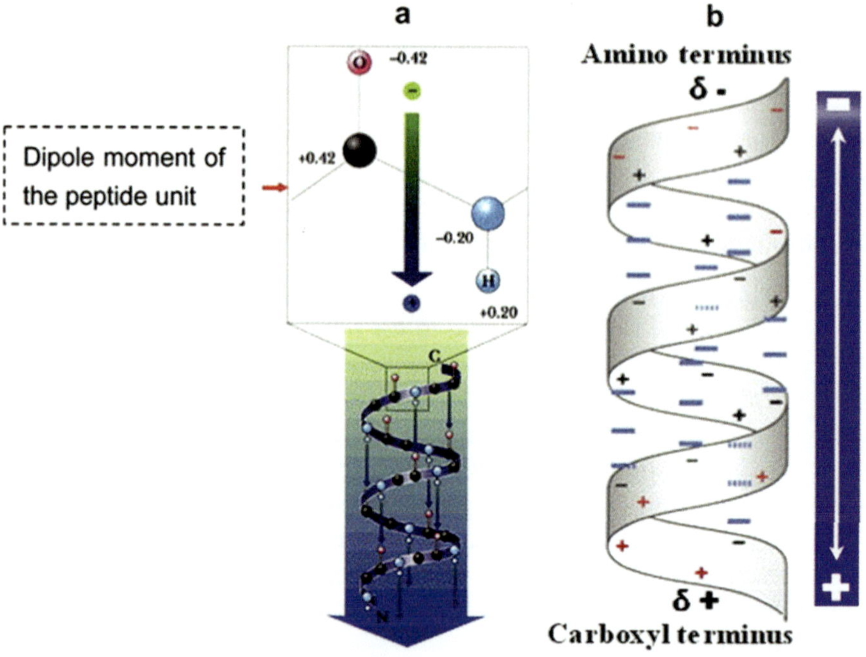

FIGURE 4.4 Schematic depiction of the geometry and dipole moment of peptide units: (a) Formation of macrodipole moment, and (b) by the alignment of peptide dipoles of helices.
Source: Reprinted with permission from Ref. [44]. © 2012 Elsevier.

- **Tertiary structure:** Certain side chains and amino acid residues present in the interior regions of native enzymes were exposed to the surface post-PEF treatment. Usually, the structural changes in proteins lead to increased surface hydrophobicity as a result of the exposure of the groups that are originally folded compactly within the intact and native protein molecule.[39] It is evident that enzymatic activity depends on the active site configuration and the surrounding protein chain conformation. Highly polar and charged regions result from a highly asymmetrical spatial distribution of charges created by amino acid residues. Additionally, the external electric fields may play a role in inducing charge separation, which may further lead to the unfolding of the protein and modification of the tertiary structure. PEF treatments initially induce the protein molecule's unfolding, which subsequently paves the way for protein interactions and aggregation. This is because the aggregation of proteins occurs with greater ease in the case of partially unfolded protein intermediates. The modification of the 3D protein structure is indicated by the movement of certain chromospheres, hydrophobic groups, and changes in disulfide linkage conformations.[44]

Although the exact mechanism of enzyme inactivation as a consequence of PEF treatment has yet to be elucidated, their conformational alterations resulting in denaturation have been suggested. Although PEF treatments are categorically and principally nonthermal, an energy-dependent increase in temperature of the treated medium (Joule or ohmic heating) is expected. This is a vital consideration with regard to food matrices containing heat-sensitive bioactive compounds.[36] Furthermore, as reports on PEF-induced enzymatic inactivation involve experimentation with long treatment time, the possibility of thermal denaturation due to ohmic heating cannot be ruled out. Moreover, a topic of interest from the industrial perspective for enzymatic application has been the suggestions of protein refolding.[28]

4.6 KINETIC MODELS FOR THE INACTIVATION OF ENZYMES BY PEF

Mathematical modeling is an essential technique for the prediction of the enzymatic inactivation and related responses under specific environmental conditions. Mathematical models are generally based on reliable experimentation, and lucid comprehension of physiological mechanisms behind the enzymatic inhibition caused due to PEF.

There are two significant parts to a predictive kinetic model; the primary model depicts the enzymatic modifications as a function of time, whereas the secondary model characterizes the primary model factors as a function of the processing and environmental conditions. A tertiary model, which arises by combining the primary and secondary models, can thereby predict the enzymatic evolution as a function of time and the processing and environmental conditions.[17]

RA is the ratio of the enzymatic activity post-treatment to the enzymatic activity observed in the untreated sample and is given by eq (4.1)

$$RA = A_t \,/\, A_0 \tag{4.1}$$

where A_t = enzymatic activity in the sample post-treatment; A_0 = enzymatic activity of the untreated sample.

The **first-order kinetics model** correlates RA as a function of the total electric energy density from the pulse generator during PEF treatment, as described below. It has been extensively employed to explain the dependency of the enzyme's RA on the electric energy density for several enzymes.

$$RA = \exp\left[-k_q * Q\right] \tag{4.2}$$

where Q = total electric energy density; k_q = constant.

The first-order fractional conversion model is most convenient to depict enzymatic inhibition by PEF,[5] as shown below.

$$\frac{RA - RA_{\infty}}{RA_0 - RA_{\infty}} = \exp\left[-k * P\right] \tag{4.3}$$

where RA_{∞} = RA after prolonged treatment; P= treatment time, pulse frequency, pulse width, electrical energy density; and k = modified rate constant.

Hülshegers equation[18] evaluates the effect of EFS as follows:

$$\ln S = \left[-b_E\right]\left[E - E_c\right] \tag{4.4}$$

where S = survival fraction at treatment time; b_E = regression coefficient; E = electric field strength; and E_c = extrapolated critical value of E for 100% survival.

However, a mathematical model in eq (4.5) was proposed after the following assumptions were taken into consideration[19]: (a) There is a linear relationship between the log survival ratio and EFS; (b) The log survival ratio and log treatment time are related in a linear fashion.

$$S = \left(t / t_c\right)^{(E-Ec)/k} \tag{4.5}$$

where t_c = extrapolated critical value of t for 100% survival; k = constant.

Weibull distribution is a continuous probability distribution that has been modified to give eq (4.6) for PEF treatment. It has been found that the scale parameter and the EFS are exponentially related.[17] It has also been suggested that the Weibull distribution explains the enzymatic inactivation at greater EFS with better accuracy than lower levels.[5]

$$Y = \exp\left[-\left(t / \delta\right)^{p}\right] \tag{4.6}$$

where t = treatment time or specific energy density; δ = scale parameter; and p =shape parameter.

Quadratic response model: The response surface methodology (RSM) is a method that decreases the number of experiments in order to obtain better statistical interpretation. It comprises mathematical and statistical protocols intended to evaluate the interactions among several chosen responses. RSM is advantageous since it analyzes the impact or influence those multiple independent variables have on PEF treatment. Modeling the effects of variables on RA is accomplished by polynomial response functions. The statistical analyses aid in indicating and informing the accuracy of the developed quadratic model with regard to the modifications in the enzymatic activity.[1,26]

Predictive modeling for enzymatic inactivation has vital practical applications for enhancing the PEF processing efficiency as well as giving insight into the analysis of certain critical parameters. The development of quadratic response models to understand the factors affecting the retention of bioactive substances during PEF is essential for a thorough understanding of the effect of various processing factors.[17]

4.7 APPLICATIONS OF PEF TECHNOLOGY ON FOOD ENZYMES

TABLE 4.1 Applications of PEF in the Inactivation of Food Enzymes.

Enzymes	Medium	Processing conditions	Observed effects	Refs.
Alkaline phosphatase, Lactoperoxidase	Apple juice	• Collinear chamber • Pulse width: 3 μs. • Flow rate: 5 L/h. • E: 34 kV/cm for lactoperoxidase and 38 (kV/cm) for Alkaline phosphatase.	• PEF treatment reduced 50% lactoperoxidase activity. • 60% inactivation of alkaline phosphatase enzyme by PEF.	[20]
Horseradish Peroxidase	Buffer solution	• E: 5–25 kV/cm. • Pulse number: 207–1242 • Temperature: 40°C.	• The decrease in peroxidase activity was ~35% post PEF 22 kV/cm for 1214 pulses. • The peroxidase inactivation was attributed to the conformational modifications of α-helix.	[45]
Lipoxygenase	Soy milk	• Frequency: 100–600 Hz • Pulse width: 1–5 μs • Treatment time: 345 μs (constant) • E: 30 kV/cm.	• 88% soybean lipoxygenase activity was reduced with an increase in treatment time, pulse strength, width, and frequency. • Optimized conditions: 42 kV/cm, 1036 μs, 400 Hz, 2 μs, 25°C.	[22]
Lipoxygenase	Tomato juice	• E: 0–35 kV/cm. • Treatment time: 20–70 μs • Temperature: 10–50°C.	• Lipoxygenase inactivation was primarily impacted by EFS.	[27]
Lipoxygenase and Peroxidase	Almond milk	• E: 7–28 kV/cm. • Frequency: 1 kHz. • Pulse width: 40 μs (unipolar square-wave) • Treatment time: 200 μs	• 50% lipoxygenase and 45% peroxidase inactivated at 28 kV/cm.	[25]

TABLE 4.1 *(Continued)*

Enzymes	Medium	Processing conditions	Observed effects	Refs.
Lipoxygenase, Hydroperoxide lyase	Tomato juice	• E: 35 kV/cm. • Treatment time: 1000 μs. • Pulse width: 1–7 μs. • Frequency: 50–250 Hz.	• RA of both enzymes decreased with an increase in frequency or pulse width. • The resistance to HIPEF was higher in Lipoxygenase than hydroxy peroxide lyase.	[2]
Pectin methylesterase	Orange juice	• E: up to 35 kV/cm at 30°C	• 90% pectin methylesterase inactivation was accomplished at 25 kV/cm, at 50°C.	[40]
Pectin methylesterase	Orange juice	• HIPEF • E: 5–35 kV/cm • Treatment time: 1500 μs. • Pulse width: 4 μs • Frequency: 200 Hz (mono- and bipolar mode).	• Pectin methylesterase inhibited when the electric field intensity, treatment time, and electrical energy density increased.	[5]
Pectin methylesterase	Tomato	• HIPEF • Generator capacitance: 3 or 10 μF • Pulse width: 0.02 or 0.04 ms. • Pulse number: 400 (mono-/ bipolar mode) • Peak voltage: 2, 2.4 kV. • E: 5–24 kV/cm.	• 93.8% inhibition of Pectin methylesterase activity with 400 pulses of 0.02-ms pulse width at 24 kV/cm. • Enzyme inactivation model fitted: Hülsheger model, classical exponential decay model, and Fermi's empirical models.	[11]

TABLE 4.1 *(Continued)*

Enzymes	Medium	Processing conditions	Observed effects	Refs.
Pectinesterase	Gazpacho vegetable soup	• Pulse width: 4 μs (mono-/bipolar square-wave). • E: 5–35 kV/cm • Treatment time: 1500 μs. • Frequency: 200 Hz. • Temperature: <40°C.	• Increased inhibition of pectinesterase activity achieved with bipolar pulses. • Greater electric field intensities and lengthier treatment times decreased pectinesterase activity significantly.	[9]
Peroxidase	Western horse-radish	• E: 4-23 kV/cm • Frequency: 50 Hz • Denaturation: 92°C, 5 min.	• Thermally denatured Peroxidase underwent refolding enhancement after 30 s of 12 kV/cm PEF at 50 Hz.	[28]
Peroxidase	Carrot juice	• E: 20–35 kV/cm • Treatment time: 300–2000 μs.	• PEF treatment (35 kV/cm/1500 μs) inactivated ~93% peroxidase activity.	[32]
Peroxidase	Carrot	• E: 1 kV/cm (capacitance: 0.5 μF; Pulse number: 20) or 1.5 kV/cm (capacitance: 1 μF; Pulse number: 20) • Blanching: 100 °C, 3 min.	• Reduction of 30–50% peroxidase activity with PEF pretreatment. • Complete enzymatic inactivation with blanching step.	[6]
Peroxidase Polyphenoloxidase	Fresh apple juice	• E: 20–40 kV/cm • Treatment time: 25–100 μs. • Pretreatment temperature: 23–50°C.	• 71% polyphenol oxidase and 68% of Peroxidase inactivated by PEF. • Optimized conditions: 50°C (preheating), 100 μs (treatment time), 40 kV/cm. • Higher enzymatic inactivation by PEF than pasteurization.	[35]

TABLE 4.1 *(Continued)*

Enzymes	Medium	Processing conditions	Observed effects	Refs.
Peroxidase Polyphenoloxidase	Mango and Papaya juice	• E: 20–40 kV/cm	• Highest PEF treatment (40 kV/cm; 360 μs) inactivated both enzymes (no Stevia present).	[4]
		• Treatment time: 100–360 μs	• Complete enzymatic inactivation of polyphenol oxidase occurred when Stevia was present.	
Polygalacturonase	Aqueous solution of commercial enzyme preparation	• E: 15–38 kV/cm • Treatment time: 300–1100 μs.	• 76.5% inactivation of polygalacturonase at 38 kV/cm for 1100 μs.	[8]
Polyphenol oxidase, Lipoxygenase	Enzyme solution in buffer	• E: 8–24 kV/cm.	• 69% polyphenol oxidase activity reduced at 24 kV/cm, 320 μs.	[23]
		• Pulse number: 6–26.	• 88% lipoxygenase activity reduction at 24 kV/cm, 962 μs.	
			• α-helix structural loss and increments of β-sheet content.	
Polyphenol oxidase, Peroxidase	Grape juice	• E: 25–35 kV/cm.	• Polyphenol oxidase activity inhibited at 35 kV/cm at 630 Hz.	[26]
		• Treatment time: 1–5 ms.	• 58.13% of Peroxidase was depleted at 30.2 kV/cm and 1000 Hz for similar treatment time.	
		• Frequency: 200–1000 Hz	• Optimized conditions: 4.94 ms, 677 Hz, 33.5 kV/cm	

4.8 SUMMARY

While extensive research has been carried out for microbial inactivation studies with PEF, enzymatic inactivation studies for the same process seem limited. The focus has rested predominantly on elucidating the role of PEF on enzymes; however, the reports are inconsistent, and in some cases, even contradictory. Nevertheless, this could be attributed to the variety of different conditions chosen for experimentation, and several other factors affecting the process cannot be disregarded. For elucidating the exact mechanism of inactivation for enzymes, more detailed studies shedding light on the action of PEF on proteins and enzymes, specifically at the molecular level, are required. The focus has to be shifted to comprehending the mechanisms of protein unfolding and refolding and, subsequently, its aggregation. Moreover, the critical PEF dosages required to induce modifications of the structure and function of protein need to be evaluated. The side effects of PEF, including the thermal effects, chemical modifications, and electrochemical reactions occurring, also need extensive research.

KEYWORDS

- **enzyme inactivation**
- **food enzyme**
- **kinetics of destruction**
- **nonthermal processing**
- **pulsed electric field**

REFERENCES

1. Aguiló-Aguayo, I.; Soliva-Fortuny, R.; Martín-Belloso, O. Impact of High-Intensity Pulsed Electric Field Variables Affecting Peroxidase and Lipoxygenase Activities of Watermelon Juice. *LWT—Food Sci. Technol.* **2010,** *43* (6), 897–902. DOI: 10.1016/j.lwt.2010.01.022 (accessed Aug 2, 2021).
2. Aguiló-Aguayo, I.; Soliva-Fortuny, R.; Martín-Belloso, O. Effects of High-Intensity Pulsed Electric Fields on Lipoxygenase and Hydroperoxide Lyase Activities in Tomato Juice. *J. Food Sci.* **2009,** *74* (8), C595–C601. DOI: 10.1111/j.1750-3841.2009.01310.x (accessed Aug 2, 2021).
3. Arshad, R. N.; Abdul-Malek, Z.; Munir, A.; Buntat, Z.; Ahmad, M. H.; Jusoh, Y. M. M.; Bekhit, A. E.-D.; Roobab, U.; Manzoor, M. F.; Aadil, R. M. Electrical Systems for

Pulsed Electric Field Applications in the Food Industry: An Engineering Perspective. *Trends Food Sci. Technol.* **2020,** *104*, 1–13. DOI: 10.1016/j.tifs.2020.07.008 (accessed Aug 2, 2021).
4. Criado, M. N.; Martínez, A.; Rodrigo, D. *Control of Polyphenoloxidase and Peroxidase Activities in Mango and Papaya Juice by Pulsed Electric Field and* Stevia Rebaudiana Bertoni *Extract Combined Process*; 2016; pp 128–131. DOI: 10.1007/978-981-287-817-5_28 (accessed Aug 2, 2021).
5. Elez-Martínez, P.; Suárez-Recio, M.; Martín-Belloso, O. Modeling the Reduction of Pectin Methyl Esterase Activity in Orange Juice by High Intensity Pulsed Electric Fields. *J. Food Eng.* **2007,** *78* (1), 184–193. DOI: 10.1016/j.jfoodeng.2005.09.014 (accessed Aug 2, 2021).
6. Gachovska, T. K.; Simpson, M. V; Ngadi, M. O.; Raghavan, G. Pulsed Electric Field Treatment of Carrots Before Drying and Rehydration. *J. Sci. Food Agric.* **2009,** *89* (14), 2372–2376. DOI: 10.1002/jsfa.3730 (accessed Aug 2, 2021).
7. Garde-Cerdán, T.; Arias-Gil, M.; Marsellés-Fontanet, A. R.; Ancín-Azpilicueta, C.; Martín-Belloso, O. Effects of Thermal and Non-Thermal Processing Treatments on Fatty Acids and Free Amino Acids of Grape Juice. *Food Control* **2007,** *18* (5), 473–479. DOI: 10.1016/j.foodcont.2005.12.004 (accessed Aug 2, 2021).
8. Giner, J.; Bailo-Ballarín, E.; Gorinstein, S.; Martín-Belloso, O. New Kinetic Approach to the Evolution of Polygalacturonase (EC 3.2.1.15) Activity in a Commercial Enzyme Preparation Under Pulsed Electric Fields. *J. Food Sci.* **2006,** *71* (6), E262–E269. DOI: 10.1111/j.1750-3841.2006.00054.x (accessed Aug 2, 2021).
9. Giner, J.; Elez-Martínez, P.; Martín-Belloso, O. Modeling Within the Bayesian Framework, the Inactivation of Pectinesterase in Gazpacho by Pulsed Electric Fields. *J. Food Eng.* **2009,** *95* (3), 446–452. DOI: 10.1016/j.jfoodeng.2009.06.006 (accessed Aug 2, 2021).
10. Giner, J.; Bailo, E.; Gimeno, V.; Martín-Belloso, O. Models in a Bayesian Framework for Inactivation of Pectinesterase in a Commercial Enzyme Formulation by Pulsed Electric Fields. *Eur. Food Res. Technol.* **2005,** *221* (3–4), 255–264. DOI: 10.1007/s00217-005-1143-2 (accessed Aug 2, 2021).
11. Giner, J.; Gimeno, V.; Espachs, A.; Elez, P.; Barbosa-Cánovas, G. V; Martín, O. Inhibition of Tomato (*Licopersicon Esculentum* Mill.) Pectin Methylesterase by Pulsed Electric Fields. *Innov. Food Sci. Emerg. Technol.* **2000,** *1* (1), 57–67. DOI: 10.1016/S1466-8564(00)00003-5 (accessed Aug 2, 2021).
12. Giteru, S. G.; Oey, I.; Ali, M. A. Feasibility of Using Pulsed Electric Fields to Modify Biomacromolecules: A Review. *Trends Food Sci. Technol.* **2018,** *72*, 91–113. DOI: 10.1016/j.tifs.2017.12.009 (accessed Aug 2, 2021).
13. Gómez, B.; Munekata, P. E. S.; Gavahian, M.; Barba, F. J.; Martí-Quijal, F. J.; Bolumar, T.; Campagnol, P. C. B.; Tomasevic, I.; Lorenzo, J. M. Application of Pulsed Electric Fields in Meat and Fish Processing Industries: An Overview. *Food Res. Int.* **2019,** *123*, 95–105. DOI: 10.1016/j.foodres.2019.04.047.
14. Gui, F.; Wu, J.; Chen, F.; Liao, X.; Hu, X.; Zhang, Z.; Wang, Z. Inactivation of Polyphenol Oxidases in Cloudy Apple Juice Exposed to Supercritical Carbon Dioxide. *Food Chem.* **2007,** *100* (4), 1678–1685. DOI: 10.1016/j.foodchem.2005.12.048 (accessed Aug 2, 2021).
15. Guionet, A.; David, F.; Zaepffel, C.; Coustets, M.; Helmi, K.; Cheype, C.; Packan, D.; Garnier, J.-P.; Blanckaert, V.; Teissié, J. *E. Coli* Electro-Eradication on a Closed Loop

Circuit by Using Milli-, Micro- and Nanosecond Pulsed Electric Fields: Comparison between Energy Costs. *Bioelectrochemistry* **2015,** *103*, 65–73. DOI: 10.1016/j.bioelechem.2014.08.021 (accessed Aug 2, 2021).

16. Han, Z.; Cai, M.; Cheng, J. H.; Sun, D. W. Effects of Electric Fields and Electromagnetic Wave on Food Protein Structure and Functionality: A Review. *Trends Food Sci. Technol.* **2018,** *75*, 1–9. DOI: 10.1016/j.tifs.2018.02.017 (accessed Aug 2, 2021).
17. Huang, K.; Tian, H.; Gai, L.; Wang, J. A Review of Kinetic Models for Inactivating Microorganisms and Enzymes by Pulsed Electric Field Processing. *J. Food Eng.* **2012,** *111* (2), 191–207. DOI: 10.1016/j.jfoodeng.2012.02.007 (accessed Aug 2, 2021).
18. Hülsheger, H.; Niemann, E.-G. Lethal Effects of High-Voltage Pulses on *E. Coli* K12. *Radiat. Environ. Biophys.* **1980,** *18* (4), 281–288. DOI: 10.1007/BF01324271 (accessed Aug 2, 2021).
19. Hülsheger, H.; Potel, J.; Niemann, E. G. Killing of Bacteria with Electric Pulses of High Field Strength. *Radiat. Environ. Biophys.* **1981,** *20* (1), 53–65. DOI: 10.1007/BF01323926 (accessed Aug 2, 2021).
20. Jaeger, H.; Meneses, N.; Moritz, J.; Knorr, D. Model for the Differentiation of Temperature and Electric Field Effects During Thermal Assisted PEF Processing. *J. Food Eng.* **2010,** *100* (1), 109–118. DOI: 10.1016/j.jfoodeng.2010.03.034 (accessed Aug 2, 2021).
21. Kovačić, Đ.; Rupčić, S.; Kralik, D.; Jovičić, D.; Spajić, R.; Tišma, M. Pulsed Electric Field: An Emerging Pretreatment Technology in a Biogas Production. *Waste Manag.* **2020**. DOI: 10.1016/j.wasman.2020.10.009 (accessed Aug 2, 2021).
22. Li, Y.-Q.; Chen, Q.; Liu, X.-H.; Chen, Z.-X. Inactivation of Soybean Lipoxygenase in Soymilk by Pulsed Electric Fields. *Food Chem.* **2008,** *109* (2), 408–414. DOI: 10.1016/j.foodchem.2008.01.001 (accessed Aug 2, 2021).
23. Luo, W.; Zhang, R. B.; Wang, L. M.; Chen, J.; Guan, Z. C. Conformation Changes of Polyphenol Oxidase and Lipoxygenase Induced by PEF Treatment. *J. Appl. Electrochem.* **2010,** *40* (2), 295–301. DOI: 10.1007/s10800-009-9973-4 (accessed Aug 2, 2021).
24. Makroo, H. A.; Rastogi, N. K.; Srivastava, B. ohmic Heating Assisted Inactivation of Enzymes and Microorganisms in Foods: A Review. *Trends Food Sci. Technol.* **2020,** *97*, 451–465. DOI: 10.1016/j.tifs.2020.01.015 (accessed Aug 2, 2021).
25. Manzoor, M. F.; Zeng, X.; Ahmad, N.; Ahmed, Z.; Rehman, A.; Aadil, R. M.; Roobab, U.; Siddique, R.; Rahaman, A. Effect of Pulsed Electric Field and Thermal Treatments on the Bioactive Compounds, Enzymes, Microbial, and Physical Stability of Almond Milk During Storage. *J. Food Process. Preserv.* **2020,** *44* (7). E-article 14541. DOI: 10.1111/jfpp.14541 (accessed Aug 2, 2021).
26. Marsellés-Fontanet, Á. R.; Martín-Belloso, O. Optimization and Validation of PEF Processing Conditions to Inactivate Oxidative Enzymes of Grape Juice. *J. Food Eng.* **2007,** *83* (3), 452–462. DOI: 10.1016/j.jfoodeng.2007.04.001 (accessed Aug 2, 2021).
27. Min, S.; Min, S. K.; Zhang, Q. H. Inactivation Kinetics of Tomato Juice Lipoxygenase by Pulsed Electric Fields. *J. Food Sci.* **2003,** *68* (6), 1995–2001. DOI: 10.1111/j.1365-2621.2003.tb07008.x (accessed Aug 2, 2021).
28. Ohshima, T.; Tamura, T.; Sato, M. Influence of Pulsed Electric Field on Various Enzyme Activities. *J. Electrost.* **2007,** *65* (3), 156–161. DOI: 10.1016/j.elstat.2006.07.005 (accessed Aug 2, 2021).
29. Pashazadeh, B.; Elhamirad, A. H.; Hajnajari, H.; Sharayei, P.; Armin, M. Optimization of the Pulsed Electric Field-Assisted Extraction of Functional Compounds from

Cinnamon. *Biocatal. Agric. Biotechnol.* **2020,** *23*. E-article 101461. DOI: 10.1016/j.bcab.2019.101461 (accessed Aug 2, 2021).

30. Pellicer, J. A.; Navarro, P.; Hernández Sánchez, P.; Gómez-López, V. M. Structural Changes Associated with the Inactivation of Lipoxygenase by Pulsed Light. *LWT* **2019,** *113*. E-article 108332. DOI: 10.1016/j.lwt.2019.108332 (accessed Aug 2, 2021).
31. Qin, S.; Timoshkin, I. V.; Maclean, M.; Wilson, M. P.; Given, M. J.; Wang, T.; Anderson, J. G.; Macgregor, S. J. Pulsed Electric Field Treatment of Saccharomyces Cerevisiae Using Different Waveforms. *IEEE Trans. Dielectr. Electr. Insul.* **2015,** *22* (4), 1841–1848. DOI: 10.1109/TDEI.2015.005152 (accessed Aug 2, 2021).
32. Quintão-Teixeira, L. J.; Soliva-Fortuny, R.; Mota Ramos, A.; Martín-Belloso, O. Kinetics of Peroxidase Inactivation in Carrot Juice Treated with Pulsed Electric Fields. *J. Food Sci.* **2013,** *78* (2), E222–E228. DOI: 10.1111/1750-3841.12019 (accessed Aug 2, 2021).
33. Reberšek, M. Beyond Electroporation Pulse Parameters: From Application to Evaluation. In *Handbook of Electroporation*; Springer International Publishing: Cham, 2017; pp 977–997. DOI: 10.1007/978-3-319-32886-7_222 (accessed Aug 2, 2021).
34. Ricci, A.; Parpinello, G. P.; Versari, A. Recent Advances and Applications of Pulsed Electric Fields (PEF) to Improve Polyphenol Extraction and Color Release During Red Winemaking. *Beverages* **2018,** *4* (1), 18–23. DOI: 10.3390/beverages4010018 (accessed Aug 2, 2021).
35. Riener, J.; Noci, F.; Cronin, D. A.; Morgan, D. J.; Lyng, J. G. Combined Effect of Temperature and Pulsed Electric Fields on Apple Juice Peroxidase and Polyphenoloxidase Inactivation. *Food Chem.* **2008,** *109* (2), 402–407. DOI: 10.1016/j.foodchem.2007.12.059 (accessed Aug 2, 2021).
36. Schottroff, F.; Gratz, M.; Krottenthaler, A.; Johnson, N. B.; Bédard, M. F.; Jaeger, H. Pulsed Electric Field Preservation of Liquid Whey Protein Formulations—Influence of Process Parameters, pH, and Protein Content on the Inactivation of Listeria Innocua and the Retention of Bioactive Ingredients. *J. Food Eng.* **2019,** *243*, 142–152. DOI: 10.1016/j.jfoodeng.2018.09.003 (accessed Aug 2, 2021).
37. Soliva-Fortuny, R.; Balasa, A.; Knorr, D.; Martín-Belloso, O. Effects of Pulsed Electric Fields on Bioactive Compounds in Foods: A Review. *Trends Food Sci. Technol.* **2009,** *20* (11–12), 544–556. DOI: 10.1016/j.tifs.2009.07.003 (accessed Aug 2, 2021).
38. Wang, Q.; Li, Y.; Sun, D.-W.; Zhu, Z. Enhancing Food Processing by Pulsed and High Voltage Electric Fields: Principles and Applications. *Crit. Rev. Food Sci. Nutr.* **2018,** *58* (13), 2285–2298. DOI: 10.1080/10408398.2018.1434609 (accessed Aug 2, 2021).
39. Xiang, B. Y.; Ngadi, M. O.; Ochoa-Martinez, L. A.; Simpson, M. V. Pulsed Electric Field-Induced Structural Modification of Whey Protein Isolate. *Food Bioprocess Technol.* **2011,** *4* (8), 1341–1348. DOI: 10.1007/s11947-009-0266-z (accessed Aug 2, 2021).
40. Yeom, H. W.; Zhang, Q. H.; Chism, G. W. Inactivation of Pectin Methyl Esterase in Orange Juice by Pulsed Electric Fields. *J. Food Sci.* **2002,** *67* (6), 2154–2159. DOI: 10.1111/j.1365-2621.2002.tb09519.x (accessed Aug 2, 2021).
41. Zhao, W.; Yang, R. Protective Effect of Sorbitol on Enzymes Exposed to Microsecond Pulsed Electric Field. *J. Phys. Chem. B* **2008,** *112* (44), 14018–14025. DOI: 10.1021/jp8062367 (accessed Aug 2, 2021).

42. Zhao, W.; Yang, R. The Effect of Pulsed Electric Fields on the Inactivation and Structure of Lysozyme. *Food Chem.* **2008,** *110* (2), 334–343. DOI: 10.1016/j.foodchem.2008.02.008 (accessed Aug 2, 2021).
43. Zhao, W.; Yang, R. Effect of High-Intensity Pulsed Electric Fields on the Activity, Conformation and Self-Aggregation of Pepsin. *Food Chem.* **2009,** *114* (3), 777–781. DOI: 10.1016/j.foodchem.2008.10.016 (accessed Aug 2, 2021).
44. Zhao, W.; Yang, R.; Zhang, H. Q. Recent Advances in the Action of Pulsed Electric Fields on Enzymes and Food Component Proteins. *Trends Food Sci. Technol.* **2012,** *27* (2), 83–96. DOI: 10.1016/j.tifs.2012.05.007 (accessed Aug 2, 2021).
45. Zhong, K.; Hu, X.; Zhao, G.; Chen, F.; Liao, X. Inactivation and Conformational Change of Horseradish Peroxidase Induced by Pulsed Electric Field. *Food Chem.* **2005,** *92* (3), 473–479. DOI: 10.1016/j.foodchem.2004.08.010 (accessed Aug 2, 2021).
46. Zhong, K.; Wu, J.; Wang, Z.; Chen, F.; Liao, X.; Hu, X.; Zhang, Z. Inactivation Kinetics and Secondary Structural Change of PEF-Treated POD and PPO. *Food Chem.* **2007,** *100* (1), 115–123. DOI: 10.1016/j.foodchem.2005.09.035 (accessed Aug 2, 2021).

CHAPTER 5

MICROWAVE PROCESSING AND OHMIC HEATING TECHNOLOGY FOR INACTIVATION OF FOOD ENZYMES

S. ABDULLAH, ABHIPRIYA PATRA, and
RAMA CHANDRA PRADHAN

ABSTRACT

Microwave processing and ohmic heating are novel food processing technologies that can be utilized as a promising alternative to traditional technologies. Both technologies have the advantage of lower energy consumption, treatment time, temperature, and higher efficiency as compared to conventional technologies. However, the recent advancement in the application of ohmic heating and microwave processing shows that both technologies are efficient in inactivating various enzymes present in the food. Utilizing these technologies over conventional thermal processing could help the food industries to develop final products with higher sensory and organoleptic properties.

5.1 INTRODUCTION

Thermal processing technologies are the most common methods used for enhancing the storage life of food. Shelf-life extension or preservation of food by thermal processing is achieved by inactivating certain enzymes and microorganisms that are liable for the spoilage of food. However, due to the involvement of high temperature, conventional thermal treatments cause certain negative impacts on the sensorial and organoleptic attributes of the

Enzyme Inactivation in Food Processing: Technologies, Materials, and Applications. Megh R. Goyal, PhD, Junaid Ahmad Malik, PhD & Ravi Pandiselvam, PhD (Eds.)

food items. These negative effects include deterioration of temperature-sensitive compounds, alteration in color, cooked off-flavor development, and so on. The consumer's demands for foods with higher quality and the undesirable effects of conventional thermal processing techniques have led to the investigation of alternative technologies for the processing of food.[47]

Novel processing technologies like microwave processing and ohmic heating, which work on volumetric heating methods using electro-heat technology, are promising alternatives to traditional processing technologies. With the help of dielectric and ionic mechanisms, ohmic and microwave processing converts electrical energy into heat energy. The basic advantages of these technologies are the faster generation of volumetric heat, which eradicates the requirement of a heating medium and minimizes fouling and overheating of surfaces.[18] Moreover, they require minimum energy and process time than conventional technologies. Furthermore, ohmic heating and microwave processing come under environment-friendly technologies since it generates a minimum volume of wastewater that has a higher energy conversion efficiency and does not require any fuel burn or steam boiler.[28]

The impact of ohmic heating and microwave on food items is mainly due to the involvement of temperature. However, in addition to thermal effects, these technologies are reported to have nonthermal effects on food products. The nonthermal effects are owing to the impact of the electric field applied on the molecules and ions with dipole moment and net charges. This may result in the destabilization of cell membranes and alteration of various biochemical reactions in food.[36] Various studies report that the nonthermal effects of these technologies can be utilized for the inactivation of enzymes from food products.

This chapter summarizes in detail the principle and applications of microwave processing and ohmic heating technology, and the impact of these technologies on numerous enzymes have been highlighted and discussed in detail.

5.2 PRINCIPLE AND APPLICATIONS OF MICROWAVE PROCESSING

Microwaves are electromagnetic radiations with the frequency between 300 MHz and 300 GHz. Generally, commercial microwave appliances work at both 2.45 GHz or 915 MHz frequencies, and domestic microwave systems work at 2.45 GHz frequency. Microwave heating occurs due to the capacity of the materials to convert the microwave radiation absorbed by it into heat energy.[14]

In food products, microwave heating is caused primarily due to ionic and dipole mechanisms. Moisture present inside the food products creates dielectric heating since water is dipolar. When the water molecules are incident by an oscillating electric field, the dipole molecules which were permanently polarized will rearrange in the electric field direction. This rearrangement happens more than a million times in a second since the applied electric field is of very high frequency. Thus internal friction is developed by the molecules which causes the generation of volumetric heating of the food products. In the ionic mechanism, the heat generation in the food products is due to the movement of ions present in the product, when the oscillating electric field is applied at a high frequency.[11] A schematic representation of the basic microwave processing system is illustrated in Figure 5.1.

Microwave processing has wider applications in the processing and preservation of food products. The major applications include blanching, baking, thawing, tempering, sterilization, pasteurization, drying, and so on. This huge popularity of microwave processing is mainly due to its various advantages, over conventional technologies, such as minimum maintenance, ease of operation, safe handling, better uniform heating, shorter cooking time, and a higher rate of heating. Moreover, microwave processing can preserve the bioactive compounds, antioxidants, color, and can reduce the antinutritional factors and microbial counts of the final product. However, the major disadvantage of microwave processing is the nonuniformity in the temperature distribution.

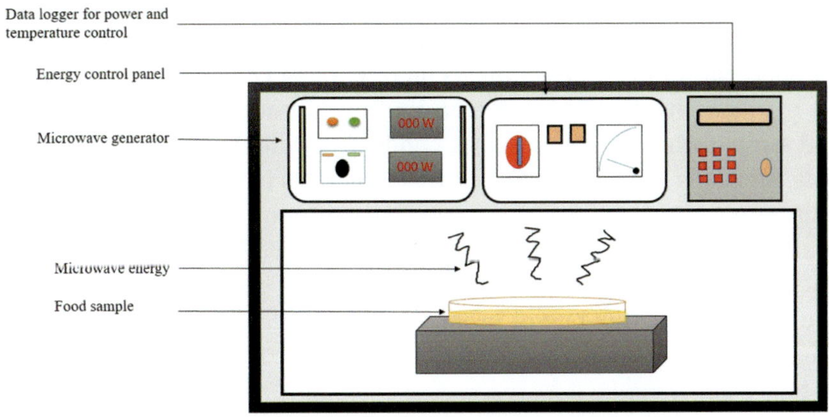

FIGURE 5.1 Schematic illustration of microwave processing setup.

The intensity of the electric field becomes higher in certain areas (hot spots) of the food products, due to the action of the microwave pattern causing nonuniformity. The nonuniformity can cause underprocessing of food products, uneven drying, irregular moisture distribution, and quality degradation of the products. Hence, uniformity is important and has to be maintained during microwave processing.[11,14,20]

5.3 PRINCIPLE AND APPLICATIONS OF OHMIC HEATING

Ohmic heating is a technology in which an alternating or direct electric current is allowed to pass through the food product for generating internal heat energy. The basic principle responsible for the generation of heat in the food materials is Joule's law. Hence, this technology is also termed as electroconductive heating, electro-heating, and Joule heating. The creation of heat energy is because of the electrical resistance produced by the food product by acting as an electrically conductive material.[26] This is a novel technology that is generally used for viscous and particulate foods. Ohmic heating follows a volumetric nature of heating which boosts the heating rate and thus shortens the processing time. Here, electrodes and food materials are in direct contact with each other, unlike in microwave. Also comparing to other traditional thermal processes, ohmic heating requires less energy and produces uniform heating throughout the process.[36]

The basic ohmic heating setup is demonstrated in Figure 5.2. The food samples, which are composed of both solids and liquids, will be in direct contact with the two electrodes located at both sides of the heating chamber. The solid part of the food sample is majorly plant cells that are bad conductors of electricity. However, the current flows through the liquid part of the food sample which is composed of conductive materials such as ions, weak acids, etc. This flow of electric current generates heat at the surface of cell walls resulting in cell lysis. Subsequently, due to the difference in electrical conductivity between the intra- and extra-cellular fluids, the current penetrates through the cell wall and thus heat is generated throughout the food matrix. Here, the rate of heating will be influenced by the distance by which the two electrodes are separated and the area of contact by them with the food sample.[26,51]

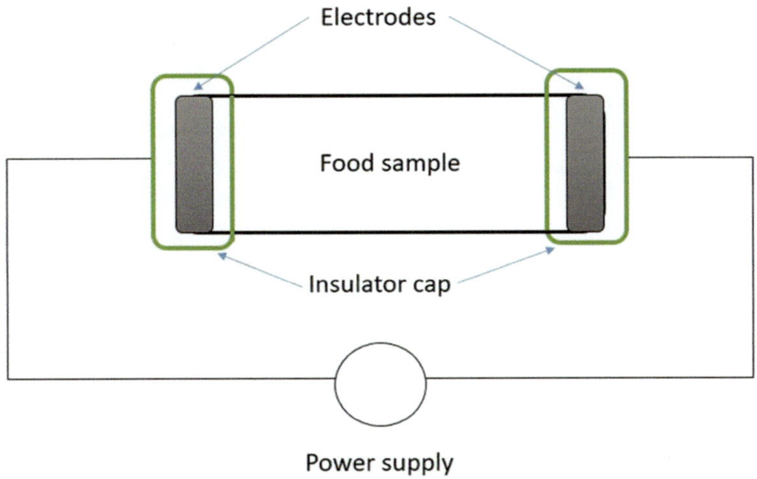

FIGURE 5.2 Schematic illustration of the basic ohmic heating system.

Ohmic heating technology possesses several advantages as well as disadvantages over other thermal technologies. A relative summary of various benefits and drawbacks of ohmic heating is given in Table 5.1. However, this technology is proved suitable for various food processing applications, such as extraction, dehydration, blanching, sterilization, evaporation, fermentation, thawing, pasteurization, preheating, cooking, drying, aseptic processing, inactivation of food enzymes, microbial inactivation, wastewater treatment, and starch galvanization. Furthermore, this technology aids to enhance the quality and expands the shelf life of food by preserving heat-sensitive compounds.[29]

TABLE 5.1 Advantages and Disadvantages of Ohmic Heating.

Advantages	Disadvantages
• Uniformity in heating and volumetric heating	• Formation of hot spots and cold spots due to the nonhomogenous food compositions
• Rapid heating and shorter processing time	• Shortest residence time for heating and holding system
• Higher energy efficiency (energy conversion from electrical energy to heat is almost 90%)	• Difficult to maintain the heating rate
• Absence of fouling due to the lack of hot surfaces	• High investment due to the high cost of electricity

TABLE 5.1 *(Continued)*

Advantages	Disadvantages
• Microbial cell damage at the lower thermal intensity	• Difficult to get appropriate inert materials to make electrodes
• Preservation of color and nutrient value	• Some of the beneficial enzymes along with protein get denatured
• Enhances the texture and inactivates enzymes	• Starch gets gelatinized and water present in the food gets vaporized
	• In some cases, the textures can be altered

5.4 ENZYME INACTIVATION MECHANISM BY OHMIC HEATING AND MICROWAVE PROCESSING METHODS

Enzymes are proteins. The major cause behind the enzyme inactivation is the denaturation of its tertiary structure by the disruption and/or the rearrangement of noncovalent bonds especially the hydrophobic interactions, ionic bonds, and hydrogen bonds. Likewise, the exposure to microwave and ohmic heating would also denature the tertiary structure of the enzymes to inactivate it. However, these exposures would not alter the primary structure of enzymes or proteins.[36]

There are two mechanisms reported by the researchers upon which an applied microwave or electric field is acting on a food enzyme, namely thermal effect and nonthermal effect. The thermal inactivation is caused by the thermal disruption of the structure of enzymes, which is similar to the inactivation caused by the conventional thermal processing technologies. However, even though the complete mechanism of action of nonthermal effect is not fully explored, studies reported that the nonthermal effect of these technologies is more dominant over the thermal effect in inactivating enzymes.[44]

When the aqueous reaction systems are exposed to microwave radiations, the molecules get excited to higher rotational and vibrational energy levels. This leads to the deterioration of chemical bonds. Additionally, researchers also report that the alteration in the tertiary and secondary structures of enzymes can also occur owing to the interaction between the polar groups of enzymes and the oscillating electric field. However, the modification in the tertiary structures of proteins and enzymes will occur only if the electric field applied is higher than 10^8 V/m. Moreover, another study proposed that the frictional energy dissipation and the oscillating electric field-induced molecular motion is the primary reason behind the modification in activity and structure of enzymes. Furthermore, another study reported that the inhibition or enhancement of the enzyme activity mainly depends on the frequency, temperature, and strength of the electric field. However, every stated hypothesis related to the mode of action of microwave and ohmic heating is not well-explored and additional investigation is required.[28,36]

5.5 IMPACT OF OHMIC HEATING ON DIFFERENT ENZYMES

5.5.1 PEROXIDASE

Peroxidase is an oxidoreductase enzyme, mainly present in plants. Peroxidase utilizes electron acceptors (mainly hydrogen peroxide) to catalyze different oxidation reactions of various inorganic and organic substrates present in plants. It is the most thermostable enzyme identified, which plays a large number of roles in plants, including lignification, rigidification, and construction of cell walls. Peroxidase also helps to strengthen the constituents of the cell walls by crosslinking the phenolic residues such as glycoprotein and polysaccharides, available in the cell walls.[7,52] However, this enzyme involves in the enzymatic browning and also generates undesirable alterations in the sensory and nutritional qualities of the foods.[5,6]

Various studies reported on the impact of ohmic heating on peroxidase enzyme are given in Table 5.2. A study reported a peroxidase inactivation of 78% during ohmic heating of sugarcane juice, in its liquid fraction, for 25 min at 75°C and 7.8 V/cm.[5] Gomes et al.[19] also reported a peroxidase inactivation of 90% while studying the effect of ohmic heating on *Tetsukabuto* pumpkin treating at 80°C for 2 min at a constant frequency (60 Hz) and voltage (180 V).

Brochier et al.[6] conducted a study on the ohmic heating of sugarcane juice at various temperatures (60–80°C) and observed that at 60°C the ohmic heating amplified the peroxidase activation, whereas at 80°C it amplified the inactivation of peroxidase. Another study on ohmic heating at various voltage gradients (0–16.7 V/cm), waveform (triangle, square, and sine), frequency (0–10^5 Hz), and temperature (60–80°C) found out that peroxidase activity was affected significantly by ohmic heating. It was also reported that the waveform and voltage gradient did not influence peroxide inactivation

TABLE 5.2 Impact of Ohmic Heating on Different Enzymes.

Enzyme	Sample	Specification	Salient observations	References
Peroxidase	Sugarcane juice	• Material: jacketed glass reactor and titanium electrode • Voltage: 3.9 and 20.5 V/cm • Waveforms: pulsed, triangle, square, and sine waves • Frequencies: 10–10^5 Hz • Temperature: 75°C and Time: 25 min	• Peroxidase activity decreased between 63% and 73%	[7]
	Sugarcane juice	• Material: jacketed glass reactor and titanium electrode • Temperature 60–80°C • The intensity of the electric field 3.57–4.39 V/cm • Voltage 25 V	• Peroxidase activity increased at 60°C • Peroxidase inactivation increased at 80°C	[6]
	Sugarcane juice	• Material: jacketed glass reactor and titanium electrode • Frequencies: 60 Hz and sine wave • Voltage gradient: 7.8 V/cm • Temperature: 75°C and Time: 25 min • Fractions of juice: solid, liquid, and concentrated	• 78% peroxidase inactivation in the liquid fraction • 100% peroxidase inactivation in the solid fraction • Intermediate inactivation rate in a concentrated fraction	[5]

TABLE 5.2 *(Continued)*

Enzyme	Sample	Specification	Salient observations	References
	Pea puree	• Temperature: 100°C • Frequency: 50 Hz • Voltage gradient: 20–50 V/cm	• Peroxidase inactivated above 30 V/cm • Inactivation time decreases with an increase in the voltage gradient • The better color was observed after 20 V/cm	[21]
	Tetsukabuto pumpkin	• Material: jacketed glass reactor and titanium electrode • Voltage: 180 V for the first 40 s and then 163 V • Frequency: 60 Hz • Time: 0–4 min • Waveform: sine, square, and triangle • Voltage gradient: 0–16.7 V/cm	• 90% decrease in peroxide activity • No significant color change	[19]
	Sugarcane juice	• Temperature: 60–80°C • Frequency: 0–105 Hz	• Peroxide inactivation was significantly affected by ohmic heating • At 60°C, the nonthermal effect of ohmic heating enhances peroxide activation • At 80°C, peroxide inactivation increased	[8]
Pectinase	Orange juice	• Material: graphite electrodes	• Pectinase activity was decreased by 9–98%	[31]

TABLE 5.2 *(Continued)*

Enzyme	Sample	Specification	Salient observations	References
		• Frequency: 50 Hz	• No retardation or regeneration of pectinase activity was observed during 100 days of storage	
		• Maximum voltage: 8 kV		
		• Storage temperature: 4°C		
	Orange juice	• Temperature: 90, 120, and 150°C	• Pectinase inactivation was 98%	[32]
		• Time: 0.68, 0.85 and 1.13 s	• Vitamin reduction was only 15%	
			• Maintained all sensory properties	
	Orange juice	• Voltages: 20–60 V/cm	• 69% reduction in pectin methylesterase	[16]
		• Frequency: 50 Hz	• 1.72–2% increase in pectin content	
		• Temperatures: 34–76°C	• 43.08–45.20 mg/100 mL of final ascorbic acid content	
Polyphenol oxidase	Elstar apple juice	• Voltage gradient: 30–40 V/cm	• 97% reduction in PPO was witnessed for ohmic heating at 80°C and 40 V/cm	[3]
		• Titanium electrode	• Better retention of ascorbic acid, polyphenol content, and carotenoid content than conventionally treated juice	
		• Temperature: 60–80°C		
		• Frequency: 60 Hz		

TABLE 5.2 *(Continued)*

Enzyme	Sample	Specification	Salient observations	References
	Sugarcane juice	• Voltage gradient: 24–48 V/cm • Temperature: 60–90 °C • Time: 5–20 min	• *D*-value and the PPO activity reduced with an increase in the voltage gradient • 97.8% PPO inactivation at 90°C, 32 V/cm for 5 min	[45]
	Watermelon juice	• Voltage: 24 V/cm • Frequency: 50 Hz • Temperature: 90°C • Time: 15, 30, 45, 60 s	• In 60 s, polyphenol oxidase activity decreased to 8.87%	[37]
	Sugarcane juice	• Temperature: 80°C • Time: 0.25, 0.50, 0.75, 1.0 and 1.25 min • Voltage gradient: 24, 32 and 48 V/cm	• PPO inactivated at 32 V/cm and 1 min	[46]
	Sugarcane juice	• Material: jacketed glass reactor and titanium electrode • Temperature: 60–80°C • Intensity of the electric field: 3.57–4.39 V/cm	• PPO is more thermo-stable than peroxide enzyme • Complete inactivation of PPO activity is possible by ohmic heating	[6]

TABLE 5.2 *(Continued)*

Enzyme	Sample	Specification	Salient observations	References
		• Voltage: 25 V		
Alkaline phosphatase	Lactose-free milk	• Stainless steel grade: 316; Titanium plate: Gr-2	• Complete inactivation of alkaline phosphatase at 63°C for 30 min	[49]
		• Temperature: 63°C and holding time: 30 min		
	Milk	• Material: glass reactor	• Inactivated alkaline phosphatase	[23]
		• Frequency: 50 Hz		
		• Time: 4–750 min	• Inactivation was majorly due to the Joule heating effect	
Urease enzyme	Soymilk	• Voltage: 160–220 V	• Inactivated urease enzyme	[33]
		• Frequency: 50–10 kHz	• Inactivation was higher than conventional processing	

greatly. However, similar to previous results, higher inactivation of peroxidase was observed at 80°C.[8]

In the same way, one more study at different gradients of voltage (3.9 and 20.5 V/cm), waveforms (pulsed, triangle, square, and sine waves), and frequencies (10–10^5 Hz) reported that during treatments there was a decrease in the peroxidase activity in between 63% and 73%.[7] These results show that during ohmic heating, though temperature influences the most in peroxidase inactivation, various other parameters, such as voltage gradient, treatment time, waveforms, and frequency also play a major role. Additionally, in all the above studies, ohmic heating was found to be superior in inactivating peroxidase than thermal treatments.

TABLE 5.3 Effect of Microwave Processing on Different Enzymes.

Enzyme	Sample	Specification	Salient observations	References
Peroxidase	Green asparagus	• Microwave pretreatment: 900 W at 30 s • Water balancing: 70, 80, and 90°C	• Microwave accelerates peroxidase inactivation	[52]
	Carrot	• Microwave powers: 70, 210, 350, and 700 W • Time: 15 s–12 min	• The microwave was more effective than thermal • Microwave showed both monophasic and biphasic behavior for peroxidase inactivation • Ascorbic acid was better preserved in microwave treatment	[48]
	Green coconut water	• Temperature: 52.5–92.9°C • Frequency: 2450 MHz	• Microwave treatment could inactivate peroxidase enzymes • Natural peroxidase showed better thermal resistance than the commercial peroxidase	[37]
	Horseradish	• Time: 15 and 30 min	• The effect of microwave treatment on the peroxidase activity was significant	[35]

TABLE 5.3 *(Continued)*

Enzyme	Sample	Specification	Salient observations	References
		• Temperature: 40–60°C	• inactivation was directly proportional to the power and temperature	
		• Power: 30 and 60 W	• Higher inactivation was reported at 60 W, 30 min, and 60°C	
	Red beetroot	• Temperature: 90°C	• 90% inactivation of peroxidase at 200 W and 5 min treatment	[30]
		• Power: 100, 150, and 200 W	• Texture and color were preserved after microwave blanching	
		• Time: 5 min		
	Sweet corn	• Frequency: 2450 MHz	• Microwave blanching was found superior for peroxidase inactivation	[25]
		• Power: 900 W	• Higher retention of geometric diameter, kernel mass, moisture content, ascorbic acid, total sugar, and the color was reported in microwave blanching	
		• Time: 40, 50, 60, 70 and 80 s		
Pectinase	Grapefruit juice	• Power: 900 W	• Pectimethylesterase activity was reduced to 10%	[22]
		• Time: 30 s	• Thermo-liable nutrients and organoleptic properties were preserved	
			• Microwave pasteurization was found superior over conventional pasteurization	
	Orange juice	• Treatments: 85°C for 2, 1, 0.5, and 0.25 min; 75°C for 5, 2.5, and 1 min; 70°C for 5, 2.5, and 1 min; 60°C for 5 and 2 min	• Pectin methyl-esterase inactivation was achieved at 70°C for 1 min with a minor reduction in vitamin C	[15]

TABLE 5.3 *(Continued)*

Enzyme	Sample	Specification	Salient observations	References
	Orange juice	• Time: 0–60 s	• 90% degradation in pectin methyl-esterase was achieved after 10 s at 60°C and 99% was achieved after 20 s at 80°C	[9]
		• Temperature: 50–95°C		
	Orange juice	• Microwave power: 540–900 W	• 93–95% reduction in pectin methylesterase was achieved after microwave treatment	[17]
		• Flow rates: 40 and 80 mL/min		
		• Temperature: 60–85°C		
Polyphenol oxidase	Loquat fruit	• Microwave power: 478 W	• Polyphenol oxidase reduction of 98, 96, and 79% was observed at E (energy intensity) = 0.28, 0.22 and 0.17 kJ/g respectively.	[12]
		• Time: 82–300 s	• Complete inactivation was obtained at 478 W (E = 0.44 kJ/g) and 210 s.	
		• Sample weight: 189–260 g		
	Mamey fruit	• The exposure time/power applied (s/W): 30/90, 165/90, 300/90, 30/590, 165/590, 300/590, 30/937, 165/937, and 300/937	• Longer microwave treatment inactivated PPO activity completely	[41]
		• Frequency: 2450 MHz	• Gentle microwave treatment enhanced PPO activity	
			• Minimum destruction to the microstructure of fruit	

TABLE 5.3 *(Continued)*

Enzyme	Sample	Specification	Salient observations	References
	Soursop fruit	• Power: 70, 140, and 210 W	• 57% of PPO activity reduced during microwave processing at 70 W in 30 s	[40]
		• Time: 30–300 s		
		• Frequency: 2.45 GHz		
	Coconut water	• Temperature: 60–100°C	• No PPO activity after 90°C	[38]
		• Frequency: 2450 MHz		
	Defatted avocado puree	• Power density: 4.4, 7.7, and 11 W/g	• 80% PPO activity reduction at 11 W/g in 80 s	[53]
		• Time: 0–90 s	• Better sensory quality and no increase in PPO residual activity during storage	
			• Increase in polyphenol content and only negligible reduction in chlorophyll content	
	Red beet	• Temperature: 90°C	• 90% inactivation of PPO at 200 W and 5 min treatment	[30]
		• Power: 100, 150, and 200 W	• Texture and color were preserved after microwave blanching	
		• Time: 5 min		
Alkaline phosphatase	Bovine milk	• Frequency: 2450 MHz	• 99.9% reduction in alkaline phosphatase activity	[50]
		• Power: 0 and 1800 W	• ALP reduced from 390,000 mU/L to 350 mU/mL after microbial treatment	
	Raw milk	• Frequency: 2450 MHz	• D value was from 128 to 13.5 s at 65°C and 70°C, respectively.	[34]
		• Power: 1000 W	• Inactivation was faster as compared to conventional heating	

TABLE 5.3 *(Continued)*

Enzyme	Sample	Specification	Salient observations	References
Lipase	Wheat germ	• Temperature: 20–60°C • Power: 10–150 W • Time: 20–1800 s	• 60% inactivation at 40°C • 100% inactivation at 60°C • Microwave showed 10% higher inactivation efficiency than conventional heating	[13]

Similar to ohmic heating, microwave processing also could affect the activity of the peroxidase enzyme. Various studies reported on the effect of microwave processing on peroxidase enzyme are summarized in Table 5.3. It is found that microwave pretreatment of asparagus, before water balancing, could successfully accelerate the peroxidase inactivation rate.[52] A similar result was also observed for red beetroot, where, along with peroxidase inactivation, organoleptic parameters such as texture and color of the final product were also preserved after microbial blanching.[30]

Kachhadiya et al.[25] compared microwave blanching with different conventional blanching methods and found that microwave blanching is superior for peroxidase inactivation since it retains more nutritional and physicochemical properties of food, such as geometric diameter, kernel mass, moisture content, ascorbic acid, total sugar, and color.

Soysal and Soylemez[48] evaluated the potential of microwave and thermal treatments on the inactivation of carrot peroxidase and reported that microwave was more efficient in inactivating peroxidase enzyme. They also found that the microwave treatment preserved more ascorbic acids than thermal treatments and it showed both monophasic and biphasic behavior on peroxide inactivation. A similar result was also obtained by Matsui et al.[39] while working on microwave processing of green coconut water. Moreover, they have added commercial peroxidase enzymes and observed that they showed lesser resistance to thermal processing than natural peroxidase present in the coconut water. Furthermore, another research on horseradish observed that microwave treatment power and temperature have a significant influence on peroxidase inactivation, as inactivation rises with an increase in temperature and treatment power.[35]

5.5.2 PECTINASE

Pectinases, generally known as pectic enzymes, are a collection of enzymes naturally present in the citrus fruit tissues. There are numerous categories of pectic enzymes; however, pectin methyl esterases and polygalacturonases are the most widely studied. These enzymes play a vital role in the ripening of fruit and increasing the yield and bioactive compounds of fruit juices. However, pectinases disturb the cloud stability of the juice by hydrolyzing pectin, one of the primary compounds for the cloud stability of juices.[2] Pectin methyl esterases act as a catalyst to de-esterify the methyl group of pectin to produce methanol and pectic acid. Whereas polygalacturonases produce D-galacturonate by hydrolyzing glycosidic linkages of polygalacturonic acid. Hence inactivation of pectinase enzymes is necessary as it enhances the cloud stability and impacts the flavor, mouthfeel, color, and viscosity of the juices.[1,4,24,42]

Several studies have reported ohmic heating as a superior technology for inactivating pectinase enzymes, which is summarized in Table 5.2. The synergistic impact of the nonthermal and thermal effects of ohmic heating is the key reason behind pectinase inactivation. Apart from that, the process intensity (frequency and electric field), final temperature, and treatment time of the ohmic treatment also influencing the inactivation.[4] Demirdöven and Baysal[16] conducted a study on the ohmic heating of orange juice to investigate its efficiency to inactivate pectin methylesterase. Process temperature (34–76°C) and voltage gradient (20–60 V/cm) were chosen as the independent variable, and the optimized condition was obtained using the response surface methodology. A pectin methylesterase inactivation of 96% was achieved using the optimized results in ohmic heated juice, which was far superior to the juice heated with the conventional technique (88%).

Leizerson and Shimoni[31] conducted a similar study on the impact of ohmic heating on orange juice and achieved a pectinase inactivation of 90–98%. This study also reported that with only a 15% reduction in vitamin C content, the ohmic heating maintained all the major sensory properties of the juice. Correspondingly, another study also reported that no retardation or regeneration of inactivated pectinase was observed in ohmic heated orange juice during the 100 days of storage.[32]

Igual et al.[22] compared the efficiency of microwave pasteurization and traditional thermal pasteurization on grape juice. The study reported that a 90% reduction in pectin methylesterase activity with no significant reduction in nutritional properties was observed in the juice after microwave pasteurization. Another experiment on orange juice reported that inactivation of pectin methyl-esterase was achieved at a microwave treatment for 1 min at 70°C.[15]

A similar result was obtained by Brugos et al.[9] while comparing the effect of conventional and microwave processing on orange juice. A pectinase (pectin methylesterase) inactivation of 99% was achieved at 80°C after 20 s and 90% was achieved at 60°C after 10 s of microwave heating. The study concluded that under the same conditions, microwave heating is more superior to traditional heating. Another experiment achieved a pectin methylesterase inactivation of 93–95% in orange juice after microwave treatment with the least quality degradation as compared to conventional treatment.[17] Details of different researches conducted to investigate the efficiency of microwave processing on pectinase inactivation are summarized in Table 5.3. The outcomes of these studies demonstrate the superiority of microwave processing over conventional techniques in the process of pectinase inactivation.

5.5.3 POLYPHENOLOXIDASE

Browning is a major concern in most vegetables and fruits. Browning occurs due to the development of benzoquinone and melanins from phenolic compounds when mechanically damaged tissues of fruits and vegetables interact with oxygen. Polyphenol oxidase (PPO) acts as a substrate for this reaction; hence, PPO is the major reason for browning in the majority of vegetables and fruits. The PPO activity enhances during cutting and peeling, therefore inactivation of PPO is a major objective in the processing of vegetables and fruits.[36] Effect of ohmic heating and microwave treatment on PPO inactivation in various vegetables and fruits are summarized in Tables 5.2 and 5.3, respectively.

Brochier et al.,[6] after an experiment on ohmic heating of sugarcane juice, reported that PPO is more thermo-stable than peroxide enzyme but still can be inactivated using ohmic heating. Abedelmaksoud et al.[3] demonstrated the impact of ohmic heating on PPO inactivation in elstar apple juice and compared it with traditional processing. The study was conducted at different combinations of voltage gradient (30–40 V/cm) and temperature (60–80°C) using an ohmic heating setup having a titanium electrode at 60 Hz. A PPO inactivation of 97% was attained for ohmic heating at 80°C and 40 V/cm, which was found better than conventional heating (91%). Moreover, better retention of ascorbic acid, polyphenol, and carotenoid content was observed for ohmic heated juice than for conventionally treated juice. A similar investigation was conducted by Saxena et al.[45] on sugarcane juice at different voltage gradients (24–48 V/cm), temperature (60–90°C), and time (5–20 min) combinations. The study achieved a PPO inactivation up to around 98%, which was obtained at 90°C, 32 V/cm for 5 min. Besides, the study has also observed that at certain combinations, the *D*-value and the PPO activity reduced with an increase in voltage gradient.

Another experiment comparing the impact of ohmic heating and hot-water treatment on watermelon juice reported that ohmic heating was found better in inactivating PPO activity (92% inactivation) as compared to hot-water treatment (64% inactivation).[37] A related experiment was conducted on sugarcane juice by Saxena et al.[46] and they proved that ohmic heating requires only less processing time for PPO inactivation as compared to conventional thermal heating.

Microwave processing also requires a shorter time for PPO inactivation than conventional processing. Palma-Orozco et al.[40] achieved a 57% reduction in PPO activity in soursop fruit after just 30 s of microwave processing at 70 W which was very shorter as compared to ultrasound (43% reduction in 220 s) and conventional processing.

Chavez-Reye et al.[12] studied the impact of microwave processing on the PPO inactivation in Loquat fruit by choosing treatment time (82–300 s) and sample weight (189–210 g) as the independent variables. Inactivation of 98, 96, and 79% was observed at *E* (energy intensity) value of 0.28, 0.22, and 0.17 kJ/g, respectively, for PPO during the study. Also, complete inactivation was observed, by preserving antioxidant activity and phenolic compounds, at 478 W ($E = 0.44$ kJ/g) and 210 s.

Similarly, another experiment on microwave treatment on mamey fruit also reported a similar result. Here, the PPO activity was completely inactivated during longer microwave treatment, whereas a gentle treatment lead to enhancement in PPO activity. However, compared to conventional thermal processing, PPO inactivity was better in microwave treatment with minimum destruction to the microstructure of the fruit (Palma-Orozco).[41]

Zhou et al.[53] investigated the impact of microwave processing on defatted avocado puree at a different times (0–90 s) and power densities (11, 7.7, and 4.4 W/g). This experiment also assessed the significance of microwave processing on the quality of treated avocado puree and compared it with the untreated puree. The study observed that the PPO activity can be reduced to 20% by the microwave treatment at 11 W/g for 80 s. The treated sample also showed better sensory quality, an increase in phenolic content, only negligible chlorophyll reduction, and no significant increase in PPO residual activity during storage. These results reveal that microwave processing is a promising alternative to conventional technologies for PPO inactivation since it required a shorter time and preserves the sensory and nutritional characteristics of the treated product.

5.5.4 ALKALINEPHOSPHATASE

Alkaline phosphatase is a glycoprotein present in the tissues of animals that necessitates two metals such as magnesium and zinc for its highest enzyme activity. At controlled pH and temperature, alkaline phosphatase reacts with disodium phenyl phosphate (its substrate) to liberate phenol. The thermal resistance of alkaline phosphatase is higher than that of the pathogenic microorganisms. Hence, its inactivation is an accepted method to determine the efficiency of the pasteurization process. For this reason, alkaline phosphatase is a significant enzyme in the milk processing industries.[23,36,50]

Castro et al.[10] studied different food-related enzymes and reported that ohmic heating does not influence the inactivation of ohmic heating. However, Suebsiri et al.[49] achieved complete inactivation of alkaline phosphatase in lactose-free milk by ohmic heating at 63°C for 30 min. Jakob et al.[23] also successfully inactivated alkaline phosphatase from milk using ohmic heating and found that the inactivation was primarily due to the Joule heating effect. All these studies demonstrate that ohmic heating can be used for the inactivation of the alkaline phosphatase enzyme.

Lin and Ramaswamy[34] compared the performance of conventional water bath heating and microwave heating on alkaline phosphatase inactivation in raw milk. Through the study, the *D* values for the alkaline phosphatase inactivation were observed lower for microwave processing, which proved that inactivation by microwave processing is much faster than conventional thermal processing technique.

Similarly, Thum et al.[50] studied the effect of microwave processing in bovine milk under different temperatures, holding time, and ramp time and compared it with conventional pasteurization. The study revealed that 99.9% of alkaline phosphatase was inactivated after microwave processing (reduction of alkaline phosphate from 390,000 to 350 mU/L). However, it was also found that the effect of microbial processing on bovine milk was similar to that of the conventional ultrahigh temperature processing.

Rahman et al.[43] reported that the degree of inactivation of alkaline phosphate using microwave treatment and conventional thermal treatment is the same. However, according to Rahman et al.[43] after considering other essential parameters, such as quality retention and microbial reduction, the microwave was found superior to conventional heating.

5.5.5 OTHER ENZYMES

Urease inactivation is often measured to determine the reduction in the activity of trypsin inhibitor in soybean products. Li et al.[33] demonstrated the impact of ohmic heating on the urease inactivation in soymilk under frequency range of 10–50 Hz and voltage gradient 160–220 V. The study observed that the ohmic heating technology successfully inactivated urease enzyme and the inactivation was much higher than the conventional processing at the same temperature. Similarly, lipoxygenase is another enzyme present in soybean and other fruits and vegetables which is responsible for the creation of off-flavors during their storage.

Kermasha et al.[27] found out that microwave treatment has the potential to activate this enzyme in a much shorter time than traditional heating methods. Additionally, Chen et al.[13] investigated the lipase inactivation potential of microwave radiation in wheat germ and observed that 60% reduction in lipase activity was achieved at 45°C and 100% reduction was achieved at 60°C of microwave treatment. Furthermore, the study reported that while comparing with conventional heating, the enzyme inactivation efficiency of the microwave was 10% higher.

5.6 CHALLENGES AND FUTURE SCOPE

The potential of ohmic heating and microwave processing technologies on the inactivation of various enzymes has been extensively studied in industrial and academic sectors. On the other hand, the successful commercialization of these technologies is so far very limited and required to resolve various challenges. The corrosive nature of the electrodes and their interaction with the food during processing is the major challenge associated with ohmic heating. Hence, electrodes and foods need to be carefully selected so that the electrochemical reactions between them should not develop any undesirable changes. Whereas, in microwave processing, the major challenge is the uneven or unpredictable temperature distribution. That is, the real temperature profile during the microwave processing cannot be measured since the heating pattern might not be constant throughout heating.

The composition of foods selected for processing is another challenge faced by both the technologies, since each component in food may have different electrochemical properties. Microwave processing or ohmic heating of these foods may lead to overheating or underheating of certain components. Another important challenge faced by both technologies is the reactivation of partially inactivated enzymes. That is, during processing, few enzymes may not completely inactivate and these partially inactivated enzymes can reactivate during storage. These problems need to be addressed in the future.

5.7 SUMMARY

Conventional thermal processing technologies have certain disadvantages such as degradation of sensory and nutritional properties of the final product. The introductory part of this chapter dealt with the problems faced by conventional thermal technologies and the need for alternative technologies in the field of food technology. Then, the chapter explored the potential of ohmic heating and microwave processing technologies and briefly discussed its principle, applications, advantages, and disadvantages. They are electro-heating technologies that utilize volumetric heating methods to process foods. Later, the chapter explained the mechanism behind enzyme inactivation during microwave and ohmic heating. The further part of the chapter discussed in detail, using recent research findings, the impact of both the technologies on specific enzymes. Finally, it was concluded that ohmic heating and microwave processing technologies have the potential to inactivate enzymes with many advantages over conventional technologies; and hence both can be a possible alternative to conventional thermal technologies.

KEYWORDS

- **alkaline phosphate**
- **enzyme inactivation**
- **microwave processing**
- **ohmic heating**
- **pectinase**
- **peroxidase**
- **polyphenol oxidase**

REFERENCES

1. Abdullah, S.; Pradhan, R. C.; Aflah, M.; Mishra, S. Efficiency of Tannase Enzyme for Degradation of Tannin from Cashew Apple Juice: Modeling and Optimization of Process Using Artificial Neural Network and Response Surface Methodology. *J. Food Process Eng.* **2020,** *43* (10), E-article 13499.
2. Abdullah, S.; Pradhan, R. C.; Pradhan, D.; Mishra, S. Modeling and Optimization of Pectinase-Assisted Low-Temperature Extraction of Cashew Apple Juice Using Artificial Neural Network Coupled with Genetic Algorithm. *Food Chem.* **2020,** *339*, E-article 127862.
3. Abedelmaksoud, T. G.; Mohsen, S. M.; Duedahl-Olesen, L.; Elnikeety, M. M.; Feyissa, A. H. Optimization of Ohmic Heating Parameters for Polyphenoloxidase Inactivation in Not-From-Concentrate Elstar Apple Juice using RSM. *J. Food Sci. Technol.* **2018,** *55* (7), 2420–2428.
4. Aghajanzadeh, S.; Ziaiifar, A. M. A Review of Pectin Methylesterase Inactivation in Citrus Juice During Pasteurization. *Trends Food Sci. Technol.* **2018,** *71*, 1–12.
5. Brochier, B.; Hertz, P. F.; Marczak, L. D. F.; Mercali, G. D. Influence of Ohmic Heating on Commercial Peroxidase and Sugarcane Juice Peroxidase Inactivation. *J. Food Eng.* **2020,** *284*, E-article 110066.
6. Brochier, B.; Mercali, G. D.; Marczak, L. D. F. Influence of Moderate Electric Field on Inactivation Kinetics of Peroxidase and Polyphenol Oxidase and on Phenolic Compounds of Sugarcane Juice Treated by Ohmic Heating. *LWT—Food Sci. Technol.* **2016,** *74*, 396–403.
7. Brochier, B.; Mercali, G. D.; Marczak, L. D. F. Effect of Ohmic Heating Parameters on Peroxidase Inactivation, Phenolic Compounds Degradation and Color Changes of Sugarcane Juice. *Food Bioprod. Process.* **2018,** *111*, 62–71.
8. Brochier, B.; Mercali, G. D.; Marczak, L. D. F. Effect of Moderate Electric Field on Peroxidase Activity, Phenolic Compounds and Color During Ohmic Heating of Sugarcane Juice. *J. Food Process. Preserv.* **2019,** *43* (12), 1–10.
9. Brugos, A. F.; Gut, J. A.; Tadini, C. C. Inactivation Kinetics of Pectin Methyl Esterase in the Microwave-Assisted Pasteurization of Orange Juice. *LWT* **2018,** *97*, 603–609.
10. Castro, I.; Macedo, B.; Teixeira, J. A.; Vicente, A. A. The Effect of Electric Field on Important Food-Processing Enzymes: Comparison of Inactivation Kinetics Under Conventional and Ohmic Heating. *J. Food Sci.* **2004,** *69* (9), C696–C701.
11. Chandrasekaran, S.; Ramanathan, S.; Basak, T. Microwave Food Processing: A Review. *Food Res. Int.* **2013,** *52* (1), 243–261.
12. Chávez-Reyes, Y.; Dorantes-Alvarez, L.; Arrieta-Baez, D.; Osorio-Esquivel, O.; Ortiz-Moreno, A. Polyphenol Oxidase Inactivation by Microwave Oven and Its Effect on Phenolic Profile of Loquat (*Eriobotrya japonica*) Fruit. *Food Nutr. Sci.* **2013,** *4* (9), 87–94.
13. Chen, Z.; Li, Y.; Wang, L.; Liu, S.; Wang, K.; Sun, J.; Xu, B. Evaluation of the Possible Non-Thermal Effect of Microwave Radiation on the Inactivation of Wheat Germ Lipase. *J. Food Process Eng.* **2017,** *40* (4), 1–11.
14. Chizoba Ekezie, F. G.; Sun, D. W.; Han, Z.; Cheng, J. H. Microwave-Assisted Food Processing Technologies for Enhancing Product Quality and Process Efficiency: A Review of Recent Developments. *Trends Food Sci. Technol.* **2017,** *67*, 58–69.

15. Cinquanta, L.; Albanese, D.; Cuccurullo, G.; Di Matteo, M. Effect on Orange Juice of Batch Pasteurization in an Improved Pilot-Scale Microwave Oven. *J. Food Sci.* **2010,** *75* (1), 46–50.
16. Demirdöven, A.; Baysal, T. Optimization of Ohmic Heating Applications for Pectin Methylesterase Inactivation in Orange Juice. *J. Food Sci. Technol.* **2014,** *51* (9), 1817–1826.
17. Demirdöven, A.; Baysal, T. Inactivation Effect of Microwave Heating on Pectin Methylesterase in Orange Juice. *Ukr. Food J.* **2016,** *5* (2), 248–261.
18. Gavahian, M.; Chu, Y. H.; Farahnaky, A. Effects of Ohmic and Microwave Cooking on Textural Softening and Physical Properties of Rice. *J. Food Eng.* **2019,** *243* (331), 114–124.
19. Gomes, C. F.; Sarkis, J. R.; Marczak, L. D. F. Ohmic Blanching of Tetsukabu to Pumpkin: Effects on Peroxidase Inactivation Kinetics and Color Changes. *J. Food Eng.* **2018,** *233*, 74–80.
20. Guo, Q.; Sun, D. W.; Cheng, J. H.; Han, Z. Microwave Processing Techniques and Their Recent Applications in the Food Industry. *Trends Food Sci. Technol.* **2017,** *67*, 236–247.
21. Icier, F.; Yildiz, H.; Baysal, T. Peroxidase Inactivation and Colour Changes During Ohmic Blanching of Pea Puree. *J. Food Eng.* **2006,** *74* (3), 424–429.
22. Igual, M.; García-Martínez, E.; Camacho, M. M.; Martínez-Navarrete, N. Effect of Thermal Treatment and Storage on the Stability of Organic Acids and the Functional Value of Grapefruit Juice. *Food Chem.* **2010,** *118* (2), 291–299.
23. Jakób, A.; Bryjak, J.; Wójtowicz, H.; Illeová, V.; Annus, J.; Polakovič, M. Inactivation Kinetics of Food Enzymes During Ohmic Heating. *Food Chem.* **2010,** *123* (2), 369–376.
24. Jiménez-Sánchez, C.; Lozano-Sánchez, J.; Segura-Carretero, A.; Fernández-Gutiérrez, A. Alternatives to Conventional Thermal Treatments in Fruit-Juice Processing. Part 1: Techniques and Applications. *Crit. Rev. Food Sci. Nutr.* **2017,** *57* (3), 501–523.
25. Kachhadiya, S.; Kumar, N.; Seth, N. Process Kinetics on Physico-Chemical and Peroxidase Activity for Different Blanching Methods of Sweet Corn. *J. Food Sci. Technol.* **2018,** *55* (12), 4823–4832.
26. Kaur, N.; Singh, A. K. Ohmic Heating: Concept and Applications—A Review. *Crit. Rev. Food Sci. Nutr.* **2016,** *56* (14), 2338–2351.
27. Kermasha, S.; Bisakowski, B.; Ramaswamy, H.; Van de Voort, F. R. Thermal and Microwave Inactivation of Soybean Lipoxygenase. *LWT—Food Sci. Technol.* **1993,** *26* (3), 215–219.
28. Kubo, M. T.; Siguemoto, É. S.; Funcia, E. S.; Augusto, P. E.; Curet, S.; Boillereaux, L.; Gut, J. A. Non-Thermal Effects of Microwave and Ohmic Processing on Microbial and Enzyme Inactivation: A Critical Review. *Curr. Opin. Food Sci.* **2020,** *35*, 36–48.
29. Kumar, J. P.; Ramanathan, M.; Ranganathan, T. V. Ohmic Heating Technology in Food Processing—A Review. *Int. J. Food Eng. Res. Technol.* **2014,** *3*, 1236–1241.
30. Latorre, M. E.; Bonelli, P. R.; Rojas, A. M.; Gerschenson, L. N. Microwave Inactivation of Red Beet (*Beta vulgaris*L. Var. Conditiva) Peroxidase and Polyphenoloxidase and the Effect of Radiation on Vegetable Tissue Quality. *J. Food Eng.* **2012,** *109* (4), 676–684.
31. Leizerson, S.; Shimoni, E. Effect of Ultrahigh-Temperature Continuous Ohmic Heating Treatment on Fresh Orange Juice. *J. Agric. Food Chem.* **2005,** *53* (9), 3519–3524.
32. Leizerson, S.; Shimoni, E. Stability and Sensory Shelf Life of Orange Juice Pasteurized by Continuous Ohmic Heating. *J. Agric. Food Chem.* **2005,** *53* (10), 4012–4018.

33. Li, F. De; Chen, C.; Ren, J.; Wang, R.; Wu, P. Effect of Ohmic Heating of Soymilk on Urease Inactivation and Kinetic Analysis in Holding Time. *J. Food Sci.* **2015,** *80* (2), E307–E315.
34. Lin, M.; Ramaswamy, H. S. Evaluation of Phosphatase Inactivation Kinetics in Milk Under Continuous Flow Microwave and Conventional Heating Conditions. *Int. J. Food Prop.* **2011,** *14* (1), 110–123.
35. Lopes, L. C.; Barreto, M. T. M.; Gonçalves, K. M.; Alvarez, H. M.; Heredia, M. F.; de Souza, R. O. M. A.; Fricks, A. T. Stability and Structural Changes of Horseradish Peroxidase: Microwave Versus Conventional Heating Treatment. *Enzym. Microb. Technol.* **2015,** *69*, 10–18.
36. Makroo, H. A.; Rastogi, N. K.; Srivastava, B. Ohmic Heating Assisted Inactivation of Enzymes and Microorganisms in Foods: A Review. *Trends Food Sci. Technol.* **2020,** *97*, 451–465.
37. Makroo, H. A.; Saxena, J.; Rastogi, N. K.; Srivastava, B. Ohmic Heating Assisted Polyphenol Oxidase Inactivation of Watermelon Juice: Effects of the Treatment on pH, Lycopene, Total Phenolic Content, and Color of the Juice. *J. Food Process. Preserv.* **2017,** *41* (6), 1–9.
38. Matsui, K. N.; Granado, L. M.; de Oliveira, P. V.; Tadini, C. C. Peroxidase and Polyphenol Oxidase Thermal Inactivation by Microwaves in Green Coconut Water Simulated Solutions. *LWT—Food Sci. Technol.* **2007,** *40* (5), 852–859.
39. Matsui; KátiaNicolau; Gut, J. A. W.; de Oliveira, P. V.; Tadini, C. C. Inactivation Kinetics of Polyphenol Oxidase and Peroxidase in Green Coconut Water by Microwave Processing. *J. Food Eng.* **2008,** *88* (2), 169–176.
40. Palma-Orozco, G.; Marrufo-Hernández, N. A.; Tobías, I.; Nájera, H. Purification and Biochemical Characterization of Polyphenol Oxidase from Soursop (*Annona muricata* L.) and its Inactivation by Microwave and Ultrasound Treatments. *J. Food Biochem.* **2019,** *43* (3), 1–11.
41. Palma-Orozco, G.; Sampedro, J. G.; Ortiz-Moreno, A.; Nájera, H. In Situ Inactivation of Polyphenol Oxidase in Mamey (*Pouteria sapota*) Fruit by Microwave Treatment. *J. Food Sci.* **2012,** *77* (4), 359–365.
42. Pradhan, D.; Abdullah, S.; Pradhan, R. C. Optimization of Pectinase Assisted Extraction of Chironji (*Buchana nialanzan*) Fruit Juice Using Response Surface Methodology and Artificial Neural Network. *Int. J. Fruit Sci.* **2020,** 1–19.
43. Rahman, M. S.; Ahmed, J.; Ramaswamy, H. S. Microwave Pasteurization and Sterilization of Foods. *Handb. Food Preserv.* **2020,** *2020*, 713–732.
44. Rifna, E. J.; Singh, S. K.; Chakraborty, S.; Dwivedi, M. Effect of Thermal and Non-Thermal Techniques for Microbial Safety in Food Powder: Recent Advances. *Food Res. Int.* **2019,** *126*, E-article 108654.
45. Saxena, J.; Ahmad Makroo, H.; Srivastava, B. Effect of Ohmic Heating on Polyphenol Oxidase (PPO) Inactivation and Color Change in Sugarcane Juice. *J. Food Process Eng.* **2017,** *40* (3), 1–11.
46. Saxena, J.; Makroo, H. A.; Srivastava, B. Optimization of Time-Electric Field Combination for PPO Inactivation in Sugarcane Juice by Ohmic Heating and its Shelf Life Assessment. *LWT—Food Sci. Technol.* **2016,** *71*, 329–338.
47. Siguemoto, É. S.; Pereira, L. J.; Gut, J. A. W. Inactivation Kinetics of Pectin Methylesterase, Polyphenol Oxidase, and Peroxidase in Cloudy Apple Juice under

Microwave and Conventional Heating to Evaluate Non-Thermal Microwave Effects. *Food Bioprocess Technol.* **2018,** *11* (7), 1359–1369.

48. Soysal, Ç.; Söylemez, Z. Kinetics and Inactivation of Carrot Peroxidase by Heat Treatment. *J. Food Eng.* **2005,** *68* (3), 349–356.
49. Suebsiri, N.; Kokilakanistha, P.; Laojaruwat, T.; Tumpanuvatr, T.; Jittanit, W. The Application of Ohmic Heating in Lactose-Free Milk Pasteurization in Comparison with Conventional Heating: The Metal Contamination and the Ice Cream Products. *J. Food Eng.* **2019,** *262*, 39–48.
50. Thum, C.; Ozturk, G.; McNabb, W. C.; Roy, N. C.; LeiteNobrega de Moura Bell, J. M. Effects of Microwave Processing Conditions on Microbial Safety and Antimicrobial Proteins in Bovine Milk. *J. Food Process. Preserv.* **2020,** *44* (3), 1–14.
51. Varghese, K. S.; Pandey, M. C.; Radhakrishna, K.; Bawa, A. S. Technology, Applications and Modelling of Ohmic Heating: A Review. *J. Food Sci. Technol.* **2014,** *51* (10), 2304–2317.
52. Zheng, H.; Lu, H. Effect of Microwave Pretreatment on the Kinetics of Ascorbic Acid Degradation and Peroxidase Inactivation in Different Parts of Green Asparagus (*Asparagus officinalis* L.) During Water Blanching. *Food Chem.* **2011,** *128* (4), 1087–1093.
53. Zhou, L.; Tey, C. Y.; Bingol, G.; Bi, J. Effect of Microwave Treatment on Enzyme Inactivation and Quality Change of Defatted Avocado Puree During Storage. *Innov. Food Sci. Emerg. Technol.* **2016,** *37*, 61–67.

CHAPTER 6

DENSE PHASE CARBON DIOXIDE AND HIGH-PRESSURE PROCESSING: EFFECTS ON FOOD ENZYME

RAVI PANDISELVAM[1], ANJINEYULU KOTHAKOTA[2], N. U. SRUTHI[3], K. B. SANKALPA[4], G. N. ASHITHA[5], and V. PRITHVIRAJ[6]

ABSTRACT

The shelf-life of fruits and vegetable products has been shortened because of the destructive or quality degrading enzymes present in these products. To alleviate this quality loss, enzyme inactivation is needed. Thermal treatment is the most commonly used conventional effective method in inactivating the enzymes and microbes present. However, thermal processing may trigger deleterious alterations in the product's sensory, quality, and nutritional characteristics. Over the past several decades, there is an increasing scope and interest in finding an alternate technology to use preservatives or the lowest possible temperature by food industries and academicians. In addition to health concerns, alternative technologies pave the way in obtaining safe and fresh-like produce with excellent quality. In this regard, a combination of various novel technologies for instance dense phase carbon dioxide, high-pressure processing, ultrasound processing, ultraviolet irradiation, and other thermal processing techniques like micro-wave and ohmic heating have been more effective in inactivating resistant enzymes than individual treatment.

Enzyme Inactivation in Food Processing: Technologies, Materials, and Applications. Megh R. Goyal, PhD, Junaid Ahmad Malik, PhD & Ravi Pandiselvam, PhD (Eds.)

6.1 INTRODUCTION

Enzymes are the form of protein that acts as biological catalysts or accelerator for promoting most of the biochemical reactions which converts the substrates into products. Enzymes consist sequence of amino acids linked with one or more polypeptides. Quality changes/loss occur in foods during storage due to the deteriorative enzymes. In this regard, the enzyme inactivation is very much needed to prevent quality loss for the ultimate end use of the produce.

In foods produce, enzymes and microorganism are the most common factors responsible for deterioration. Enzymes can breakdown the essential nutrients present in foods. For instance, enzymes such as lipase and protease break the fats and proteins in the foods, respectively. The original form of the enzyme needs to be maintained to prevent it from denaturation. The three-dimensional molecular structure of the globular protein can be stabilized by hydrophobic interactions, hydrogen bonding, intermolecular bonding, ion pairing, Coulomb interaction, and steric constraints.

Ozone technology for the sterilization process does the same by getting the sulfhydryl group to get oxidized forming shorter peptides from enzymes which causes antimicrobial action.[88,92] But care should be taken to decide the correct dosage and conditions to avoid negative effects.[91] The influence of ozone can reduce degradations such as the yellowing of broccoli by inhibiting the chlorophyll degrading enzymes.[89] The ozone treatment can modify the crosslinking of amylose and amylopectin molecules thereby imparting specific modifications to starch molecules.[90] The inactivation of enzymes and their activity should be reduced for stabilizing the food produced. Inactivation mechanisms of different processing techniques are given in Table 6.1.

Even the heat treatment gives rise to enzyme inactivation but higher heat resistance enzyme will result in negative effect. This may alter the food quality attributes and nutritional properties. Whitaker[110] and Lamikanra[69] have reported about the enzymes responsible for that nutritional loss, textural degradation, taste and odor loss, and color and flavor changes which are given in Table 6.2. To prevent such a loss, an alternative combination of novel methods is becoming the current research interest in enzyme inactivation.

This chapter analyzes the effectiveness of DPCD and HPP on enzyme inactivation, their mechanism of action, and the combination of these techniques with other novel technologies for controlling or inactivating the quality degrading enzymes which are naturally present in foods.

6.2 DENSE PHASE CARBON DIOXIDE (DPCD)

DPCD can be termed as cold pasteurization and ages back to the 1950s where the activity of enzymes and microorganisms are influenced through the inhibitory potencies of molecular carbon dioxide (CO_2) at a lower pressure of less than 50 MPa. The foods are not exposed to adversarial heat and thus its freshness is well retained along with other sensory parameters.[6,60] DPCD denotes that phase of matter which remains fluid but is dense with respect to its gaseous state.[114] Also, the subsequent reuses of the CO_2 gaseous after the process classify this technique as green and sustainable technologies. In brief, DPCD utilizes pressure in combination with CO_2 for achieving food preservation by both enzyme and microbial inactivation.

The CO_2 is an innocuous, nonflammable, and low-cost gas in industrial quantities. CO_2 is also risk-free and can easily be removed by mere depressurization or simply out-gassing. The drop in pH caused by the dissolved CO_2 could be useful in inactivating enzymes, decreasing microorganisms, as well as preserving and stabilizing the nutrients. The physiognomies of CO_2 allow an extensive inactivation of microbial enzymes at reasonably moderate treatment conditions, where thermal processing is nominal or ineffective.[54] These characteristics of CO_2 when combined with the properties of supercritical fluid (SCF) prove to be more advantageous in enzyme inactivation. In contrast to the behavior of liquids and gases, the physical characteristics of SCFs could be attuned over a varied span with only a comparatively minor modification in pressure or temperature.[19]

TABLE 6.1 Disintegration Mechanism of Processing Techniques on Enzymes.

Processing technique	Disintegration mechanism on enzymes	References
Cold plasma	Mutation of enzyme secondary structure.	[103]
Dense phase carbon dioxide	Decline in the media pH and variation of ionic equilibrium causes destruction.	[87]
High-pressure processing	Destruction or reduction of enzymes is related to the conformational changes which modify the functions of enzymes either by enhancing or decreasing their biological activity and modifying their substrate specificity.	[52]
Ozone processing	Attacks OH radicals, specifically the double bonds, causing the formation of unstable ozonides resulting in disintegration.	[32]

TABLE 6.1 *(Continued)*

Processing technique	Disintegration mechanism on enzymes	References
Pulsed electric field	Electric field results in association or dissociation of protein functional groups, charged side chains, and alters the arrangement of helices. Also causes conformational changes by disrupting the electrostatic interactions of peptide chains.	[11, 108, 115]
Ultraviolet light processing	Photo oxidation is explained to be the inactivation mechanism for PPO affecting the protein native structure. Very poor penetration on turbid and colored liquid beverages. Still under study.	[36, 99]
Ultrasonication	Water sonolysis, cavitation produced after sonication, denaturation of proteins, and mechanical force exerted by the collapse of bubble formation lead to the formation of free radicals.	[78]

TABLE 6.2 Enzymes Responsible for Degradation in Fruits and Vegetables.

Nutritional loss	Textural degradation	Taste and odor loss	Color and flavor loss
1. Thiaminase 2. Ascorbic acid oxidase 3. Polyphenol oxidase 4. Polyphenol oxidase 5. Lipoxygenase	1. Pectinase 2. Cellulase 3. Hemi cellulose 4. Peroxidase 5. β-glucosidases	1. Lipooxygenase 2. Lipases 3. proteases	1. Polyphenol oxidase 2. Chlorophyllase 3. Peroxidase 4. Lipoxygenase 5. Lipase 6. Anthocyanases

The intrinsic low viscosities coupled with high diffusivities of SCFs as similar to gases considerably intensify the pace of substrate mass flow into the enzyme. Additionally, their liquid-like densities give rise to greater dissolution compared to that of gases. The solubility of substances can be regulated with ease by merely changing the pressure and temperature of SCFs which in turn varies its solvent power.[80] This can be continuously done at the reactor outlet, allowing a combined reaction and separation stage, which simplifies the downstream processing in addition to the recycling of the solvent. Separation of the reaction products and water in an incessantly operated reactor is also possible due to the difference in solubility in the media.

Another advantage of the DPCD technique is that it can be operated at mild temperature ranges from 30 to 50°C. DPCD has the ability to inactivate specific enzymes at very low temperatures, whereas conventional thermal treatment is not effective comparatively.[101] In the case of enzyme inactivation, DPCD treatment is potentially more beneficial than conventional heat pasteurization. The quality traits, nutrients, and other favorable components including anthocyanins and polyphenols are well retained as the exposure to heat are very mild. When compared to the required pressures in high hydrostatic pressure (HHP), DPCD uses only about 1/10th (generally 20 MPa).[5] Consequently, this makes handling, regulation, and management much easier in the DPCD technique. Additionally, the exclusion of oxygen is advantageous in preventing oxidation reactions.[6]

6.2.1 ENZYME INACTIVATION IN FRUITS AND FRUIT JUICES

DPCD is a prospectus in handling fruits and beverages and processing, where enzymes if not inactivated might cause adverse effects on the quality. Niu et al.[85] indicated that the residual activity (RA) of polyphenol oxidase (PPO) present in cloudy apple juice reduced with arise in temperature from 25 to 65°C when the pressure was kept at 20 MPa and was completely inactivated at 65°C. Its minimum RA, when given mild heat (MH) treatment was 38.6% at 65°C. In the same study, the RA of pectin methylesterase (PME) upon DPCD treatment showed a considerably lower value as compared to MH-treated juice, where the least value obtained was 18%.

Xu et al.[111] used a continuous DPCD system and observed that 22 MPa and 60°C, PPO was completely inactivated and PME was reduced by 54.3% when the processing time was raised to 10 min indicating that PME is more resilient to DPCD than PPO. Yu et al.[113] found that RA of PPO in banana juice did not decrease significantly when MH-treated at a temperature lower than 50°C and merely decreased by 20.5% at 60°C showing banana pulp PPO has greater heat endurance and further increasing the temperature may cause adverse detrimental effects. On DPCD treatment, the RA of PPO was 11.6% at 60°C temperature and 20 MPa pressure, and the change in juice color was found to be slower than that from MH treatment. The juice was also found to have increased clarity as DPCD pretreatment stemmed in a decline in viscosity of the juice. On treating Hami melon juice with DPCD (8–35 MPa, 5–60 min, 35–65°C), the minimum RA of PPO, lipoxygenase (LOX), and peroxidase (POD) were 25.26, 0.02, and 38.46%, respectively, at 35 MPa.[26] PPO and POD also showed restoration of activity after storage for 4 weeks showing that DPCD completely inactivated LOX. The aroma and flavor of the treated juice were comparable to those of the fresh juice when stored for 4 weeks at 4°C. PPO was completely inactivated and activity of

POD was decreased by 85% when strawberry juice was processed for 30 min at 45°C temperature and 60 MPa pressure.[76] The treatment also ensured good microbial quality with no visible color changes.

Liu et al.[71] evaluated the outcomes of combined DPCD-MH treatment on watermelon (10–30 MPa, 50°C, 5–60 min) and concluded that the RA of PPO, POD, and PME reduced significantly with pressure and processing time, and PME showed the least activity of 14.5% at 30 MPa and 60 min. Thus, longer processing time, augmented pressure, and raised temperature strengthens the inactivation effect of DPCD treatment.

6.2.2 ENZYME INACTIVATION IN VEGETABLES

Spilimbergo et al.[102] studied the influence of DPCD on the quality aspects of fresh-cut carrot. The authors reported that PME and polygalacturonase (PG) were entirely inactivated and PPO and POD showed 70% and 65% activity, respectively, at treatment conditions of 12 MPa and 40°C with 30 min exposure. The enzyme activities were not expressively dissimilar at the termination of the 4-week storage study. The treatment also displayed no detrimental influences on antioxidant capacity and carotenoids of the carrot samples.

The indigenous enzyme activity of carrot slices was also studied by Bi et al.[17] and found that an initial increase in RA of POD, PPO, and PME which gradually decreased with the minimum values of 75.8, 90.9, and 52.8%, respectively, at a treatment condition of 5 MPa pressure, 20°C temperature, and 15 min exposure. The less significant loss in hardness can be accredited to the reduced temperature used in the treatment.

A systematic inactivation study was performed by Benito-Román et al.[15] using tyrosinase from mushrooms as a prototypical PPO enzyme with DPCD (5–20 MPa, 25–45°C, 2–15 min) and the activity showed a sudden decrease within the initial 2 min which was then trailed by slow decay. The whole experiments performed at 20 MPa virtually revealed complete inactivation irrespective of the temperature but the inactivation rate intensified with an increase in temperature. DPCD processing was used for beetroot juice processing to inactivate the tissue enzymes.[75] The shortest decimal reduction time was noticed for PG (256 min) while the longest for POD (384 min). Overall, it can be observed that enzymes in vegetables showed higher resistance to DPCD treatment than fruits which could be accredited to the naturally higher pH in vegetables than fruit products. This resistance can be explained by the innate cellular self-defense mechanism to different biotic and abiotic stresses.[102]

6.2.3 MECHANISM OF ENZYME INACTIVATION

Each food processing approach, including DPCD, covers dual objectives: to destroy food-borne pathogens or detrimental microbes and immobilize the enzymes present in the foodstuff that can cause objectionable modifications in the flavor, texture, or color. The magnitude of enzyme inactivation by means of DPCD depends on the kind and origin of the same and treatment operating conditions, in addition to the properties of the treatment medium. The mechanism of inactivation of enzyme activity by DPCD can be ascribable to many reasons, for instance, lowering of pH, conformational modifications of the enzyme, repressive effects of molecular CO_2, or due to the depressurization effects.

When CO_2 gets dissolved in the hydration layer, it combines with water and a series of reactions take place resulting in a number of by-products including carbonic acid, bicarbonate, carbonate, and H^+ ions, depicted as below[40]:

$$CO_2 + H_2O \leftrightarrow H_2CO_3 \leftrightarrow HCO_3^- + H^+$$

$$HCO_3^- \leftrightarrow H^+ + CO_3^{2-}$$

This causes a temporary drop in the pH value by surpassing the cell's buffering capacity which in turn causes intensive attraction and repulsion among protein side chains that cause folding as well as the unfolding of peptide chains.[54] Beneath pH 2, the proteins titrated against HCl get converted to an acid-denatured form which is preceded by an acid unfolded state and molten globule state at pH 3–4. Thus, a decrease in pH brought by DPCD is very well accounted for by the enzyme inactivation mechanism. Enzyme inactivation by DPCD is also said to be caused by the molecular CO_2, where surplus CO_2 forms covalent compounds along with free amino groups resulting in the formation of carbamates superficially on the enzyme.[43] These carbamates have the potential to bring about charge elimination at histidine and lysine remnants, and consequently pave the way to enzyme inactivation. CO_2 also undergoes hydrophobic interactions at enzyme active sites which could furthermore influence enzyme inactivation.[112]

The density of CO_2 increases at a higher pressure and temperature and dense CO_2 acts as a hydrophobic organic solvent. This leads to the alteration of poise between the interactions of water and protein eliciting the hydrophobic units to get imperiled to the external hydrophilic media. Enzyme inactivation by DPCD can also be elucidated by the subunit dissociation and modifications in protein backbone structure induced by the pressure. Supercritical CO_2 has the ability to penetrate easily through large molecules and ooze out vital water molecules within the enzyme microenvironment, which results in enzyme inactivation, owing to the water apportioning the enzymatic hold-up and the solvent.[29]

The activity loss of enzyme noticed during depressurization can be clarified incidentally that in the course of the depressurization stage, with the discharge of CO_2 dispersed in the bound water over the enzyme protein, protein hydration by water molecules by means of noncovalent bonds gets diminished.[43]

6.2.4 QUALITY CHANGES IN FOOD PRODUCE TREATED WITH DPCD

DPCD even at low pressures caused severe tissue damage in some fruits.[50] Beer treated with DPCD in a continuous approach was compared with heat-treated beer for its physical and sensory attributes.[34] Heat-treated beer differed significantly from fresh beer in taste and aroma than the DPCD-treated beer. Also, DPCD-treated beer was less cloudy and had less foam opacity compared with heat pasteurized beer.

Ferrentino et al.[42] treated red grapefruit juice in a continuous DPCD system (13.8–34.5 MPa, 5–9 min, 40°C) and observed that the mean cloud values of the treated samples showed 1.9 times greater value than the untreated juice. The better cloud stability of the DPCD-treated samples can be endorsed to the depressurization which might have caused liquid homogenization resulting in smaller composite particles of the juice colloid. The impact of DPCD treatment (0.6 MPa, 35°C, 6.8 min, 8 CO_2) on total antioxidant activity (TAA) and total phenolic content (TPC) of guava pulp were investigated, where TAA showed less significant changes in organic fractions.[95] Comparable results were observed for TPC. In a study performed by Chen et al.[26] on Hami melon juice, better preservation of β-carotene was observed in DPCD (55°C, 35 MPa, 60 min) treated samples as compared to typical high-temperature short-time pasteurization.

Bi et al.[17] reported that hardness was well retained when carrot slices were treated with DPCD with a maximum loss of 7.9% achieved for samples exposed for 15 min at 5 MPa and 20°C. On treating button mushrooms with high CO_2 (95–100%)[70] at the time of sealing the package significantly increased the antioxidant ability during poststorage. The decline in sensory scores of the button mushroom was reduced with CO_2 treatment during storage. The authors noticed that a high concentration of CO_2 could also result in injury to the surface tissue of the mushroom cap, nevertheless it also significantly repressed the browning in the course of cold storage. Moreover, compared with the untreated sample, the 1-octen-3-ol content of button mushroom was maintained in the CO_2-treated sample.

The physicochemical and sensory analyses on DPCD-treated coconut water revealed that DPCD had no significant effect on the dry matter as well as soluble solid content.[22] Moreover, out of three water-soluble vitamins present in coconut water (B_1, B_2, B_5), DPCD slightly reduced only vitamin B_2, which shows its less intense influence on the nutritional qualities of the final product. It was also found that the CO_2-treated samples were liked equally to the controlled samples while heat-treated samples showed a significantly higher off-flavor.

6.3 HIGH-PRESSURE PROCESSING

HPP is recognized as one of the popular techniques for eliminating microbes and inactivation of food spoilage enzymes. To attain the balance between enzymatic and microbial safety, the HPP technique gave the most promising results. One of the crucial principles governing HPP is Le-Chatelier's principle, which depicts that the chemical reactions which cause a decrease in total volume (negative activation volume) are heightened with the application of pressure. The pressure applies during HPP is instantaneous and uniform all through the chamber. Food does not undergo any crushing or compression action due to this isostatic pressure. It is any how advantageous compared to the traditional thermal treatments where the product temperature is gradually increased.[83]

For the high-pressure processing of food, the product needs to be hermetically sealed in a flexible packaging material and processed under high pressure ranging from 100 to 600 MPa. Water, food-grade glycerol, glycerol and water solution, castor oil, and ethanol can be used as a pressure transfer medium to achieve uniform pressure distribution.[7,83] Low molecular weight compounds including minerals, aroma compounds, and vitamins undergo no significant changes upon application of high pressure due to the least compressibility of covalent bonds; however, proteins and starch molecules alter the inherent structure.[10] Due to these reasons, HPP is recognized as a novel and emerging technology that can be useful in controlling food spoilage enzymes and microorganisms in the processing and industrial sector.[9,14] Apart from the inactivation of enzymes, HPP is known for the texturization of food. For instance, improved firmness in fruit by the activity of pectin methylesterase is due to the application of high pressure.[86]

6.3.1 ENZYME INACTIVATION BY HPP

Enzyme inactivation during HPP is primarily due to modification of intermolecular structure and conformational alteration at enzyme active sites.[9] Enzymes found in horticultural produce present a greater variation in enzymatic activity and resistance to high pressure. These characteristics make it impossible to generalize pressure, exposure time, and treatment temperature combination for the inactivation of enzymes.[59]

6.3.1.1 OXIDATIVE ENZYMES

The conversion of phenol to quinones by enzymatic oxidation is the initiation of the browning reaction. These enzymes also cause the degradation of anthocyanins and end up with a loss of antioxidant activity in fruit products.[14,105] The activity of PPO is mainly found in apple juice, pineapple puree, blueberry puree, strawberry puree, and in many fruit juices and smoothies. Complete inactivation (100%) of these enzymes is reported in pear puree at treatment conditions of 600 MPa and 90°C when treated for 5 min.[116] and in avocado puree at 450 MPa, 40°C, and 60 min.[16] In other studies, the maximum inactivation achieved was 92%.

According to Jesus et al.,[59] POD is more resistant than PPO; hence, POD requires higher pressure and temperature for inactivation. The result is in agreement with the report of Kaushiket al.[65] and Marszałeket al.[76] in mango and strawberry puree, respectively. In contrast, a higher reduction of POD than PPO was reported by Terefe et al.[105] in strawberry puree. Also, it is clear from the available literature that pressure alone is not enough for the inactivation of food enzymes and it needs synergy effect of both temperature and high pressure.

6.3.1.2 PECTIC ENZYMES

Pectic enzymes including PG and PME are important enzymes that are to be inactivated so as to maintain the quality of horticultural produces. These PG and PME are responsible for the pectin network breakdown in the cell wall of fruits and fruit products.[65] PME is mainly reported in orange[84,96] and tomato.[31,94] In orange juice, PME was inactivated up to 95% at a pressure of 500–800 MPa along with MH.[84,96] In the case of tomatoes, 95–98% inactivation of PG was attained at a pressure ranging from 700 to 800 MPa and 60 to 90°C temperature. The PG in avocado and tomato puree was inactivated at 450 MPa pressure with 20°C and 700 MPa with 80°C, respectively.[16,68] PG requires relatively less pressure and temperature for inactivation when compared to PME and oxidative enzymes.[68] It has been reported that mild temperature and moderate pressure processes can cause an antagonistic effect in many enzymes including PPO, POD, PME, and PG.[24,49,76,105]

6.3.2 MECHANISM OF INACTIVATION

HPP causes various effects on enzyme activities and microbial contaminants. The major changes that take place during HPP are the distraction of cell membrane and cell wall, chemical reactions, conversion processes catalyzed by enzymes, and alterations in biopolymers such as the formation of a gel and denaturation of protein.[83,86] From the earlier available literature database, we understood that the application of pressure induces the inactivation of enzyme activity by permanently destroying the active site and brings alterations in the structure of protein by conformational changes. Furthermore, enzyme–substrate interactions will boost up due to the liberation of intracellular enzymes as a result of the high-pressure application, thus resulting in discoloration.[66] Therefore, optimum pressure and holding time are the most important factors that needed to be standardized depending on the food composition and enzyme type.

According to Abe,[1] at a reduced pressure of about 50 MPa, synthesis of proteins in the microorganism cause inhibition, ribosomes level gets reduced at a moderate pressure of 100 MPa, and partial protein denaturation occurs at 200 MPa pressure leads to damage of internal cell structure and membrane. The irreversible changes occur in enzymes and proteins at high pressure (>300 MPa).

6.3.3 ACTIVATION OF ENZYMES UNDER HPP

Despite various advantages of high-pressure processing, a major demerit with the system is that it can activate certain enzymes. For instance, while processing fruits and vegetables in HPP mode, enzymatic reaction associated with browning occurs due to the activation of enzyme PPO.[3] Enzymes were activated during high-pressure treatment due to limited conformation change, availability of an active enzyme from a complex group and inhibitor, and proteolysis.[3] The high-pressure application may enhance the reaction rate by increasing the enzyme sensitivity toward the substrate.[39] Application of high pressure causes membrane interruption and conformational modification in enzyme structure, which leads to the development of the enzyme catalytic activity and its interface with the substratum.[104]

In raspberries and strawberries, even at 600 and 400 MPa pressure, the activity of β-glucosidase, peroxidase, and PPO was increased.[44] Moreover, for apple treated at 600 MPa and blueberry treated at 400 MPa, the pressure was not sufficient for enzyme inactivation.[97,106] The contrasting impacts of pressure and temperature can even lead to stabilization and activation of enzymes by the development or disintegration of intramolecular interactions among the protein functional groups and water.[105] This interaction will change the functionality, native structure, and stability of the protein.[105] At the initial stage of thermal inactivation, water molecules are loosened by protein and it results in the structural rearrangement. In contrast, the application of high pressure inhibits the structural rearrangement due to the hydration of nonpolar and polar groups.[82] Hence, certain enzymes are activated in HPP at specific operating conditions.

6.4 HYBRID TECHNOLOGIES RELATED TO HPP

Hybrid technologies (combining two technologies) result in a more concerted action which helps to reduce processing time, lower the levels of application, and can make the technology more energy-efficient.[47,67]

6.4.1 COMBINED HIGH PRESSURE AND THERMAL TECHNOLOGIES

HPP has been producing high-quality products with maximum preservation of low-molecular-weight compounds like vitamins, minerals, and other nutraceuticals. Even after the treatment, some baro-resistant spores and enzymes may get retained in the product which can cause further quality deterioration during storage.[51,86]

Extensive studies have been conducted in regard to horticultural produce for the effective inactivation of enzymes through HHP.[23,46,87,94,105] Previous studies pointed out that the inactivation of enzymes induced by HPP involves a complex mechanism that relays on the source of the enzyme and interaction with the constituents of the products. High pressure actuates structural changes such as the reversible, irreversible, partial, or complete unfolding of enzymes.[65] Same enzymes present in different fruits and vegetables react differently with HPP, some of the enzymes were found to be inactivated, and some were found to get activated or stabilized during application of pressure.[23] Hence, thermal processing combined with HPP will help to produce a better-quality product even at lower ranges of temperature and pressure.[23]

HPP combined with mild to elevated temperature is effectively applied as a method of inactivation of quality degrading enzymes in fruits and vegetables. The commercial application level of elevated pressure (>500 MPa) along with temperature showed a synergistic action in the inactivation of enzymes.[72,104,105] The combined high-pressure thermal (HPT) process could retain flavor, color, and texture of product than the equivalent thermal process because heating during pressurization (adiabatic compression) and cooling during compression help to decrease the overall thermal load of treatment. In addition to that, instantaneous and uniform transmission of pressure results in enhanced uniformity of temperature during pressurization, which assists the reduction of heat-induced damages.[61,79]

High pressure and temperature induce the enzyme inactivation through a complex mechanism and it is hard to describe when enzymes and substrate get pressurized simultaneously.[23,64] High pressure and elevated temperature treatments work synergistically as it causes irreversible unfolding of enzymes in their tertiary and quaternary structures leading to complete enzyme denaturation and inactivation.[8] Hydrogen bonds are found to be the key factor in maintaining the structure of enzymes, especially the helical structure of peptides, and those relations are favored at lower pressures (below 400 MPa). Whereas the interaction of hydrogen bonds is disrupted at higher pressures (500–600 MPa) and finally leads to an unstable enzyme structure with less functionality.[52] Higher temperature of around 50–70°C causes disruption of covalent and noncovalent bonds in the enzyme configuration and also opens the outer hydration layer.[18,23]

Manassero et al.[74] explored the consequence of combined high pressure and thermal treatments (HPT) on lipoxygenase inactivation in calcium-added soy milk. The authors observed a complete inactivation of lipoxygenase activity as a function of the processing parameters. Fang et al.[41] observed that kiwifruit treated with more than 400 MPa pressure and moderate heat (<50°C) could increase the levels of peroxidase inactivation. The authors noticed that an increase in the processing time (>15 min) did not show any significant difference in the inactivation rate.

The RA of the enzymes may be due to the presence of POD isozymes and their resistance to pressure. A complete inactivation of PPO and POD was achieved in coconut water at a pressure of 400–600 MPa and temperature 90°C, treated for 300 s.[28] Both enzymes showed an increase in inactivation rate with respect to temperature when increased from 40 to 90°C.[28] The authors reported that the POD enzymes of coconut water displayed better resistance to HPT than PPO. Zhou et al.[116] also observed similar results in acidified pear puree treated at 600 Mpa /90°C for 5 min.

The inactivation of the PPO enzyme was found to increase with a rise in pressure. The resistance of the endogenous enzymes to HHP recommends combining the treatments with other technologies.[45] PPO and POD enzymes of pineapple puree were treated beyond 400 MPa and 50–70°C, which also resulted in oxidoreductase inactivation.[24] Kaur et al.[62] researched on the influence of HPT treatments at 300–600 MPa /30–60°C for 0–15 min and noticed a synergetic influence of pressure and temperature on the inactivation rate of PPO.

Terefe et al.[107] studied endogenous enzyme inactivation in pear under HPT. The HPT process of 600 MPa pressure and 20°C temperature for 5 min showed an inactivation rate of 32, 74, and 51% for PPO, POD, and PME, respectively. Treatments in the temperature range of 40–60°C resulted in a drop in enzyme inactivation and a rise in residual activities. The HPT processing at 600 MPa over 80–100°C temperature attained the inactivation of POD and PPO as 92 and 90%, respectively. But the treatment showed undesirable changes in the textural attributes of the product.

Kaushik et al.[65] studied the effects of combined HPT treatments on the inactivation of enzymes including PME, POD, and PPO of mango and 600 MPa/70°C for 20 min was found to be the most effective processing condition in the reduction of activities to 52, 15, and 26%, respectively. The authors noticed that PME is the most baro-resistant enzyme followed by POD and PPO. The synergistic action of pressure and temperature was shown above 50°C for the inactivation of enzymes. PG enzymes in avocado observed a complete inactivation even after the mild thermal and pressure treatment (10 min/70°C/0.1 MPa or 15 min/20°C/450 MPa).[16]

Generally, POD is known as a more thermostable enzyme and is normally recognized as a pointer for the effectiveness of blanching treatments. Several findings have, however, proved that POD is not essentially the most heat-stable enzyme.[27,28,105] In contrast, a higher baro-resistance of PPO than POD was reported in pineapple puree,[23] strawberry puree,[105] strawberry pulp,[21] and feijoa puree.[87] These studies point out that the presence of several isoforms in fruits induces a higher baro-resistivity in PPO than POD. During applied stress like pressure and temperature, PPO is more stable than POD.[109] The resistance of enzymes under high pressure and temperature changes according to the inherent characters of each fruit and vegetable, hence it is required to conduct studies specific to the product.

6.4.2 HIGH PRESSURE COMBINED WITH NONTHERMAL TECHNOLOGIES

Although the combination of HPT processing could be used for enzyme inactivation to an acceptable level.[72] But the heat-sensitive product cannot be processed by this method due to heat-induced loss of color, nutrients, and flavor of the food during processing. Inflated demand for high-quality fresh-like foods has piloted the development of enormous novel technologies. A new concept called hurdle technology proposed a combination of nonthermal technologies for food preservation with higher safety and quality. HPP in combination with DPCD and HPP in combination with ultrasound (US) processing are found to be some of the effective technologies for enzyme inactivation.

6.4.2.1 HIGH PRESSURE COMBINED DENSE PHASE CO_2

Dense phase carbon dioxide (DPCD) and HHP systems operate with different processes even though the pressure has a direct effect on DPCD.[53] The operation of combined technologies could be influenced by different constraints like temperature, pressure, initial pH, processing time, CO_2 state and density, and sample media. During the pressurization and depressurization stages of HHP, the occurrence of CO_2 may cause different actions on the enzymes.[53] At the time of pressurization, CO_2 dissolved into the food material interacts with the water molecules resulting in the formation of carbonic acid, and hence the surrounding pH gets reduced. The reduction in pH leads to the denaturation of proteins to some extent. Whereas the CO_2 molecules react with protein chains of enzymes and their active sites which causes a hydrophobic effect.[35,53,57]

A gas–liquid interface was formed due to the rapid depressurization which causes a change in the phase of CO_2 molecules to the gaseous state from its dense phase. During this phase, the orientation of polar and nonpolar remnants to the aqueous phase and gaseous phase would instigate the unfolding of protein chains. A homogenization effect could be created by the abrupt discharge of CO_2.[54]

Limited studies have been done on enzyme inactivation using HHP combined with DPCD and most of them showed a potential reduction in the active enzymes of food. The combination of high pressure and CO_2 causes a minimum of three stresses that can ideally alter the enzymatic activity of the food. In HHP, higher pressure is required for enzyme inactivation, which is noticeably greater than the pressures used in the combined HHP and DPCD process. Hence, the pressure alone would not justify the inactivation of enzymes during combined HHP and DPCD processing.

Similarly, only a few studies have expressed the synergistic action on the inactivation of PPO, LOX, and PME enzymes as in carrot and orange juice.[93,30] Park et al.[93] narrated that DPCD treatment of 4.9 MPa followed by HHP treatment at 300–600 MPa resulted in RA less than 11, 35, and 9% for enzymes PPO, PME, and LOX in carrot juice, respectively. The reduction in residual activities was 2–3 folds less than that of the HHP process or DPCD process alone but it needs more processing time. The addition of 0.2% CO_2 to orange juice before HHP treatment (500 and 800 MPa) has significantly reduced the activities of PME at 25°C and PPO at 50°C. Also, this combined technique at all pressure levels was more effective against microbial activity than the HHP process alone.[30]

Ortuno et al.[87] experimented on enzyme inactivation by filling CO_2 in the headspace before the application of HHP. Results showed a substantial reduction in POD, PPO, and PME enzyme activities by combined treatment of CO_2 with HHP processing than HHP treatment(300–600 MPa) alone. The occurrence of CO_2 saturation in foods during the combined treatment helped PPO and POD inactivation compared to HPP alone. The synergic action of HPP and dissolved CO_2 might have resulted in the changes of α-helix structure and 3D folding of β-sheet arrangements of POD and PPO enzymes.[25,48,55] The studies on the conformational changes in the enzymes showed both HHP and combined treatments resulted in changes of secondary confirmation, but more extensive alterations were observed in the combined treatments.[38]

Duong and Balaban[37] studied the effect of combined HHP and DPCD on three enzymes, namely, peroxide, PPO, and PME in feijoa puree. The study concluded that optimal treatment conditions of 600 MPa pressure, pH of 3, CO_2 saturation, and processing time of 13 min attained the RA of 74.3, 70.9, and 53.9% for the POD, PPO, and PME, respectively. It was observed that an increase in processing time and lowering of pH decreases all enzyme activities. The increase in CO_2 showed major effects on PME and the increase in pressure on PPO.

Duong et al.[38] examined the impact of combined HHP and DPCD on activity and secondary confirmation of pure PPO in mushrooms with the processing conditions being 600 MPa, 20°C, for 1–9 min. The combined treatment significantly reduced the RA to 12% after 9 min treatment. No recovery in the activity of samples was observed after 1-month storage. The significant loss in activity corresponds to the extensive decrease of α helix fractions and enlargement of protein size thus altering the protein secondary conformations. It is suggested that more research needs to be conducted on different food products to evaluate the enzyme inactivation in specified products by the combined HHP and DPCD treatments.

6.4.2.2 HIGH-PRESSURE PROCESSING IN COMBINATION WITH SONICATION

Among general sonication, thermosonication, and manothermosonication treatments, manothermosonication is more efficient in inactivating enzymes.[98]

A frequency range of 20–100 kHz was applied for physical and chemical alterations in US processing. When a US wave passes through a medium, it alternates the compression and rarefaction cycles. Cavitation of bubbles occurs in high-pressure conditions when rarefaction surpasses the adhesive forces between the liquid molecules. During the cavitation process, each bubble experiences a localized field due to neighboring bubbles, which makes them unstable and ultimately leads to collapse. Microstreaming and bubble collapse create high shear stress along with the generation of localized high temperature and pressure. These results in different changes and denaturation of proteins in the material according to the characteristics of material and eventually lead to the inactivation of enzymes.[56] The cavitation process and the collapse during transient and repetitive transient cavitation causes a localized increase in temperature up to 5000 K and pressure, this induces different chemical changes and the formation of free radicals.[4]

The combinations of HHP and US processes were efficient methods of enzyme inactivation in apple juice. Different enzymes (such as polyphenolase, pectin methylesterase, and peroxidase) in apple juice showed discoloration and browning of the product.[2] A complete inactivation of enzymes was observed in combined US and HHP treatment. The individual treatment of US and HHP could not achieve a complete inactivation of enzymes.

6.4.2.3 MULTIPLE PULSE AND HIGH-PRESSURE PROCESSING COMBINATION

The HHP process includes pressurization to the required pressure, holding for a predetermined time, and decompression to atmospheric pressure. Whereas multipulse HHP consists of a repetition of the HHP process, that is, multiple compression, holding, and decompression cycles. In HHP processing, the important process parameters are pressure, temperature, and holding time. But in the case of multipulsed HHP, besides the parameters of HHP, some other factors also affect the treatment which includes pulse duration, number of pulses, the duration between pulses (off pressure time), compression and decompression rates, and phase shape.[20]

Several studies on pulse pressure treatment have been reported that showed significant influence on the inactivation of various enzymes. It was also found that multipulse HHP treatment was more proficient than a single pulse HHP treatment for the same holding time.[33,73] The dynamic changes in pressure during pulsed pressurization could influence the different actions such as hydration of charged groups, intramolecular interactions, hydrogen bond stabilization, and distraction of bound water during high-pressure inactivation of enzymes.[39,63]

During pulse pressure processing, enzyme inactivation due to disruption of the secondary or tertiary structure of proteins occurs by compression produced during pressurization and rapid expansion owing to sudden depressurization.[13] Pulse pressurization in pineapple puree[24] and litchi-based mixed fruit beverage[58] reduced the activity of the PPO enzyme. The enzyme inactivation using pulse processing is due to the structural breakage, or by modifications in the tertiary or quaternary conformation of the enzyme.[8,100] The inactivation of orange juice PME using pulse pressure was influenced by the constraints such as pH, pressure level, holding time, and soluble solids. It was also observed that a drop in enzyme activity was found due to instantaneous pressure and the effects of cyclic pressure were established to be more efficient at reduced pH and higher pressure levels.[12]

Pulse pressurization of orange juice reported around 90% and 37% reduction in PME activity for application of 400 MPa at pH 3.2 and 3.7, respectively. Concentrated orange juice observed a 79% and 21% reduction of PME with three pulse pressure treatments of 400 MPa at pH of 3.2 and 3.7, respectively. The orange juice PME was found to be more tolerant in concentrated juice and the inactivation rate increased with the number of pulses, pressure level, and lowering of pH.[13]

Kaushik et al.[63] reported that cycle pressure application on litchi juice resulted in 39.7% and 29.1% inactivation of PPO and POD using triple pulse at 600 MPa for 15 min holding time per cycle, respectively. Multiple cycles were found to be an added hurdle in addition to the single-cycle HHP for enzyme inactivation. The cycle pressure process resulted in better inactivation rates of enzymes as compared to static pressure. Even though there is no industrial application of multipulse HHP treatment till now due to its monetary value, longer treatment time, the requirement of more energy, and complicated HHP equipment.

6.5 CHALLENGES AND FUTURE SCOPE

The application of nonthermal techniques has a few concerns that limit their usage considerably. There are various researches regarding the inactivation of microbes using the cold-plasma but researches on enzyme inactivation are very limited. The studies so far revealed successful application in different products; however, practical application of these techniques is compulsory to display the enzyme inactivation in real food systems to craft a clearer picture of the protein interactions at a molecular level.

The current knowledge on cold plasma–protein interactions between the model system and enzyme in foods could be beneficial in evaluating the product quality and shelf-life. Anyhow, precautions are needed to depict the results from the model system. Enzymes in the biological system are generally found attached to the subcellular organelles under the surface.[81] Therefore, before drawing conclusions, valid proofs are required for the model system of enzyme inactivation for applying to the real food systems. An idea about enzyme recovery is also necessary to improve the efficacy of the plasma treatment during storage. Though understanding plasma chemistry is complex, it is necessary to transform the existing and forthcoming knowledge for applying cold-plasma protein interactions into real-world applications.

During HPP, the sensitivity of highly resistant enzymes like PPO and PME is dependent on their origin and other environmental factors. At economically feasible conditions, the high pressure could only achieve partial inactivation of these enzymes.[104] It is obvious that enzyme-mediated quality degradations take place during the storage period and often it is controlled using enzyme inhibitors, oxygen scavengers, appropriate packaging, and refrigerated storage. Therefore, looking into the additional capital and operating costs, its future application would be confined to those high-value products where expenditure on capital costs is justified because of assured returns.

The limitation with DPCD is that it is not commercialized yet. Moreover, DPCD is more expensive in terms of parts used and requires specialized high-pressure equipments, which amounts to a high initial cost. Looking into the current issue of global warming, there is a necessity for seizing and recycling CO_2 during the operation.[3,5] The treating of solids is also one of the limitations of solids, as they cannot run on a continuous process and the CO_2 diffusion to the solid is slow, which leads to cellular damage and textural defects which is a great concern. Therefore, a reduced level of superficial free water may restrain CO_2 solubility in the food with its entanglement to enzyme inactivation.

6.6 SUMMARY

Novel nonthermal techniques such as HPP and DPCD described in the chapter potentially inactivates the enzymes and retained the texture, enhanced the shelf-life and quality of the product, thereby retaining the nutritional factors. The chapter might motivate the scientists for additional research to obtain more theoretical and practical insights into the various nonthermal techniques and their interaction with enzymes that are significant in maintaining food quality in the industries. However, future perspectives with regard to cold plasma, HPP, and DPCD are:

- In order to understand the plasma–protein interactions at a molecular level, with better monitoring tools and also to control the plasma chemistry, more adequate research on the plasma is required. More research on the activity of the enzyme during storage of pressured fruits and vegetables is required for the claim on good stability of the products. Detailed studies are to be done in the future for interpreting the mechanism of stimulated inactivation of fruit enzymes. During pressurization, the identification of changes in the native enzyme composition is also to be done in the future. An economical and productive high-pressure process pertaining to enzyme inactivation in fruit purees and juices must be optimized, considering the consumers and their accessibility to the high-pressure processed fruit products.

- To commercialize the DPCD technology, additional information such as mathematical models to predict the extent of microbe reduction, enzyme inactivation, and other conducive conditions, such as pressure, temperature, time, CO_2 ratio, and other synergistic processes are pertinent. Also, the application of this technology would be enhanced with the availability of continuous solids treatment processes.

KEYWORDS

- **enzyme inactivation**
- **food quality**
- **horticultural produce**
- **pectinmethylesterase**
- **peroxidase**
- **polyphenolase**
- **ultrasound**

REFERENCES

1. Abe, F. Exploration of the Effects of High Hydrostatic Pressure on Microbial Growth, Physiology and Survival: Perspectives from Piezo Physiology. *Biosci. Biotechnol. Biochem.* **2007,** *71* (10), 2347–2357.
2. Abid, M.; Jabbar, S.; Hu, B.; Muhammad, M.; Wu, T.; Lei, S.; Zeng, X. Thermo-Sonication as a Potential Quality Enhancement Technique of Apple Juice. *Sonochemistry* **2014,** *21*, 984–990.
3. Asaka, M.; Aoyama, Y.; Nakanishi, R.; Hayashi, R. Purification of a Latent Form of *Polyphenoloxidase* from La France Pear Fruit and Its Pressure-Activation. *Biosci. Biotechnol. Biochem.* **1994,** *58* (8), 1486–1489.
4. Ashokkumar, M.; Sunartio, D.; Kentish, S.; Mawson, R.; Simons, L.; Vilkhu, K.; Versteeg, C. Modification of Food Ingredients by Ultrasound to Improve Functionality:

A Preliminary Study on a Model System. *Innov. Food Sci. Emerg. Technol.* **2008,** *9*, 155–160.

5. Balaban, M. O.; Duong, T. Dense Phase Carbon Dioxide Research: Current Focus and Directions. *Agric. Agric. Sci. Procedia* **2014,** *2*, 2–9.
6. Balaban, M. O.; Ferrentino, G., Eds. *Dense Phase Carbon Dioxide: Food and Pharmaceutical Applications*; John Wiley and Sons: New York, 2012; p 309.
7. Balasubramaniam, V. M.; Martínez-Monteagudo, S. I.; Gupta, R. Principles and Application of High Pressure-Based Technologies in the Food Industry. *Annu. Rev. Food Sci. Technol.* **2015,** *6*, 435–462.
8. Balny, C.; Masson, P. Effects of High Pressure on Proteins. *Food Rev. Int.* **1993,** *9*, 611–628.
9. Bang, W. S.; Chung, H. J. Effect of High Hydrostatic Pressure on the Enzyme Activities in *Saccharomyces cerevisiae* and *Escherichia coli*. *New Biotechnol.* **2010,** *27* (4), 440–444.
10. Barba, F. J.; Terefe, N S.; Buckow, R.; Knorr, D.; Orlien, V. New Opportunities and Perspectives of High-Pressure Treatment to Improve Health and Safety Attributes of Foods. *Food Res. Int.* **2015,** *77*, 725–742.
11. Barsotti. L.; Cheftel, J. Food Processing by Pulsed Electric Fields. II. *Biological Aspects. Food Res. Int.* **1999,** *15* (2), 181–213.
12. Basak, S.; Ramaswamy, H. S. Ultra High Pressure Treatment of Orange Juice: A Kinetic Study on Inactivation of Pectin Methyl Esterase. *Food Res. Int.* **1996,** *29 (7)*, 601–607.
13. Basak, S.; Ramaswamy, H. S. Pulsed High Pressure Inactivation of Pectin Methyl Esterase in Single Strength and Concentrated Orange Juices. *Can. Biosyst. Eng.* **2001,** *43*, 3–25.
14. Bayindirli, A.; Alpas, H.; Bozoglu, F.; Hizal, M. Efficiency of High-Pressure Treatment on Inactivation of Pathogenic Microorganisms and Enzymes in Apple, Orange, Apricot and Sour Cherry Juices. *Food Control* **2006,** *17* (1), 52–58.
15. Benito-Román, Ó.; Sanz, M. T.; Melgosa, R.; de Paz, E.; Escudero, I.; Beltrán, S., Studies of Polyphenol Oxidase Inactivation by Means of High Pressure Carbon Dioxide (HPCD). *J. Supercrit. Fluids* **2018,** *147*, 310–321.
16. Bermejo-Prada, A.; Buggenhout, S. V.; Otero, L.; Houben, K.; Loey, V.; Hendrickx, M. E. Kinetics of Thermal and High-Pressure Inactivation of Avocado Polygalacturonase. *Innov. Food Sci. Emerg. Technol.* **2014,** *26*, 51–58.
17. Bi, X.; Wu, J.; Zhang, Y.; Xu, Z.; Liao, X. High Pressure Carbon Dioxide Treatment for Fresh-cut Carrot Slices. *Innov. Food Sci. Emerg. Technol.* **2011,** *12* (3), 298–304.
18. Boonyaratanakornkit, B. B.; Park, C. B.; Clark, D. S. Pressure Effects on Intra-and Intermolecular Interactions within Proteins. *BBA—Prot. Struct. Mol. Enzymol.* **2002,** *1595* (1), 235–249.
19. Brunner, G. Supercritical Fluids: Technology and Application to Food Processing. *J. Food Eng.* **2005,** *67* (1–2), 21–33.
20. Buzrul, S. Multi-Pulsed High Hydrostatic Pressure Treatment of Foods. *Foods*, **2015,** *2015*, 173–183.
21. Cao, X.; Zhang, Y.; Zhang, F.; Wang, Y.; Yi, J.; Liao, X. Effects of High Hydrostatic Pressure on Enzymes, Phenolic Compounds, Anthocyanins, Polymeric Color and Color of Strawberry Pulps. *J. Sci Food Agric.* **2011,** *91*, 877–885.
22. Cappelletti, M.; Ferrentino, G.; Endrizzi, I.; Aprea, E.; Betta, E.; Corollaro, M. L.; Charles, M.; Gasperi, F.; Spilimbergo, S. High Pressure Carbon Dioxide Pasteurization

of Coconut Water: A Sport Drink with High Nutritional and Sensory Quality. *J. Food Eng.* **2015,** *145*, 73–81.
23. Chakraborty, S.; Kaushik, N.; Rao, P. S.; Mishra, H. N. High Pressure Inactivation of Enzymes: A Review on Its Recent Applications on Fruit Purees and Juices. *Compreh. Rev. Food Sci. Food Saf.* **2014,** *13* (4), 578–596.
24. Chakraborty, S.; Rao, P. S.; Mishra, H. N. Kinetic Modeling of Polyphenol oxidase and Peroxidase Inactivation in Pineapple (*Ananas comosus* L.) Puree during High-pressure and Thermal Treatments. *Innov. Food Sci. Emerg. Technol.* **2015,** *27*, 57–68.
25. Chen, J. S.; Balaban, M. O.; Wei, C. I.; Marshall, M. R.; Hsu, W. Y. Inactivation of *Polyphenol Oxidase* by High-Pressure Carbon Dioxide. *J. Agric. Food Chem.* **1992,** *40* (12), 2345–2349.
26. Chen, J.; Zhang, J.; Song, L.; Jiang, Y.; Wu, J.; Hu, X. Changes in Microorganism, Enzyme, Aroma of Hami Melon (*Cucumis melo* L.) Juice Treated with Dense Phase Carbon Dioxide and Stored at 4°C. *Innov. Food Sci. Emerg. Technol.* **2010,** *11* (4), 623–629.
27. Chisari, M.; Barbagallo, R. N.; Spagna, G. Characterization of Polyphenol Oxidase and Peroxidase and Influence on Browning of Cold Stored Strawberry Fruit. *J. Agric. Food Chem.* **2007,** *55*, 3469–3476.
28. Chourio, A. M.; Salais-fierro, F.; Mehmood, Z.; Martinez-Monteagudo, S. I. Inactivation of *Peroxidase* and *Polyphenoloxidase* in Coconut Water using Pressure-Assisted Thermal Processing. *Innov. Food Sci. Emerg. Technol.* **2018,** *49*, 41–50.
29. Clifford, A. A.; Williams, J. R. Introduction to Supercritical Fluids and Their Applications. In *Supercritical Fluid Methods and Protocols*, **2000,** 1–16.
30. Corwin, H.; Shellhammer, T. H. Combined Carbon Dioxide and High Pressure Inactivation of Pectin Methylesterase, Polyphenol Oxidase, Lactobacillus Plantarum and *Escherichia coli*. *J. Food Sci.* **2002,** *67* (2), 697–503.
31. Crelier, S.; Robert, M. C.; Claude, J.; Juillerat, M. A. Tomato (*Lycopersicon esculentum)* Pectin Methylesterase and Polygalacturonase Behaviors Regarding Heat- and Pressure-Induced Inactivation. *J. Agric. Food Chem.* **2001,** *49* (11), 5566–75.
32. Criegee. Mechanism of Ozonolysis. *Angewan. Chem. Int. Ed.* **1975,** *14* (11), 745–752.
33. Curl, A. L.; Jansen. E. F. Effect of High Pressure on Pepsin and Chymotrypsinogen. *J. Biol. Chem.* **1950,** *185*, 713–723.
34. Dagan, G. F.; Balaban, M. O. Pasteurization of Beer by a Continuous Dense Phase CO_2 System. *J. Food Sci.* **2006,** *71* (3), 164–169.
35. Damar, L.; Murat, O. B. Review of Dense Phase CO_2 Technology: Microbial and Enzyme Inactivation. *Rev. Food Sci.* **2006,** *71* (1), 1–11.
36. Davies. M. J.; R. J. Truscott. Photo-oxidation of Proteins and its Role in Cataractogenesis. *J. Photochem. Photobiol. B Biol.* **2001,** *63* (1), 114–125.
37. Duong, T.; Balaban, M. Optimization of the Process Parameters of Combined High Hydrostatic Pressure and Dense Phase Carbon Dioxide on Enzyme Inactivation in Feijoa (*Acca sellowiana*) puree using Response Surface Methodology. *Innov. Food Sci. Emerg. Technol.* **2014,** *26*, 93–101.
38. Duong, T.; Balaban, M.; Perera, C. Effects of Combined High Hydrostatic Pressure and Dense Phase Carbon Dioxide on the Activity, Structure and Size of Polyphenoloxidase. *J. Food Sci.* **2015,** *80* (11), 2486–2494.

39. Eisenmenger, M. J.; Reyes-De-Corcuera, J. I. High Hydrostatic Pressure Increased Stability and Activity of Immobilized Lipase in Hexane. *Enzyme Microbiol. Technol.* **2009,** *45,* 118–125.
40. England, A. H.; Duffin, A. M.; Schwartz, C. P.; Uejio, J. S.; Prendergast, D.; Saykally, R. J. On the Hydration and Hydrolysis of Carbon Dioxide. *Chem. Phys. Lett.* **2011,** *514* (4–6), 187–195.
41. Fang, L.; Jiang, B.; Zhang, T. Effect of Combined High Pressure and Thermal Treatment on Kiwifruit Peroxidase. *Food Chem.* **2008,** *109* (4), 802–807.
42. Ferrentino, G.; Plaza, M. L.; Ramirez-Rodrigues, M.; Ferrari, G.; Balaban, M. O. Effects of Dense Phase Carbon Dioxide Pasteurization on the Physical and Quality Attributes of a Red Grapefruit Juice. *J. Food Sci.* **2009,** *74* (6), 333–341.
43. Fricks, A. T.; Souza, D. P. B.; Oestreicher, E. G.; Antunes, O. A. C.; Girardi, J. S.; Oliveira, D.; Dariva, C. Evaluation of Radish (*Raphanus sativus* L.) Peroxidase Activity after High-Pressure Treatment with Carbon Dioxide. *J. Supercrit. Fluids* **2006,** *38* (3), 347–353.
44. Garcia-Palazon, A.; Suthanthangjai, W.; Kajda, P.; Zabetakis, I. The Effects of High Hydrostatic Pressure on β-glucosidase, Peroxidase and Polyphenoloxidase in Red Raspberry (*Rubus idaeus*) and Strawberry (*Fragaria ananassa*). *Food Chem.* **2004,** *88,* 7–10.
45. Garcia-Parra, J.; Gonzalez-Cebrino, F.; Delgado, J.; Ramirez, C. High Pressure Assisted Thermal Processing of Pumpkin Puree: Effect on Microbial Counts, Color, Bioactive Compounds and Polyphenoloxidase Enzyme. *Food Bioprod. Process.* **2016,** *98,* 124–132.
46. Gonzalez-Cebrino, F.; García-Parra, J.; Contador, R.; Tabla, R.; Ramírez, R. Effect of High-Pressure Processing and Thermal Treatment on Quality Attributes and Nutritional Compounds of Songold Plum Purée. *J. Food Sci.* **2012,** *77 (8)*, C866–C873.
47. Gould, G. W.; Sale, A. J. H. Initiation of Germination of Bacterial Spores by Hydrostatic Pressure. *Microbiology* **1970,** *60* (3), 335–346.
48. Gui, F.; Chen, F.; Wu, J.; Wang, Z.; Liao, X.; X. Inactivation and Structural Change of Horseradish Peroxidase Treated with Supercritical Carbon Dioxide. *Food Chem.* **2006,** *97* (3), 480–489.
49. Guiavarch, Y.; Segovia, O.; Hendrickx, M.; Van Loey, A. Purification, Char- acterization, Thermal and High-Pressure Inactivation of a Pectin Methylesterase from White Grapefruit (*Citrus paradisi*). *Innov. Food Sci. Emerg. Technol.* **2005,** *6* (4), 363–371.
50. Haas, G. J.; Prescott Jr, H. E.; Dudley, E.; Dik, R.; Hintlian, C.; Keane, L. Inactivation of Microorganisms by Carbon Dioxide under Pressure. *J. Food Saf.* **1989,** *9* (4), 253–265.
51. Heinz, V.; Buckow, R. Food Preservation by High Pressure. *J. Consump. Protect. Food Saf.* **2010,** *5* (1), 73–81.
52. Hendrickx, M.; Ludikhuyze, L.; Broeck, I. Van Den; Weemaes, C. Effects of High Pressure on Enzymes Related to Food Quality. *Trends Food Sci. Technol.* **1998,** *9* (5), 197–203.
53. Hong, S. I.; Pyunb, Y. R. Membrane Damage and Enzyme Inactivation of *Lactobacillus Plantarum* by High-Pressure CO_2 Treatment. *Int. J. Food Microbiol.* **2001,** *63,* 19–28.
54. Hu, W.; Zhou, L.; Xu, Z.; Zhang, Y.; Liao, X. Enzyme Inactivation in Food Processing using High Pressure Carbon Dioxide Technology. *Crit. Rev. Food Sci. Nutr.* **2011,** *53,* 145–161.

55. Ishikawa, H.; Shimoda, M.; Yonekura, A.; Osajima, Y. Inactivation of Enzymes and Decomposition of α-helix Structure by Supercritical Carbon Dioxide Microbubble Method. *J. Agric. Food Chem.* **1996,** *44* (9), 2646–2649.
56. Islam, N.; Zhang, M.; Adhikari, B. The inactivation of Enzymes by Ultrasound—A Review of Potential Mechanisms. *Food Rev. Int.* **2013,** *30* (1), 37–41.
57. Ismaya, W. T.; Rozeboom, H. J.; Weijn, J. J.; Fusetti, F.; Wichers, H. J.; Dijkstra, B. W. Crystal Structure of *Agaricus Bisporus* Mushroom Tyrosinase: Identity of the Tetramer Subunits and Interaction with Tropolone. *Biochemistry* **2011,** *50* (24), 5477–5486.
58. Jayachandran, L. E.; Chakraborty, S.; Rao, P. S. Inactivation Kinetics of the Most Baro-Resistant Enzyme in High Pressure Processed Litchi-Based Mixed Fruit Beverage. *Food Bioprocess Technol.* **2016,** *9* (7), 1135–1147.
59. Jesus, A. L. T.; Leite, T. S.; Cristianini, M. High Isostatic Pressure and Thermal Processing of Açaí Fruit (*Euterpe oleracea Martius*): Effect on Pulp Color and Inactivation of Peroxidase and Polyphenol Oxidase. *Food Res. Int.* **2018,** *105*, 853–862.
60. Ji, H.; Zhang, L.; Liu, S.; Qu, X.; Zhang, C.; Gao, J. Optimization of Microbial Inactivation of Shrimp by Dense Phase Carbon Dioxide. *Int. J. Food Microbiol.* **2012,** *156* (1), 44–49.
61. Juliano, P.; Knoerzer, K.; Barbosa-Cánovas, G. V. High Pressure Thermal Processes: Thermal and Fluid Dynamic Modeling Principles. In *Engineering Aspects of Thermal Food Processing*; Simpson, R., Ed.; CRC Press/Taylor & Francis: Boca Raton, FL, 2009; pp 93–161.
62. Kaur, B.; Rao, P. S.; Nema, P. K. Effect of Hydrostatic Pressure and Holding Time on Physicochemical Quality and Microbial Inactivation Kinetics of Black Tiger Shrimp (*Penaeus monodon*). *Innov. Food Sci. Emerg. Technol.* **2016,** *33*, 47–55.
63. Kaushik, N.; Kaur, B. P.; Rao, P. S. Inactivation of Polyphenol Oxidase and Peroxidase Enzymes during Pulsed, Static and Cyclic Pressurization of Litchi (*Litchi chinensis*) Juice. *Food Bioprod. Process.* **2016,** *100*, 412–423.
64. Kaushik, N.; Kaur, B. P.; Rao, P. S.; Mishra, H. N. Effect of High Pressure Processing on Color, Biochemical and Microbiological Characteristics of Mango Pulp (*Mangifera indica*). *Innov. Food Sci. Emerg. Technol.* **2014,** *22*, 40–50.
65. Kaushik, N.; Rao, P. S.; Mishra, H. N. Effect of High Pressure and Thermal Processing on Spoilage-Causing Enzymes in Mango (*Mangifera indica*). *Food Res. Int.* **2017,** *100*, 885–893.
66. Keenan, D. F.; Roble, C.; Gormley, R.; Butler, F.; Brunton, N. P. Effect of High Hydrostatic Pressure and Thermal Processing on The Nutritional Quality and Enzyme Activity of Fruit Smoothies. *Food Sci. Technol.* **2012,** *45*, 50–57.
67. Knorr, D. Effects of High-Hydrostatic-Pressure Processes on Food Safety and Quality. *Food Technol.* **1993,** *47*, 156–161.
68. Krebbers, B.; Matser, A. M.; Hoogerwerf, S. W.; Moezelaar, R.; Tomassen, M. M.; van den Berg, R. W. Combined High-pressure and Thermal Treatments for Processing of Tomato Puree: Evaluation of Microbial Inactivation and Quality Parameters. *Innov. Food Sci. Emerg. Technol.* **2003,** *4* (4), 377–385.
69. Lamikanra, O. Enzymatic Effects on Flavor and Texture of Fresh-Cut Fruits and Vegetables. In *Fresh Cut Fruits and Vegetables: Science, Technology and Market*; Lamikanra, O., Ed.; Ed. CRC Press: Washington, DC, 2002.
70. Lin, Q.; Lu, Y.; Zhang, J.; Liu, W.; Guan, W.; Wang, Z. Effects of High CO_2 in-Package Treatment on Flavor, Quality and Antioxidant Activity of Button Mushroom (*Agaricus bisporus*) during Postharvest Storage. *Postharv. Biol. Technol.* **2017,** *123,* 112–118.

71. Liu, Y.; Hu, X.; Zhao, X.; Song, H. Combined Effect of High Pressure Carbon Dioxide and Mild Heat Treatment on Overall Quality Parameters of Watermelon Juice. *Innov. Food Sci. Emerg. Technol.* **2012,** *13*, 112–119.
72. Ludikhuyze, L.; Van, A.; Denys, I. S.; Hendrickx, M. E. Effects of High-Pressure on Enzymes Related to Food Quality. In *Ultra high Pressure Treatments of Foods*; Springer: New York, 2003; pp 115–166.
73. Ludikhuyze, L. R.; Van den Broeck, I.; Weemaes, C. A.; Hendrickx, M. E. Kinetic Parameters for Pressure- Temperature Inactivation of *Bacillus subtilis* α-amylase Under Dynamic Conditions. *Biotechnol. Progress* **1997,** *13*, 617–623.
74. Manassero, C. A.; Vaudagna, S. R.; Sancho, A. M.; Añón, M. C.; Speroni, F. Combined High Hydrostatic Pressure and Thermal Treatments Fully Inactivate Trypsin Inhibitors and Lipoxygenase and Improve Protein Solubility and Physical Stability of Calcium-Added Soymilk. *Innov. Food Sci. Emerg. Technol.* **2016,** *35*, 86–95.
75. Marszałek, K.; Krzyżanowska, J.; Woźniak, Ł.; Skąpska, S. Kinetic Modelling of Polyphenol Oxidase, Peroxidase, Pectin Esterase, Polygalacturonase, Degradation of the Main Pigments and Polyphenols in Beetroot Juice during High Pressure Carbon Dioxide Treatment. *LWT-Food Sci. Technol.* **2017,** *85*, 412–417.
76. Marszałek, K.; Mitek, M.; Skąpska, S. The Effect of Thermal Pasteurization and High Pressure Processing at Cold and Mild Temperatures on The Chemical Composition, Microbial and Enzyme Activity in Strawberry Purée. *Innov. Food Sci. Emerg. Technol.* **2015,** *27*, 48–56.
77. Marszałek, K.; Skąpska, S.; Woźniak, Ł.; Sokołowska, B. Application of Supercritical Carbon Dioxide for the Preservation of Strawberry Juice: Microbial and Physicochemical Quality, Enzymatic Activity and the Degradation Kinetics of Anthocyanins During Storage. *Innov. Food Sci. Emerg. Technol.* **2015,** *32*, 101–109.
78. Mason, T. J.; Lorimer, J. P.; Bates, D. M.; Zhao, Y. Dosimetry in Sonochemistry: The Use of Aqueous Terephthalate Ion as a Fluorescence Monitor. *Ultrasonication Sonochem.* **1994,** *1*, 91–95.
79. Matser, A. A.; Krebbers, B., Van den Berg, R. W.; Bartels, P. V. Advantages of High Pressure Sterilization on Quality of Food Products. *Trends Food Sci. Technol.* **2004,** *15* (2), 79–85.
80. McHugh, M.; Krukonis, V., Eds. *Supercritical Fluid Extraction: Principles and Practice*; Elsevier: New York, 2013; pp 135–143.
81. Misra, N. N.; Pankaj, S. K.; Segat, A.; Ishikawa, K. Cold Plasma Interactions with Enzymes in Foods and Model Systems. *Trends Food Sci. Technol.* **2016,** *55*, 39–47.
82. Mozhaev, V. V.; Heremans, K.; Frank, J.; Masson, P.; Balny, C. High-Pressure Effects on Protein Structure and Function. *Protein Struct. Funct. Bioinform.* **1996,** *24* (1), 81–91.
83. Muntean, M. V.; Marian, O.; Barbieru, V.; Catunescu, G.; Ranta, O.; Drocas, I.; Terhes, S. High Pressure Processing in Food Industry: Characteristics and Application. *Agric. Sci. Procedia* **2016,** *10*, 377–383.
84. Nienaber, U.; Shellhammer, T. High-pressure Processing of Orange Juice: Kinetics of *Pectinmethylesterase* Inactivation. *J. Food Sci.* **2001,** *66* (2), 328–331.
85. Niu, S.; Xu, Z.; Fang, Y.; Zhang, L.; Yang, Y.; Liao, X.; Hu, X. Comparative Study on Cloudy Apple Juice Qualities from Apple Slices Treated by High Pressure Carbon Dioxide and Mild Heat. *Innov. Food Sci. Emerg. Technol.* **2010,** *11* (1), 91–97.
86. Oey, I.; Van der Plancken, I.; Van Loey, A.; Hendrickx, M. Does High Pressure Processing Influence Nutritional Aspects of Plant Based Food Systems. *Trends Food Sci. Technol.* **2008,** *19* (6), 300–308.

87. Ortuno, C.; Duong, T.; Balaban, M.; Benedito, J. Combined High Hydrostatic Pressure and Carbon Dioxide Inactivation of Pectin Methylesterase, Polyphenol Oxidase and Peroxidase in Feijoa Puree. *J. Supercrit. Fluids* **2013,** *82*, 56–62.
88. Pandiselvam, R.; Mayookha, V. P.; Kothakota, A.; Sharmila, L.; Ramesh, S. V.; Bharathi, C. P.; Gomathy, K.; Srikanth, V. Impact of Ozone Treatment on Seed Germination-A Systematic Review. *Ozone Sci. Eng.* **2019,** 1–16.
89. Pandiselvam, R.; Kaavya, R.; Jayanath, Y.; Veenuttranon, K.; Lueprasitsakul, P.; Divya, V.; Kothakota, A.; Ramesh, S. V. Ozone as a Novel Emerging Technology for the Dissipation of Pesticide Residues in Foods–a Review. *Trends Food Sci. Technol.* **2020,** *97*, 38–54.
90. Pandiselvam, R.; Manikantan, M. R.; Divya, V.; Ashokkumar, C.; Kaavya, R.; Kothakota, A.; Ramesh, S. V. Ozone: An Advanced Oxidation Technology for Starch Modification. *Ozone Sci. Eng.* **2019,** *41*, 491–507.
91. Pandiselvam, R.; Subhashini, S.; Banuu Priya, E. P.; Kothakota, A.; Ramesh, S. V.; Shahir, S. Ozone Based Food Preservation: a Promising Green Technology for Enhanced Food Safety. *Ozone Sci. Eng.* **2018,** *41* (1), 17–34.
92. Pandiselvam, R.; Sunoj, S.; Manikantan, M. R.; Kothakota, A.; Hebbar, K. B. Application and Kinetics of Ozone in Food Preservation. *Ozone Sci. Eng.* **2017,** *39* (2), 115–126.
93. Park, S. J.; Lee, J. I.; Park, J. Effects of a Combined Process of High Pressure Carbon-Dioxide and High Hydrostatic Pressure on the Quality of Carrot Juice. *J. Food Sci.* **2002,** *65* (5), 1827–1834.
94. Plaza, L.; Muñoz, M.; Ancos, B.; Cano, M. P. Effect of Combined Treatments of High-Pressure, Citric acid and Sodium Chloride on Quality Parameters of Tomato Puree. *Eur. Food Res. Technol.* **2003,** *216* (6), 514–519.
95. Plaza, M. L.; Ramirez-Rodrigues, M. M.; Balaban, M. O.; Balaban, M. O. Quality Improvement of Guava Puree by Dense Phase Carbon Dioxide (DP-CO_2) Pasteurization (Paper #236–09). In *IFT Annual Meeting of Food Expo*, Chicago, IL, July 20, 2010; p 16.
96. Polydera, A.; Galanou, E.; Stoforos, N.; Taoukis, P. Inactivation Kinetics of Pectin Methylesterase of Greek Navel Orange Juice as a Function of High Hydrostatic Pressure and Temperature Process Conditions. *J. Food Eng.* **2004,** *62* (3), 291–298.
97. Prestamo, G.; Arabas, J.; Fonberg-Broczek, M.; Arroyo, G. Reaction of B- cereus Bacteria and Peroxidase Enzymes under Pressures >400 MPa. *J. Agric. Food Chem.* **2001,** *49* (6), 2830–2834.
98. Raviyan, P.; Zhang, Z.; Feng, H. Ultrasonication for Tomato Pectin methylesterase Inactivation: Effect of Cavitation Intensity and Temperature on Inactivation. *J. Food Eng.* **2005,** *70*, 189–196.
99. Rhim. J. W.; A. Gennadios, D. Fu, C. L. Weller; M. A. Hanna. Properties of Ultraviolet Irradiated Protein Films. *LWT— Food Sci. Technol.* **1999,** *32* (3), 129–133.
100. Riahi, E.; Ramaswamy, H. S. High-pressure Processing of Apple Juice: Kinetics of Pectin Methyl Esterase Inactivation. *Biotechnol. Progress* **2003,** *19* (3), 908–914.
101. Shimoda, M.; Yamamoto, Y.; Cocunubo-Castellanos, J.; Tonoike, H.; Kawano, T.; Ishikawa, H.; Osajima, Y. Antimicrobial Effects of Pressured Carbon Dioxide in a Continuous Flow System. *J. Food Sci.* **1998,** *63* (4), 709–712.
102. Spilimbergo, S.; Komes, D.; Vojvodic, A.; Levaj, B.; Ferrentino, G. High Pressure Carbon Dioxide Pasteurization of Fresh-cut Carrot. *J. Supercrit. Fluids* **2013,** *79*, 92–100.

103. Surowsky, B.; Fischer, A.; Schlueter, O.; Knorr, D. Cold Plasma Effects on Enzyme Activity in a Model Food System. *Innov. Food Sci. Emerg. Technol.* **2013,** *19*, 146–152.
104. Terefe, N. S.; Buckow, R.; Versteeg, C. Quality-Related Enzymes in Fruit and Vegetable Products: Effects of Novel Food Processing Technologies, Part 1: High- Pressure Processing. *Crit. Rev. Food Sci. Nutr.* **2014,** *54* (1), 24–63.
105. Terefe, N. S.; Yang, Y. H.; Knoerzer, K.; Buckow, R.; Versteeg, C. High Pressure and Thermal Inactivation Kinetics of Polyphenol Oxidase and Peroxidase in Strawberry Puree. *Innov. Food Sci. Emerg.ng Technol.* **2010,** *11* (1), 52–60.
106. Terefe, N. S.; Delon, A.; Versteeg, C. Thermal and High Pressure Inactivation Kinetics of Blueberry Peroxidase. *Food Chem.* **2017,** *232*, 820–826.
107. Terefe, N. S.; Tepper, P.; Ullman, A.; Knoerzer, K.; Juliano, P. High Pressure Thermal Processing of Pears: Effect on Endogenous Enzyme Activity and Related Quality Attributes. *Innov. Food Sci. Emerg. Technol.* **2016,** *33*, 56–66.
108. Tsong. T. Y.; R. D. Astumian. Electro Conformational Coupling and Membrane Protein Function. *Progr. Biophys. Mol. Biol.* **1987,** *50* (1), 1–45.
109. Vamos-Vigyazo, L.; Haard Polyphenol Oxidase and Peroxidase in Fruits and Vegetables. *Crit. Rev. Food Sci. Nutr.* **1981,** *15*, 49–127.
110. Whitaker. J. R. Enzymes: Monitors of Food Stability and Quality. *Trends Food Sci. Technol.* **1991,** *2*, 94–97.
111. Xu, Z.; Zhang, L.; Wang, Y.; Bi, X.; Buckow, R.; Liao, X. Effects of High Pressure CO_2 Treatments on Microflora, Enzymes and Some Quality Attributes of Apple Juice. *J. Food Eng.* **2011,** *104* (4), 577–584.
112. Yoshimura, T.; Furutera, M.; Shimoda, M.; Ishikawa, H.; Miyake, M.; Matsumoto, K.; Osajima, Y.; Hayakawa, I. Inactivation Efficiency of Enzymes in Buffered System by Continuous Method with Microbubbles of Supercritical Carbon Dioxide. *J. Food Sci.* **2002,** *67 (9)*, 3227–3231.
113. Yu, Y.; Xiao, G.; Wu, J.; Xu, Y.; Tang, D.; Chen, Y.; Wen, J.; Fu, M. Comparing Characteristic of Banana Juices from Banana Pulp Treated by High Pressure Carbon Dioxide and Mild Heat. *Innov. Food Sci. Emerg. Technol.* **2013,** *18*, 95–100.
114. Zhang, J.; Davis, T. A.; Matthews, M. A.; Drews, M. J.; LaBerge, M.; An, Y. H. Sterilization Using High-pressure Carbon Dioxide. *J. Supercrit. Fluids* **2006,** *38* (3), 354–372.
115. Zhao. W.; R. Yang. Effect of High-Intensity Pulsed Electric Fields on the Activity, Conformation and Self-Aggregation of Pepsin. *Food Chem.* **2009,** *114* (3), 777–781.
116. Zhou, L.; Liua, W.; Stockmann, R.; Terefe, N. S. Effect of Citric Acid and High-Pressure Thermal Processing on Enzyme Activity and Related Quality Attributes of Pear Puree. *Innov. Food Sci. Emerg. Technol.* **2018,** *45*, 196–207.

CHAPTER 7

RADIOFREQUENCY TECHNOLOGY FOR INACTIVATION OF FOOD ENZYMES

N. U. SRUTHI and YASHASWINI PREMJIT

ABSTRACT

Food enzymes are primarily responsible for the development of off-flavor and color changes in foods due to food contact with substrates, such as lipids, proteins, polyphenols, etc., during processing or in the course of storage. The use of thermal treatments to inactivate enzymes and improve the quality of different products is a common practice in food processing. Enzyme inactivation is conventionally achieved through blanching, which has several disadvantages, such as slow conduction of heat, significant loss of soluble nutrients, and huge water usage. Radiofrequency heating is an alternative to conventional hot-water blanching techniques for improving safety and maintaining food quality. Radiofrequency energy is a nonionizing form of electromagnetic energy, but with a longer wavelength (1–100 MHz). Radiofrequency heating has proved to be an efficient and promising mild treatment for blanching different foods and can transform the structure of proteins and can inactivate enzymes, such as lipase, lipoxygenase, polyphenol oxidases, peroxidase, pectin esterase, catalase, etc. The chapter provides a comprehensive review of the principle of radiofrequency heating, its application in inactivating food enzymes, associated advantages and limitations.

Enzyme Inactivation in Food Processing: Technologies, Materials, and Applications. Megh R. Goyal, PhD, Junaid Ahmad Malik, PhD & Ravi Pandiselvam, PhD (Eds.)

7.1 INTRODUCTION

Food quality can be defined as the aggregate features significant to the characteristics of a product that fulfills consumer requirements. These factors can be both external (size, shape, color, morphology, etc.) and internal that include other complex aspects.[49] In a thermodynamic consensus, foods are unstable systems exhibiting the propensity to shift from a less disordered, high energy state toward a high entropy, low enthalpy state; and such a thermodynamic instability causes quality deterioration in foods due to different external factors, such as chemical constituents, microbial features, biochemical, and physical reactions occurring within food matrix during processing or storage.[7]

Enzymatic spoilage of food is one such instability, which occurs as a result of the biochemical changes, where enzymes catalyze certain reactions leading to browning, proteolysis, lipolysis, etc.[3] The use of enzymes to modify, transform, or produce a food product is widespread and is often desired; however, endogenous enzymes inside the food are of paramount interest as far as spoilage is concerned. The activity of such deteriorative enzymes together with microbial growth and other oxidative reactions substantially shorten the shelf-life and reduce the acceptability of horticultural produce. Enzymes work best at a specific temperature and pH ranges, and its activity is greatly dependent on the concentration of both enzyme and substrate and the presence of any inhibitors or activators.

The major enzyme accountable for a specific quality loss may vary from one food to another. However, it can be generalized that color and flavor changes are associated with peroxidase, polyphenol oxidase, chlorophyllase, lipase, and lipoxygenase; textural degradation is caused by pectinases, cellulase, and hemicellulose; and enzymes, such as thiaminase, ascorbic acid oxidase, etc., bring about nutritive loss.[46] Hence, arresting the activity of such enzymes is of utmost importance in food processing steps and enzyme inactivation is one of the prime goals of food processing (thermal processing).

Enzymes are globular proteins and their inactivation is a chemical process involving rearrangement, dissociation, or aggregation of noncovalent bonds, for instance, hydrogen and ionic bonds and associated conformational changes.[33] Processing by thermal techniques is prevalent physical approach to inactivate enzymes in food products owing to their easiness in usage and nonchemical characteristics. Conventional heating process, under both steady and unsteady conditions, causes a heterogeneous distribution of heat as a result of internal resistance developed within different sections of the food material causing thermal deterioration to the product quality.[54] Adverse effects of traditional thermal processing techniques in line with consumer demands for high-quality products have encouraged researchers to explore innovative processing technologies to replace traditional heating methods.

This chapter explores the principle, mechanism, and equipment of radio-frequency (RF) heating as an alternative to traditional heat processing for enzyme inactivation and subsequent preservation of different foods. It also summarizes recent applications and research activities employing RF heating for food enzyme inactivation and the associated limitations and challenges.

7.2 OVERVIEW OF RADIO-FREQUENCY HEATING

RF heating was initially employed as a medical treatment method and later for textile drying, curing glue in plywood manufacturing, and welding plastics.[14] It was first explored in the food industry for blanching and subsequently for cooking, dehydration, thawing frozen foods, postdrying of bakery products, postharvest disinfestation of agricultural produce, and pasteurization and sterilization of foods.[23] More recent studies show the potential of RF heating as a promising replacement for traditional heating in countless applications, such as roasting, enzyme inactivation, microbial decontamination, as well as to minimize sensory and nutritional losses in the processed foods.

RF heating is a volumetric heating technique involving nonionizing electromagnetic waves, similar to microwave processing, covering a broad span of high frequencies stretching from 1 to 100 MHz.[27] However, only band designations of 13.56, 27.12, and 40.68 MHz remain in use today for industrial, research, and therapeutic treatments.[39] When the dielectric food matrices are subjected to an electric field produced by alternating current at frequencies corresponding to radio waves, the electrical energy gets wholly converted into heat. RF heating technique comprises a set of parallel electrodes to which a high voltage alternating current signal is applied to generate longer and more penetrating waves interacting with ions, atoms, and molecules producing extensive internal heat within the food sandwiched between the electrodes.[23] This capacitive character together with wave characteristics makes RF treatments much more uniform than microwave processing, preventing surface overheating, or hot or colds spots. Therefore, RF heating can rapidly raise food temperatures by using less processing times which can guarantee an improved final product quality with a reduced treatment cost as compared to conventional thermal processing.

Many factors contribute to the RF heating proficiency in agricultural commodities, such as the food dielectric properties, the energy density in the material, and the design of RF heating equipment.[11] All fresh foods and their products possess dielectric property, which enables them to accumulate or dissipate energy resembling a capacitor or else a resistance when placed in an electromagnetic field. The properties of treated material in RF processing are permeability, permittivity, and the material electrical conductivity.[40] Since permeability of food is often pondered to be near to that of vacuum, it shows no or very little influence on the dielectric heating, whereas the material permittivity renders a significant contribution in the dielectric heating of foods.[27] The ability to store or dissipate electrical energy is determined by defining the relative dielectric constant, as follows[44]:

$$\varepsilon_r = \varepsilon' - j\varepsilon'' \tag{7.1}$$

The dielectric constant is related to the food's capacitance characteristics and symbolizes the capacity of a substance to accumulate electromagnetic energy. While the dielectric loss factor is linked to the resistance characteristics and indicates the extent of heat generation in the course of dielectric heating.[43] Dielectric properties are dependent on many factors, such as temperature, moisture, density, and food composition. Additionally, different classes of fresh food materials have distinct and varied dielectric properties. The properties also differ in different parts of the same food material.

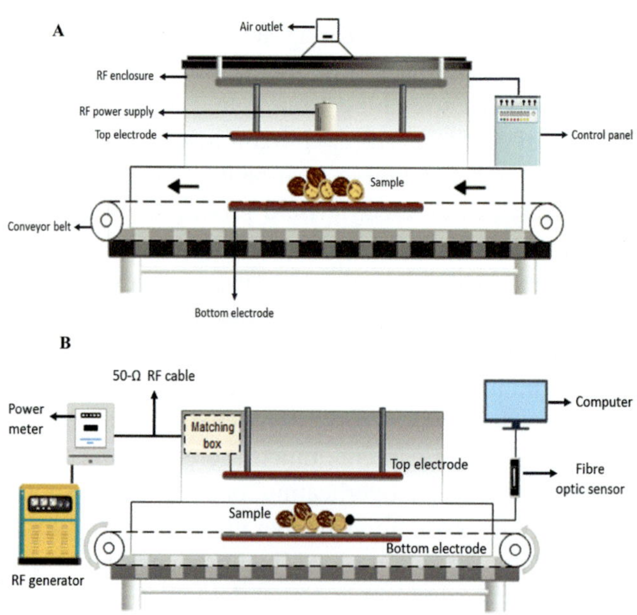

FIGURE 7.1 Major RF heating apparatus: (A) free running oscillator type; and (B) 50-Ω RF heating system.

However, a part of the electromagnetic energy gets obstructed and reflected from the food surface, and the residual energy which gets absorbed by the sample is responsible for generating heat (Fig. 7.1). The energy intensity generally decreases gradually with increasing penetration depth, which is calculated using the following equation[40]:

$$d_p = \frac{c}{2\pi f \sqrt{2\varepsilon' \left[\sqrt{1 + \frac{\varepsilon''^2}{\varepsilon'}} - 1 \right]}} \tag{7.2}$$

It can be inferred from the equation that the penetration depth will get reduced at higher frequencies. To be exact, a shorter frequency is sufficient to attain larger penetration than microwave heating. Also, RF energy is a potential heating method for low moisture foods in bulk.

The major RF heating systems widely used for research and industrial application include free running oscillator type and the 50 Ω technology. Of these, the parallel plate configuration of the free-running oscillator system is the most commonly used design for food processing (Fig. 7.1). In an open-circuit RF heating system, RF energy is produced by a typical oscillator circuit constituted of triode tubes. A tuning circuit is formed between the loaded sample and the electrodes which is inductively joined with the output circuit generating heat.[38] A continuous heating is achieved by a movable top electrode which ensures proper adjustment of the energy coupled within the product and a conveyor belt through which products can be moved. This type of system offers various advantages such as high efficiency, simple structure, and economic feasibility.[56] In 50-Ω RF heating system, there exist an impulsive and automatic impedance matching set-up, which spontaneously maintains the total impedance of the operating circuit at 50 Ω to be on a par with that of the generator.[57] This provides a steady connection of RF energy during heating, and a preset frequency specifically controls the power and the feedback. However, these systems have a high production cost, which limits their use to an experimental level.

RF heating is presently employed in different fresh food processing operations, such as cooking, microbial decontamination, insect disinfestation, thawing and tempering, enzyme inactivation, and more recently on structural modification of functional food components. The mechanism of action for each application is different and can depend on different operating parameters.

7.3 MECHANISM OF ENZYME INACTIVATION BY RF HEATING

The basic mechanism of enzyme inactivation through RF is associated with the thermal denaturation of enzyme proteins. Thermal inactivation of enzymes accountable for quality deterioration in agricultural produce mandates accurate ascertainment of temperature resistances of indicator enzymes prior to performing a proper blanching process. Processing involving heat is accompanied by a number of reactions in series, parallel, and competing. Such interacting and complex reactions result in quality loss and an appropriate first-step kinetic modeling can be applied to analyze such simple food matrices. The kinetics of chemical reactions is used to compute the enzyme activity of any standard food system similar to a conventional rate law[42]:

$$\frac{dP}{dt} = \pm kP^n \tag{7.3}$$

The basics of studies on the kinetics of quality alterations in food owing to thermal treatments are to enumerate the change as a function of time of heating by applying reaction rate constants that depend on temperature. The *D*-value defines the heating time required in minutes to attain a one-log change in the specific food quality at a constant temperature and is given by the following equation.[5]

$$\mathrm{D} = \frac{2.303}{\mathrm{k}} \tag{7.4}$$

The Arrhenius expression gives the accurate relationship to explain the effect of absolute temperature (*T*) on the chemical reaction rate constant (*k*):

$$k = k_0 e^{-\frac{E_a}{RT}} \tag{7.5}$$

The activation energy describes the least energy required to commence a chemical reaction. For a reaction to continue at a moderate rate, the system temperature should be sufficient such that there exists a significant sum of molecules with translational energy less than or equivalent to the activation energy. The reaction rate improves with the increase in temperature as a result of the increase in the sum total of molecules with energy exceeding its activation energy.

The rate of change of a biological or chemical system is further quantified by the use of Q_{10} value, which is the ratio of the rate of change in particular quality at temperature $(T + 10)$°C to that at T, and is expressed as follows:

$$Q_{10} = \frac{k(T + 10°C)}{k(T)} \tag{7.6}$$

The *Z*-values can be correlated to the activation energy, when the Arrhenius relationship persists, and the following relationship at a given temperature range can be employed for $T_1 < T_2$:

$$Z = \frac{2.303\,R T_1 T_2}{E_a} \tag{7.7}$$

Therefore, the theoretical basis of enzyme inactivation by thermal processing can be established by *D*-value, *Z*-value, E_a, *k*, Q_{10}, and *T*, and individual factors can be very well converted by any two of the other known model parameters.

In RF heating, when dielectric foods are positioned in the alternating electromagnetic field, two important phenomena take place: ionic depolarization and dipole rotation (Fig. 7.2). The former explains the forward and backward oscillation of positive and negative ions between the negative and positive regions of the electric field under the continued reversal of polarity, which contributes to the resistance heating. The latter describes the alignment of dipole molecules present in foods, for instance, water molecules, to become steady with the path of the alternating fields, which causes friction between the molecules leading to heat generation.[38] This internal generation of heat, as a result of continuous and rapid motion of ions and dipoles, avoids any temperature lag between the surface and the interior of the product.[35] Although both the phenomenon exists concurrently in the course of dielectric heating, and ionic depolarization is normally the main cause for generating heat at the RF frequencies.[23]

The food moisture plays a vital function in dielectric heating techniques as the conformation flexibility of the enzyme is largely dependent on moisture content.[1] When the moisture of food increases with a simultaneous increment in temperature, the content and kinesis of free water increase within the food matrix which causes a subsequent upsurge in the dielectric loss factor. This facilitates the food to absorb more electromagnetic energy attaining a higher heating rate suggesting that preconditioning the food before the treatment can be an effective practice to convalesce the efficiency of enzyme inactivation by RF heating, especially in food grains. In fresh foods with high water activity, enzyme inactivation is generally attributed to the protein denaturation, as the process require enough water for initially hydrating the protein.[27]

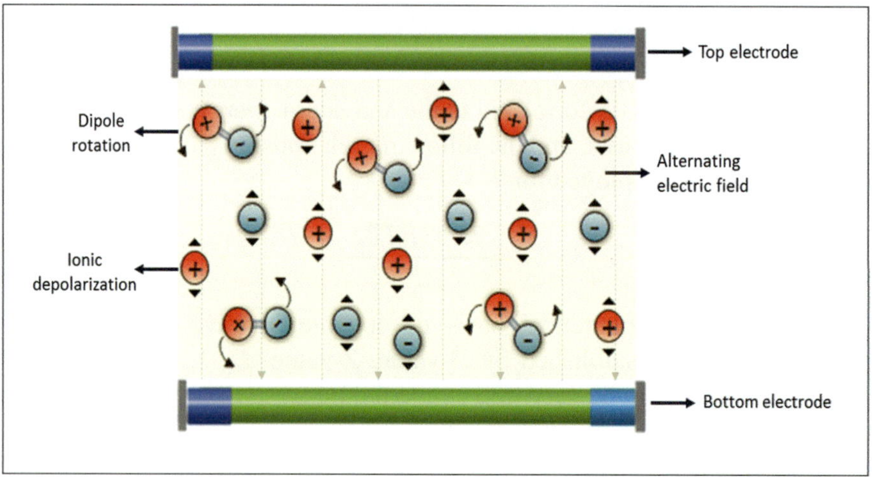

FIGURE 7.2 Dipole rotation and ionic depolarization in an alternating electric field.

In cereals and their products with comparatively low water activity, the mechanism of enzyme inactivation is mostly correlated with the damaged membranes of cells via lipid fluidization, since protein denaturation cannot take place due to inadequate water. As soon as the factors such as temperature, nutrient content, and water content become appropriate, the sublethally damaged cell membrane of the enzymes gets recovered, and turn out to be metabolically active. Partly inhibition of inactive endogenous enzymes in food can only be achieved if the moisture content of the food is augmented to be in equilibrium with the atmosphere.[29]

Figure 7.3 shows important changes in the enzyme protein confirmation when a food is subjected to RF processing. Different studies on RF heating of food products have shown the ability of RF energy to modify the protein confirmation of target enzymes.[12,32] Protein folding into its right shape is a key to whether a globular protein or a membrane protein performs its function perfectly. However, the hydrogen bonds, which play a major role in folding, are considerably weak, and it does not take much heat, acidity, or other stress, causing protein denaturing. Circular dichroism spectroscopy studies conducted on RF-treated samples for analyzing the secondary structure contents of enzyme showed a decrease in α-helix and increase in parallel, antiparallel, β-turn, and random coil contents, confirming the change in protein confirmation.[55]

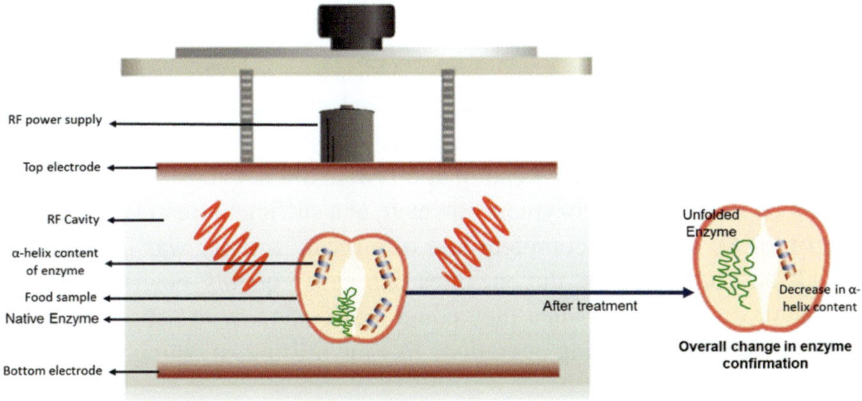

FIGURE 7.3 Mechanism of enzyme inactivation by RF heating.

Thermal inactivation of enzymes can be well thought-out as a first-order monophasic model, or a second-order biphasic model.[24] The former is authenticated based on the hypothesis that to inactivate a particular enzyme, a single bond cleavage is adequate. The latter is centered on the assessment of thermal inactivation kinetics of enzyme mode made by two factions that vary in their thermal stability. POD is known to show a higher resistance during thermal processing. Two isozymes having dissimilar thermo-stabilities have been described in POD, which results in biphasic inactivation. The temperature required for deactivation is correlated to the donor matrix provided of each isoenzyme.[53]

The matrix includes both heat-labile fractions and heat resistant portions. The interaction between these fractions and the time–temperature combinations of RF heating plays a crucial role in POD inactivation. Also, the thermal inactivation of POD is able to be demonstrated with two distinct kinetic models attributable to the existence of certain isoenzymes. The heat-labile fraction can be commendably inactivated at reduced temperatures, while the heat-resistant portion requires a sufficient amount of heat energy. The following section gives a deeper insight to the various mechanisms of interaction of RF energy with different food components that result in enzyme inactivation and quality preservation. Table 7.1 gives an overview of different applications of RF heating technology inactivation of enzymes in foods.

7.4 APPLICATIONS OF RF PROCESSING FOR FOOD ENZYME INACTIVATION

Lopez et al.[32] stated that the influence of RF electric fields on POD, PPO, PE, catalase, and AA activity was analyzed in apple juice, milk, and sweet potato extract. They found that RF heating at 60 MHz did not show any significant impact on the analyzed enzymes; however, at a sufficient treatment temperature of 70°C, partial or complete loss of enzyme activity was observed. In cereals and other grains, the enzymes can cause quality deterioration and off-flavor development when they come into contact with active substrates during different processing operations like the milling, crushing, or grinding process.

The LA, LOX, POD, urase, and myrosinase are key enzymes accountable for off-flavor development in agricultural produce. Rice bran, an important by-product during rice milling is extremely liable to rancidity brought by the endogenous LA and LOX. During storage, a rapid increase in activities of these enzymes is observed mainly because of the increase in microflora that is responsible for the production of these endogenous enzymes. In this context, hot air-assisted RF treatment was studied on LA and LOX inactivation and subsequent stability of rice bran against lipolysis.[29]

It was found that the average residual activities of LA and LOX significantly reduced with an increase in treatment temperature and holding time, with LA displaying a superior thermal resistance than LOX. It was also noted that rice bran with low moisture content possesses difficulty in irreversibly inactivating LA and LOX, and the activity can be partially inhibited on the condition that the moisture content of the bran is rose to the atmospheric equilibrium in the course of storage. Similarly, a commercial by-product of wheat flour, the germ, is also prone to lipase-induced rancidity and thus has limited incorporation into food products. However, wheat germ has a low moisture content and a flake-like structure allowing it to trap air, which together deliver a low thermal conductivity.[10]

Ling et al.[29] evaluated the LA activities in wheat germ employing a series of RF heating settings and found that heating to a temperature of 90°C exclusives of holding could not entirely inactivate LA in the sample. It was also documented that LA activity at the center of the sample bulk decreased significantly when the container's top and bottom surfaces were covered with polyetherimide plates, owing to the increased temperature caused at the cold spot and thus lowering the variation of LA activities throughout the hot and cold spots. The uniformity of RF heating was also enhanced with an increase in bulk density of the sample, but decreased with an increase in sample moisture content.

The widespread application of soy bean in food industry is limited by the presence of antinutrients such as LOX, urease, and trypsin inhibitor. Jiang et al.[19] researched the effectiveness of RF treatment to inactivate the antinutritional factors present in soybean, and a rapid heating of the sample was achieved within 180 s. Thus, the rising rate of temperature in RF heating was found higher to that of traditional heating as a consequence of the displacement of space-charges and the mutual friction that exists between the polar molecules. However, a nonuniform distribution of the electric field was observed caused by the difference in dielectric properties of the treated sample and the container. A higher rate of inactivation LOX was accomplished with RF treatment at lower temperatures and shorter treatment and holding times (113°C, 150, and 120 s) as compared to conventional thermal treatment. The degree of inactivation was influenced by the heating rate and the LOX distribution within the bean. A similar trend was observed for urease where 93.4% reduction in the activity was observed within 300 s and the residual activity became 14.1% at the end of 600 s.

Aviles-Gaxiola et al.[4] reported that the during dry heat treatment, the enzymes located on the seed coat gets quickly inactivated as the surface temperature rises sharply, and a longer treatment duration is required for inactivating the enzymes present in the internal tissues. On the contrary, during RF treatment, lower temperatures are attained by the seed coats and internal tissues during the initial stages, and with increase in the heating time, the internal temperature rapidly reach the target temperature which explains the sharp decline of enzyme activity in RF-treated foods.[29]

Enzymes cause color degradation in foods, specifically fruits and vegetables. The major enzymes which are of concern in degradation of color in plants and other foods are PPO, POD, LOX, chlorophylase, and β-glucosidase. The occurrence of enzymatic browning is the major reason for the observed color changes in plant-derived foods, apart from other process parameters.[45] PPOs, which are commonly distributed in animals, plants, fungi, and bacteria, catalyze the phenolic compounds into extremely reactive o-quinones that results in color changes. An RF processing technique was proposed for heating apple tissues for inactivating PPO in a distinct state from phenolic substrates.[46] A subsequent PEF processing was also accompanied on the juice squeezed from RF preprocessed apple to improve the juice quality. It was observed that as a result of the differences in dielectric properties of apple juice and apple slice, the rate of heating with RF treatment was

TABLE 7.1 Summary of Recent Findings on Enzyme Inactivation in Foods by RF Heating.

Food matrix	Process mode	Power/frequency	Target enzymes	Residual activity	Quality attributes	Reference
Apple juice	Continuous system	3.5 kW/27.12 MHz	PPO	13.57%/10 min	↑ L^*	[46]
					↔Total phenols	
Apple tissue	Batch mode (200 g)	3.5 kW/27.12 MHz	PPO	Membrane-bound PPO: 13%/10 min	↔ TSS, pH	[47]
					↑ L^*,ΔE	
					↓ BI	
				Soluble PPO: 45%/10 min	↓ Total phenols	
Carrot cube	Batch mode	12 kW/27.12 MHz	POD	2.99%/3 min	↓ Hardness	
					↓ L^*, a^*, b^*	
Coconut water	Batch mode (100 mL)	10 kW/40.68 MHz	POD	10.08%/8.2 min	↑ TSS	[8]
					↔ Total phenols	
			PPO	17.69%/8.2 min	↔ Total sugar	
					↓ L^*	
Mustard seed	Batch mode	13.5 MHz	Myrosinase	10–15%/112°C	↔ Total phenols	[17]
					↔ Scavenging Potential	
					↑ OAC, WAC	
Perilla seed	Batch mode	–	LA	5.7% decrease/90°C/10 min	↓ PV	[50]
			LOX	5.94% decrease/90°C/10 min	↓ TBA	
			APX	7.62% decrease/90°C/10 min		
			SOD	12.089% decrease/90°C/10 min		

TABLE 7.1 *(Continued)*

Food matrix	Process mode	Power/frequency	Target enzymes	Residual activity	Quality attributes	Reference
Potato	Batch mode (20 g)	6 kW/27.12 MHz	PPO	Enzyme extract: 0.19%/85°C	↑ Weight loss	[55]
				Potato cuboids: 3.24%/85°C	↑ Δ E	
				Mashed potato: 3.54%/85°C	↓ Hardness	
Rice bran	Batch mode (200 g)	6 kW/27.12 MHz	LA	19%/100°C/15 min	↔ FFA, PV	[29]
			LOX	5.5%/100°C/15 min		
Soybean	Batch mode (37 g)	2.1 kW/27.12 MHz	LOX	4.8%/113°C/270 s	↑ FC, FS	[19]
			Urease	14.1%/113°C/300 s	↑ Flavor	
Soybean	Continuous system	1.8 kW/27.12 MHz	LOX	5.7%/210 s	↑ FC, FS	[20]
					↓ L^*, a^*	
					↑ b^*	
Stem lettuce	Batch mode	6 kW/27.12 MHz	POD	66.03%/65°C	↔ Weight loss	[53]
				6.46%/85°C	↓ L^*, a^*, b^*	
					↑ Vitamin C	
Sweet potato	Batch mode	6 kW/27.12 MHz	POD	23.79%/90°C/4 min	↑ Weight loss	[18]
				5.45%/90°C/6 min	↓ L^*	
				0.00%/90°C/8 min	↑ a^*, b^*	
					↓ Hardness	

TABLE 7.1 *(Continued)*

Food matrix	Process mode	Power/frequency	Target enzymes	Residual activity	Quality attributes	Reference
Wheat germ	Batch mode (650 g)	6 kW/27.12 MHz	LA	18.2%/100°C/15 min	↑ Color	[31]
				22.5%/110°C/5 min	↑ WAC, OAC	
					↓ PS	
Wheat germ	Batch mode	12 kW/27.12 MHz	LA	9.1%/14.5% MC/110°C/6 min	↓ *L**	[25]
			POD	4.8%/20% MC/110°C/6 min	↑ *b**	
			PPO	1.7%/14.5% MC/110°C/6 min	↔ *a**	
			LOX	2.3%/20% MC/100°C/15 min	↔ PV, FFA	
Wheat	Batch mode (900 g)	12 kW/27.12 MHz	SOD	30.3% decrease/70°C/10 min	↓ Seed vigor	[21]
			CAT	14.6% decrease/70°C/10 min	↔ Germination rate	
			POD	43.8% decrease/70°C/10 min	↑ *b**	
Corn	Batch mode (900 g)	12 kW/27.12 MHz	SOD	25.9% decrease/70°C/10 min	↔ Δ E	
			CAT	34.3% decrease/70°C/10 min	↓ TPC, M&Y	
			POD	29% decrease/70°C/10 min		

↓, decrease in the level; ↑, increase in the level; ↔, insignificant change.

different for both the food matrix. Also, the PPO in apple tissue displayed a reduced activity and gets easily inactivated than the PPO present in the juice; specifically, PPO in tissue presented lower heat stability than that present in juice, and RF treatment can efficiently inactivate the PPO in apple than that in the extracted form.

Mishra et al.[36] indicated that PPO is generally found bound to the organelle membrane, forming mPPO, and when such cells get damaged, mPPO is freed into the juice resulting in high catalytic activity, and that mPPO is easily deactivated compared to the released PPO.

Zhang et al.[55] examined the performance of RF heating on the denaturation of both pure mushroom PPO and potato PPO by a mild application of heat (25–85°C) by varying the gap between the electrodes. The findings indicated that a fastest rate of heating rate was attained when the electrode gap was maintained at 110 mm for mushroom PPO and 120 mm for potato PPO. The rates gradually decreased with increase in temperature. The relative activity of both mushroom PPO and potato PPO was found to decrease with increase in treatment temperature. The relative activity also varied with different potato matrices given that enzyme extract exhibited the lowest activity followed by cuboids and mashed potato at a treatment temperature of 75°C.

Yao et al.[53] confirmed heat-induced changes in the protein confirmation of POD in RF-treated stem lettuce and stated that the varying degree of POD inactivation with temperature can be attributed to the different thermal resistance characteristics of their isoenzymes. A different study by the same team observed that the residual activity of POD presented an abrupt decrease when the center temperature of the sample ascended to around 85°C. The heat-labile fraction of POD was speculated to be commendably reduced at lower temperatures while heat-resistant fraction requires a slightly higher temperature for getting inactivated. It is to be noted that providing the RF heating did not reach the required temperature for denaturation of the particular target enzyme, no significant reduction in activity will be detected; however, the same rate of enzyme inactivation can be accomplished under RF heating employing different electric field intensities but at the same treatment temperature.

7.5 QUALITY CHANGES IN THE RF HEATED FOODS

The quality of food processed through RF heating is a crucial component in regard to shelf-life stability, quality of processing, market acceptability, and inclusive consumer approval. The present section describes different physicochemical and functional changes in food during RF treatment.

Though enzyme activity is a fundamental aspect in quality control of fresh agricultural produce and their products, yet other important factors such as moisture content, a_w, weight loss, FFA, PV, color properties, functional properties, and structural characteristics also play significant role in the quality check. For dried food products, there exists a critical value of a_w, below which quality loss can be shrugged off and product stability can be ensured throughout its shelf-life.[6] Water act as both reactant and solvent, and in foods containing fats, there exist a strong constructive relationship between the product moisture content and the rate of hydrolysis reactions. Thus, reducing MC could essentially hinder the hydrolysis of lipids and can prevent rancidity. However, optimal a_w for retarding rancidity is different for different food products. Hot air assisted-RF heating caused a significant decrease in moisture content and a_w value with increasing treating temperature and holding time, when rice bran was subjected to RF heating.[29]

In the same study, though there occurred an increase in the FFA content during storage, yet the accumulations of FFA in RF-treated samples were all appreciably lower than those of the control ones.[29] A similar result was obtained for PV, where PV build-up in the samples decreased pointedly with rise in treatment temperature and holding time.

Manzocco et al.[34] observed that apples blanched by RF method exhibited an enhanced resistance to browning, and apple puree was found to have better sweetness than that made from hot water-blanched samples. The retention of water-soluble proteins during RF treatment can thus produce high-quality products as compared to conventional methods.

RF treatment was found to significantly improve the foaming properties of grain flour.[19] These properties are of importance to the grain flour for its utilization in diverse food products like biscuits, ice creams, cookies, and cakes. The noticeable improvement can be attributed to the RF-induced protein denaturation and the subsequent unfolding of its native form that stimulated the protein molecules to both adsorption and diffusion at air or water interface.[30] Conversely, conventional thermal treatment results in the formation of certain insoluble polymers with comparatively greater molecular weight that might not get adsorbed or diffused into the two-phase interface.

Similarly, an increase in both emulsifying stability and emulsifying activity index was observed in the RF-treated sample. The responsible phenomena can be possibly associated with the alterations in protein spatial structures. This signifies that RF processing directed a partial unfolding of proteins and instigated the uncovering of hydrophobic amino acids masked within the protein inner core, which resulted in better assimilation of proteins onto the surface of oil droplets and thus enhanced the emulsifying capacity.[13]

The effect of hot air assisted-RF processing on SOD, CAT, and POD activities of both wheat and corn grains was studied by Jiao et al.[21] An initial increase in the enzyme activities was found when the treatment was carried out at 60–65°C and held for 10 min. Podlesny et al.[41] described this unusual increase in enzyme activity as a result of the enhancement of seed metabolism and biological activity of cell molecules when they interact with electromagnetic field and mild heat energy. When treatment temperature was gradually increased beyond 70°C, the enzyme activities started reducing significantly on account of the heat-induced enzyme inactivation. However, these enzymes have the potential or scavenging free radicals and reactive oxygen species, exhibiting a protective function and having a positive correlation with seed vigor. Hence, mild RF treatment is recommended for such enzymes when the process is aimed to improve seed quality. RF treatment study on potato cuboids with varied electrode gaps showed no significant differences on weight loss, whereas treatment temperature showed a significant effect.[55]

The weight loss was found to increase with increase in attained temperature. With increase in temperature, water evaporation and cellular damage occur leading to nutrients outflow. The microstructure of the sample also gets destroyed contributing to weight loss. Similar result was found in RF blanched sweet potato where a decline in both moisture content and a_w of samples was also obtained.[18] Here, the weight loss was explained as a result of greater weight of lost water-soluble substances than that of the absorbed water during blanching. When compared with other blanching techniques, RF heating presented higher values of weight loss contrasted to steam blanching and PEF treatments.[16,37]

Scanning electron microscopy studies on RF-treated samples also corroborated that the microstructure of the sample was steadily damaged with increasing temperature. RF treatment caused changes in the cell structure and the links between the adjacent cells. The cell wall folds get sternly destroyed and debris gets sunken to the bottom of the cell-wall cavity making the cell wall rough.[53]

Pectin degradation is also observed which degrades cell-to-cell adhesion and contact, causing a collapse of the cell wall and a decrease in hardness. The value of L^* and b^* reduced with increase in temperature due to the cell collapse and liquid release causing a lower reflectance.[48] However, a^* and ΔE values showed a slight increase with increasing temperature which might be on account of Maillard reaction, nonenzymatic browning, and possible alterations in heat-sensitive components. However, varied results are observed with different food products treated which can be attributed to the product composition and structure.

Vitamin C is an indispensable micronutrient which is mandatory for basic metabolic functioning of the human system among other vitamins and minerals. It is broadly established that fruits and vegetables are the best sources of vitamin C. RF heating was found to decrease the loss of vitamin C as compared to hot water blanching although a slight decreasing trend was observed in RF heating rate.[52] The results proposed that the vitamin C being water soluble, resulted in a weighty proportion of overall loss during hot water blanching, and that dry-blanching with no contact with water was advantageous for retaining vitamin C. It is seen that RF blanching treatments can be an impending technique for moderate blanching process that can inactivate the deteriorative enzymes present in food without influencing product quality. Nevertheless, from the perception of food safety and quality, it is indispensable to conduct further studies on the detailed mechanism of the interaction of RF electromagnetic waves with endogenous food enzymes that lead to their complete or partial inactivation.

7.6 ADVANTAGES AND LIMITATIONS OF RF HEATING

The major advantages of RF heating in the food industry are its ability to overcome many heat transfer problems that are prevalent in traditional thermal processing. RF easily targets the product without affecting the air surrounding it, and this manages to drive moisture from inside out, substantially reducing the processing time. Sample geometry is a prominent factor influencing the enzyme inactivation level and the efficacy difference of conventional and RF heating system.[53] In traditional hot blanching, the time taken for heat to transfer from the surface to the sample core from the hot water is more and subsequently, the temperature within the samples fails to reach the anticipated temperature for specific enzyme inactivation within the preset time. In contrast, the entire sample gets uniformly heated during RF treatment, thus averting an extreme temperature gradient. In general, prolonged blanching may not only cause leaching out of essential nutrients to the heating medium but also results in textural degradation.[26] A variation in the membrane function of the phospholipid bimolecular layer is caused as a result of thermal stress and subsequent alteration in the morphological configuration. Further, the permeability of cell membrane gets increased with extended thermal processing, which facilitates seepage of intracellular fluids.[52]

A longer RF treatment can cause a deliberate destruction of lipid and protein confirmations of the cell membrane. Such damage can possibly be diminished through reducing the heating or holding times and employing elevated heating rate. Rapid RF processing at an accelerated rate may be additionally appropriate for both dry and solid foods like quick-frozen or deep processed foods, since in this way, a better texture and improved nutritional value can be maintained even with enhanced enzyme inactivation rates. Conversely, extended blanching with a gradual heating rate is beneficial when liquid fruit and vegetable juice are to be processed since damage to the cell structure elevates the suspension of nutrients in the medium. The use of RF heating is also reported to result in reduced energy consumption, which can be pondered to be of great advantage over conventional methods of heating.[41]

In RF processing, the electromagnetic field is mostly created in a directional fashion in between the electrodes, unlike the field approaching the sample from several directions as in microwave processing.[15] Also, RF processing employs lower frequencies which results in longer penetration depths thus making it useful to process bulk food materials having larger dimensions.

In spite of plentiful academic research experimentation and productive findings, the progress toward commercializing the applications of RF heating or sterilization or decontamination purposes still remain in a nascent stage. The reason or such a slow progress includes lack of knowledge among the equipment manufacturers and suppliers who hesitate the development of the technique, the lack of proper funding required to assist methodical research and scale-up, and the lack of familiarity on the impending benefits of RF heating in the food applications.[23] Additionally, the evident nonuniform heating is even now a main impediment to adopting the RF technique in the food industry. The corners and edges of the rectangular slabs in the RF processing system commonly absorbs extra energy as compared to the other portions of the sample holder and thus initiate nonuniform heating. It is also instigated by the varied dielectric properties of food stuffs and the encompassing medium which results in an asymmetrical distribution of the field in the treatment chamber.[35]

The local high temperatures thus created can give rise to relentless quality loss, for example, lipolysis or lipid oxidation or the reactions involving nonenzymatic browning.[2] Therefore, RF heating uniformity must be perfected prior to scaling up the technique to the industrial level. Various approaches have been espoused to enhance the uniformity of RF heating, which include substituting the adjacent heating media with water or plastic material[51] or fending the sample with a plastic material having comparable ε'[22] or combining the heating with product movement and rotation or sample agitation or moving on top of a conveyor belt.[9,28]

Another potential drawback in RF processing is the selection of optimum frequency from the band designations permitted within the country of operation. Presently, the frequencies allocated for various treatments normally comprise those positioned at 6.78, 13.56, 27.12, and 40.68 MHz. Consequently, any devices of telecommunication applications that use these same frequencies will have the ability to endure RF interferences from other appliances, hindering their performance. These mandates shielding of RF energy in local applications to prevent such interferences with other frequencies devoted to telecommunication equipments. Providentially, the longer wavelengths make it easier for shielding RF waves than microwaves.

7.7 SUMMARY

RF processing is a volumetric heating for treating agriculture and food products and holds the prospective to focus on budding food safety associated with both high and low moisture foods. The results of the present chapter summarizes recent findings on applications of RF heating for enzyme inactivation in different foods and the associated quality changes. It was seen that RF resulted in the effective inactivation of different food enzymes at shorter treatment times as compared to conventional hot air blanching. Also, the treatment was accompanied by better preservation of sensory, physicochemical, and functional characteristics of the product. The precedent decade has seen a considerable amount of research works and publications in the field of radiofrequency heating and a sustained development in RF applications in food processing is anticipated in the years ahead.

However, voluminous amount of promising usages of RF processing in foods are not studied and cautiously examined yet, and there still is a demand for future studies and developments that can take into account the latest technological advancements after considering a thorough economic analysis. Broad future studies would be needed on investigating different systems or alternatives to properly design an RF processing device that can ensure homogenous heating throughout the bulk of the treated product. Also, there is insufficient data on changes in dielectric properties of various foods associated with RF that gets varied depending on the food composition, water content, viscosity, RF frequency, and processing temperature. Careful consideration of these factors can help in designing suitable and efficient RF systems for heating foods. Researches can also focus on estimating optimal RF frequencies, holding time, electrode gap, and other configurations for processing a specific food or a group of comparable foods. Such studies can also cogitate the influence of RF processing on food quality and sensory assessment with a detailed clarification of different interaction mechanisms between RF energy wave and individual food components. Future studies should also focus on hurdle processing, which is a combined method utilizing the synergistic effects of individual processing methods to improve the overall quality attributes of foods by reducing the high treatment intensities.

KEYWORDS

- **dielectric properties**
- **food quality**
- **functional properties**
- **protein denaturation**
- **radio frequency**
- **sensory attributes**

REFERENCES

1. Adams, J. B. Review: Enzyme Inactivation During Heat Processing of Food-Stuffs. *Int. J. Food Sci. Technol.* **1991,** *26* (1), 1–20. DOI: 10.1111/j.1365-2621.1991.tb01136.x (accessed Aug 02, 2021).
2. Alfaifi, B.; Tang, J.; Jiao, Y.; Wang, S.; Rasco, B.; Jiao, S.; Sablani, S. Radio Frequency Disinfestation Treatments for Dried Fruit: Model Development and Validation. *J. Food Eng.* **2014,** *120* (1), 268–276. DOI: 10.1016/j.jfoodeng.2013.07.015 (accessed Aug 02, 2021).
3. Amit, S. K.; Uddin, M. M.; Rahman, R.; Islam, S. M. R.; Khan, M. S. A Review on Mechanisms and Commercial Aspects of Food Preservation and Processing. *Agric. Food Secur.* **2017,** *6* (1), 51. DOI: 10.1186/s40066-017-0130-8 (accessed Aug 02, 2021).

4. Avilés-Gaxiola, S.; Chuck-Hernández, C.; Serna-Saldívar, S. O. Inactivation Methods of Trypsin Inhibitor in Legumes: A Review. *J. Food Sci.* **2018,** *83* (1), 17–29. DOI: 10.1111/1750-3841.13985 (accessed Aug 02, 2021).
5. Awuah, G. B.; Ramaswamy, H. S.; Economides, A. Thermal Processing and Quality: Principles and Overview. *Chem. Eng. Process.: Process Intensif.* **2007,** *46* (6), 584–602. doi: 10.1016/j.cep.2006.08.004; (accessed Aug 02, 2021).
6. Bell, L. N. Moisture Effects on Food's Chemical Stability. In *Water Activity in Foods: Fundamentals and Applications*; Barbosa-Cánovas, G. V., Eds.; Black-well Publishing Ltd.: Oxford, 2008; pp 173–198. DOI: 10.1002/9780470376454.ch7 (accessed Aug 02, 2021).
7. Van Boekel, M. A. J. S. Kinetic Modeling of Food Quality: A Critical Review. *Compr. Rev. Food Sci. Food Saf.* **2008,** *7* (1), 144–158. DOI: 10.1111/j.1541-4337.2007.00036.x (accessed Aug 02, 2021).
8. Eswaran, G. M.; Anand, T.; Hema, V. Exploration on the Effectiveness of Radio Frequency Treatment and Flash Pasteurization on Enzyme and Microbial Activity for Coconut Water Preservation. *Biotechnol. J. Int.* **2019,** 1–7. DOI: 10.9734/bji/2019/v23i230073 (accessed Aug 02, 2021).
9. Garbati-Pegna, F.; Sacchetti, P.; Canuti, V.; Trapani, S.; Bergesio, C.; Belcari, A.; Zanoni, B.; Meggiolaro, F. Radio Frequency Irradiation Treatment of Dates in a Single Layer to Control *Carpophilus Hemipterus*. *Biosyst. Eng.* **2017,** *155*, 1–11. DOI: 10.1016/j.biosystemseng.2016.11.011 (accessed Aug 02, 2021).
10. Gili, R. D.; Torrez Irigoyen, R. M.; Penci, M. C.; Giner, S. A.; Ribotta, P. D. Physical Characterization and Fluidization Design Parameters of Wheat Germ. *J. Food Eng.* **2017,** *212*, 29–37. DOI: 10.1016/j.jfoodeng.2017.05.011 (accessed Aug 02, 2021).
11. Guo, C.; Mujumdar, A. S.; Zhang, M. New Development in Radio Frequency Heating for Fresh Food Processing: Review. *Food Eng. Rev.* **2019,** *11* (1), 29–43. DOI: 10.1007/s12393-018-9184-z (accessed Aug 02, 2021).
12. Guo, C.; Zhang, Z.; Chen, J.; Fu, H.; Subbiah, J.; Chen, X.; Wang, Y. Effects of Radio Frequency Heating Treatment on Structure Changes of Soy Protein Isolate for Protein Modification. *Food Bioprocess Technol.* **2017,** *10* (8), 1574–1583. DOI: 10.1007/s11947-017-1923-2 (accessed Aug 02, 2021).
13. Hassan, A. B.; von Hoersten, D.; Mohamed-Ahmed, I. A. Effect of Radio Frequency Heat Treatment on Protein Profile and Functional Properties of Maize Grain. *Food Chem.* **2019,** *271*, 142–147. DOI: 10.1016/j.foodchem.2018.07.190 (accessed Aug 02, 2021).
14. Hou, L.; Johnson, J. A.; Wang, S. Radio Frequency Heating for Postharvest Control of Pests in Agricultural Products: A Review. *Postharvest Biol. Technol.* **2016,** *113*, 106–118. DOI: 10.1016/j.postharvbio.2015.11.011 (accessed Aug 02, 2021).
15. Huang, Z.; Marra, F.; Subbiah, J.; Wang, S. Computer Simulation for Improving Radio Frequency (RF) Heating Uniformity of Food Products: A Review. *Crit. Rev. Food Sci. Nutr.* **2018,** *58* (6), 1033–1057. DOI: 10.1080/10408398.2016.1253000 (accessed Aug 02, 2021).
16. Ignat, A.; Manzocco, L.; Brunton, N. P.; Nicoli, M. C.; Lyng, J. G. The Effect of Pulsed Electric Field Pre-Treatments Prior to Deep-Fat Frying on Quality Aspects of Potato Fries. *Innov. Food Sci. Emerg. Technol.* **2015,** *29*, 65–69. DOI: 10.1016/j.ifset.2014.07.003 (accessed Aug 02, 2021).

17. Ildikó, S. G.; Klára, K. A.; Marianna, T. M.; Ágnes, B.; Zsuzsanna, M. B.; Bálint, C. The Effect of Radio Frequency Heat Treatment on Nutritional and Colloid-Chemical Properties of White Mustard (*Sinapis alba* L.). *Innov. Food Sci. Emerg. Technol.* **2006,** *7* (1–2), 74–79. DOI: 10.1016/j.ifset.2005.06.001 (accessed Aug 02, 2021).
18. Jiang, H.; Ling, B.; Zhou, X.; Wang, S. Effects of Combined Radio Frequency with Hot Water Blanching on Enzyme Inactivation, Color and Texture of Sweet Potato. *Innov. Food Sci. Emerg. Technol.* **2020,** *66*, E-article 102513. DOI: 10.1016/j.ifset.2020.102513 (accessed Aug 02, 2021).
19. Jiang, Y.; Li, L.; He, F.; Yan, W.; Tang, Y.; Yang, R.; Zhao, W. Highly Effective Inactivation of Anti-Nutritional Factors (Lipoxygenase, Urease and Trypsin Inhibitor) in Soybean by Radio Frequency Treatment. *Int. J. Food Sci. Technol.* **2020,** *2020*, E-article 14605. DOI: 10.1111/ijfs.14605 (accessed Aug 02, 2021).
20. Jiang, Y.; Wang, S.; He, F.; Fan, Q.; Ma, Y.; Yan, W.; Tang, Y.; Yang, R.; Zhao, W. Inactivation of Lipoxygenase in Soybean by Radio Frequency Treatment. *Int. J. Food Sci. Technol.* **2018,** *53* (12), 2738–2747. DOI: 10.1111/ijfs.13885 (accessed Aug 02, 2021).
21. Jiao, S.; Zhong, Y.; Deng, Y. Hot Air-Assisted Radio Frequency Heating Effects on Wheat and Corn Seeds: Quality Change and Fungi Inhibition. *J. Stored Prod. Res.* **2016,** *69*, 265–271. DOI: 10.1016/j.jspr.2016.09.005 (accessed Aug 02, 2021).
22. Jiao, Y.; Tang, J.; Wang, S. A New Strategy to Improve Heating Uniformity of Low Moisture Foods in Radio Frequency Treatment for Pathogen Control. *J. Food Eng.* **2014,** *141*, 128–138. DOI: 10.1016/j.jfoodeng.2014.05.022 (accessed Aug 02, 2021).
23. Jiao, Y.; Tang, J.; Wang, Y.; Koral, T. L. Radio-Frequency Applications for Food Processing and Safety. *Ann. Rev. Food Sci. Technol.* **2018,** *9*, 105–127. DOI: 10.1146/annurev-food-041715-033038 (accessed Aug 02, 2021).
24. Kaur, K.; Kaushal, P. Enzymes as Analytical Tools for the Assessment of Food Quality and Food Safety. In *Biomass, Biofuels, Biochemicals: Advances in Enzyme Technology*; Singh, R. S., Singhania, R. R., Pandey, A., Larroche, C., Eds.; Elsevier, 2019; pp 273–292. DOI: 10.1016/B978-0-444-64114-4.00010-8 (accessed Aug 02, 2021).
25. Liao, M.; Damayanti, W.; Zhao, Y.; Xu, X.; Zheng, Y.; Wu, J.; Jiao, S. Hot Air-Assisted Radio Frequency Stabilizing Treatment Effects on Physicochemical Properties, Enzyme Activities and Nutritional Quality of Wheat Germ. *Food Bioprocess Technol.* **2020,** *13* (5), 901–910. DOI: 10.1007/s11947-020-02446-5 (accessed Aug 02, 2021).
26. Liburdi, K.; Benucci, I.; Esti, M. Effect of Microwave Power and Blanching Time in Relation to Different Geometric Shapes of Vegetables. *LWT* **2019,** *99*, 497–504. DOI: 10.1016/j.lwt.2018.10.029 (accessed Aug 02, 2021).
27. Ling, B.; Cheng, T.; Wang, S. Recent Developments in Applications of Radio Frequency Heating for Improving Safety and Quality of Food Grains and their Products: A Review. *Crit. Rev. Food Sci. Nutr.* **2020,** *60* (15), 2622–2642. DOI: 10.1080/10408398.2019.1651690 (accessed Aug 02, 2021).
28. Ling, B.; Hou, L.; Li, R.; Wang, S. Storage Stability of Pistachios as Influenced by Radio Frequency Treatments for Postharvest Disinfestations. *Innov. Food Sci. Emerg. Technol.* **2016,** *33*, 357–364. DOI: 10.1016/j.ifset.2015.10.013 (accessed Aug 02, 2021).
29. Ling, B.; Lyng, J. G.; Wang, S. Effects of Hot Air-Assisted Radio Frequency Heating on Enzyme Inactivation, Lipid Stability and Product Quality of Rice Bran. *LWT—Food Sci. Technol.* **2018,** *91*, 453–459. DOI: 10.1016/j.lwt.2018.01.084 (accessed Aug 02, 2021).

30. Ling, B.; Ouyang, S.; Wang, S. Effect of Radio Frequency Treatment on Functional, Structural and Thermal Behaviors of Protein Isolates in Rice Bran. *Food Chem.* **2019,** *289*, 537–544. DOI: 10.1016/j.foodchem.2019.03.072 (accessed Aug 02, 2021).
31. Ling, B.; Ouyang, S.; Wang, S. Radio-Frequency Treatment for Stabilization of Wheat Germ: Storage Stability and Physicochemical Properties. *Innov. Food Sci. Emerg. Technol.* **2019,** *52*, 158–165. DOI: 10.1016/j.ifset.2018.12.002 (accessed Aug 02, 2021).
32. Lopez, A.; Baganis, N. A. Effect of Radio-Frequency Energy at 60 MHz on Food Enzyme Activity. *J. Food Sci.* **1971,** *36* (6), 911–914. DOI: 10.1111/j.1365-2621.1971.tb15558.x (accessed Aug 02, 2021).
33. Makroo, H. A.; Rastogi, N. K.; Srivastava, B. Ohmic Heating Assisted Inactivation of Enzymes and Microorganisms in Foods: A Review. *Trends Food Sci. Technol.* **2020,** *97*, 451–465. DOI: 10.1016/j.tifs.2020.01.015 (accessed Aug 02, 2021).
34. Manzocco, L.; Anese, M.; Nicoli, M. C. Radiofrequency Inactivation of Oxidative Food Enzymes in Model Systems and Apple Derivatives. *Food Res. Int.* **2008,** *41* (10), 1044–1049. DOI: 10.1016/j.foodres.2008.07.020.
35. Marra, F.; Zhang, L.; Lyng, J. G. Radio Frequency Treatment of Foods: Review of Recent Advances. *J. Food Eng.* **2009,** *91* (4), 497–508. DOI: 10.1016/j.jfoodeng.2008.10.015 (accessed Aug 02, 2021).
36. Mishra, B. B.; Gautam, S.; Sharma, A. Purification and Characterization of Polyphenol Oxidase (PPO) from Eggplant (Solanum melongena). *Food Chem.* **2012,** *134* (4), 1855–1861. DOI: 10.1016/j.foodchem.2012.03.098 (accessed Aug 02, 2021).
37. Mukherjee, S.; Chattopadhyay, P. K. Whirling Bed Blanching of Potato Cubes and Its Effects on Product Quality. *J. Food Eng.* **2007,** *78* (1), 52–60. DOI: 10.1016/j.jfoodeng.2005.09.001 (accessed Aug 02, 2021).
38. Orsat, V.; Jusoh, Y. M. M. Electrical Conductivity Effect on Dielectric Properties and Radio-Frequency Heating. In *Radio-Frequency Heating in Food Processing: Principles and Applications*; Tang, J., Awuah, G. B., Eds.; CRC Press: Boca Raton, 2014; pp 73–90. DOI: 10.1201/b17740 (accessed Aug 02, 2021).
39. Ozturk, S.; Kong, F.; Trabelsi, S.; Singh, R. K. Dielectric Properties of Dried Vegetable Powders and their Temperature Profile During Radio Frequency Heating. *J. Food Eng.* **2016,** *169*, 91–100. DOI: 10.1016/j.jfoodeng.2015.08.008 (accessed Aug 02, 2021).
40. Piyasena, P.; Dussault, C.; Koutchma, T.; Ramaswamy, H. S.; Awuah, G. B. Radio Frequency Heating of Foods: Principles, Applications and Related Properties—A Review. *Crit. Rev. Food Sci. Nutr.* **2003,** *43* (6), 587–606. DOI: 10.1080/10408690390251129 (accessed Aug 02, 2021).
41. Podleśny, J.; Stochmal, A.; Podleśna, A.; Misiak, L. E. Effect of Laser Light Treatment on some Biochemical and Physiological Processes in Seeds and Seedlings of White Lupine and Faba Bean. *Plant Growth Regul.* **2012,** *67* (3), 227–233. DOI: 10.1007/s10725-012-9681-7 (accessed Aug 02, 2021).
42. Prigogine, I. Chemical Kinetics and Dynamics. *Ann. NY Acad. Sci.* **2003,** *988* (1), 128–132. DOI: 10.1111/j.1749-6632.2003.tb06091.x (accessed Aug 02, 2021).
43. Routray, W.; Orsat, V. Recent Advances in Dielectric Properties–Measurements and Importance. *Curr. Opin. Food Sci.* **2018,** *23*, 120–126. DOI: 10.1016/j.cofs.2018.10.001 (accessed Aug 02, 2021).
44. Sosa-Morales, M. E.; Valerio-Junco, L.; López-Malo, A.; García, H. S. Dielectric Properties of Foods: Reported Data in the 21st Century and their Potential Applications.

LWT—Food Sci. Technol. **2010,** *43* (8), 1169–1179. DOI: 10.1016/j.lwt.2010.03.017 (accessed Aug 02, 2021).
45. Terefe, N. S.; Buckow, R.; Versteeg, C. Quality-Related Enzymes in Fruit and Vegetable Products: Effects of Novel Food Processing Technologies, Part 1: High-Pressure Processing. *Crit. Rev. Food Sci. Nutr.* **2014,** *54* (1), 24–63. DOI: 10.1080/10408398.2011.566946 (accessed Aug 02, 2021).
46. Tian, Y.; Wang, S.; Yan, W.; Tang, Y.; Yang, R.; Zhao, W. Inactivation of Apple (*Malus domestica Borkh*) Polyphenol Oxidases by Radio Frequency Combined with Pulsed Electric Field Treatment. *Int. J. Food Sci. Technol.* **2018,** *53* (9), 2054–2063. DOI: 10.1111/ijfs.13781 (accessed Aug 02, 2021).
47. Tian, Y.; Yan, W.; Tang, Y.; Yang, R.; Zhao, W. Inactivation of Membrane—Bound and Soluble Polyphenol Oxidases in Apple (*Malus domestica Borkh*) by Radio Frequency Processing for Improved Juice Quality. *J. Food Process Eng.* **2018,** *41* (8), online. DOI: 10.1111/jfpe.12923 (accessed Aug 02, 2021).
48. Tribst, A. A. L.; Leite Júnior, B. R. D. C.; De Oliveira, M. M.; Cristianini, M. High Pressure Processing of Cocoyam, Peruvian Carrot and Sweet Potato: Effect on Oxidative Enzymes and Impact in the Tuber Color. *Innov. Food Sci. Emerg. Technol.* **2016,** *34*, 302–309. DOI: 10.1016/j.ifset.2016.02.010 (accessed Aug 02, 2021).
49. Trimigno, A.; Marincola, F. C.; Dellarosa, N.; Picone, G.; Laghi, L. Definition of Food Quality by NMR-Based Foodomics. *Curr. Opin. Food Sci.* **2015,** *4*, 99–104. DOI: 10.1016/j.cofs.2015.06.008 (accessed Aug 02, 2021).
50. Vearasilp, S.; Thobunluepop, P.; Thanapornpoonpong, S.; Pawelzik, E.; von Hörsten, D. Radio Frequency Heating on Lipid Peroxidation, Decreasing Oxidative Stress and Aflatoxin B1 Reduction in Perilla frutescens L. Highland Oil Seed. *Agric. Agric. Sci. Procedia* **2015,** *5*, 177–183. DOI: 10.1016/j.aaspro.2015.08.027 (accessed Aug 02, 2021).
51. Wang, S.; Birla, S. L.; Tang, J.; Hansen, J. D. Postharvest Treatment to Control Codling Moth in Fresh Apples Using Water Assisted Radio Frequency Heating. *Postharvest Biol. Technol.* **2006,** *40* (1), 89–96. DOI: 10.1016/j.postharvbio.2005.12.005 (accessed Aug 02, 2021).
52. Yao, Y.; Sun, Y.; Cui, B.; Fu, H.; Chen, X.; Wang, Y. Radio Frequency Energy Inactivates Peroxidase in Stem Lettuce at Different Heating Rates and Associate Changes in Physiochemical Properties and Cell Morphology. *Food Chem.* **2020,** 128360. DOI: 10.1016/j.foodchem.2020.128360 (accessed Aug 02, 2021).
53. Yao, Y.; Wei, X.; Pang, H.; Wang, K.; Liu, Q.; Fu, H.; Chen, X.; Wang, Y. Effects of Radio-Frequency Energy on Peroxidase Inactivation and Physiochemical Properties of Stem Lettuce and the Underlying Cell-Morphology Mechanism. *Food Chem.* **2020,** *322*, 126753. DOI: 10.1016/j.foodchem.2020.126753 (accessed Aug 02, 2021).
54. Yildiz-Turp, G.; Sengun, I. Y.; Kendirci, P.; Icier, F. Effect of Ohmic Treatment on Quality Characteristic of Meat: A Review. *Meat Sci.* **2013,** 93 (3), 441–448. DOI: 10.1016/j.meatsci.2012.10.013 (accessed Aug 02, 2021).
55. Zhang, Z.; Wang, J.; Zhang, X.; Shi, Q.; Xin, L.; Fu, H.; Wang, Y. Effects of Radio Frequency Assisted Blanching on Polyphenol Oxidase, Weight Loss, Texture, Color and Microstructure of Potato. *Food Chem.* **2018,** *248*, 173–182. DOI: 10.1016/j.foodchem.2017.12.065 (accessed Aug 02, 2021).
56. Zhao, Y.; Flugstad, B.; Kolbe, E.; Park, J. W.; Wells, J. H. Using Capacitive (Radio Frequency) Dielectric Heating in Food Processing and Preservation—A Review. *J.*

Food Process Eng. **2000,** *23* (1), 25–55. DOI: 10.1111/j.1745-4530.2000.tb00502.x (accessed Aug 02, 2021).

57. Zhou, H.; Guo, C.; Wang, S. Performance Comparison Between the Free Running Oscillator and 50 Ω Radio Frequency Systems. *Innov. Food Sci. Emerg. Technol.* **2017,** *39*, 171–178. DOI: 10.1016/j.ifset.2016.12.003 (accessed Aug 02, 2021).

CHAPTER 8

ULTRAVIOLET AND LIGHT-EMITTING DIODE TECHNOLOGIES FOR INACTIVATION OF FOOD ENZYMES

TAMANNA RAMESH[1] and JATINDRA K. SAHU[2]

ABSTRACT

Among the nonthermal technologies, ultraviolet (UV), and light-emitting diode (LED) methods utilize the light energy from 200 to 760 nm of wavelength to inactivate pathogenic microorganisms and undesirable enzymes in foods. These technologies have been predominantly utilized in treating beverages and solid food surfaces, and have received due acknowledgment by processors and consumers. The effectiveness of inactivation under UV and LED treatments depends largely on the source of enzymes and type of food matrices under investigation. Low penetration depth of light poses a major challenge to the utilization of the light-based methods in solid and/or semisolid foods. This chapter focuses on the scope of UV and LED technologies for inactivation of food enzymes.

8.1 INTRODUCTION

The rising demand from the consumers for minimally processed foods has accelerated the development and utilization of several nonthermal technologies for the enhancement of safety against pathogenic microorganisms with minimal impact on the quality foods.[95] Among several nonthermal technologies, authors have discussed ultraviolet (UV), pulsed-UV (PUV),

Enzyme Inactivation in Food Processing: Technologies, Materials, and Applications. Megh R. Goyal, PhD, Junaid Ahmad Malik, PhD & Ravi Pandiselvam, PhD (Eds.)

and light-emitting diode (LED) and their effects on the undesirable food enzymes.

The food processing industry has utilized UV-light for water treatment, treatments of solid surfaces, and packaging materials.[34,42] UV light is further classified into three categories based on its wavelength ranges: UV-A (315–400 nm), UV-B (280–315 nm), and UV-C (200–280 nm). At 253.7 nm, UV-C has proven to be the most germicidal and is effectively used for air, surface, water, and food disinfection.[7] U.S. Food and Drug Administration (US FDA) has permitted the utilization of UV light for processing and treating food as long as the source of light is from low-pressure mercury lamps discharging 90% of the light at 253.7 nm.[85] UV light is allowed by the FDA to be deployed for reducing pathogenic microbes in fruit juices, sterilizing water used for food processing or production and controlling the surface microbes. One of the key benefits of UV-based treatment is lack of residues or by-products after processing.[40] UV is used widely in the food industry for pasteurizing liquid foods including but not limited to fruit juices, cider, milk, and egg whites.[4,23,25,63]

PUV contains pulses of broad-spectrum radiation at high frequency of UV, visible light, or infrared at wavelengths of 100–400, 400–700, and 700–1100 nm, respectively.[39] For every second, there are usually 1–20 flashes from these pulses with a surface level dosage of 0.01–50 J cm^{-2}.[93] Similar to UV, the application of PUV has been recognized as safe and nontoxic due to its nonionizing nature.[27] The US FDA has permitted the utilization of PUV[86] for the control of surface microbes as long as the total PUV dosage does not surpass 12.0 J/cm^2 and the period of each pulse is <2 ms.[64–66,68]

Recently, the UV disinfection using LEDs has been explored actively. UV-LEDs are formed using the combination of p- and n-type semiconductors.[10] This moves the electrons into holes that are positively charged between both semiconductors. The electrons and holes rejoin to produce light. This light's wavelength depends on the semiconductor material (e.g., indium gallium nitride produces visible light, aluminum gallium nitride and aluminum nitride produces UV light).[11] LEDs have compact design and are shock-resistant.

Unlike the traditional UVC mercury lamps, UV-LEDs do not require time for warming up and are energy efficient and longer life of >10^5 h.[74] Similar to UV and PUV, LEDs do not form any toxic or nontoxic residues as well. At present, UV-LEDs are used for inactivation of pathogens, that is, safety of food products, sterilization of surfaces in production facilities, and post-harvest storage.[17] Several researchers have verified the benefits of LEDs in slowing down senescence in vegetables,[60] quickening secondary metabolites during ripening[27] and slowing down the loss of nutritional aspects of plants and fresh produce after harvest.[14,91]

Although light-based technologies have been predominantly explored for inactivation of pathogens in foods, yet researchers have demonstrated the inactivation of food enzymes[48,49,81,94] that can deteriorate the quality of fresh produce and beverages.

This chapter includes discussions on the effect of light-based technologies on four major undesirable enzymes (e.g., lipoxygenase (LOX), polyphenol oxidase (PPO), pectin methylesterase (PME), and peroxidase (POD)), which deteriorate the quality of fresh produce and beverages through development of undesirable flavor attributes and enzymatic browning, oxidation of lipids, and bleaching of pigments.[87] Traditionally, these enzymes are inactivated using heat treatment.[41] However, growing customer demand for minimally processed foods with no artificial preservatives has led to several nonthermal methods being deployed for minimizing or eliminating undesirable changes to the color, flavor, texture, and sensory aspects of food and to prevent the loss of heat-sensitive nutrients.

This chapter discusses mechanism of enzyme inactivation using UV and LED technologies; factors affecting the inactivation mechanism; applications in food processing; use of light-based technologies as a hurdle technology in combination with other thermal and/or nonthermal methods to inactivate the food enzymes; and future prospectives of the light-based technologies.

8.2 MECHANISM OF ENZYME INACTIVATION USING LIGHT-BASED TECHNOLOGIES

Light energy from UV and LED has the ability to modify the native structure of the proteins on which the enzymatic activity is dependent.[13,18,69,88] This process is known as protein photo-oxidation and it occurs through two major ways: direct photo-oxidation and indirect protein oxidation (Fig. 8.1). Direct oxidation of enzymes is effective only when UV light is absorbed by the protein/amino acids. Tryptophan (Trp), tyrosine (Tyr), phenylalanine (Phe), histidine (His), cysteine (Cys), and cystine are the some of the major chromophoric amino acids while the rest of the amino acids typically do not captivate UV light over 230 nm.

- In the direct photo-oxidation, the light energy is directly absorbed by the protein structure (or chromophores) through a phenomenon called photo-ionization, which generates free radicals or excited states (singlet or triplet).[12]

- In the indirect protein oxidation, the light radiation is facilitated by development and consequent reactions of the singlet oxygen. The singlet oxygen is produced by energy transfer to the ground level by either the protein or chromophores.[12]

The functional activity of the enzymes is thus compromised due to backbone fragmentation or formation of aggregates, cross-links, and side-chain oxidation.[90] Several studies indicate that the reactive protein peroxides developed on the residues of His, Tyr, Trp typically lead to secondary inactivation. The primary structure of proteins is affected by photooxidation modifying the individual amino acids to form protein carbonyls[31] and leading to loss of aromatic amino acids (Trp, Tyr, Phe) as they have the absorption maxima between 280 and 290 nm.[43]

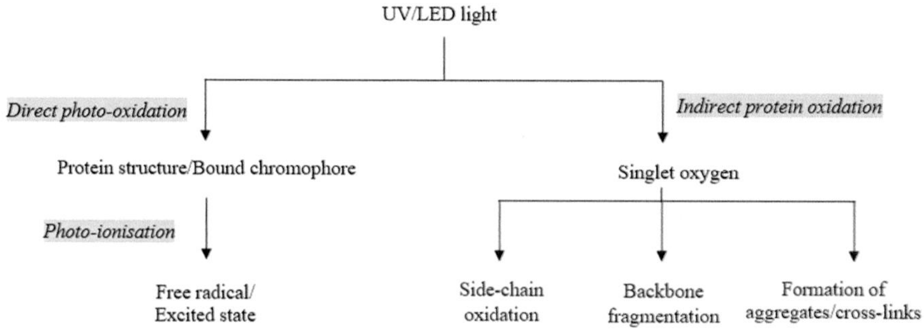

FIGURE 8.1 Flow Diagram for UV-LED based Direct and indirect photo-oxidation.

UV light causes the direct photochemical changes to these proteins due to the limitation of absorption above 310 nm.[62] The concentration of protein carbonyls could indicate the level of protein oxidation.[30]

8.3 FACTORS AFFECTING LIGHT-BASED ENZYME INACTIVATION

Enzyme inactivation using light-based technologies is affected primarily by the light parameters, such as dosage, wavelength, and the type of lamps. For instance, UV radiation produced by medium-pressure mercury vapor lamps are more effective in enzyme inactivation of beverages in comparison to low-pressure mercury vapor lamps.[31] In addition, it will be affected by the food product and its composition, concentration of enzymes, and penetration depth of light in the food matrix. In addition, the pigments, organic acids, and sugars available in these foods affect the ability of light to be absorbed by the food matrices leading to decreased inactivation of enzymes.

Also, external factors (such as: temperature, pH and water activity) directly affect the efficacy of deactivation of undesirable food enzymes. The temperature of the food or processing conditions directly affects the stability of enzymes and tend to accelerate the reaction. When the temperature increases, the enzyme activity increases as well. Denaturation of these undesirable enzymes occur when the temperature rises above 60–70°C.[56]

The pH of the product distresses the enzymes. On the basis of the enzyme, its source and concentration, and depending on the availability of enzyme inhibitors, every enzyme has an optimal pH (LOX has an optimal pH of 7–9). Water activity and moisture content plays a significant role in enzyme activity. When the moisture increases, the enzyme activity tends to increase. Enzyme activity decreases at low water activity. To inactivate an enzyme, the combination of dosage and wavelength of light for each food matrix needs to be optimized along with external factors to disrupt the optimal thriving conditions for these enzymes and inactivate them effectively.

8.4 LIGHT-BASED INACTIVATION OF FOOD ENZYMES

PPO, a group of copper-containing proteins, is one of the primary causes of enzymatic browning in fresh produce when some fresh produces are cut open and come in contact with air.[53] With the oxygen from the air, PPO serves as a catalyst for the hydroxylation of monophenols and the oxidation of o-diphenols to o-quinones. This further generates orthoquinones, which are polymerized and form the undesirable brown pigments. This causes loss of nutritional and sensory attributes in foods.[55]

POD, a heme-containing pectic enzyme, is commonly found in vegetables and acts as a catalyst for the oxidation reactions of peroxides or oxygen, leading to formation of undesirable flavors and colors in unblanched raw vegetables as well as frozen vegetables.[54] POD is typically inactivated using blanching as it is highly heat-stable enzymes and must be inactivated for increased shelf-life of vegetables on freezing.[83]

PME, a pectic enzyme, acts as a catalyst in pectin resulting in a calcium pectate gel after hydrolysis. This further causes instability of beverages and deposits of suspended solids in juices.[82] LOX acts as a catalyst for the oxidation of unsaturated fatty acids and generates hydroperoxides and free radicals. These further cause discoloration through bleaching of chlorophylls and carotenoids in vegetables that are not subject to blanching.[5,75] LOX produces volatile flavor components and free radicals in beverages causing flavor concerns. UV light has demonstrated effectiveness in mitigating the levels of undesirable micro-organisms and enzymes in various foods.[7,20,21,26]

Recently, UV has been predominantly deployed for enzyme inactivation in beverages and fresh produce (fruits and vegetables). Tables 8.1 and 8.2 list the food matrix, processing condition, and the rate of enzyme inactivation in each of these food matrices. Overall, extended treatment time is one of the major challenges in utilizing UV light for enzyme inactivation. However, when combined with mild heat treatment, the treatment time is reduced drastically providing an additional hurdle to the enzyme inactivation process. Table 8.1 focuses on fruit juices and vegetable juices; and Table 8.2 focuses on fresh produce.

TABLE 8.1 UV and LED Light-Based Inactivation of Food Enzymes in Fruit Juices.

Food matrix	Processing conditions	Enzyme inactivation	Reference
Apple juice	HPM lamp of 400 W—Energy was 3.88×10^{-7} E min^{-1}. The clarified apple juice was at 2.4 cm depth.	PPO and POD activity—completely inactivated in 100 and 15 mins, respectively.	[22]
	UV-C light (254 nm, 0–53 kJ/L) in a dean flow reactor with a 9 W UV-C lamp at 253.7 nm.	No significant reduction of PPO and PME even at the maximum dosage of 53 kJ/L.	[57]
	UV Source of 36 W LPM lamps with peak radiation of UV-C (253.7 nm) and a 18.3 W UV-B lamps (290–315 nm); Flow rate was 30 L h^{-1}. Energy UV-C was 4.019 kJ L^{-1} and for UV-B was 2.043 kJ L^{-1} for every cycle.	UVC: At 100.48 kJ L^{-1} (25 cycles), the PPO level was reduced by over 20%; UVB: No impact on PPO level	[58]
	UV-LED lamps (254 and 280 nm)	PPO level was reduced to 32.58% and 34.38% respectively at 280/365 and 280/405 nm respectively. However, thermal treatment at 70°C for 2 min reduced enzyme activity to 22.34%.	[3]

TABLE 8.1 *(Continued)*

Food matrix	Processing conditions	Enzyme inactivation	Reference
	LPM lamps with UVC light (254 nm)	UVC light at 55°C reduced PPO activity by 39.39% while UVC along reduced it by 22.72% demonstrating the synergistic effect of UVC and heat.	[24]
Carrot juice	LPM lamp at 8 W (254 nm) with total dose being 3.92 J/mL in one pass that is, 3.6 min.	UV (3.92 J/mL, 3.6 min) with a mild heat of 60 °C reduced PME activity by 69.3% whereas heating at 60 °C without heat reduced the PME activity by 56.8% after 18.1 min; Heating at 80 °C completely inactivated PME activity.	[28]
		PPO activity was reduced by 51.3% with UV (3.92 J/mL, 3.6 min) and heat; heating at 60 °C for 18.1 reduced PPO by 58.9% and at 80 °C, PPO activity was undetected.	
Carrot and orange juice	30 W LPM UV lamp was used (2.66–53.10 J/cm^2)	PME was reduced by 78% using UVC treatment (10.6 J/cm^2). However, heating at 72 °C reduced PME by 92%.	[9]
Coconut water	UV-C at 4.8 J/mL was compared against thermal pasteurization.	PPO activity was decreased by 46.4% using UVC and by 70.3% using heat treatment (90 °C, 1.7 min).	[16]
Grape juice	HPM lamp of 400 W with 3.88×10^{-7} E min^{-1} energy. White and Pink Grape Juice were at 2.4 cm depth.	PPO was not eliminated entirely during 140 min treatment; Reduced only 80% and 50% of the original levels in white grapes and pink grapes, respectively. POD was completely inactivated.	[19]
Mango nectar	UV light (75–450 kJ/m^2) with mercury lamps of intensity of 25 mW/cm^2	PPO activity was eliminated in 156–204 min.	[29]

TABLE 8.1 *(Continued)*

Food matrix	Processing conditions	Enzyme inactivation	Reference
Nectarine juice	Medium pressure mercury lamps of 460 W was used 2.40×10^{-5} W cm^2 at 25 °C and 45 °C.	Both PPO and POD were eliminated by 60% at 25 °C and completely at 45°C after 100 min for both juices.	[2]
Orange juice	LPM UV lamp at 6 W of UVC (254 nm). The depth of orange juice was between 0.21 and 0.48 mm.	PME level was eliminated by 70% after heating at 70°C for 2 s; However, UVC (73.8 mWs·cm^{-2}) did not eliminate PME effectively.	[84]
Peach juice	HPM lamp of 460 W with 4.49 10^{-2} W cm^2 at 25°C and 45°C.	Effectively inactivated PPO and POD when subject to heat in addition to UV light; At 25°C and 45°C, PPO activity was mostly eliminated in 120 min and 60 min, respectively.	[1]
Pineapple juice	UV pasteurizer (FPE Inc, New York, USA) and flow rates (0.043, 0.064 and 0.086 L·s^{-}).	PME activity was reduced by 60.53% at 55°C in 10 min. However, UVC (5.61–11.23 mJ cm^{-2}) did not eliminate PME effectively.	[72]
Watermelon juice	LPM lamp at 9 W (254 nm) was used	PME was reduced by 65% at 9.69 J/mL dosage (15 min)	[92]

TABLE 8.2 Summary of Light-based Inactivation of Enzymes in Fresh Produce.

Food matrix	Processing conditions	Enzyme inactivation	Reference
Apple slices	15 W lamps with max emission at 253.7 nm; Apple surfaces were exposed to irradiance of 13.8 Wm^{-2} UV-C	10× decreased POD activity in 33 min	[35, 51, 52]
Banana	UV-C at 0.03 and 0.04 KJ/m^2	PPO activity was completely eliminated at 0.03 kJ/m^2	[15, 32]
Lotus root	UV-C lamp at 75 W placed 30 cm away from the samples; 1.5–3 KJ/m^2	Noticeable reduction of PPO, and POD. Prolonged exposure enhanced the enzyme activity.	[89]

TABLE 8.2 *(Continued)*

Food matrix	Processing conditions	Enzyme inactivation	Reference
Pineapple	LED light + photosensitizer (curcumin)	Reduced PPO by 33.5% and POX by 25.7%; Preserved bromelain activity	[6]
Potato slices	Irradiation levels from 2.28 to 13.68 KJ/m^2	UV-C at 6.84 KJ/m^2 reduced PPO, and POD	[80]
Shiitake mushroom	UV-C lamp placed 60 cm away; 4 KJ/m^2; Fluence rate is 2.82 ± 0.44 $mWcm^{-2}$	Enhanced the enzyme activities during storage studies	[37]
Soy	Pulsed UV light placed 6 cm away; Sample temperature was at 33.5 ± 1.8°C at 16 s	99.95% reduction of LOX activity	[36]
Tomato	UV-C, 1.5–4.5 kJ m^{-2}	60% decrease in PME activity	[50]
Zucchini	16 W UV-C lamps, at 10 cm from the sample (11 J/m^2)	POD inactivation followed first order kinetics; Synergistic effect upon combining blanching with UV	[61]

8.5 HURDLE TECHNOLOGY

In order to warrant safety of liquid foods, fruits, and vegetables, hurdle technology employs various sublethal stressors either individually in a sequence or in combination to obtain an increasing inhibitory effect on the pathogenic organisms in the foods.[78,79] These sublethal stressors could be physical, chemical, or biological. Combination of hurdles could be modification of pH with combination of pressure or UV methods to obtain pasteurization or sterilization.

Some of the most common physicochemical hurdles are water activity, pH level, redox potential, salt, ozone, lactic acid, spices, herbs, etc. Examples of physical hurdles include sterilization, pasteurization, UV, photodynamic inactivation, vacuum packaging, chilling, freezing, aseptic packaging, etc. Examples of microbiologically derived hurdles could be bacteriocins and antibiotics.[46,47]

UV and LED light based inactivation of enzymes is still at a stage,[8,11,33,38,44,45,59,76,77] where only few researchers[70] have explored the extent of hurdle technology's impact in this area. Till today, combination of light treatment with mild heat is commonly used hurdle method to inactivate enzymes.[71–73] In the future, other hurdles listed in this section could be combined appropriately with UV or LED light treatment to establish combinations of lethal factors for effective enzyme inactivation.[67] In terms of future scope, this is a prime area for exploration and establishment of parameters for elimination of undesirable enzymes in both beverages and fresh produce.

8.6 SUMMARY

Light-based technologies (such as UV, pulsed light (PL), and LED) are being actively explored for their efficacy against undesirable and pathogenic microorganisms in liquid foods (beverages, egg white, etc.). However, their efficacy against undesirable enzymes is noteworthy and is an area of interest for food and beverage manufacturers. In addition, several researchers have demonstrated increased retention of desirable nutrients, which improve the overall quality of these food products and make them more appealing to the consumers.

UV and PL are already being used commercially for pasteurization of beverages and can be easily expanded to inactivation of enzymes in industrial scale after validation of the process. UV and LED light technologies have the potential to inactivate undesirable enzymes in beverages and on fresh produce. However, they are limited by the penetration depth of the light and pose serious challenges to the effectiveness of inactivation based on the food matrix that is subjected to this treatment. Hurdle Technology could be deployed to inactivate the undesirable enzymes. In this chapter, based on existing research work, a combination of mild heat and UV/LED light would prove to be effective in inactivating the food enzymes through the stress reactions caused by the combination of heat and light. UV-LED technology is still at its infant stage and additional research is needed to evaluate the effectiveness of LED on enzyme inactivation as opposed to that of the traditional mercury lamps.

ACKNOWLEDGMENTS

Special thanks to Dr. Kaavya Rathnakumar and Ms. Bharathi Ramesh for their encouragement and continued support. In addition, the authors want to sincerely thank the editors for their exceptional support and patience.

KEYWORDS

- **enzyme inactivation**
- **food enzymes**
- **light-based**
- **light-emitting diode (LED)**
- **nonthermal technology**
- **ultraviolet (UV)**

REFERENCES

1. Aguilar, K.; Garvín, A.; Ibarz, A. Effect of UV Processing on Enzymatic Activity and the Physicochemical Properties of Peach Juices for Different Varieties. *Innov. Food Sci. Emerg. Technol.* **2018,** *48*, 83–89.
2. Aguilar, K.; Ibarz, R.; Garvín, A.; Ibarz, A. Effect of UV–Vis Irradiation on Enzymatic Activities and the Physicochemical Properties of Nectarine Juices from Different Varieties. *LWT—Food Sci. Technol.* **2016,** *65*, 969–977.
3. Akgün, M. P.; Ünlütürk, S. Effects of Ultraviolet Light Emitting Diodes (LEDs) on Microbial and Enzyme Inactivation of Apple Juice. *Int. J. Food Microbiol.* **2017,** *260*, 65–74.
4. Basaran, N.; Quintero-Ramos, A.; Moake, M. M.; Churey, J. J.; Worobo, R. W. Influence of Apple Cultivars on Inactivation of Different Strains of *Escherichia Coli* O157: H7 in Apple Cider by UV Irradiation. *Appl. Environ. Microbiol.* **2004,** *70* (10), 6061–6065.
5. Baysal, T.; Demirdöven, A. Lipoxygenase in Fruits and Vegetables: A Review. *Enzyme Microb. Technol.* **2007,** *40* (4), 491–496.
6. Bhavya, M. L.; Shewale, S. R.; Rajoriya, D.; Hebbar, H. U. Impact of Blue LED Illumination and Natural Photosensitizer on Bacterial Pathogens, Enzyme Activity and Quality Attributes of Fresh-Cut Pineapple Slices. *Food Bioprocess Technol.* **2021,** *14* (2), 362–372.
7. Bintsis, T.; Litopoulou-Tzanetaki, E.; Robinson, R. K. Existing and Potential Applications of Ultraviolet Light in the Food Industry—A Critical Review. *J. Sci. Food Agric.* **2000,** *80* (6), 637–645.
8. Bonciu, E. Food Processing, a Necessity for the Modern World in the Context of Food Safety: A Review. *Ann. Univ. Craiova-Agric. Montanol. Cadastre Ser.* **2018,** *47* (1), 391–398.
9. Caminiti, I. M.; Noci, F.; Morgan, D. J.; Cronin, D. A.; Lyng, J. G. The Effect of Pulsed Electric Fields, Ultraviolet Light or High Intensity Light Pulses in Combination with Monothermo sonication on Selected Physicochemical and Sensory Attributes of an Orange and Carrot Juice Blend. *Food Bioprod. Process.* **2012,** *90* (3), 442–448.
10. Cassar, J. R.; Ouyang, B.; Krishnamurthy, K.; Demirci A. Microbial Decontamination of Food by Light-Based Technologies: Ultraviolet (UV) Light, Pulsed UV Light (PUV),

and UV Light-Emitting Diodes (UV-LED). In *Food Safety Engineering*; Demirci, A., Feng, H., Krishnamurthy K., Eds.; Springer: Cham, 2020; pp 493–521.

11. Chevremont, A. C.; Farnet, A. M.; Coulomb, B.; Boudenne, J. L. Effect of Coupled UV-A And UV-C LEDs on Both Microbiological and Chemical Pollution of Urban Wastewaters. *Sci. Total Env.* **2012,** *426*, 304–310.
12. Davies, M. J.; Truscott, R. J. Photo-Oxidation of Proteins and Its Role in Cataractogenesis. *J. Photochem. Photobiol. B: Biol.* **2001,** *63* (1–3), 114–125.
13. Davies, M. J. Protein Oxidation and Peroxidation. *Biochem. J.* **2016,** *473* (7), 805–825.
14. Dhakal, R.; Baek, K. H. Short Period Irradiation of Single Blue Wavelength Light Extends the Storage Period of Mature Green Tomatoes. *Postharvest Biol. Technol.* **2014,** *90*, 73–77.
15. Ding, P.; Ling, Y. S. Browning Assessment Methods and Polyphenol Oxidase in UV-C Irradiated Banana Fruit. *Int. Food Res. J.* **2014,** *21* (4), 1667–1674.
16. Donsingha, S.; Assatarakul, K. Kinetics Model of Microbial Degradation by UV Radiation and Shelf-Life of Coconut Water. *Food Control* **2018,** *92*, 162–168.
17. D'Souza, C.; Yuk, H. G.; Khoo, G. H.; Zhou, W. Application of Light-Emitting Diodes in Food Production, Postharvest Preservation, and Microbiological Food Safety. *Compr. Rev. Food Sci. Food Saf.* **2015,** *14* (6), 719–740.
18. Edwards, A. M.; Silva, E. Effect of Visible Light on Selected Enzymes, Vitamins and Amino Acids. *J. Photochem. Photobiol. B: Biol.* **2001,** *63* (1–3), 126–131.
19. Falguera, V.; Garza, S.; Pagán, J.; Garvín, A.; Ibarz, A. Effect of UV–Vis Irradiation on Enzymatic Activities and Physicochemical Properties of Four Grape Musts from Different Varieties. *Food Bioprocess Technol.* **2013,** *6* (8), 2223–2229.
20. Falguera, V.; Pagán, J.; Garza, S.; Garvín, A.; Ibarz, A. Inactivation of Polyphenol Oxidase by Ultraviolet Irradiation: Protective Effect of Melanins. *J. Food Eng.* **2012,** *110* (2), 305–309.
21. Falguera, V.; Pagán, J.; Garza, S.; Garvín, A.; Ibarz, A. Ultraviolet Processing of Liquid Food—A Review, Part 2: Effects on Microorganisms and on Food Components and Properties. *Food Res. Int.* **2011,** *44* (6), 1580–1588.
22. Falguera, V.; Pagán, J.; Ibarz, A. Effect of UV Irradiation on Enzymatic Activities and Physicochemical Properties of Apple Juices from Different Varieties. *LWT—Food Sci. Technol.* **2011,** *44* (1), 115–119.
23. Gayán, E.; Condón, S.; Álvarez, I. Biological Aspects in Food Preservation by Ultraviolet Light: A Review. *Food Bioprocess Technol.* **2014,** *7* (1), 1–20.
24. Gayán, E.; Serrano, M. J.; Monfort, S.; Álvarez, I.; Condón, S. Pasteurization of Apple Juice Contaminated with *Escherichia Coli* by a Combined UV—Mild Temperature Treatment. *Food Bioprocess Technol.* **2013,** *6* (11), 3006–3016.
25. Geveke, D. J. UV Inactivation of *E. Coli* in Liquid Egg White. *Food Bioprocess Technol.* **2008,** *1* (2), 201–206.
26. Gómez-López, V. M.; Koutchma, T.; Linden, K. Ultraviolet and Pulsed Light Processing of Fluid Foods. In *Novel Thermal and Non-Thermal Technologies for Fluid Foods*; Cullen, P. J., Tiwari, B. K., Valdramidis, V. P., Eds.; Academic Press: New York, 2012; pp 185–223.
27. Gomez-Lopez, V. M.; Ragaert, P.; Debevere, J.; Devlieghere, F. Pulsed Light for Food Decontamination: A Review. *Trends Food Sci. Technol.* **2007,** *18* (9), 464–473.
28. Gouma, M.; Álvarez, I.; Condón, S.; Gayán, E. Pasteurization of Carrot Juice by Combining UV-C and Mild Heat: Impact on Shelf-Life and Quality Compared to

Conventional Thermal Treatment. *Innov. Food Sci. Emerg. Technol.* **2020,** *64*, E-article 102362.
29. Guerrero-Beltren, J. A.; Barbosa-Cenovas, G. V. Inactivation of *Saccharomyces cerevisiae* and Polyphenoloxidase in Mango Nectar Treated with UV Light. *J. Food Prot.* **2006,** *69* (2), 362–368.
30. Hellwig, M. The Chemistry of Protein Oxidation in Food. *Angewandte Chemie Int. Edit.* **2019,** *58* (47), 16742–16763.
31. Hinds, L. M.; O'Donnell, C. P.; Akhter, M.; Tiwari, B. K. Principles and Mechanisms of Ultraviolet Light Emitting Diode Technology for Food Industry Applications. *Innov. Food Sci. Emerg. Technol.* **2019,** *56*, E-article 102153.
32. Huang, J. Y.; Xu, F.; Zhou, W. Effect of LED Irradiation on the Ripening and Nutritional Quality of Postharvest Banana Fruit. *J. Sci. Food Agric.* **2018,** *98* (14), 5486–5493.
33. Ibrahim, M. A.; McAdam, J.; Autin, O.; Jefferson, B. Evaluating the Impact of LED Bulb Development on the Economic Viability of Ultraviolet Technology for Disinfection. *Environ. Technol.* **2014,** *35* (4), 400–406.
34. Iqbal, A.; Murtaza, A.; Hu, W.; Ahmad, I.; Ahmed, A.; Xu, X. Activation and Inactivation Mechanisms of Polyphenol Oxidase During Thermal and Non-Thermal Methods of Food Processing. *Food Bioprod. Process.* **2019,** *117*, 170–182.
35. Jang, J. H.; Moon, K. D. Inhibition of Polyphenol Oxidase and Peroxidase Activities on Fresh-Cut Apple by Simultaneous Treatment of Ultrasound and Ascorbic Acid. *Food Chem.* **2011,** *124* (2), 444–449.
36. Janve, B. A; Yang, W.; Marshall, M. R.; Reyes-De-Corcuera, J. I.; Rababah, T. M. Nonthermal Inactivation of Soybean (*Glycine max* sp.) Lipoxygenase by Pulsed Ultraviolet Light. *J. Food Sci.* **2014,** *79* (1), C8–C18.
37. Jiang, T.; Jahangir, M. M.; Jiang, Z.; Lu, X.; Ying, T. Influence of UV-C Treatment on Antioxidant Capacity, Antioxidant Enzyme Activity and Texture of Postharvest Shiitake (*Lentinus edodes*) Mushrooms During Storage. *Postharvest Biol. Technol.* **2010,** *56* (3), 209–215.
38. Kebbi, Y.; Muhammad, A. I.; Sant'Ana, A. S.; do Prado-Silva, L., Liu, D.; Ding, T. Recent Advances on the Application of UV-LED Technology for Microbial Inactivation: Progress and Mechanism. *Compr. Rev. Food Sci. Food Saf.* **2020,** *19* (6), 3501–3527.
39. Keklik, N. M.; Krishnamurthy, K.; Demirci, A. Microbial Decontamination of Food by Ultraviolet (UV) and Pulsed UV Light. In *Microbial Decontamination in the Food Industry*; Demirci, A., Ngadi, M. O., Eds.; Woodhead Publishing: London, 2012, pp 344–369.
40. Keyser, M.; Műller, I. A.; Cilliers, F. P.; Nel, W.; Gouws, P. A. Ultraviolet Radiation as a Non-Thermal Treatment for the Inactivation of Microorganisms in Fruit Juice. *Innov. Food Sci. Emerg. Technol.* **2008,** *9* (3), 348–354.
41. Klibanov, A. M. Stabilization of Enzymes against Thermal Inactivation. *Adv. Appl. Microbiol.* **1983,** *29* (1), 28–32.
42. Koutchma, T. Advances in Ultraviolet Light Technology for Non-Thermal Processing of Liquid Foods. *Food Bioprocess Technol.* **2009,** *2* (2), 138–155.
43. Koutchma, T. Principles and Applications of UV Light Technology. In *Ultraviolet Light in Food Technology: Principles and Applications*; Vol. 2, 2nd ed.; CRC Press: Boca Raton, Florida, USA, 2019; pp 29–42.
44. Koutchma, T.; Popović, V. UV Light-Emitting Diodes (LEDs) and Food Safety. In *Ultraviolet LED Technology for Food Applications*; Koutchma, T., Ed.; Academic Press: Oxford, UK, 2019; pp 91–117.

45. Lante, A.; Tinello, F.; Nicoletto, M. UV-A Light Treatment for Controlling Enzymatic Browning of Fresh-Cut Fruits. *Innov. Food Sci. Emerg. Technol.* **2016,** *34*, 141–147.
46. Leistner, L.; Gorris, L. Food Preservation by Hurdle Technology. *Trends Food Sci. Technol.* **1995,** *6* (2), 41–46.
47. Leistner, L; Basic Aspects of Food Preservation by Hurdle Technology. *Int. J. Food Microbiol.* **2000,** *55* (1–3), 181–186.
48. Li, S.; Yang, X.; Yang, S.; Zhu, M.; Wang, X. Technology Prospecting on Enzymes: Application, Marketing and Engineering. *Comput. Struct. Biotechnol. J.* **2012,** *2* (3), E-article 201209017.
49. Lv, C., Zhao, G., & Ning, Y; Interactions Between Plant Proteins/Enzymes and Other Food Components, and Their Effects on Food Quality. *Crit. Rev. Food Sci. Nutr.* **2017,** *57* (8), 1718–1728.
50. Mansourbahmani, S.; Ghareyazie, B.; Kalatejari, S.; Mohammadi, R. S.; Zarinnia, V. Effect of Post-Harvest UV-C Irradiation and Calcium Chloride on Enzymatic Activity and Decay of Tomato (*Lycopersiconesculentum* L.) Fruit During Storage. *J. Integr. Agric.* **2017,** *16* (9), 2093–2100.
51. Manzocco, L.; Dri, A.; Quarta, B. Inactivation of Pectic Lyases by Light Exposure in Model Systems and Fresh-Cut Apple. *Innov. Food Sci. Emerg. Technol.* **2009,** *10* (4), 500–505.
52. Manzocco, L.; Quarta, B.; Dri, A. Polyphenoloxidase Inactivation by Light Exposure in Model Systems and Apple Derivatives. *Innov. Food Sci. Emerg. Technol.* **2009,** *10* (4), 506–511.
53. Marshall, M. R.; Kim, J.; Wei, C. Enzymatic Browning. In *Fruits, Vegetables and Seafoods*; Vol. 49; FAO: Rome, 2000; pp 259–312.
54. Matheis, G.; Whitaker, J. R. Modification of Proteins by Polyphenol Oxidase and Peroxidase and Their Products. *J. Food Biochem.* **1984,** *8* (3), 137–162.
55. Mayer, A. M.; Harel, E. Phenoloxidases and Their Significance in Fruit and Vegetables. *Food Enzymol.* **1991,** *1*, 373–398.
56. Morales-Blancas, E. F.; Chandia, V. E.; Cisneros-Zevallos, L. Thermal Inactivation Kinetics of Peroxidase and Lipoxygenase from Broccoli, Green Asparagus and Carrots. *J. Food Sci.* **2002,** *67* (1), 146–154.
57. Müller, A.; Briviba, K.; Gräf, V.; Greiner, R.; Herrmann, C.; Kuballa, T.; Stahl, M. R. UV-C Treatment Using a Dean Vortex Technology: Impact on Apple Juice Enzymes and Toxicological Potential. *Innov. Food Sci. Emerg. Technol.* **2013,** *20*, 238–243.
58. Müller, A.; Noack, L.; Greiner, R.; Stahl, M. R.; Posten, C. Effect of UV-C And UV-B Treatment on Polyphenol Oxidase Activity and Shelf-Life of Apple and Grape Juices. *Innov. Food Sci. Emerg. Technol.* **2014,** *26*, 498–504.
59. Muramoto, Y.; Kimura, M.; Nouda, S. Development and Future of Ultraviolet Light-Emitting Diodes: UV-LED Will Replace the UV Lamp. *Semicond. Sci. Technol.* **2014,** *29* (8), E-article 084004.
60. Nassarawa, S. S.; Abdelshafy, A. M.; Xu, Y.; Li, L.; Luo, Z. Effect of Light-Emitting Diodes (LEDs) on the Quality of Fruits and Vegetables During Postharvest Period: A Review. *Food Bioprocess Technol.* **2020,** 1–27.
61. Neves, F. I.; Vieira, M. C.; Silva, C. L. Inactivation Kinetics of Peroxidase in Zucchini (*Cucurbita pepo* L.) by Heat and UV-C Radiation. *Innov. Food Sci. Emerg. Technol.* **2012,** *13*, 158–162.

62. Neves-Petersen, M. T.; Gajula, G. P.; Petersen, S. B. UV Light Effects on Proteins: from Photochemistry to Nanomedicine. *Mol. Photochem. Var. Asp.* **2012,** *2012*, 125–158.
63. Oteiza, J. M.; Peltzer, M.; Gannuzzi, L.; Zaritzky, N. Antimicrobial Efficacy of UV Radiation on *Escherichia Coli* O157: H7 (EDL 933) in Fruit Juices of Different Absorptivities. *J. Food Prot.* **2005,** *68* (1), 49–58.
64. Pimputkar, S.; Speck, J. S.; DenBaars, S. P.; Nakamura, S. Prospects for LED Lighting. *Nat. Photonics* **2009,** *3* (4), 180–182.
65. Popović, V.; Koutchma, T. Characterizing the Performance of a Continuous-Flow UV-LED System for Treatment of Juices and Beverages Using Multiple Wavelengths. *Food Eng. Rev.* **2020,** *2020*, 1–10.
66. Prasad, A.; Du, L.; Zubair, M.; Subedi, S.; Ullah, A.; Roopesh, M. S. Applications of Light-Emitting Diodes (LEDs) in Food Processing and Water Treatment. *Food Eng. Rev.* **2020,** *12* (3), 268–289.
67. Putnik, P.; Pavlić, B.; Šojić, B.; Zavadlav, S.; Žuntar, I.; Kao, L.; Kitonić, D.; Kovačević, D. B. Innovative Hurdle Technologies for the Preservation of Functional Fruit Juices. *Foods* **2020,** *9* (6), 699.
68. Regulations [42 FR 14635, March 15, 1977]; Us Printing Press: Washington, DC, 2000. https://www.federalregister.gov/documents/2000/11/29/00–30453/irradiation-in-the-production-processing-and-handling-of-food (accessed May 19, 2021).
69. Rhim, J. W.; Gennadios, A.; Fu, D.; Weller, C. L.; Hanna, M. A. Properties of Ultraviolet Irradiated Protein Films. *LWT—Food Sci. Technol.* **1999,** *32* (3), 129–133.
70. Sampedro, F.; Fan, X. Inactivation Kinetics and Photoreactivation of Vegetable Oxidative Enzymes After Combined UV-C and Thermal Processing. *Innov. Food Sci. Emerg. Technol.* **2014,** *23*, 107–113.
71. Scheidegger, D.; Pecora, R. P.; Radici, P. M.; Kivatinitz, S. C. Protein Oxidative Changes in Whole and Skim Milk After Ultraviolet or Fluorescent Light Exposure. *J. Dairy Sci.* **2010,** *93* (11), 5101–5109.
72. Sew, C. C.; Ghazali, H. M.; Martín-Belloso, O.; Noranizan, M. A. Effects of Combining Ultraviolet and Mild Heat Treatments on Enzymatic Activities and Total Phenolic Contents in Pineapple Juice. *Innov. Food Sci. Emerg. Technol.* **2014,** *26*, 511–516.
73. Seyderhelm, I.; Boguslawski, S.; Michaelis, G.; Knorr, D. Pressure Induced Inactivation of Selected Food Enzymes. *J. Food Sci.* **1996,** *61* (2), 308–310.
74. Singh, S. C. Basics of Light Emitting Diodes, Characterizations and Applications. *Handb. Light Emitting Schottky Diode Res.* **2009,** *2009*, 45–55.
75. Smith, W. L.; Lands, W. E. Oxygenation of Unsaturated Fatty Acids by Soybean Lipoxygenase. *J. Biol. Chem.* **1972,** *247* (4), 1038–1047.
76. Song, K.; Mohseni, M.; Taghipour, F. Application of Ultraviolet Light-Emitting Diodes (UV-LEDs) for Water Disinfection: A Review. *Water Res.* **2016,** *94*, 341–349.
77. Stranieri, S.; Ricci, E. C.; Banterle, A. Convenience Food with Environmentally-Sustainable Attributes: A Consumer Perspective. *Appetite* **2017,** *116*, 11–20.
78. Tapia, M. S.; Welti-Chanes, J. Hurdle Technology Principles Applied in Decontamination of Whole and Fresh-Cut Produce. Chapter 24; In *Decontamination of Fresh and Minimally Processed Produce*; Gomez-Lopez, V. M., Ed.; Online John Wiley & Sons, Inc.: New York, 2012, pp 417–449.
79. Tapia, M. S.; Alzamora, S. M.; Chirife, J. Effects of Water Activity (a_w) on Microbial Stability as a Hurdle in Food Preservation. Chapter 14; In *Water Activity in Foods: Fundamentals and Applications*; Barbosa-Canovas, G. V., Ed.; Online John Wiley & Sons, Inc.: New York, 2020; pp 323–355.

80. Teoh, L. S.; Lasekan, O.; Adzahan, N. M.; Hashim, N. The Effect of Ultraviolet Treatment on Enzymatic Activity and Total Phenolic Content of Minimally Processed Potato Slices. *J. Food Sci. Technol.* **2016,** *53* (7), 3035–3042.
81. Terefe, N. S.;Buckow, R.; Versteeg, C. Quality-Related Enzymes in Fruit and Vegetable Products: Effects of Novel Food Processing Technologies, Part 1: High-Pressure Processing. *Crit. Rev. Food Sci. Nutr.* **2014,** *54* (1), 24–63.
82. Tiwari, B. K.; Muthukumarappan, K.; O'donnell, C. P.; Cullen, P. J. Inactivation Kinetics of Pectin Methylesterase and Cloud Retention in Sonicated Orange Juice. *Innov. Food Sci. Emerg. Technol.* **2009,** *10* (2), 166–171.
83. Tomás-Barberán, F. A.; Espín, J. C. Phenolic Compounds and Related Enzymes as Determinants of Quality in Fruits and Vegetables. *J. Sci. Food Agric.* **2001,** *81* (9), 853–876.
84. Tran, M. T. T.; Farid, M. Ultraviolet Treatment of Orange Juice. *Innov. Food Sci. Emerg. Technol.* **2004,** *5* (4), 495–502.
85. US Food and Drug Administration. 21 CFR 179.39. Ultraviolet Radiation for the Processing and Treatment of Food. https://www.accessdata.fda.gov/scripts/cdrh/cfdocs/cfcfr/CFRSearch.cfm?fr=179.39 (accessed May 26, 2021).
86. US Food and Drug Administration. Sec. 179.41. Pulsed Light for the Treatment of Food. https://www.accessdata.fda.gov/scripts/cdrh/cfdocs/cfcfr/CFRSearch.cfm?fr=179.41 (accessed May 26, 2021).
87. Van Loey, A.; Smout, C.; Hendrickx, M. Inactivation of Enzymes. In *Handbook of Food Enzymology*; Alphons, G. J. V., Dominic W. S., Whitaker, J. R., Eds.; CRC Press: Boca Raton, Florida, 2002; pp 49–58.
88. Von Tappeiner, H. About the Effect of Photodynamic (Fluorescing) Substances on Protozoans and Enzymes. *Deutsches Archivfürklinische Medizin (Ger. Arch. Clin. Med.)* **1904,** *80*, 427–487.
89. Wang, D.; Chen, L.; Ma, Y.; Zhang, M.; Zhao, Y.; Zhao, X. Effect of UV-C Treatment on the Quality of Fresh-Cut Lotus (*Nelumbo nucifera*) Root. *Food Chem.* **2019,** *278*, 659–664.
90. Wright, A.; Hawkins, C. L.; Davies, M. J. Photo-Oxidation of Cells Generates Long-Lived Intracellular Protein Peroxides. *Free Rad. Biol. Med.* **2003,** *34* (6), 637–647.
91. Xu, F.; Cao, S.; Shi, L.; Chen, W.; Su, X.; Yang, Z. Blue Light Irradiation Affects Anthocyanin Content and Enzyme Activities Involved in Postharvest Strawberry Fruit. *J. Agric. Food Chem.* **2014,** *62* (20), 4778–4783.
92. Zhang, C.; Trierweiler, B.; Li, W.; Butz, P.; Xu, Y.; Rüfer, C. E.; Zhao, X. Comparison of Thermal, Ultraviolet-C, and High-Pressure Treatments on Quality Parameters of Watermelon Juice. *Food Chem.* **2011,** *126* (1), 254–260.
93. Demirci, A.; Krishnamurthy, K. Pulsed Ultraviolet Light. In *Nonthermal Processing Technologies for Food*; Zhang, H. Q., Barbosa-Cánovas, G. V., Balasubramaniam, V. B., Dunne, C. P., Farkas, D. F., Yuan, J. T., Eds.; Wiley-Blackwell Publishing: Oxford, 2011; pp 249–261.
94. Zhang, Y.; He, S.; Simpson, B. K. Enzymes in Food Bioprocessing: Novel Food Enzymes, Applications, and Related Techniques. *Curr. Opin. Food Sci.* **2018,** *19*, 30–35.
95. Zink, D. L. The Impact of Consumer Demands and Trends on Food Processing. *Emerg. Infect. Dis.* **1997,** *3* (4), 467–472.

CHAPTER 9

APPLICATIONS OF ULTRASOUND TECHNOLOGY FOR INACTIVATION OF FOOD ENZYMES

ANANDU CHANDRA KHANASHYAM[1], M. ANJALY SHANKER[2], PRIYAMVADA THORAKKATTU[3], T. JAYASREE JOSHI[4], RAVI PANDISELVEM[5], and ANJINEYULU KOTHAKOTA[6]

ABSTRACT

Many endogenous enzymes found in food samples degrade the quality and nutritional value of raw and processed foods. To inactivate enzymes found in food, the food industry has long relied on traditional heat treatments. However, the intensity of these heat processes with respect to time and temperature can cause negative impacts in foods, such as nutritional losses, the formation of undesirable flavors, color darkening, and functional property deterioration. Ultrasound technology has been considered as an alternative processing option to traditional thermal techniques for food preservation in recent years. The application of ultrasound to liquid medium can produce cavitation, which results in the production of shock waves and reactive oxygen species. These mechanical and chemical phenomenal changes could alter the secondary and tertiary enzyme structure, rendering enzyme inactivation. Ultrasound treatment is reported effective on inactivating enzymes from both plant and animal sources. The plant endogenous enzymes that are particularly significant for fruit and vegetable products include pectin methylesterase, polyphenol oxidase, polygalacturonase, and peroxidase. Furthermore, enzymes such as lactoperoxidase and alkaline phosphatase

Enzyme Inactivation in Food Processing: Technologies, Materials, and Applications. Megh R. Goyal, PhD, Junaid Ahmad Malik, PhD & Ravi Pandiselvam, PhD (Eds.)

are also addressed in this chapter with respect to milk and milk products. The effect and efficacy of sonication on inactivating food enzymes are also outlined in this chapter.

9.1 INTRODUCTION

Enzymes are group of proteins that catalyze biochemical reactions. The action of certain enzymes on food matrix could lead to loss of quality and nutritional parameters in processed foods and leading to food wastage and economic losses. Nowadays, consumer awareness toward healthy and minimally processed foods are gaining more momentum, reducing the demand for traditional thermally processed food products. This led the industry to adapt to alternative nonthermal processing methods such as ultrasound technology.

Ultrasound is a form of sound (pressure) waves of high frequency that are difficult to be detected by the human ears. The frequency of the ultrasound range above 20 kHz, and this ultrasonic spectrum can be broken down into two distinct zones. Power ultrasound that has a frequency range from 20 kHz to 1 MHz and diagnostic ultrasound that have a frequency range above 1 MHz. Diagnostic ultrasonography, which has a frequency of over 1 MHz, is mostly utilized for medical and industrial imaging.

Today ultrasound in combination with pressure (manosonication), heat (thermosonication), or both (manothermosonication (MTS)) is used for various applications in food (such as microbial inactivation, pesticide degradation, desensitization of food allergens, enzyme inactivation, toxin degradation, meat tenderization, homogenization, etc.), even though there are wide use for ultrasound in food, this chapter focuses on power ultrasound and its application in inactivation of food enzymes.

9.2 ENZYME INACTIVATION: MECHANISM AND KINETICS

9.2.1 INACTIVATION MECHANISM

As ultrasound waves need a medium to propagate, region of alternating high and low pressures is generated in the medium during their propagation. When the medium of propagation is compressible in nature, such as air, the medium will be capable of accommodating the pressure variations. But, however in the case of inelastic medium likes most liquids, the medium literally gets torn apart due to the high-pressure variations. This will lead to the formation microbubbles.[60] Moreover, liquids already include cavities of gas or nanobubbles, which act as nuclei and aid in the production of microbubbles.[117]

The cavitation or blake threshold is the lowest acoustic pressure at which it is seen, and it is a function of the solution nanobubble radius, vapor pressure, surface tension, and system pressure. This gas bubble will expand from a small gas pocket in the rarefraction phase then grow until it hits the compression phase, creating bubble oscillation. There are two kinds of acoustic cavitation: inertial and noninertial. The bubbles oscillate regularly in a low amplitude sound field in noninertial cavitation. Inertial cavitation, on the other hand, causes the bubbles to rapidly expand until it crosses the threshold energy needed to maintain the bubble's vapor pressure, causing it to become unstable and collapse. This causes the gas bubbles to develop and burst repeatedly, resulting in the formation of shock waves which create regions with high temperature (5500°C) and pressure (50,000 KPa).[78]

Furthermore, cavitation can result in the development of high-energy shear waves, which can generate turbulence. High pressure, turbulence, and temperature can cause water molecules to decompose, resulting in the creation of free radicals (Fig. 9.1) ($H_2O \rightarrow H + OH^*$). These chemical (free radical production) and physical (temperature and pressure) changes caused by cavitation can interact with food enzymes, causing structural breakdown, and eventually inactivation.[82]

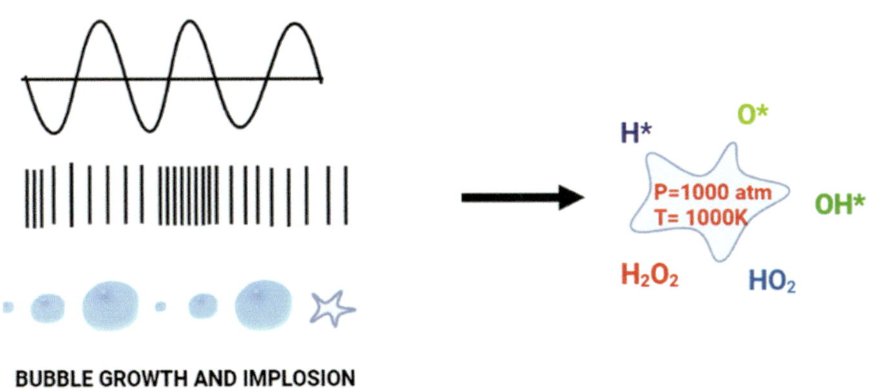

FIGURE 9.1 Cavitation and free radical formation.

Enzymes are indeed a group of proteins, which serve as a biocatalyst. Enzyme activity has a significant impact on the final quality of food products. The presence of endogenous enzymes in food products can impair color, flavor, and nutritional value. Enzymes are easily denatured by minor changes in environmental parameters such as pressure, temperature, pH, etc. Power ultrasound, either alone or in conjunction with other thermal and nonthermal approaches, has already been shown to be effective in inactivating food enzymes.[48] The chemical and physical effects of cavitation are primarily responsible for the enzyme inactivation, which includes:

- The high micro-zone temperature and pressure created during cavitation.
- The formation of free radicals.
- The mechanical and shear stress resulting from micro-jets, microstreaming, and shock waves.

These extreme conditions resulting from cavitation could break down the weak intermolecular forces such as VanDer Waals forces and hydrogen bonds in the peptide chain resulting in an irreversible change to the tertiary and secondary protein conformation. The three-dimensional structure of the enzyme active sites will be further modified by changes in tertiary protein structure, altering the enzyme–substrate interaction leading to loss of enzymes biological activity.[48] The interaction of free radicals with amino-acids can also cause an enzyme's biological activity to be lost, compromising its stability, catalytic activity, and substrate binding capacity. The enzyme inactivation mechanism of ultrasound is shown in Figure 9.2.

The spatial arrangement and amino acid content of the targeting enzyme, ultrasonic power and frequency, pH, time, etc. plays a significant role in enzyme inactivation by ultrasound. The factors affecting the enzyme inactivation by ultrasound are discussed in detail in the sections below.

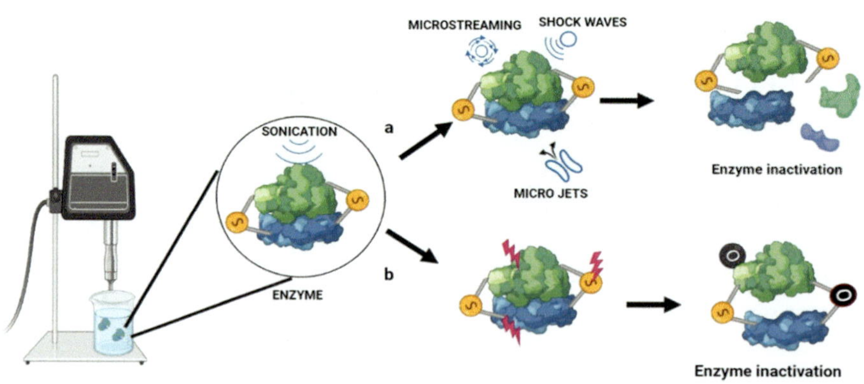

FIGURE 9.2 Enzyme inactivation by ultrasound through mechanisms such as physical (a) and chemical (b).

The hydrophobic interaction and the hydrogen bonds are widely known to play a crucial role in protein stability, when water molecules come into contact with hydrophobic protein areas, the protein forms hydrogen-bonded pentagonal structures.[16] Any factor that causes damage to this molecular structure could affect protein stability. Cavitation bubbles will expose the protein molecules to the water–air interface, which can likely cause the hydrophobic interaction to weaken and the protein to denature. Depending on their nature, different types of enzymes, such as mono-, dimeric, and polymeric enzymes, have different ultrasonic inactivation mechanisms. When monomeric enzymes are denatured, they either produce aggregates or fragments of the enzyme molecule, whereas, polymeric enzymes typically fragment into monomeric molecules after sonication. Breaking polymeric globular proteins into subunits, de-lipidation or removal of lipids from lipoproteins, dissociation or splitting polypeptides into fragments, and oxidation of aromatic residues are some of the impacts of sonication to enzymes.[7]

Trypsin inactivation by ultrasound has been linked to two separate mechanisms: alternations to molecular confirmations and damage to molecular structures.[95] The generation of a huge interfacial area between water and air during cavitation caused by ultrasound, will disrupts hydrogen bonds and hydrophobic interactions in the molecule resulting in structural alternations.[95] The free radicals generated during cavitation are also responsible for trypsin's alteration and damage. When free radicals react with trypsin, the charge distribution on the protein surface is altered, resulting in damage to the enzyme active sites.

Huang et al.[46] studied the inhibitory action of ultrasound on trypsin inhibitors and reported conformational changes after sonication. The disulfide bond of Kunitz trypsin inhibitor was affected by free radicals produced by ultrasound, resulting in the formation of a terminal thiol group. This resulted in a change in enzyme structure and, eventually led to inactivation. Whereas, in the case of peroxidase (PD) inactivation, the mechanism was suggested due to the splitting of prosthetic heme group from the enzyme.[62]

9.2.2 INACTIVATION MECHANISM AND KINETICS

Mathematical models and kinetic parameters for enzyme inactivation are widely used to evaluate the efficacy of ultrasound, and combined ultrasound treatments such as thermosonication, and MTS processes, for developing novel procedures and improving food quality.[66] Most of these calculations are based on enzyme residual activity (R_A), which is calculated by the following equation:

$$R_A = \frac{A}{Ao} \tag{9.1}$$

where, R_A = Residual enzyme activity; A = Final enzyme activity; Ao = Initial enzyme activity.

Although most of the enzymes follow a linear trend when plotted against natural logarithm of R_A versus time, yet few enzymes follow a nonlinear trend. For those enzymes that follow a linear trend, first-order kinetics model is widely used to describe its inaction kinetics. This equation is applicable for most of the pressure and thermal-assisted enzyme inactivation by ultrasound. For ultrasound of a specific energy, temperature or pressure the inactivation kinetics can be explained by using the following model.

$$\ln \frac{A}{Ao} = K_T t \tag{9.2}$$

Where, A = Final enzyme activity after time t; Ao = Initial enzyme activity; K_T = Inactivation rate; t = Time.

The first-order enzyme inactivation kinetics can also be explained by using decimal reduction time (D_T) with the help of Bigelow model as follows:

$$\log \frac{A}{Ao} = -t / D_T \tag{9.3}$$

For enzymes that follow a nonlinear fashion, a first-order biphasic model can be used which can be represented as follows:

$$A = A_s \exp\left(-K_s t\right) + A_L \exp\left(-K_L t\right) \tag{9.4}$$

Where, A_S = Enzyme activity stable fraction; A_L = Enzyme activity liable fraction; k_S = Inactivation rate constants of the stable fraction; K_L = Inactivation rate constants of the labile fraction.

9.3 APPLICATION OF ULTRASOUND FOR INACTIVATION OF FOOD ENZYMES

9.3.1 ENZYME INACTIVATION IN FRUITS AND FRUIT JUICES

Likelihood of applications of emerging technologies is allied to the quality retention properties that they hold in comparison with the conventional techniques. Among the food strata, the application of conventional techniques was most erratic in quality properties due to the tissue structure possessed. Pioneering ultrasound (US) for the control of enzymes in fruits is a viable choice, which helps in plummeting the quality glitches to a greater extent. Being rich in enzymes, studies on ultrasound treatment was a matter of importance and has gained momentum during the recent years. Chen et al.[23] studied the outcome of ultrasound in regulating the pericarp browning that is allied with the activity of polyphenol oxidase (PPO) and PD enzyme in litchi fruit. The treatment was observed to be operative in controlling the activity of PPO and PD in the early storage period itself which can be correlated with the browning inhibition during later storage stages. Adding up to the inhibitory activities, it was observed that the treatment also constructively shaped the anthocyanin retention, which shows the maintenance of quality matrix conversed.

Suitability of US treatment as a substitute to thermal and other treatments were evaluated in cantaloupe melon juice with emphasis given on quality parameters and enzymes like PPO, POD, and ascorbate PD.[36] Treatment intensity around 376 W/cm^2 for a time period of 10 min was operative in overall inactivation of ascorbate PD and noteworthy activity diminutions in PPO and POD while upholding the color and homogeneity of the juice during 6 weeks of low-temperature storage. Analogous to these results, effectiveness of the equivalent treatment conditions was observed in PPO inactivation of pineapple juice.[30] Higher intensity and long exposure timings (376 W/cm^2 for 10 min) induced 20% experiential reduction in PPO activity. Enhanced color properties and cloud stability of juice samples along 42 days of storage sums up the quality retention complementing the inactivation kinetics. Enzyme inactivation using US treatment is mainly correlated to attributes including cavitation, mechanical effects, micro-mechanical shocks, free radical formations, etc.[48]

TABLE 9.1 Scope of Ultrasound on Enzyme Inactivation in Fruit Juices.

Product	Enzyme	Treatment parameters	Remarks	Reference
Apple juice	Polyphenol oxidase, peroxidase and pectin methylesterase	Thermosonication via ultrasound in-bath (25 kHz, 0.06 W cm^{-3}) for 30 min and ultrasound with-probe sonicator (0.3 Wcm^{-3}, 20 kHz) for 5 and 10 min at 20°C, 40°C and 60°C	Lowest residual activity was observed with probe sonicator at 60°C and was around 6.15%, 9.0% and 7.10% for PPO, peroxidase and the latter one	[1]

TABLE 9.1 *(Continued)*

Product	Enzyme	Treatment parameters	Remarks	Reference
Bayberry juice	Polyphenol oxidase and peroxidase	Sonication treatment at intensity levels 452, 362, 271, 181 and 90 W/cm^2 for time period of 1–12 min at frequency of 20 kHz.	Minimum residual activity was around 9.49% for POD at 452 W/cm^2 whereas PPO was completely inactivated at same intensity.	[20]
Blackberry juice	Pectin methylesterase and polyphenol oxidase	Treatment at 1500 W at amplitude level-28 µm with 2 s on and 4 s off pulse durations at a constant frequency of 20 kHz an	Percentage residual activity of pectin methylesterase and poly phenol oxidase was around 53.5 and 1.5%	[21]
Melon juice	Peroxidase, polyphenol oxidase and ascorbate peroxidase	Sonication treatment at power levels of 100, 300 and 500 W and ultrasound intensities of 75, 226 and 376 W/cm^2 at frequency of 19 KHz	Residual activity got reduced to 23.29% for PD, 70.48% for PPO by treatment at 376 W/cm^2 for 10 min and complete inactivation for the latter enzyme	[36]
Pear juice	Peroxidase, Polyphenol oxidase and pectin methylesterase	Ultrasound treatment at 65°C, 45°C and 25°C for 10 min at frequency 20 kHz and amplitude 70%	Highest enzyme inactivation was at treatment at 65°C with residual activities of 4.3%, 3.25% and 1.91% respectively.	[83]
Pineapple juice	Polyphenol oxidase	Sonication treatment at power levels of 100, 300 and 500 W and intensities of 75, 226 and 376 W/cm^2	376 W/cm^2 treatment for time period of 10 min reduced enzyme activity by 20%	[30]
Quince juice	Polyphenol oxidase	Ultrasonic treatments at different times-5–30 min at intensity 200 W and ultrasound intensities-100–400 W for 20 min	Activity was lowered to 35% at treatment intensity 400 W for 20 min. 30 min treatment at 200 W reduced activity up to 38%.	[47]

TABLE 9.1 *(Continued)*

Product	Enzyme	Treatment parameters	Remarks	Reference
Sour soup nectar	Polyphenol oxidase	Sonication at 24 kHz, at temperature-34°C, 44°C and 54°C, and for time periods-2, 6 and 10 min	Initial PPO activity was reduced by 34–67% by temperature between 34–44°C and 91–99% decrease was attained by treatment at 54°C.	[6]
Sugarcane juice	Peroxidase	US treatment at 20 kHz frequency, 50 W cm^{-2}, amplitude 140 µm at 50, 60, 70, 75 and 80 °C.	High inactivation rates were attained at 70°C and 75°C.	[32]

Cavitation being a primary factor, emancipates immense extent of energy with upsurge in temperature and pressure. This heat proliferation was found to have an unambiguous impact on the inactivation kinetics of enzymes. Cooling with incidence of ice bath all through US treatment was performed to examine the discrete validities of cavitation and heat on inactivation in bay berry juice.[20]

Complete inactivation of PPO was observed at treatment intensity of 452 W/cm^2, whereas a residual activity of 3.28% was observed in case of US treatment with associated cooling. Time entailed with 90% reduction of initial enzyme activity indicated as D value, was 41.44–2.59 min in ultrasound and 55.92–7.05 min in US treatment assisted with cooling. The disparity in the kinetics can be associated with the synergistic effect in ultrasound compared with the latter. Analogous outcomes were noted in case of PD enzyme where the residual activity of enzyme is lower in US treatment than with cooling.

With PD enzyme owning lesser kinetic parameters than PPO, Cao et al.[20] proposed the likelihood of using PD as an index enzyme for ultrasonication treatment. Author also infers the impracticability of thermal treatment over sonication for limiting the activity of both the enzymes. In line with reading matter, the competence of inactivation was found to have an explicit connection with numerous treatment aspects like product parameters (e.g., pH, antioxidants present, composition, etc.), process parameters (e.g., treatment period, intensity, etc.) and mechanism of inactivation.[22] Authority of fruit type/cultivar on PPO enzyme inactivation was reviewed by Sulaiman et al.[91] on fruit samples like pear, apple, and strawberry. Sonication treatment at 32°C with divergent conditions was effectual in demoting the PPO activity, having pear PPO being utmost resilient followed by apple and strawberry. Hence in certain cases accomplishment of anticipated benchmarks of inactivation kinetics comprises the involvement of extended contact timings and higher parameter values. This description that is comparable with the microbial stability of food stuffs could affect the quality and nutritional matrix of food stuffs making it undesirable.

Proposal of combination technologies gains magnitude in this scenario, where the efficacy is improved by combining the technique with other existing technologies like combination with heat and pressure, combination with other technologies, etc. The degree of effectiveness diverges according to aspects like food material under consideration, different processing parameters, etc. and can be understood only by detailed literature studies. Amalgamation of ultrasound and heat as discretion to thermal treatments employed by and large named as thermosonication, is reviewed for its efficacy in a variety of fruit juice samples.

Saeeduddin et al.[83] premeditated the impact of ultrasound pasteurization on the inactivation of pear juice enzymes. Temperature dependency of the ultrasound treatment with definite process parameters (frequency 20 kHz, amplitude 70% and time 10 min at temperatures 25, 45, 65 °C) was amply evident in the residual activity of enzymes. The unveiled residual activities at 65 °C of PD, PPO, and pectin methylesterase enzyme was 4.3%, 3.25%, and 1.91%, and that of treatment at 45 °C was 43.2%, 37.83% and 40.22% which clearly shows the superiority of temperature adjustments. Comparable inactivation profile was observed in case of thermo ultrasound treatment in bay berry juice.[21] Treatment was successfully in inactivating pectin methylesterase and PPO by 63.71% and 98.28% with maintaining the quality attributes of the juice sample.

Influence of temperature profile on inactivation was recorded in sour soup nectar by Anaya-Esparza et al.[6] at three different temperature range and three different time periods. Findings illustrate that thermo-ultrasound at 34 °C and 44 °C reduced the initial activity of PPO by 34–67% whereas treatments at higher temperature (54 °C) lessened the activity between 91% and 99%. Here the efficacy of inactivation can be ascribed to both the parameters, the fragmentation and impairment of protein structure by the pressure gradient engendered by cavitation forces and to the protein denaturation caused by the temperature divergences. In a similar way, the combination of ascorbic acid and US was found to develop a synergistic upshot on the activity of PPO in apple.[53]

MTS is a combination technique merging pressure and temperature treatments with sonication procedure, where moderate pressures are employed to permit cavitation at temperatures bordering on the boiling point.[101] Literature studies have been documented showing the positive effects of MTS on controlling the enzyme populations in orange and lemon juice. Kuldiloke[58] reported the possibility of just about total inactivation of one of the heat stable enzymes pectinesterase in lemon and other citrus products. Study states that treatment at 70 °C, 300 kPa, and 2 min was effective in achieving 94% inactivation and a small change in treatment conditions (80 °C, 200 kPa, 5 min) yielded a jump of 96% in inactivation of pectinesterase enzyme. Whether ultrasound is applied as stand-alone technology or in combinations, the efficacy of the technique over conventional inactivation methods like thermal treatments is assured and the extension of ground-level application of idem is allied with the comprehensiveness of literature findings engrossing diverse fruit cultivars and process parameters.

9.3.2 ENZYME INACTIVATION IN VEGETABLES

Vegetables are vital fragment of day-to-day food matrix with prolific quantities of vitamins, dietary fibers, minerals, and other nutritious components. The multifarious structure and components make it vulnerable to different mechanical damages and thereby concerning the postharvest quality of vegetables. Anomalous enzyme activity also has an imperative hand on metabolic disorders associated with postharvest quality. This elucidates the magnitude of enzyme inactivation in vegetables, implicating both thermal and nonthermal methods. Ultrasound is an imperative procedure amongst nonthermal methods, which have been a subject matter for exploration and advancement for several years.[54] Continually the application of ultrasound in vegetables rather earned weightage over relentless studies and progressions.

Impact of ultrasound on peroxidase (POD) inactivation in tomato was studied by Ercan and Soysal[34] under five different ultrasonic powers. Compared with thermal treatment, the application of ultrasound caused inactivation of POD at short time intervals. Total inactivation was detected at 75% power for a period of 90 s and 50% power for 150 s, which plainly portrays the authority of ultrasound power on inactivation rate.

Practicality of ultrasound for carrot blanching with emphasis on PD and pectin methylesterase inactivation was researched by Gamboa-Santos et al.[38] and was compared with conventional blanching treatment. Highest inactivation rate was around 90% for PD and 50% for pectin methylesterase by means of US treatment. Author elucidates the likelihood of employing the US treatments for fractional inactivation of enzyme and can be a well-thought-out alternative to conventional blanching treatments. Applicability matrix of US treatment is not only curbed to fresh produce or core product but also to their by-products and processed forms. Analogous research encompasses US for the enzyme stability of green coconut water at treatment conditions of 286 W/L, 20 kHz for 0–30 min.[82] PD inactivation of about 27% was attained after 30 min of exposure and was operative in inactivation of enzyme. On gauging the worth of sonication as a pretreatment for thermal technique, PD enzyme post US treatment displays lower thermal resistance consenting the application of lower processing temperature and time combinations.

TABLE 9.2 Scope of Ultrasound Treatment on Enzyme Inactivation in Vegetables.

Product	Enzyme	Treatment parameters	Remarks	Reference
Carrot	Peroxidase	Thermosonication treatments at 7 frequency modes at 60°C for 12 min: mono-frequency (40, 33 and 22 kHz); dual-frequency (22/40, 33/40 and 22/33 kHz); and tri-frequency (22/33/40 kHz)	Residual activity was reduced to 22% which was 65% inferior than thermally treated samples	[42]
Carrot juice	Polyphenolase, peroxidase, pectin methylesterase and lipoxygenase	Thermosonication at 70% amplitude, 20 kHz, and 48 W cm^{-2} for 5–10 min at 20 °C, 40°C and 60 °C with pulse cycle-5 s on/off.	Thermosonication at 60°C for 10 min was effective among the treatments	[49]
Coconut water	Peroxidase	Treatment conditions of 286 W/L, 20 kHz for 0–30 min	Peroxidase inactivation of about 27% was attained after 30 min of exposure.	[82]

TABLE 9.2 *(Continued)*

Product	Enzyme	Treatment parameters	Remarks	Reference
Fresh cut potato	Polyphenol Oxidase	Ultrasound treatment conditions-200 W, 40 kHz, time periods-1, 5 and 10 min	PPO enzyme inactivation of about 50% was achieved by 5 min of sonication	[5]
Horse radish	Peroxidase	Treatments at 1175, 1144, 995, 862, 583, 378 & 20 kHz for 60 min using power points between 2.1 and 64 W.	Complete inactivation was achievable at 378 and 583 KHz.	[9]
Pakchoi (Chinese cabbage)	Poly phenol oxidase and peroxidase	Ultrasound treatment at 30 kHz, for 5, 10 and 15 min followed by modified atmospheric packaging with 5% O_2 + 10% CO_2 + 85% N_2	Ultrasound treatment for 10 min followed by modified atmosphere packaging was effective in reducing the activity of both enzymes	[116]
Tomato	Polygalacturonase and pectin methylesterase	Thermo ultrasound treatment at 65 μm, 20 kHz, and temperatures 50–75 °C.	Enhanced inactivation rates; increased by 1.5–6 times for pectin methylesterase and by 2.3–4 for polygalacturonase at 60–75°C	[94]
Tomato	Peroxidase	Ultrasound power at 15, 25, 40, 50, 75% for 20–150 s	Total inactivation was detected at 75% ultrasound power for 90 s and 50% power for 150 s	[34]
Tomato	Pectin methylesterase	US treatment at intensities in the series of 0.004–0.020 mg L^{-1} min^{-1} and at 50 °C, 61 °C, and 72 °C	Ultrasound resulted in variation of D-values between 24.0 and 240.6 min Reduction of D value up to 0.3 min achieved after thermosonication at 0.008 mg L^{-1} min^{-1} at 72 °C.	[80]

TABLE 9.2 *(Continued)*

Product	Enzyme	Treatment parameters	Remarks	Reference
Tomato	Polygalacturonase and pectin methylesterase	Manothermosonication treatments at 117 μm amplitude, 2 kg pressure, 20 kHz and 70 °C for 1 min.	Untraceable levels of pectin methylesterase and inactivated 62% of the latter enzyme after MTS treatment	[101]
Tomato juice	Polygalacturonase and pectin methylesterase	Thermosonication treatments at 25, 50 and 75 μm, 24 kHz and at temperatures 60 °C, 65 °C and 70 °C	Reduction up to 90% of initial activity of enzyme was achieved. Beneficial inactivation kinetics was evident at 60 °C and 65 °C, but was not prominent at 70 °C.	[107]

Sonication as a stand-alone technology and as a combination or a pretreatment technique in refining the conventional techniques safeguarding quality is of greater significance and has better application potentials. Parallel to the inactivation kinetics of whole vegetable, literature studies also depicted the efficacy of sonication in fresh-cut whole potato[5] and fresh-cut sweet potato[75] augmenting an added approach of treatment. Thermo ultrasound is a reliable substitute to conventional thermal treatments for enzyme inactivation with retention of quality attributes of vegetable products. Thermo ultrasound treatment was operative in upholding the rheological properties of tomato products along with high enzyme inactivation rate.[94]

Complete inactivation of pectin methylesterase was achievable at 75 °C in tomato juice under US treatment (20 kHz, 75 μm) and 72% deactivation of polygalacturonase was achievable after 4 min exposure. Enhanced inactivation rates 1.5–6 times for the former and 2.3–4 times for the latter was pragmatic in assessment with the thermal treatments. Impression of contemplating thermo sonication as an alternative for the thermal processes which safeguards the quality characteristics of produce was proposed in several literature evaluations.[49,116]

Analogous to routine acclimatization allied with the parameters of treatment and product under consideration, here also various combinations of product and process parameters should be considered when aiming for higher efficacy. Upper hand of thermo sonicated produce over US as a standalone technique and conventional treatments[42] defends the likelihood of augmented investigations in this trail. MTS is also found to be an operative method in monitoring the enzyme activity of vegetable produce. Undetectable levels of pectin methylesterase in MTS-treated tomato juice contrary to 38% activity displayed after thermal treatments shows the clear hand of MTS over other techniques.[101]

Outcomes also illustrate inactivation 62% of total polygalacturonase activity over thermal treatments in tomato juice. Usage of ultrasound as a pretreatment technique for fresh produce prior to storage and other processing methods is still in the expansion phase. Usage of ultrasound treatment (30 kHz, for 5, 10, and 15 min) in combination with modified atmospheric packaging was found to be operational in extending the freshness of pakchoi. Among the treatment combinations, US treatment for 10 min followed by MAP was successful in lessening the activity of poly phenoloxidase and PD activity.[116] Application of US prior to dehydration of carrot was operative in reducing the activity by 90% for PPO and 50% for pectin methylesterase after 15 min of treatment give wider range entails optimization of process and product parameters to develop more competent treatments and superior quality produce.

9.3.3 INACTIVATION OF ENZYMES IN OTHER FOOD PRODUCTS

With accumulative apprehensions and modifications in consumption pattern of consumers, the need and upkeep of a healthy diet that embraces major food groups like lean proteins, whole grains, healthy fats, fruits and vegetables, milk and dairy foods, etc. is gaining momentum. Prominence of the consideration and progress of novel approaches of preservation of every food groups. Application of ultrasound on administering and preservation of different food groups were reviewed and premeditated under different prerequisites. Food systems trails a distinct inactivation pattern due to the presence of divergent components. These patterns should be clearly assessed to safeguard the efficacy of the process.

Apart from divergent components of the food system, they are also reliant on the process parameters like amplitude, time duration, etc. One typical example for the same is the inactivation kinetics followed by enzymes in milk where there is inactivation dependency in parameters including amplitude,

TABLE 9.3 Efficacy of Ultrasound in Controlling the Enzymes in Divergent Food Matrixes.

Product	Enzyme	Treatment parameters	Remarks	Reference
Barley	Alpha amylase	Sonication treatment at temperatures 30, 50 and 70 °C for 5–15 min at intensities around 20, 60 and 100% of total power of device.	Temperature dependency of the treatment was clearly distinguishable with low rate of inactivation at lower temperatures	[110]
Chicken meat	Proteases: Calpains and cathepsins	Ultrasound and LTST treatment for 15 min at 0.2 W/cm^2 and 40 kHz at 55 °C	Calpains and cathepsins retained 50% and 17.6% of the total activity after 2 min of treatment.	[61]
Egg	Lysozyme	Manothermosonication at 117 μm, 200 kPa, 60–80 °C	Helped in 10 fold decrease in total lysozyme activity	[65]
Garlic slices	Polyphenol oxidase	Ultrasound for 40 min, at 40 kHz, 600 W power, at 30 °C and pulsating period of 10 s followed by osmo dehydration and vacuum pressure at 100 mbar	Combination treatment exhibited a residual activity of 13.21%	[4]
Milk	Lactoperoxidase and alkaline phosphate	Sonication at amplitude levels 0, 20, 40, 60, and 80% at temperatures 20–40 °C for timings around 30, 60, 90, and 120 s	Treatment at 80% amplitude and 40 °C exhibited maximum inactivation rate of both enzymes	[35]
Mushroom	Polyphenol oxidase	Thermosonication treatments 25 kHz and 50% power at temperatures between 55 and 75 °C	Complete inactivation at 75 °C. Residual activity of 1% was observed at treatment at 60 °C	[26]
Skim Milk	Alkaline phosphate, γ-glutamyltranspeptidase and lactoperoxidase	Sonication and thermosonication treatment at amplitude level of 120 μm and frequency 20 kHz at temperature between 55–75 °C	Thermosonication at temperature around 75 °C leads to total inactivation of former two enzymes and a residual activity of 47.2% for the latter	[102]

TABLE 9.3 *(Continued)*

Product	Enzyme	Treatment parameters	Remarks	Reference
Wheat bran	Polyphenol oxidase	High intensity ultrasound treatments for 5,10, 15 min at 24 kHz, amplitude levels of 60%. 80%, and 100%	Effective in controlling the PPO by inactivating 93% of enzyme	[43]
Whole Milk	Alkaline phosphate, γ-glutamyltranspeptidase and lactoperoxidase	Sonication and thermosonication treatment at amplitude level of 120 μm and frequency 20 kHz at temperature between 55–75 °C	Thermosonication at 61 °C leads to complete inactivation of former two enzymes and a residual activity of 30.8% for the latter	[102]
Whole wheat flour	Polyphenol oxidase	Ultrasound treatment at 750 W power, 20 KHz using a probe setup followed by chemical treatment with ascorbic acid	Combination treatment brought down the activity below 20%.	[49]

time, and temperature.[35] Highest inactivation rate of lactoperoxidase and alkaline phosphate (AP) was observed at 80% amplitude level and 40 °C, whereas the lowest rate is at amplitude level of 20% at temperature 20 °C.

The effect of ultrasound on inactivation kinetics of alpha-amylase during barley germination was evaluated by Yaldagard et al.[110] Temperature dependency of inactivation kinetics was evident, which exhibited low rate of inactivation at lower temperatures. Effectiveness of ultrasound in controlling the activity of enzymes in grains are also reported. Habuš et al.[43] illustrates the efficacy of treatment on controlling the PPO by inactivating 93% of enzyme at the end of treatment. Time period and power dependency of the inactivation pattern of PPO in wheat bran is in consent with the prior discussions. Utility of heat and ultrasound combinations in regulating the enzyme fractions is something that does not entails any substantiation and was considerably described in fresh produce. Thermosonication treatments 25 kHz and 50% power at temperatures oscillating between 55 °C and 75 °C in mushrooms showed a residual activity of 1.0% for treatment at 60 °C.[25]

With increase in temperature there is an apparent reduction in the residual activity of PPO in these mushrooms. Complete inactivation of PPO was attained at 75 °C with no predominant influence on the quality parameters. Potential of amalgamation of heat (55–75 °C) and ultrasound (20 kHz frequency, 120 μm amplitude) in food produce has been stated in the inactivation of enzymes in milk sample.[102] Complete inactivation of AP and γ-glutamyl transpeptidase were displayed at 75 °C in skim milk and 61 °C in whole milk whereas lactoperoxidase exhibited a minimum residual activity of 30.8% and 47.2% at given temperatures in whole and skim milk.

Divergence of food components didn't act as limiting factor in the temperature dependency and efficacy of treatment. Superiority of thermosonication over ultrasound as an individual technology is clear and evident with the results amalgamated from these literature studies. Upper hand inactivation kinetics of proteases enzyme was exhibited by collective outcome of ultrasound and low-temperature short time heating on chicken meat.[61] Cathepsins and calpains that are the major proteases retained 50% and 17.6% of the total activity after 2 min of treatment. Analogs to the upper hand of heat and ultrasound combinations, the efficacy of combined pressure, temperature, and ultrasound technique; manothermosonication is well proven and is gaining momentum with creditable outcomes. The enhancement of intensity of ultrasound treatment when applied under moderate pressure frames the drive of higher effectiveness of the process. Confrontation of enzymes in milk to manothermosonication and heat treatment reported advantage of former overheat treatment with reduction in protease activity by 6% and lipase activity by 7%.[100] Manothermosonication has also being found effective in inactivation of egg lysozyme by 10-fold within a period of 3.5 min.[65] Besides information on usefulness of heat and pressure treatments with ultrasound, there is only limited resources that validates the competence of combination of ultrasound with other chemical and technologies of preservation.

Niu et al.[49] reviewed such a possibility by combining the efficacy of ultrasound and ascorbic acid in controlling the activity of PPO in wheat flour and enzymatic darkening in whole wheat noodles. Combinations of ultrasound and ascorbic acid was effective in plummeting the activity of PPO below 20% whereas individual treatment was only competent to devise a reduction of about 40–45% within same treatment period. Comparable evaluations were found in treatment, which involves additive influence of vacuum pressure and osmosonication as a pretreatment in maintaining the quality of garlic slices.[4] Comparisons between ultrasound as an individual treatment and combination of ultrasound and osmotic dehydration and influence of vacuum pressure andosmosonication in totality was made in controlling the PPO activity, showing a residual activity of 30.16%, 19.64%, and 13.21% at the end of treatment.

The least activity of PPO enzyme after combination treatment of vacuum pressure and osmosonication can be attributed to the harmonious effect of vacuum pressure, osmotic pressure, and cavitation procedure on the normal activity of enzyme. Change in conditions ascribes the degradation of structure of enzyme and thereby controlling the activity. As this current approach lack detailed reviews and studies, general implications on the effectiveness of such combinations are very difficult to gather.

9.4 FACTORS AFFECTING PROCESSING EFFICIENCY OF ULTRASOUND

Ultrasound treatment can inactivate enzymes. Although, enzymes are susceptible to heat or thermal treatment, the exposure of food for longer holding times at high temperatures can adversely affect food quality. Recently, ultrasound has been gaining interest for enzyme inactivation. Sonication, thermosonication (ultrasound with heat), manosonication (ultrasound with pressure), and MTS (ultrasound with pressure and heat) at various frequencies and power can all be used to inactivate enzymes.[72] The reaction of an enzyme to ultrasound treatment may be influenced by physiological characteristics of the enzyme as well as operational parameters such as input power or intensity of ultrasonic irradiation, ultrasonic frequency, duty cycle, and other factors. Horn-type and bath-type systems are the most common experimental designs for creating acoustic cavitation. When an ultrasonic horn is submerged in liquid, an acoustic wave significantly shorter than the acoustic wavelength is emitted from the horn tip. An ultrasonic transducer is mounted to the exterior surface of a liquid tank for bath-type.

Özbek and Ülgen[74] carried out a study on six different enzymes under sonication at low frequency and observed that operational parameters like acoustic power, time of exposure, wave duty cycle, processing time, and viscosity of the enzyme solution had a significant effect on enzyme stability. These factors that affect the processing efficiency of ultrasound are discussed in detail.

9.4.1 ULTRASONIC POWER

It has long been known that increasing ultrasonic power leads to a higher concentration of cavitation bubbles with a bigger average bubble size. Remarkably, high-intensity sonication has the potential to harm enzymes. The thermal, mechanical, and chemical impacts generated by the cavitation and acoustic streaming caused by the ultrasound shock waves are sufficient to inactivate enzymes. When the power of ultrasound irradiation increased from 100 to 500 W, the biological activity of trypsin reduced, indicating that trypsin was inactivated and denatured.[92] Similarly, when the acoustic power was increased from 20 to 100 W, the lactase enzyme activity reduced from 35.9% to 21% (~41.5% decrease).[86] Interestingly, ultrasound can both activate and inactivate enzymes.

According to several investigations, enzyme inactivation happens when the optimal intensity is exceeded. For example, while using a low-intensity ultrasound of 17.33 W, cellulase activity increased by 23.5%, whereas when using a high-intensity ultrasound of 43.4 W, cellulase activity reduced by 25%.[93] Tian et al.[95] found that when the ultrasonic intensity surpassed the optimal intensity, with an increase in ultrasonic intensity, enzyme activity decreased steadily. This could be attributed to increase in cavitation effects by high-intensity ultrasound, which generates shear in the medium, causing damage to polypeptide chains and consequently causing deactivation of the enzyme.[85]

Jadhav and Gogate[50] noted that enzyme activity grew constantly until it reached an optimum intensity of 12.22 W/cm^2, after which it decreased as the intensity was increased further. They also observed a total loss of enzyme activity at 16.30 W/cm^2. Similarly, Ma et al.[63] disclosed that the highest Alcalase activity was found at an ultrasonic intensity of 80 W, resulting in a 4.9% increase in enzyme activity. However, the Alcalase activity decreased gradually when the ultrasonic intensity exceeded 80 W. Wang et al.[106] observed that ultrasonic intensity is an important factor for cellulase activity. The free cellulase activity rose as the ultrasonic intensity was increased up to 15 W, but as the ultrasonic intensity was increased to 35 and 40 W, the cellulase activity dropped. Cho et al.[27] found that as the power increased from 50 to 200 W, the effective diameters of the bubbles increased.

9.4.2 *DUTY CYCLE*

Along with the ultrasonic intensity, the duty cycle is a crucial operating parameter. The duty cycle is an important factor in determining how long the enzyme is exposed to ultrasound. The duty cycle of an enzyme refers to how often it is sonicated. Continuous sonication should, in theory, result in faster intensification. Continuous and sustained irradiation at a greater duty cycle, on the other hand, induces increased heat generation around the enzyme solution, which could contribute to enzyme breakdown.[50] With increasing duty cycle, the rate of reaction is reduced due to enzyme inactivation caused by ultrasonic irradiation. The authors also stated that enzyme activity decreased ~49.5% and hydrolysis degree decreased ~20.7% when duty cycle was changed from 10% to 90%.[103]

9.4.3 *FREQUENCY*

Coakley et al.[28] examined the deactivation of alcohol dehydrogenase, lysozyme, and catalase in a model solution by exposing them to 20 kHz ultrasound for varied periods. The authors also discovered that alcohol dehydrogenase and lysozyme were inactivated exponentially, although catalase had only moderate impacts. Yasuda, Matsushima, and Asakura[112] recently discovered that the mode diameter of cavitation bubbles produced by ultrasound has a lower frequency dependence.

Brotchie et al.[14] found that as the ultrasound frequency was increased, the mean value of the cavitation bubbles distribution reduced by 60%, and the bubble dispersion became much smaller. With an increase in ultrasonic frequency from 18 to 26 kHz, the catalytic activity of immobilized cellulase increased from 6.56% to 10.45%. At 29 kHz, however, the immobilized cellulase activity dropped by 1.02%. The ability of ultrasound to enhance the surface area of immobilized cellulase particles could be the cause for this. However, when too much ultrasonic energy was absorbed by the immobilized cellulase particles, the activity was reduced because the surface was degraded.[106]

9.4.4 ULTRASONICATION TIME

Yasuda et al.[112] previously demonstrated that for various ultrasonic powers and frequencies, the concentration of bubbles formed by acoustic cavitation increases with irradiation time. The length of time exposed in treatment can have a favorable or negative impact on an enzyme's activity. Although, most of the studies suggests that the enzyme activity decreases beyond optimum sonication time.

Ma et al.[63] stated that alkalase activity rose by 5% and 8% after 4 min of treatment with an intensity of 80 W. When the treatment time was longer than 10 min, alcalase activity decreased. This could be due to the severe conditions (increased pressure, temperature, or shear force caused by ultrasonic treatment) as well as the formation of free radicals during water sonolysis.

Wang et al.[106] revealed that increasing the pretreatment duration boosted cellulase activity significantly within the first 15 min. Cellulase activity was reduced when the treatment time exceeded 15 min. In another research by Subhedar and Gogate[90], the authors noted that an ultrasonic exposure of 30 min increased the cellulase enzyme activity by 23.4%, Enzymatic activity reduced when the treatment time was increased to 55 min. Short-term ultrasonic treatment only modifies the configuration in a positive way, resulting in increased enzyme activity. When exposed to ultrasonic treatment for longer periods of time, however, the protein structure may be disrupted. This could even happen under ideal ultrasonic intensity settings.[63] Another factor suggested by the authors is that of extreme temperature circumstances, which might result in the production of free radicals or shear force because of extended exposure.[92]

9.4.5 ENZYME CONCENTRATION

In determining the effects of ultrasound, the concentration of enzymes is critical. Enzymatic molecules are all so distributed randomly and have a really low density at very low enzymatic concentrations, that they cannot effectively interact with the micro streams generated by cavitation bubbles. Excess enzyme molecules, on the other hand, obstruct the energy-transfer mechanism, reducing the amount of energy available for the acoustic cavitation events beyond the optimum enzyme concentration. Acoustic cavitation may also result in the formation of enzyme aggregates, which could result in a decreased degree of intensification. Similar effects can be seen in enzyme kinetics, where there is a competition for substrate at increasing concentrations, slowing down the rate of reaction.[50] This was in line with the findings of a study conducted on milk lactose.[87] They looked at the effect of different lactase enzyme concentrations (0.2–2 mL/L) at the same temperature and pH and discovered that enzyme activity increases with enzyme concentration until it reaches the optimum enzyme concentration of 1 mL/L, after which the enzyme may have saturated the lactose with active enzymes.

9.4.6 TEMPERATURE

Temperature is a key factor in reaction rate control. According to the Arrhenius theory, the activity of enzymes increases as the temperature rises. But studies have reported that the enzyme undergoes thermal denaturation beyond the optimum temperature and hence causes decreased enzyme activity. Scudino et al.[87] observed that AP in milk was only inactivated by high-intensity ultrasound from 3 KJ/mL, and the temperature was lesser to 57 °C, which alone could not be able to cause enzyme inactivation. The high-intensity ultrasound was not able to inactivate AP for the treatments with low processing temperature <36 °C. They also discovered that lactose PD in milk could only be inactivated by combining high-intensity ultrasound treatment with adequate temperature.

9.5 EFFECT OF ULTRASOUND TECHNOLOGY ON FOOD QUALITY

Ultrasound treatment has a significant impact on a variety of food quality parameters. Table 9.4 summarizes recent studies on the impact of ultrasonic treatment on food quality. The next sessions look at the effects of sonication on certain key quality criteria.

9.5.1 PROTEIN

Ultrasound treatments have shown to induce significant changes on the structural and functional attributes of many proteins. The treatment has modified the bulk functionality of bovine serum albumin protein owing to the structural changes, increased hydrophobicity, and sulfhydryl group reduction.[41] Ultrasound treatment has improved the surface hydrophobicity and emulsion stability of egg white proteins. But it has reduced the foam stability the same.[8] High-intensity ultrasound treatments results in partial protein unfolding and as a result sulfhydryl group gets exposed to the surface. Shear and turbulence created during the cavitation phenomenon will further break the disulfide bonds. This conversion of S–S bonds to S–H bonds results changes the tertiary structure of proteins.[45] Xiong et al.[109] also reported the changes in tertiary structure and particle size increase of ovalbumin after ultrasound application.

Jambrak et al.[53] studied the effect of ultrasonication on functional properties of whey protein. They found that foaming capacity and stability of whey, improved after the treatments (20 and 40 kHz). The increased in foaming properties may be due to the hydrolysis of whey protein hydrolysate.[15] O'sullivan et al.[71] studied the influence of sonication on emulsifying properties of different proteins of animal and plant origin. They observed a reduction in size of protein due to the cavitation-induced hydrodynamic shear force. They also observed a reduction in the intrinsic viscosity which has contributed to increase in the hydrophobicity of protein.

9.5.2 PH AND ELECTRICAL CONDUCTIVITY

Most of the sonicated food materials have not shown any significant change in values of pH.[2,18,26,113] Changes in electrical conductivity are highly correlated with the initial state and properties of food sample. Studies show that there was an increased electrical conductivity after ultrasound treatment in some food materials.[27] The increase is due to formation of hydroxyl radicals during sonication.[64,77] Ultrasound treatment can also result in the release of mineral elements which are entrapped inside the tissues. This can also contribute to conductivity changes.[119] Also at higher frequencies, electrical conductivity decreases due to viscosity changes and the formation of ion aggregates.[89]

9.5.3 TOTAL SOLUBLE SOLIDS (TSS) AND SUGAR CONTENT

Sonication has significant effect in the TSS and sugar content. Zou and Jiang[118] conducted a study using carrot juice and they observed a slight increase in TSS (3%) and total sugars (2.6%), respectively. The increase may be due to the destruction of tissues and cell walls due to ultrasound treatment.[119]

Similar results of increased sugar content were reported for cantaloupe melon, apple juice, and sirkencubin syrup.[2,36,113] Aday et al.[3] observed that as the ultrasound power is increasing, the TSS was decreasing in strawberry samples. Nadeem et al.[68] reported an increase in soluble solids in sonicated carrot-grape mixed juice. They also found that the TSS content was gradually decreasing during the storage period. Hashemi[44] has studied the influence of continuous and pulsed ultrasound treatment on Mirabelle plum. He found that the plums treated with pulsed ultrasound (60 min) have higher content of TSS, than the continuous treated samples.

9.5.4 ASCORBIC ACID AND PHENOLIC COMPOUNDS

Retention of ascorbic acid content in ultrasonicated samples was found to be significantly higher than thermal processed samples.[25,118] This could be due to the reduction in dissolved oxygen content during cavitation.[24] However, the prolonged exposure of ultrasound waves also results in reduction of ascorbic acid content due to the free radical formation in cavitation bubbles.[96] Better retention of phenolic compounds also observed in sonicated samples than in thermal processed ones.[2,11,68,113]

According to Zafra-Rojas et al.[114] the rise in phenolic content after ultrasound treatment in samples could be due to the disruption of cell walls and successive release of phenolic compounds entrapped inside the cell matrix. A second hydroxyl group gets added to the ortho/para-positions of phenolic compounds during sonication which leads to increase in antioxidant potential.[9] Lafarga et al.[59] reported that bio accessibility of phenolic compounds varies within different fruits and vegetables. They observed increased bio accessibility in green pepper and lettuce, and no effect for red pepper, tomato, and zucchini samples after ultrasound treatment at a frequency of 40 kHz for 20 min. Literatures show an increase in anthocyanin content of strawberries after sonication.[29]

Birmpa et al.[13] reported that most of the vegetables, retained their bright color after ultrasonication at low treatment time (up to 20 min) and after which it started decreasing. This increase is due to the release of anthocyanins, which are glucosidically linked with sugar molecules[37] and the reduction might be due to degradation of anthocyanin on exposure to high shear force during cavitation.[97]

9.5.5 SENSORY ATTRIBUTES

Impact of ultrasound treatment on sensory attributes of different food samples have been investigated by many researchers. Peña-Gonzalez et al.[76] studied the effects of high intensity ultrasound on sensory characteristics of bovine meat. They found that the treatment increased meat tenderness and fragmentation. No significant change in the sensory score of ultrasound treated sirkencubin syrup sample.[113]

Similar results were observed for mango juice,[105] apple juice,[88] and orange juice.[84] Jalilzadeh et al.[51] studied the effect of ultrasonication on organoleptic properties of feta cheese. The treatment resulted, an increase in color and appearance of the cheese. Bermúdez-Aguirrersteeg et al.[10] observed a modification in the color of milk samples treated with ultrasound.

9.5.6 MICROBIAL COUNT

Literatures show that acoustic cavitation directly inactivates microorganisms.[40,108] The mechanical shear effects due to cavitation and formation of high reactive radicals and other compounds contributes to microbial destruction.[56,57,67] Gao et al.[40] studied the influence of ultrasound on different bacteria and they observed 4.5-log cycle reduction of *Enterobacter aerogenes* and *Bacillus subtilis*. No significant effect was observed for Staphylococcus species. They also reported that bacteria with a soft and thick capsule are more resistant to ultrasonication.

Yang et al.[111] observed that sonication reduces spoilage microbes and has shown no significant effect on blue mold. Studies show that gram-negative bacteria have higher sensitivity than gram-positive ones when exposed to ultrasound waves.[17,33] Pokhrel et al.[79] found that ultrasound treatment of 24

KHz at 58 °C for 2 min has resulted 5-log reduction of *Escherichia coli* in carrot juice samples with no effect on its physicochemical properties. Gani et al.[39] observed significant decrease in microbial load with increased treatment time.

TABLE 9.4 Effects of Ultrasound Treatment on Food Quality.

Food Sample	Processing Conditions	Key changes in quality attributes	Reference
Apple juice	Ultrasound Frequency:25 kHz Treatment time: 0, 30, 60 and 90 min Treatment Temperature: 20 °C	No significant changes observed for pH, TSS and titratable acidity. Treated samples showed higher values for phenolic compounds, ascorbic acid, antioxidant potential	[2]
Avocado puree	Ultrasound Frequency:20 kHz Treatment time: 1–10 min	PPO activity showed an increasing trend	[12]
Calcot	Ultrasound Frequency: 40 kHz Treatment time: 0, 10, 25 and 45 min Treatment Temperature: 17 °C	No significanteffect on the antioxidant properties. Treated samples retained firmness, color and weight.	[120]
Cape goose-berry juice	Ultrasound Frequency: 42 kHz Treatment time: 10, 20, 40 min Treatment Temperature: 30 °C	Increase in phenolic compounds and carotenoids observed	[73]
Carrot	Ultrasound frequency: 21 and 35 kHz Treatment time: 10, 20, and 30 min Treatment temperature: 25 °C	Treated samples showed an increase in carotenoids content	[70]
Carrot juice	Ultrasound frequency:40 kHz Treatment time: 20, 40, 60 min Ultrasound intensity: 0.5 W/cm^2	Significantly increase in TSS, total sugars, ascorbic acid and total carotenoids observed	[118]

TABLE 9.4 *(Continued)*

Food Sample	Processing Conditions	Key changes in quality attributes	Reference
Carrot-grape juice blend	Ultrasound frequency: 20 kHz Pulse duration: 5 s on. 5 s off for 5 min Treatment temperature: 15 °C	Significant increase was observed in total phenolic contents after sonication	[68]
Fresh-Cut Potatoes	Ultrasound frequency: 40 kHz Treatment time: 1, 5, 10 min Storage period: 12 days Storage temperature: 3 °C	Polyphenoloxidase activity showed a decreasing trend from 4th day and it increased again toward the end of storage period	[5]
Green asparagus	Ultrasound frequency: 40 kHz Treatment time: 10 min Storage temperature: 4 °C	Higher levels of total phenolic content, ascorbic acid and TSS observed.	[104]
Red wine	Ultrasound frequency: 45, 80 and 100 kHz Treatment time: 20 min Treatment temperature: 20 °C	Showed a significant reduction in total phenolic content	[115]
Safflower protein isolate	Ultrasound frequency: 40 kHz Treatment time: 15 and 30 min Treatment temperature: 20 °C	Protein solubility increased and oil absorption capacity decreased in the samples	[121]
Sirkencubin syrup	Ultrasound frequency: 26 kHz Treatment time: 2, 5, 8, 10 and 15 min Storage temperature: −18 °C	Significant increase was observed in total phenolic content and antioxidant activity. No significant change for pH and TSS	[113]
Strawberry	Ultrasound frequency:33 kHz Treatment time: 0, 10, 20, 30, 40, 60, min Storage period: 15 days Storage temperature: 4 °C	Better retention of pH, vitamin C and TSS was obtained for 30 and40 min. Higher exposure time showed detrimental effect on all quality attributes.	[39]

TABLE 9.4 *(Continued)*

Food Sample	Processing Conditions	Key changes in quality attributes	Reference
Strawberry	Ultrasound frequency: 0, 25, 28, 40 or 59 kHz Treatment time: 10 min Treatment temperature: 20 °C Storage period: 8 days Storage temperature: 5 °C	Treatment resulted decrease in firmness of the fruit. Higher values of TSS, acidity and vitamin C was observed	[19]
Strawberry juice	Ultrasound frequency: 40 kHz Treatment time: 10, 30 min Storage temperature: 5 °C	Significant increase in antioxidant activity and polyphenol content, and no change in total acidity and °Brix values	[98]
Sweet potato and wheat flours	Ultrasound frequency: 20 kHz Pulse duration: 3 s on, 1 s off Treatment temperature: 20 °C	Reduced the particle size, gel hardness, total phenolic contents and pasting viscosities observed. Flour digestibility was found to be improved.	[31]

9.6 SUMMARY

Inactivation of enzymes prevailing in different states as a response to processing is an imperative prerequisite that safeguards stabilization and thereby preservation of food quality. There is a general acceptability about the effectiveness of ultrasound process as an individual treatment in achieving food stability by controlling the activity of enzymes. Literature evaluations and studies put forward the idea of combination of heat and pressure with ultrasound technique which is found to have a positive disposition toward inactivation kinetics. However, the percentages of these positive inclinations are reliant on factors like type of enzyme associated, processing factors including amplitude, frequency, power, etc., method of sonication, temperature, and pressure associated and also to the composition of food matrix that is taken care of. One of the major constraints emanates in this segment is the shortcoming of standardized reporting in terms of factors influencing the process. Lack of these standardized reports confines techno-economic assessment which is a requisite in endorsing industrial application. Another key limiting factor regarding industrial application is the deficiency of energy-efficient ultrasound processors which guarantees the economic viability of the procedure. In spite of the favorable results associated with ultrasound as an individual or combination technology, the real-world application is very limited. Added learning and trials are desirable to amend the restraints associated with the hands-on application of the technology. Deficient in realistic revisions and experiments, application of ultrasound in food matrix and its promising upshots will be restricted to the workshops.

KEYWORDS

- **enzymes**
- **inactivation**
- **polyphenol oxidase**
- **sonication**
- **thermosonication**

REFERENCES

1. Abid, M.; Jabbar, S.; Hu, B.; Hashim, M. M.; Wu, T.; Lei, S.; Zeng, X. Thermosonication as a Potential Quality Enhancement Technique of Apple Juice. *Ultrason. Sonochem.* **2014**, *21* (3), 984–990.
2. Abid, M.; Jabbar, S.; Wu, T.; Hashim, M. M.; Hu, B.; Lei, S.; Zeng, X. Effect of Ultrasound on Different Quality Parameters of Apple Juice. *Ultrason. Sonochem.* **2013,** *20* (5), 1182–1187.
3. Aday, M. S.; Temizkan, R.; Büyükcan, M. B.; Caner, C. An Innovative Technique for Extending Shelf-life of Strawberry: Ultrasound. *LWT—Food Sci. Technol.* **2013,** *52* (2), 93–101.
4. Alolga, R. N.; Osae, R.; Essilfie, G.; Saalia, F. K.; Akaba, S.; Chikari, F. Sonication, Osmosonication and Vacuum-assisted Osmosonication Pretreatment of Ghanaian Garlic Slices: Effect on Physicochemical Properties and Quality Characteristics. *Food Chem.* **2021,** *343*, E-article 128535.
5. Amaral, R. D. A.; Benedetti, B. C.; Pujola, M.; Achaerandio, I.; Bachelli, M. L. B. Effect of Ultrasound on Quality of Fresh-Cut Potatoes During Refrigerated Storage. *Food Eng. Rev.* **2015,** *7 (2)*, 176–184.
6. Anaya-Esparza, L. M.; Velázquez-Estrada, R. M.; Sayago-Ayerdi, S. G.; Sánchez-Burgos, J. A.; Ramírez-Mares, M. V.; de Lourdes García-Magana, M.; Montalvo-González, E. Effect of Thermosonication on Polyphenol Oxidase Inactivation and Quality Parameters of Soursop Nectar. *LWT* **2017,** *75*, 545–551.
7. Arroyo, C.; Lyng, J. G. The Use of Ultrasound for the Inactivation of Microorganisms and Enzymes. *Ultrasound Food Process.: Recent Adv.* **2017,** *2017*, 258–265.

8. Arzeni, C.; Pérez, O. E.; Pilosof, A. M. Functionality of Egg White Proteins as Affected by High Intensity Ultrasound. *Food Hydrocoll.* **2012,** *29* (2), 308–316.
9. Ashokkumar, M.; Sunartio, D.; Kentish, S.; Mawson, R.; Simons, L.; Vilkhu, K.; Versteeg, C. K. Modification of Food Ingredients by Ultrasound to Improve Functionality: A Preliminary Study on a Model System. *Innov. Food Sci. Emerg. Technol.* **2008,** *9* (2), 155–160.
10. Bermúdez-Aguirre, D. A. N. I. E. L. A.; Mawson, R.; Versteeg, K.; Barbosa-Cánovas, G. V. Composition Properties, Physicochemical Characteristics and Shelf-Life of Whole Milk After Thermal and Thermo-Sonication Treatments. *J. Food Qual.* **2009,** *32* (3), 283–302.
11. Bhat, R.; Kamaruddin, N. S. B. C.; Min-Tze, L.; Karim, A. A. Sonication Improves Kasturi Lime (*Citrus Microcarpa*) Juice Quality. *Ultrason. Sonochem.* **2011,** *18* (6), 1295–1300.
12. Bi, X.; Hemar, Y.; Balaban, M. O.; Liao, X. The Effect of Ultrasound on Particle Size, Color, Viscosity and Polyphenol Oxidase Activity of Diluted Avocado Puree. *Ultrason. Sonochem.* **2015,** *27*, 567–575.
13. Birmpa, A.; Sfika, V.; Vantarakis, A. Ultraviolet Light and Ultrasound as Non-Thermal Treatments for the Inactivation of Microorganisms in Fresh Ready-to-Eat Foods. *Int. J. Food Microbiol.* **2013,** *167* (1), 96–102.
14. Brotchie, A.; Grieser, F.; Ashokkumar, M. Effect of Power and Frequency on Bubble-size Distributions in Acoustic Cavitation. *Phys. Rev. Lett.* **2009,** *102* (8), E-article 084302.
15. Caessens, P. W.; De Jongh, H. H.; Norde, W.; Gruppen, H. The Adsorption-induced Secondary Structure of β-Casein and of Distinct Parts of its Sequence in Relation to Foam and Emulsion Properties. *Biochimica et Biophysica Acta—Protein Struct. Mol. Enzymol.* **1999,** *1430* (1), 73–83.
16. Cameron, I. L.; Kanal, K. M.; Keener, C. R.; Fullerton, G. D. A Mechanistic View of the Non-ideal Osmotic and Motional Behavior of Intracellular Water. *Cell Biol. Int.* **1997,** *21* (2), 99–113.
17. Cameron, M.; McMaster, L. D.; Britz, T. J. Electron Microscopic Analysis of Dairy Microbes Inactivated by Ultrasound. *Ultrason. Sonochem.* **2008,** *15* (6), 960–964.
18. Campelo, P. H.; Junqueira, L. A.; Resende, J. V. D.; Zacarias, R. D.; Fernandes, R. V. D. B.; Botrel, D. A.; Borges, S. V. Stability of Lime Essential Oil Emulsion Prepared Using Biopolymers and Ultrasound Treatment. *Int. J. Food Prop.* **2017,** *20*, S564–S579.
19. Cao, S.; Hu, Z.; Pang, B.; Wang, H.; Xie, H.; Wu, F. Effect of Ultrasound Treatment on Fruit Decay and Quality Maintenance in Strawberry After Harvest. *Food Control* **2010,** *21* (4), 529–532.
20. Cao, X.; Cai, C.; Wang, Y.; Zheng, X. The Inactivation Kinetics of Polyphenol Oxidase and Peroxidase in Bayberry Juice during Thermal and Ultrasound Treatments. *Innov. Food Sci. Emerg. Technol.* **2018,** *45*, 169–178.
21. Cervantes-Elizarrarás, A.; Piloni-Martini, J.; Ramírez-Moreno, E.; Alanís-García, E.; Güemes-Vera, N.; Gómez-Aldapa, C. A.; del Socorro Cruz, N. Enzymatic Inactivation and Antioxidant Properties of Blackberry Juice after Thermoultrasound: Optimization Using Response Surface Methodology. *Ultrason. Sonochem.* **2017,** *34*, 371–379.
22. Chandrapala, J.; Oliver, C.; Kentish, S.; Ashokkumar, M. Ultrasonics in Food Processing–Food Quality Assurance and Food Safety. *Trends Food Sci. Technol.* **2012,** *26* (2), 88–98.

23. Chen, Y.; Jiang, Y.; Yang, S.; Yang, E.; Yang, B.; Prasad, K. N. Effects of Ultrasonic Treatment on Pericarp Browning of Postharvest Litchi Fruit. *J. Food Biochem.* **2012,** *36* (5), 613–620.
24. Cheng, L. H.; Soh, C. Y.; Liew, S. C.; Teh, F. F. Effects of Sonication and Carbonation on Guava Juice Quality. *Food Chem.* **2007,** *104* (4), 1396–1401.
25. Cheng, X.-F.; Zhang, M.; Adhikari, B. Changes in Quality Attributes of Strawberry Purees Processed by Power Ultrasound or Thermal Treatments. *Food Sci. Technol. Res.* **2014,** *20* (5), 1033–1041.
26. Cheng, X. F.; Zhang, M.; Adhikari, B. The Inactivation Kinetics of Polyphenol Oxidase in Mushroom (*Agaricusbisporus*) During Thermal and Thermosonic Treatments. *Ultrason. Sonochem.* **2013,** *20* (2), 674–679.
27. Cho, S. H.; Kim, J. Y.; Chun, J. H.; Kim, J. D. Ultrasonic Formation of Nanobubbles and Their Zeta-potentials in Aqueous Electrolyte and Surfactant Solutions. *Coll. Surf. A: Physicochem. Eng. Asp.* **2005,** *269* (1–3), 28–34.
28. Coakley, W. T.; Brown, R. C.; James, C. J.; Gould, R. K. The Inactivation of Enzymes by Ultrasonic Cavitation at 20 kHz. *Arch. Biochem. Biophys.* **1973,** *159* (2), 722–729.
29. Cordenunsi, B. R.; Genovese, M. I.; do Nascimento, J. R. O.; Hassimotto, N. M. A.; dos Santos, R. J.; Lajolo, F. M. Effects of Temperature on the Chemical Composition and Antioxidant Activity of Three Strawberry Cultivars. *Food Chem.* **2005,** *91* (1), 113–121.
30. Costa, M. G. M.; Fonteles, T. V.; de Jesus, A. L. T.; Almeida, F. D. L.; de Miranda, M. R. A.; Fernandes, F. A. N.; Rodrigues, S. High-Intensity Ultrasound Processing of Pineapple Juice. *Food Bioprocess Technol.* **2013,** *6 (4)*, 997–1006.
31. Cui, R.; Zhu, F. Effect of Ultrasound on Structural and Physicochemical Properties of Sweetpotato and Wheat Flours. *Ultrasonics Sonochemistry*, **2020,** *66*, E-article 105118.
32. de Medeiros, J. K.; Sarkis, J. R.; Jaeschke, D. P.; Mercali, G. D. Thermosonication for Peroxidase Inactivation in Sugarcane Juice. *LWT* **2021,** *140*, E-article 110730.
33. Drakopoulou, S.; Terzakis, S.; Fountoulakis, M. S.; Mantzavinos, D.; Manios, T. Ultrasound-Induced Inactivation of Gram-Negative and Gram-Positive Bacteria in Secondary Treated Municipal Wastewater. *Ultrason. Sonochem.* **2009,** *16* (5), 629–634.
34. Ercan, S. Ş.; Soysal, Ç. Effect of Ultrasound and Temperature on Tomato Peroxidase. *Ultrason. Sonochem.* **2011,** *18* (2), 689–695.
35. Ertugay, M. F.; Şengül, M. Effect of Ultrasound Treatment on Milk Homogenisation and Particle Size Distribution of Fat. *Turk. J. Vet. Anim. Sci.* **2004,** *28* (2), 303–308.
36. Fonteles, T. V.; Costa, M. G. M.; de Jesus, A. L. T.; de Miranda, M. R. A.; Fernandes, F. A. N.; Rodrigues, S. Power Ultrasound Processing of Cantaloupe Melon Juice: Effects on Quality Parameters. *Food Res. Int.* **2012,** *48* (1), 41–48.
37. Galoburda, R.; Boca, S.; Skrupskis, I.; Seglina, D. Physical and Chemical Parameters of Strawberry Puree. In *9th Baltic Conference on Food Science and Technology; Food for Consumer Well-Being; FOODBALT Conference Proceedings*; Jelgava, LLU, 2014; pp 172–178.
38. Gamboa-Santos, J.; Montilla, A.; Soria, A. C.; Villamiel, M. Effects of Conventional and Ultrasound Blanching on Enzyme Inactivation and Carbohydrate Content of Carrots. *Eur. Food Res. Technol.* **2012,** *234* (6), 1071–1079.
39. Gani, A.; Baba, W. N.; Ahmad, M.; Shah, U.; Khan, A. A.; Wani, I. A.; Gani, A. Effect of Ultrasound Treatment on Physico-Chemical, Nutraceutical and Microbial Quality of Strawberry. *LWT—Food Sci. Technol.* **2016,** *66,* 496–502.

40. Gao, S.; Lewis, G. D.; Ashokkumar, M.; Hemar, Y. Inactivation of Microorganisms by Low-Frequency High-Power Ultrasound, Part I: Effect of Growth Phase and Capsule Properties of the Bacteria. *Ultrason. Sonochem.* **2014,** *21* (1), 446–453.
41. Gülseren, İ.; Güzey, D.; Bruce, B. D.; Weiss, J. Structural and Functional Changes in Ultrasonicated Bovine Serum Albumin Solutions. *Ultrason. Sonochem.* **2007**, *14* (2), 173–183.
42. Guo, Y.; Wu, B.; Guo, X.; Liu, D.; Wu, P.; Ma, H.; Pan, Z. Ultrasonication and Thermosonication Blanching Treatments of Carrot at Varying Frequencies: Effects on Peroxidase Inactivation Mechanisms and Quality Characterization Evaluation. *Food Chem.* **2021,** *343*, E-article 128524.
43. Habuš, M.; Novotni, D.; Gregov, M.; Štifter, S.; Čukelj Mustač, N.; Voučko, B.; Ćurić, D. Influence of Particle Size Reduction and High-Intensity Ultrasound on Polyphenol Oxidase, Phenolics, and Technological Properties of Wheat Bran. *J. Food Process. Preserv.* **2021,** *45* (3), E-article 15204.
44. Hashemi, S. M. B. Effect of Pulsed Ultrasound Treatment Compared to Continuous Mode on Microbiological and Quality of Mirabelle Plum During Postharvest Storage. *Int. J. Food Sci. Technol.* **2018,** *53* (3), 564–570.
45. Hu, H.; Wu, J.; Li-Chan, E. C.; Zhu, L.; Zhang, F.; Xu, X.; Pan, S. Effects of Ultrasound on Structural and Physical Properties of Soy Protein Isolate (SPI) Dispersions. *Food Hydrocoll.* **2013,** *30* (2), 647–655.
46. Huang, H.; Kwok, K. C.; Liang, H. H. Inhibitory Activity and Conformation Changes of Soybean Trypsin Inhibitors Induced by Ultrasound. *Ultrason. Sonochem.* **2008,** *15* (5), 724–730.
47. Iqbal, A.; Murtaza, A.; Marszałek, K.; Iqbal, M. A.; Chughtai, M. F.; Hu, W.; Xu, X. Inactivation and Structural Changes of Polyphenol Oxidase in Quince (*Cydonia oblonga Miller*) Juice Subjected to Ultrasonic Treatment. *J. Sci. Food Agric.* **2020,** *100* (5), 2065–2073.
48. Islam, M. N.; Zhang, M.; Adhikari, B. The Inactivation of Enzymes by Ultrasound—A Review of Potential Mechanisms. *Food Rev. Int.* **2013,** *30* (1), 1–21.
49. Islam, M. N.; Zhang, M.; Adhikari, B. The Inactivation of Enzymes by Ultrasound—A Review of Potential Mechanisms. *Food Rev. Int.* **2014,** *30 (1)*, 1–21.
50. Jabbar, S.; Abid, M.; Hu, B.; Hashim, M. M.; Lei, S.; Wu, T.; Zeng, X. Exploring the Potential of Thermosonication in Carrot Juice Processing. *J. Food Sci. Technol.* **2015,** *52* (11), 7002–7013.
51. Jadhav, S. H.; Gogate, P. R. Intensification in the Activity of Lipase Enzyme Using Ultrasonic Irradiation and Stability Studies. *Ind. Eng. Chem. Res.* **2014,** *53* (4), 1377–1385.
52. Jalilzadeh, A.; Hesari, J.; Peighambardoust, S. H.; Javidipour, I. The Effect of Ultrasound Treatment on Microbial and Physicochemical Properties of Iranian Ultrafiltered Feta-Type Cheese. *J. Dairy Sci.* **2018,** *101* (7), 5809–5820.
53. Jambrak, A. R.; Mason, T. J.; Lelas, V.; Paniwnyk, L.; Herceg, Z. Effect of Ultrasound Treatment on Particle Size and Molecular Weight of Whey Proteins. *J. Food Eng.* **2014,** *121,* 15–23.
54. Jang, J. H.; Moon, K. D. Inhibition of Polyphenol Oxidase and Peroxidase Activities on Fresh-cut Apple by Simultaneous Treatment of Ultrasound and Ascorbic acid. *Food Chem.* **2011,** *124* (2), 444–449.

55. Jiang, Q.; Zhang, M.; Xu, B. Application of Ultrasonic Technology in Post Harvested Fruits and Vegetables Storage: A Review. *Ultrason. Sonochem.* **2020,** E-article 105261.
56. Jin, J.; Ma, H.; Wang, K.; Yagoub, A. E. G. A.; Owusu, J.; Qu, W.; Ye, X. Effects of Multi-frequency Power Ultrasound on the Enzymolysis and Structural Characteristics of Corn Gluten Meal. *Ultrason. Sonochem.* **2015,** *24,* 55–64.
57. Kadkhodaee, R.; Povey, M. J. Ultrasonic Inactivation of Bacillus Alpha-Amylase. I. Effect of Gas Content and Emitting Face of Probe. *Ultrason. Sonochem.* **2008,** *15* (2), 133–142.
58. Knorr, D.; Zenker, M.; Heinz, V.; Lee, D. U. Applications and Potential of Ultrasonics in Food Processing. *Trends Food Sci. Technol.* **2004,** *15* (5), 261–266.
59. Kuldiloke, J. Effect of Ultrasound, Temperature and Pressure Treatments on Enzyme Activity and Quality Indicators of Fruit and Vegetable Juices. Doctoral Thesis, Technische Universität Berlin, Fakultät III—Prozesswissenschaften, 2002. https://www.depositonce.tu-berlin.de/handle/11303/835.
60. Lafarga, T.; Rodríguez-Roque, M. J.; Bobo, G.; Villaró, S.; Aguiló-Aguayo, I. Effect of Ultrasound Processing on the Bio accessibility of Phenolic Compounds and Antioxidant Capacity of SelectedVegetables. *Food Sci. Biotechnol.* **2019,** *28* (6), 1713–1721.
61. Lauterborn, W.; Mettin, R. Acoustic Cavitation: Bubble Dynamics in High-Power Ultrasonic Fields. *In Power Ultrason.* **2015,** 37–78.
62. Li, P.; Sun, L.; Wang, J.; Wang, Y.; Zou, Y.; Yan, Z.; Xu, W. Effects of Combined Ultrasound and Low-Temperature Short-Time Heating Pretreatment on Proteases Inactivation and Textural Quality of Meat of Yellow-Feathered Chickens. *Food Chem.* **2021,** *355*, E-article 129645.
63. Lopez, P.; Burgos, J. Peroxidase Stability and Reactivation after Heat Treatment and Manothermosonication. *J. Food Sci.* **1995,** *60* (3), 451–455.
64. Ma, H.; Huang, L.; Jia, J.; He, R.; Luo, L.; Zhu, W. Effect of Energy-Gathered Ultrasound on Alcalase. *Ultrason. Sonochem.* **2011,** *18* (1), 419–424.
65. Makino, K.; Mossoba, M. M.; Riesz, P. Chemical Effects of Ultrasound on Aqueous Solutions. Formation of Hydroxyl Radicals and Hydrogen Atoms. *J. Phys. Chem.* **1983,** *87* (8), 1369–1377.
66. Manas, P.; Munoz, B.; Sanz, D.; Condon, S. Inactivation of Lysozyme by Ultrasonic Waves Under Pressure at Different Temperatures. *Enzyme Microb. Technol.* **2006,** *39* (6), 1177–1182.
67. Marques Silva, F. V.; Sulaiman, A. Advances in Thermosonication for the Inactivation of Endogenous Enzymes in Foods. *Ultrasound: Adv. Food Process. Preserv.* **2017,** 101–130.
68. Mason, T. J.; Riera, E.; Vercet, A.; Lopez-Buesa, P. Application of Ultrasound. *Emerg. Technol. Food Process.* **2005,** 323–351.
69. Nadeem, M.; Ubaid, N.; Qureshi, T. M.; Munir, M.; Mehmood, A. Effect of Ultrasound and Chemical Treatment on Total Phenol, Flavonoids and Antioxidant Properties on Carrot-Grape Juice Blend During Storage. *Ultrason. Sonochem.* **2018,** *45*, 1–6.
70. Niu, M.; Hou, G. G.; Li, X.; Wang, L.; Chen, Z. Inhibitory Effects of Ultrasound Combined with Ascorbic Acid or Glutathione on Enzymatic Darkening of Whole-wheat Raw Noodles. *LWT—Food Sci. Technol.* **2014,** *59* (2), 901–907.
71. Nowacka, M.; Wedzik, M. Effect of Ultrasound Treatment on Microstructure, Color and Carotenoid Content in Fresh and Dried Carrot Tissue. *Appl. Acoust.* **2016,** *103*, 163–171.

72. O'donnell, C. P.; Tiwari, B. K.; Bourke, P.; Cullen, P. J. Effect of Ultrasonic Processing on Food Enzymes of Industrial Importance. *Trends Food Sci. Technol.* **2010,** *21* (7), 358–367.
73. Ordóñez-Santos, L. E.; Martínez-Girón, J.; Arias-Jaramillo, M. E. Effect of Ultrasound Treatment on Visual Color, Vitamin C, Total Phenols, and Carotenoids Content in Cape Gooseberry Juice. *Food Chem.* **2017,** *233*, 96–100.
74. O'sullivan, J.; Murray, B.; Flynn, C.; Norton, I. The Effect of Ultrasound Treatment on the Structural, Physical and Emulsifying Properties of Animal and Vegetable Proteins. *Food Hydrocoll.* **2016,** *53*, 141–154.
75. Özbek, B.; Ülgen, K. Ö. The Stability of Enzymes after Sonication. *Process Biochem.* **2000,** *35* (9), 1037–1043.
76. Pan, Y.; Chen, L.; Pang, L.; Chen, X.; Jia, X.; Li, X. Ultrasound Treatment Inhibits Browning and Improves Antioxidant Capacity of Fresh-Cut Sweet Potato during Cold Storage. *Rsc Adv.* **2020,** *10* (16), 9193–9202.
77. Peña-Gonzalez, E.; Alarcon-Rojo, A. D.; Garcia-Galicia, I.; Carrillo-Lopez, L.; Huerta-Jimenez, M. Ultrasound as a Potential Process to Tenderize Beef: Sensory and Technological Parameters. *Ultrason. Sonochem.* **2019,** *53*, 134–141.
78. Petrier, C.; Jeunet, A.; Luche, J. L.; Reverdy, G. Unexpected Frequency Effects on the Rate of Oxidative Processes Induced by Ultrasound. *J. Am. Chem. Soc.* **1992,** *114* (8), 3148–3150.
79. Piyasena, P.; Mohareb, E.; McKellar, R. C. Inactivation of Microbes using Ultrasound: A Review. *Int. J. Food Microbiol.* **2003,** *87* (3), 207–216.
80. Pokhrel, P. R.; Bermúdez-Aguirre, D.; Martínez-Flores, H. E.; Garnica-Romo, M. G.; Sablani, S.; Tang, J.; Barbosa-Cánovas, G. V. Combined Effect of Ultrasound and Mild Temperatures on the Inactivation of E. coli in Fresh Carrot Juice and Changes on Its Physicochemical Characteristics. *J. Food Sci.* **2017,** *82* (10), 2343–2350.
81. Raviyan, P.; Zhang, Z.; Feng, H. Ultrasonication for Tomato Pectinmethylesterase Inactivation: Effect of Cavitation Intensity and Temperature on Inactivation. *J. Food Eng.* **2005,** *70* (2), 189–196.
82. Riesz, P.; Kondo, T. Free Radical Formation Induced by Ultrasound and its Biological Implications. *Free Rad. Biol. Med.* **1992,** *13* (3), 247–270.
83. Rojas, M. L.; Trevilin, J. H.; dos Santos Funcia, E.; Gut, J. A. W.; Augusto, P. E. D. Using Ultrasound Technology for the Inactivation and Thermal Sensitization of Peroxidase in Green Coconut Water. *Ultrason. Sonochem.* **2017,** *36*, 173–181.
84. Saeeduddin, M.; Abid, M.; Jabbar, S.; Wu, T.; Hashim, M. M.; Awad, F. N.; Zeng, X. Quality Assessment of Pear Juice under Ultrasound and Commercial Pasteurization Processing Conditions. *LWT—Food Sci. Technol.* **2015,** *64* (1), 452–458.
85. Samani, B. H.; Khoshtaghaza, M. H.; Lorigooini, Z.; Minaei, S.; Zareiforoush, H. Analysis of the Combinative Effect of Ultrasound and Microwave Power on *Saccharomyces cerevisiae* in Orange Juice Processing. *Innov. Food Sci. Emerg. Technol.* **2015,** *32*, 110–115.
86. Scudino, H.; Silva, E. K.; Gomes, A.; Guimarães, J. T.; Cunha, R. L.; Sant'Ana, A. S.; Cruz, A. G. Ultrasound Stabilization of Raw Milk: Microbial and Enzymatic Inactivation, Physicochemical Properties and Kinetic Stability. *Ultrason. Sonochem.* **2020,** *67,* E-article 105185.
87. Şener, N.; Apar, D. K.; Özbek, B. A Modelling Study on Milk Lactose Hydrolysis and β-galactosidase Stability Under Sonication. *Process Biochem.* **2006,** *41* (7), 1493–1500.

88. Shen, Y.; Zhu, D.; Xi, P.; Cai, T.; Cao, X.; Liu, H.; Li, J. Effects of Temperature-controlled Ultrasound Treatment on Sensory Properties, Physical Characteristics and Antioxidant Activity of Cloudy Apple Juice. *LWT* **2021,** *142,* E-article 111030.
89. Southall, J. P.; Hubbard, H. S. A.; Johnston, S. F.; Rogers, V.; Davies, G. R.; McIntyre, J. E.; Ward, I. M. Ionic Conductivity and Viscosity Correlations in Liquid Electrolytes for Incorporation into PVDF Gel Electrolytes. *Solid State Ionics* **1996,** *85* (1–4), 51–60.
90. Subhedar, P. B.; Gogate, P. R. Enhancing the Activity of Cellulase Enzyme Using Ultrasonic Irradiations. *J. Mol. Catal. B: Enzymatic* **2014,** *101*, 108–114.
91. Sulaiman, A.; Soo, M. J.; Farid, M.; Silva, F. V. Thermosonication for Polyphenoloxidase Inactivation in Fruits: Modeling the Ultrasound and Thermal Kinetics in Pear, Apple and Strawberry Purees at Different Temperatures. *J. Food Eng.* **2015,** *165*, 133–140.
92. Sun, Y.; Zeng, L.; Xue, Y.; Yang, T.; Cheng, Z.; Sun, P. Effects of Power Ultrasound on the Activity and Structure of β-D-Glucosidase with Potentially Aroma-Enhancing Capability. *Food Sci. Nutr.* **2019,** 7 (6), 2043–2049.
93. Szabó, O. E.; Csiszár, E. The Effect of Low-frequency Ultrasound on the Activity and Efficiency of a Commercial Cellulase Enzyme. *Carbohydr. Polym.* **2013,** *98* (2), 1483–1489.
94. Terefe, N. S.; Gamage, M.; Vilkhu, K.; Simons, L.; Mawson, R.; Versteeg, C. The Kinetics of Inactivation of Pectin Methylesterase and Polygalacturonase in Tomato Juice by Thermosonication. *Food Chem.* **2009,** *117* (1), 20–27.
95. Tho, P.; Manasseh, R.; Ooi, A. Cavitation Microstreaming Patterns in Single and Multiple Bubble Systems. *J. Fluid Mech.* **2007,** *576*, 191–233.
96. Tian, Z. M.; Wan, M. X.; Wang, S. P.; Kang, J. Q. Effects of Ultrasound and Additives on the Function and Structure of Trypsin. *Ultrason. Sonochem.* **2004,** *11* (6), 399–404.
97. Tiwari, B. K.; O'Donnell, C. P.; Muthukumarappan, K.; Cullen, P. J. Ascorbic Acid Degradation Kinetics of Sonicated Orange Juice During Storage and Comparison with Thermally Pasteurised Juice. *LWT—Food Sci. Technol.* **2009,** *42* (3), 700–704.
98. Tiwari, B. K.; O'Donnell, C. P.; Patras, A.; Cullen, P. J. Anthocyanin and Ascorbic Acid Degradation in Sonicated Strawberry Juice. *J. Agric. Food Chem.* **2008,** *56* (21), 10071–10077.
99. Tomadoni, B.; Cassani, L.; Viacava, G.; Moreira, M. D. R.; Ponce, A. Effect of Ultrasound and Storage Time on Quality Attributes of Strawberry Juice. *J. Food Process Eng.* **2017,** *40* (5), E-article 12533.
100. Tsikrika, K.; Chu, B. S.; Bremner, D. H.; Lemos, M. A. The Effect of Different Frequencies of Ultrasound on the Activity of Horseradish Peroxidase. *LWT* **2018,** *89*, 591–595.
101. Vercet, A.; Lopez, P.; Burgos, J. Inactivation of Heat-Resistant Lipase and Protease from Pseudomonas Fluorescens by Manothermosonication. *J. Dairy Sci.* **1997,** *80* (1), 29–36.
102. Vercet, A.; Lopez, P.; Burgos, J. Inactivation of Heat-Resistant Pectinmethylesterase from Orange by Manothermosonication. *J. Agric. Food Chem.* **1999,** *47* (2), 432–437.
103. Villamiel, M.; de Jong, P. Influence of High-Intensity Ultrasound and Heat Treatment in Continuous Flow on Fat, Proteins, and Native Enzymes of Milk. *J. Agric. Food Chem.* **2000,** *48* (2), 472–478.
104. Waghmare, G. V.; Rathod, V. K. Ultrasound Assisted Enzyme Catalyzed Hydrolysis of Waste Cooking Oil Under Solvent Free Condition. *Ultrason. Sonochem.* **2016,** *32,* 60–67.

105. Wang, J.; Fan, L. Effect of Ultrasound Treatment on Microbial Inhibition and Quality Maintenance of Green Asparagus During Cold Storage. *Ultrason. Sonochem.* **2019,** *58*, E-article 104631.
106. Wang, J.; Liu, Q.; Xie, B.; Sun, Z. Effect of Ultrasound Combined with Ultraviolet Treatment on Microbial Inactivation and Quality Properties of Mango Juice. *Ultrason. Sonochem.* **2020,** *64*, E-article 105000.
107. Wang, Z.; Lin, X.; Li, P.; Zhang, J.; Wang, S.; Ma, H. Effects of Low Intensity Ultrasound on Cellulase Pretreatment. *Bioresour. Technol.* **2012,** *117,* 222–227.
108. Wu, J.; Gamage, T. V.; Vilkhu, K. S.; Simons, L. K.; Mawson, R. Effect of Thermosonication on Quality Improvement of Tomato Juice. *Innov. Food Sci. Emerg. Technol.* **2008,** *9* (2), 186–195.
109. Wu, J.; Nyborg, W. L. Ultrasound, Cavitation Bubbles and Their Interaction with Cells. *Adv. Drug Deliv. Rev.* **2008,** *60* (10), 1103–1116.
110. Xiong, W.; Wang, Y.; Zhang, C.; Wan, J.; Shah, B. R.; Pei, Y.; Li, B. High Intensity Ultrasound Modified Ovalbumin: Structure, Interface and Gelation Properties. *Ultrason. Sonochem.* **2016,** *31*, 302–309.
111. Yaldagard, M.; Mortazavi, S.; Tabatabaie, F. The Effects of Ultrasound on the Activity of Alpha-Amylase During Barley Germination. *Afr. J. Biotechnol.* **2008,** *7*, 2465–2471.
112. Yang, Z.; Cao, S.; Cai, Y.; Zheng, Y. Combination of Salicylic Acid and Ultrasound to Control Postharvest Blue Mold Caused by *Penicillium expansum* in Peach Fruit. *Innov. Food Sci. Emerg. Technol.* **2011,** *12* (3), 310–314.
113. Yasuda, K.; Matsushima, H.; Asakura, Y. Generation and Reduction of Bulk Nanobubbles by Ultrasonic Irradiation. *Chem. Eng. Sci.* **2019,** *195,* 455–461.
114. Yikmiş, S. Effect of Ultrasound on Different Quality Parameters of Functional Sirkencubin Syrup. *Food Sci. Technol.* **2019,** *40*, 258–265.
115. Zafra-Rojas, Q. Y.; Cruz-Cansino, N.; Ramírez-Moreno, E.; Delgado-Olivares, L.; Villanueva-Sánchez, J.; Alanís-García, E. Effects of Ultrasound Treatment in Purple Cactus Pear (*Opuntia Ficus-Indica*) Juice. *Ultrason. Sonochem.* **2013,** *20* (5), 1283–1288.
116. Zhang, Q. A.; Shen, Y.; Fan, X. H.; Garcia Martin, J. F. Preliminary Study of the Effect of Ultrasound on Physicochemical Properties of Red Wine. *CyTA—J. Food* **2016,** *14* (1), 55–64.
117. Zhang, X. T.; Zhang, M.; Devahastin, S.; Guo, Z. Effect of Combined Ultrasonication and Modified Atmosphere Packaging on Storage Quality of Pakchoi (*Brassica chinensis L.*). *Food Bioprocess Technol.* **2019,** *12* (9), 1573–1583.
118. Zhang, X. H.; Quinn, A.; Ducker, W. A. Nanobubbles at the Interface between Water and a Hydrophobic Solid. *Langmuir* **2008,** *24* (9), 4756–4764.
119. Zou, Y.; Jiang, A. Effect of Ultrasound Treatment on Quality and Microbial Load of Carrot Juice. *Food Sci. Technol.* **2016,** *36,* 111–115.
120. Zou, Y.; Xie, C.; Fan, G.; Gu, Z.; Han, Y. Optimization of Ultrasound-Assisted Extraction of Melanin from Auricularia Auricula Fruit Bodies. *Innov. Food Sci. Emerg. Technol.* **2010,** *11* (4), 611–615.
121. Zudaire, L.; Lafarga, T.; Viñas, I.; Abadias, M.; Brunton, N.; Aguiló-Aguayo, I. Effect of Ultrasound Pre-Treatment on the Physical, Microbiological, and Antioxidant Properties of Calçots. *Food Bioprocess Technol.* **2019,** *12* (3), 387–394.
122. Zuñiga-Salcedo, M. R.; Ulloa, J. A.; Bautista-Rosales, P. U.; Rosas Ulloa, P.; Ramírez-Ramíez, J. C.; Silva-Carrillo, Y.; Hernández, C. Effect of Ultrasound Treatment on Physicochemical, Functional and Nutritional Properties of a Safflower (*Carthamus tinctorius L.*) Protein Isolate. *Italian J. Food Sci.* **2019,** *31* (3), 18–23.

CHAPTER 10

POTENTIAL OF ENZYMES FOR BIODEGRADATION OF PLASTIC WASTE

ROSHAN ARA and ANAND KUMAR PANDEY

ABSTRACT

Widespread increment in consumption and the virtue of its enduringness has become the worst vice of plastic and other synthetic polymers, posing a major threat to the environment. According to CPCB (Central Pollution Control Board, India) estimations, 8% of total solid waste is plastic lying on the edge of recalcitrance. The issue of their degradation remains a challenge to the researcher still and all methods discovered. Various cost-efficient technologies and eco-amiable treatments have been tried to reduce or eliminate plastic wastes like landfilling, incineration, and recycling. The former two tools have their own drawbacks related to the environment, and the latter one throws a high cost upon the treater. Among all, the enzymatic degradation method releases some relief of capital expenses and can be considered an innocuous option for ecological health. Many fungal and bacterial enzymes have been reported to have plastic degrading activity, prominently in fungi. This chapter accumulates a brief matter about enzymes capable of degrading plastic filth such as cutinase, laccase, protease, esterase, and urease, and accountable microbial species of *Bacillus, Streptococcus, Pseudomonas, Aspergillus, Phaenarochaete, Pestaloptiopsis,* and others. A conclusive outcome is made on the basis of gathered information to alter the status of accumulated plastic heaps on the land and oceans and provide alternative options to the consumers.

Enzyme Inactivation in Food Processing: Technologies, Materials, and Applications. Megh R. Goyal, PhD, Junaid Ahmad Malik, PhD & Ravi Pandiselvam, PhD (Eds.)

10.1 INTRODUCTION

Plastics are contrived, long synthetic organic polymers backboned by carbon and hydrogen. Generally, eight types of atoms (Carbon, Hydrogen, Oxygen, Nitrogen, Fluorine, Sulphur, Chlorine, Silicon) are used to synthesize a thousand varieties of plastic. The synthesis of plastic generally involves addition polymerization or condensation polymerization.[5] The examples include typically high density poly ethylene (HDPE), low density poly ethylene (LDPE), polyvinyl chloride (PVC), polypropylene (PP), and polystyrene (PS). The consumption of plastics is greater than that of steel per year as per the records.[1] They have replaced the usage of many materials like metal, glass, wood, leather, and many other kinds of stuff.[51] They are utile as they have good features of durability, strength, and resistance to corrosiveness.[8] Therefore, they are very stable and not easily degradable; hence, their accumulation in the terrestrial and aquatic bodies is becoming a threat to the environment.[57]

Plastic wastes can be categorized into thermosetting and thermoplastic, which comprise huge heaps of litter[1]: (1) *Thermosetting plastics* are networked and highly branched polymers. After heating, they are set into a solid state (curing), so they are called thermosetting plastics. The solid-state results in cross bonds between the adjacent molecules and chains, forbidding the slippage between the chains; (2) In *thermoplastics,* the atoms and molecules are joined, and there is no cross linking between the chains. They can be molded multiple times under the effect of heat and cold. They become softer at higher temperatures and stiffer at cold temperatures.

The recalcitrance nature of the accumulated polymer wastes has driven greater attention toward its removal or degradation. Many approaches are being applied, but among all, the enzymatic route proves to be the best in terms of cost, efficiency, and environmental concern. The researchers have reported an array of microbial strains (examples: *Pseudomonas chlororaphis*, *Rhodococcus erythropolis*, *Bacillus brevis*, *Rhizopus delemar*, *Streptomyces spp,* and others) possessing the enzymes having plastic degrading capacity. Most of them are cutinase, laccase, lipase, etc.; these enzymes modify the surface of the plastics resulting in their degradation.[27] Apart from using an enzymatic approach of plastic wastes degradation, the alternate option of bioplastics can also be preferred. The bioplastics contain a functional group that makes them compatible for the degradation naturally, generating the least toxic pollutants. The research on the scale up and development of various economic bioplastic is in trend nowadays.

This chapter thus elaborates different physical, chemical, and biological plastic degrading techniques with a lead focus on potent enzymes and their associated microbial species in detail. Further alternate strategies to reduce synthetic plastic loads like biodegradable plastic have also been discussed.

10.2 STATISTICAL DATA OF PLASTIC WASTES

According to the global survey of "Statista" on "Global Plastic Production 1950–2019," the production of plastic has touched the line of 368 million metric tons worldwide. China is the leading producer of plastic appliances and wastes, followed by Germany, the USA, Italy, and France. The data in Figure 10.1 depicts the global plastic production (in a million metric tons) trend during past four decades.[42]

As per the present trend of production and waste management, 63 million metric tons of plastic have been generated, out of which 12% was managed by incineration, 79% disposed in landfills, and 9% was recycled.[23,42] The different types of plastic produced worldwide are illustrated in Figure 10.2.

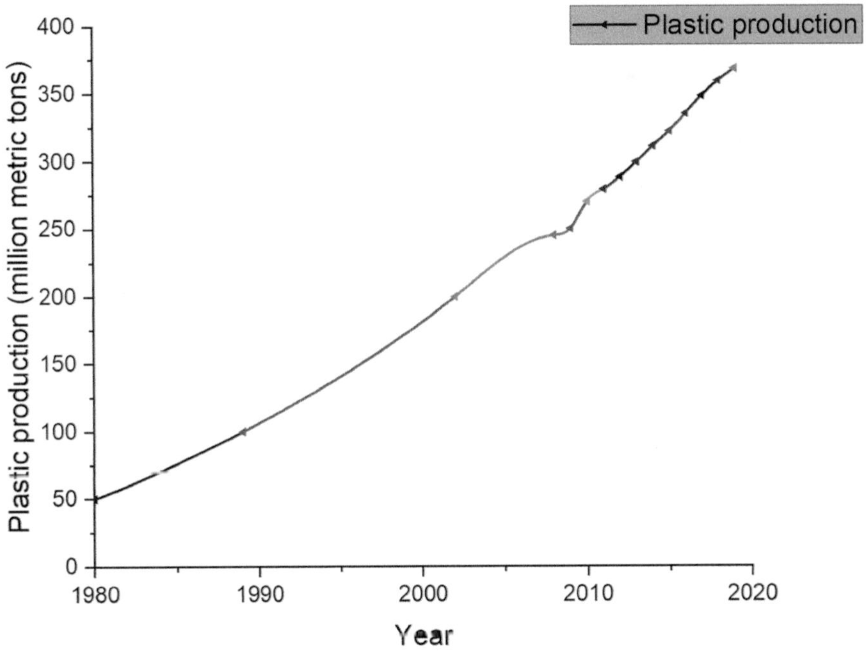

FIGURE 10.1 Trend of Worldwide production of plastic in the past 40 years.

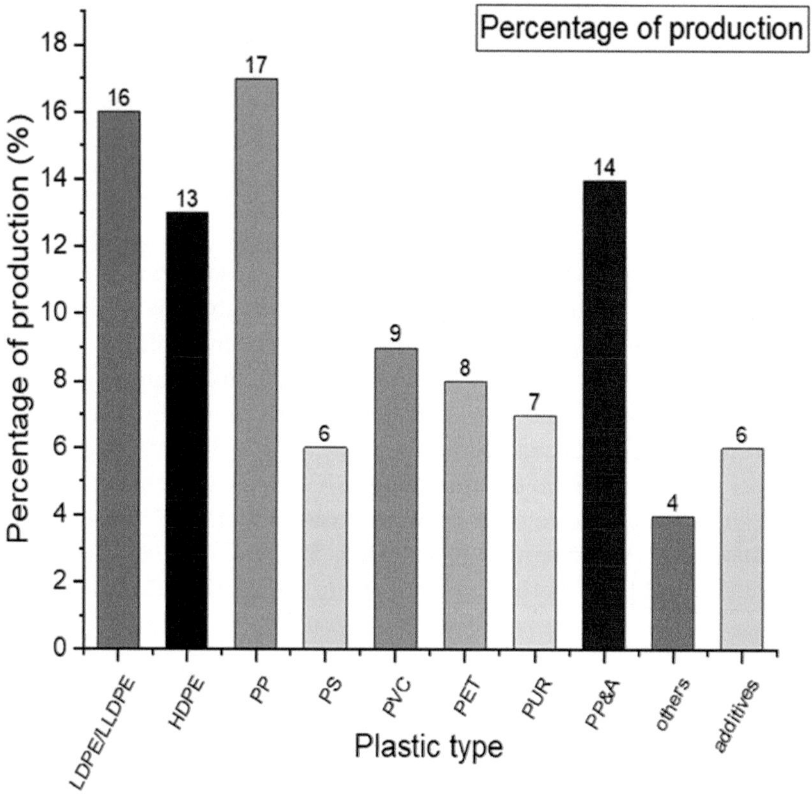

FIGURE 10.2 Global plastic production: LDPE/LLDPE—Low density polyethylene/ Linear low density polyethylene; HDPE—High density polyethylene; PP—Polypropylene; PS—Polystyrene; PVC—Polyvinyl chloride; PET—polyethylene terephthalate; PUR—Polyurethane; PP&A—polyster, polyamide and acrylic fiber.

10.3 APPROACHES TO DEGRADE PLASTIC WASTES

Degradation of plastic wastes and inadequate management of wastes has arisen as a great concern for environmental health. Around 300 million metric tons of plastics are made throughout the world per annum, and 50% of them are used as disposables. The characteristic features of plastics, such as their resistance and durability, are both boon and bane.[50] Their resistance capacity against any impact leads to their accumulation in different natural bodies.

FIGURE 10.3 Biotechnological approach to deal with plastic waste.

According to the report "Plastics: A Global Outlook" of Global Industry Analysts, Inc. (GIA), a "garbage patch (a ring of marine trash)" has been developed in the Asia Pacific due to the accumulation of plastic wastes mixed with chemical sludge and debris.[5] Various approaches have been developed to deal with such a huge waste load. Keeping concerns with the conventional methods of waste management, plastic waste can be treated in the following ways:

i. Landfills
ii. Incineration
iii. Recycling
iv. Modification

Apart from these, biotechnological approaches to treat and manage plastic wastes have been proven as a remedy to the concern. These approaches can serve as easy and eco-amiable approaches to remove or convert the polymer litters into another type of value-added product or biodegradable form.[5] These bio-based approaches are the microbial or enzymatic treatment of plastic trashes or the development of alternate biodegradable polymers. The management of plastic wastes can be achieved in ways as depicted in Figure 10.3.

10.3.1 LANDFILLS

Landfills include a wide range of unregulated and regulated practices and vary depending on different geographical regions. The waste of landfills may be sorted or may contain industrial, construction, commercial, mining, and forestry components or may have been disposed of in monitored settings. These landfills contain complex ecosystems of microbes ranging from obligate anaerobes to aerobes. All the landfills change microbiologically or geochemically with time and with changing conditions of the surrounding environment. A landfill can be broadly defined as a landscape feature of large size that contains millions of megagrams of waste made up of natural and anthropogenic organic and inorganic matter and local buried soils. Disposal of plastic wastes or any kind of wastes in landfilling has remained as a traditional practice to manage the wastes. Land scarcity is the main factor that does not nourish this practice. Dry tomb or wet cell types of the landfill are used for this purpose. Wet cell land fill results in landfill leachate, which can contaminate the environment. In addition, the different toxic compounds used in plastic making, such as plasticizers, organotin, phenolics can leak out into the ground contaminating the groundwater or other water bodies.[5,52]

10.3.2 INCINERATION

Incineration is an exothermic process in which combustion of organic material is done and leads to the production of bottom ash, particulates, heat, and flue gases. Decomposition of plastic wastes through incineration or pyrolysis can extenuate a load of wastes in landfills and can further be utilized to recover energy from these wastes' heap. The energy released during incineration can be stored for electricity generation, too.[56] Energy repossession with incineration offers following benefits[5]:

- Waste mass is reduced by 90%.
- The harmful chemicals in the waste stream are demolished.
- The inorganic fraction of the waste is converted to an inert slag with mineralization in incineration, which can then be used for different applications.
- It can serve as an ideal route to treat heterogeneous and highly polluted wastes.

Pyrolysis is an endothermic process in which the waste is converted to a mixture of gases consisting of hydrogen, carbon monoxides, and carbon dioxide by subjecting it to high temperatures greater than 700°C. The decomposition is performed under anaerobic conditions in the presence of high pressure and heat, which contribute to breaking the bonds of waste compounds. It can also be used as a method of recycling. Through pyrolysis, the plastic wastes can be converted into usable gases, pyrolytic oil (mixture of hydrocarbon fraction), and solid residues.[7] Different types of products can be obtained based on the treatment in different thermal ranges given to the plastic wastes which are listed in Table 10.1. To treat wastes through pyrolysis, different types of reactors have been used (generally fluidized bed reactors).[26]

TABLE 10.1 Different Chemical Products Generated from Thermal Treatment of Plastic Resins.

Plastic resins	Products at low temperature	Products at high temperature
Polycaprolactam	Caprolactam	–
Polyethylene	Olefins, oils, waxes	Oils and gases
Polyethylene terephthalate	Vinyl terephthalate	–
Polypropylene	Vaseline	Oils and gases
Polystyrene	Styrene	Styrene
Polytetrafluoroethylene	Monomer	Tetrafluoroethylene
Polyvinyl chloride	Benzene, HCl	Toluene

10.3.3 RECYCLING

Recycling plastic is a better approach to reuse the wasted resources and materials. Although it might be helpful in waste reduction, all the plastics are not recyclable. Recycled plastics are of somewhat lower quality which again is not a desirable trait for their market demand and consumer satisfaction. Recycling can be done in many ways and for many objectives, either to convert plastics into other usable forms through mechanical or chemical means, or valorizing the plastic wastes to get different forms of energy. There are following categories for the recycling and recovery of plastic wastes[21,30]:

Primary recycling is also referred to as "mechanical recycling." This type of processing is done to convert plastic scraps into a product that is characteristically similar to original products. Such recycling type is also called closed-loop recycling and can possibly used for thermoplastics but as commercially used packaging plastics also contain other polymers in amalgamated form and materials like paper, metals, pigments, adhesives and inks it is not so easy to recycle via primary recycling approaches. This type of recycling can be utilized when the polymer could be separated effectively with high purity from source plastic and could be made stable against any degradation while reprocessing and successive use. Thus, this approach is generally used for clear PET bottles and HDPE (High density polyethylene) milk bottles.[49]

Secondary recycling is the processing of plastic scraps into a product having different characteristics than that of the original product. Downgrading or secondary recycling is utilized when the plastic recovered is transformed for an application that will not make use of the virgin resin of the polymer like lumber plastic. Recycling activities like cleaning, reduction of size, regranulating, and sorting are considered under this type of recycling process.[30]

Tertiary recycling is the processing of plastics wastes to obtain or recover chemicals constituents of plastic polymer. These recovered constituents can then either be utilized to synthesize chemicals or to again manufacture the plastic. Though this approach sounds technically feasible but is uneconomic as the cost of petrochemical feedstock is lower than the process and plant cost required to generate monomer forms from plastic waste. PET plastic is the common ones that have been recycled by this approach as its de-polymerization is possible under mild conditions. It can be broken into monomers by methanolysis, glycolysis, or hydrolysis.[21]

Quaternary recycling is a valorization technique in which the plastic wastes are processed to convert them into a usable form of energy. The technique utilized for such recycling is incineration. This approach considers the high energy value present in the polymers of plastic waste and also reduces the volume of waste. It is a sustainable method for recovering organics content from the plastic polymers as well as for preserving the oil and petroleum source. The major associated disadvantage is emission of byproducts like dioxins. Example of energy form produced from quaternary recycling is Tire-derived fuel.[21]

10.3.4 MODIFICATION

Surface modifications by rearranging the main chain, adding functional groups, and additives are done for easy degradation or reconstruction into another product. In the reconstruction of the surface of plastics, the glass transition temperature with no crosslinking plays an essential role in promoting the event in a forward direction.[58]

10.4 REDUCTION OF PLASTIC WASTES BY BIOLOGICAL METHODS

Each type of plastic has its own kind of application, such as polyethylene (commonly used in packaging materials, bags, water, and milk bottles), polyurethane (used to make medical disposables, automobile materials, and industrial objects), polyethylene terephthalate (used for manufacturing single-use PET bottles), etc. All these polymers have high molecular weight and complex structures that offer a great resistance toward their deterioration, thus causing harm to the environment. The reduction in plastic wastes can be achieved by different routes like mechanical, chemical, and biological (using enzymes and microbes) methods scissoring the bonds, transformation of molecules, and formation of new bonds.[47] The chemical route of degradation is again a too toxic and heavy route to follow up. Physical methods require large infrastructures, which increase the cost of processes. Therefore, the biological method that utilizes easy and fast-growing microbes and their enzymes to reduce the plastic ravages offers a clean, attractive, and safe mode to be considered.

There are different types of living organisms (bacteria, fungi, and plants) present in the environment which reduce or degrade the plastic litters but anaerobic digestion in the landfills and leads to the formation of greenhouse gases (GHGs), resulting in global warming.[15] The species of *Streptomyces, Corynebacterium, Pseudomonas, Micrococcus, Anthrococcus*, etc. are spectacular microbes exemplified in Table 10.2, which produce specific enzymes including laccase, peroxidases, hydrolases, cutinases to effectively degrade the plastic polymers into their monomer forms without the production of GHGs.[11,15,25].

TABLE 10.2 List of Microbial Species and Enzymes Responsible for Polymer Degradation.

Microorganism	Enzyme	Polymer
Pseudomonas chlororaphis, P. fluorscens, Rhodococcus erythropolis, R. rhodochrous, Streptococcus epidermidis, S. cohnii, Streptomyces badius, S. viridosporous, Bacillus brevis, B. circulancs, B.mycoides, B. thuringiensis, Anthrobacter viscous, A. paraffineus, Asprgillus niger, Micrococcus leuteus, Fusarium spp., Verticellium lecanii	Laccase, manganese peroxidase, lignin peroxidase, alkane hydroxylase,	Polyethylene

TABLE 10.2 *(Continued)*

Microorganism	Enzyme	Polymer
Bacillus brevis, Rhizopus delemar, and others	Lipase, esterase, proteinase K	Polylactic acid
*Curvularia senegalensis, Fusarium solani, Trichoderma spp. a*nd others	Cholesterol esterase, urease	Polyurethane
Pseudomonas chlororaphis, Aspergillus niger, and others		Polyvinyl chloride
Alcaligenis faecalis, penicillium funiculosum, Aspergillus fumigatus, Pseudomonas lemoignie and others	Serine hydrolases, esterase, lipases	Poly(3-hydroxybutyrate)
Clostridium botulinum, Fusarium solani, Bacillus brevis, and others	Glucosidases, cutinase	Polycaprolactone
Streptomyces spp., Clostridium acetobutylicum, and others	Lipases	Poly(3-hydroxybutyrate-co-3-hydroxyvalerate)

10.4.1 LACCASE

Laccases are a kind of oxidoreductase enzyme (EC 1.10.3.2) that oxidizes substrates of wide varieties while conducting dioxygen electron reduction to water and are thermally stable. A study done by Santo et al. proved that laccase produced by an actinomycetes *Rhodococcus ruber* have the capacity to degrade the polyethylene films (both in cell-free extracts and in suspension) in the presence of copper. The amorphous part of polyethylene plays an important role in its degradation by increasing the rate of degradation. The laccase oxidizes the bonds and disunites the links between the polymeric chains. This enzymatic activity reduced the molecular weight of the polyethylene after 2 weeks of incubation at 37°C resulting in an increased amount of carbonyl functional groups.[42]

Sumathi et al. (2016) studied a fungus of *Cochliobolus sp.* for effective degradation of low molecular weight polyvinyl chloride, which secretes laccase enzyme. The degradative effects of fungus on the plastic polymer were studied by using techniques like gas chromatography-mass spectrometry, FTIR, and SEM, and significant differences were discovered between the control sample and the sample which was subjected to the fungus.[53]

Ojha et al. (2017) isolated two fungal strains, which probably produced laccase enzyme, including Penicillium chrysogenum-NS10 and Penicillium oxalicum-NS4 capable of degrading high- and low-density polyethylene sheets and performed an optimization study for a duration of 90 days. They utilized different techniques like AFM, FTIR, and FE-SEM to analyze the degradation and noticed significant outcomes.[38]

In another experiment, laccase was obtained from *Trametes versicolor*, having a high reactivity against polyethylene. This enzyme oxidizes the phenolic compounds and converts the molecular O_2 to water, losing a single electron in the reaction. It could oxidize the nonphenolic compounds in the presence of laccase mediator systems (LMS) such as 1-hydroxybenzotriazol (HBT) or 2,2′-azinobis(3-ethylbenzothiazoline-6-sulfonate) (ABTS). The degradation of polyethylene was ascertained by monitoring change in elongation, tensile strength, and distribution of its molecular weight. It resulted in the decrease in molecular weight of plastic film from 242,000 to 28,300 after 3 days of treatment.[20]

10.4.2 MANGANESE PEROXIDASE (E.C. 1.11.1.13)

According to the study of Liyoshi et al., the strain of *Phanerochaete chrysosporium* and an isolate of IZU-154 produce the enzymes that could degrade the polyethylene films of high molecular weight very effectively. The use of surfactants like CHAPSO, Tween 80, and Tween 20 promotes the event. These strains were reported to degrade the polyethylene membrane by 70% under incubation of 15 days.[20]

Sack et al., in 1997, performed a degradation study on different polycyclic aromatic hydrocarbons by manganese peroxidase of Nematoloma frowardii crude preparation and found significant degradation. They further evaluated the effect of glutathione on the degradation process and found an increase in degradation, thus making it evident that glutathione mediates the activity of manganese peroxidase.[45]

10.4.3 ALKANE HYDROXYLASES

It catalyzes the degradation of oligomeric hydrocarbon chains oxidizing terminal groups. A study by Yoon et al. showed that the enzyme produced by recombinant *Pseudomonas* sp. degraded the polyethylene film of low molecular weight to CO_2 by 20% after incubating it for 80 days at 37°C.[16]

Muhonja et al. (2018) performed biochemical and molecular characterization of fungi and bacteria of Dandora dumpsite Nairobi, Kenya, and isolated bacterial species, namely D4 yn-Brevibacillus borstelensis, Bacillus cereus strain A5,a and B1,1-Pseudomonas putida that were possessing alkane hydroxylase enzyme genes. These bacteria were able to degrade low-density polyethene effectively. Of these, Bacillus cereus strain A5 a was one of the highest degrading bacterial strains.[34]

10.4.4 LIPASE/ESTRASE

Akira et al. studied that commercial lipases can be used to degrade the polylactic acid completely. Although few enzymes were also reported earlier, it took them approximately 1 year to degrade the polycaprolactone. They monitored the degradation by the enzyme by analyzing the weight loss of the film and molecular weight distribution (after 100 days of incubation at 37°C) using gel permeation chromatography (GPC). Among all the various types of lipases used, the enzymes, named lipase QL, lipase Asahi, and lipase F, were the most efficient candidates resulting in the degradation of the film in 4–17 days of incubation.[22,64]

A fusion of enzymes of esterase class (lipase and cutinase; Lip-Cut) resulted in efficient degradation of polycaprolactone. Results showed that the fusion enzymes had a greater capability of degradation and resulted in weight loss by 13.3, 11.8, and 5.7 times higher than lipase, cutinase, and lipase–cutinase mixture, respectively. The results were analyzed by GC-MS (Gas Chromatography-Mass Spectroscopy).[33]

Another study carried out by Christenson et al. showed the effect of cholesterol esterase against poly(ether urethane) and poly(carbonate urethane) films. Both the films were degraded by oxidation rather than hydrolysis. The films were incubated in 400 U/mL of enzyme for 36 days. The enzymatic activity resulted in an $8 \pm 5\%$ loss in the weight of the films. The results were analyzed by attenuated total reflectance Fourier transform infrared and gel permeation chromatography. The film characterization was obtained by scanning electron microscopy.[9]

10.4.5 PROTEINASE K

First, this enzyme was used to study the characteristic of degradation of polylactic acid and copolymers of polylactic acid. During degradation, the grooves/pits are formed on the surface of the film observed by scanning electron microscope. The results showed that the amorphous region degrades faster than the crystalline region.[57] Yamashita et al. studied the degradation efficiency of proteinase K on polylactic acid amorphous film. They analyzed the results by QCM (Quartz Crystal Microbalance) and reported that the amorphous film having 20-nm thickness was completely degraded by 100 μg/L of proteinase K. The results were confirmed by AFM (Atomic Force Microscopy).[63]

Huang et al. utilized proteinase K enzyme of Tritirachium album and embedded it into a poly(L-lactic acid film. A degradation or loss of weight of 78% was observed after 96 h, thus proving the high potential of the enzyme for the degradation of this plastic type.[23]

10.5 MECHANISM OF PLASTIC DEGRADATION THROUGH ENZYMES

The degradation of plastics using enzymes is proven to be the easiest and attractive route. Each type of polymer has its own mechanism of degradation. Generally, the degradation of polyethylene by enzymes takes place in the following steps[51,65]:

- The microbes adhere to the surface of the polymer and deteriorate the surface by releasing enzymes. This process is known as biodeterioration
- The deteriorated surface is further acted upon by the enzyme and also by natural forces and get fragmented by the process of bio fragmentation
- The fragments of polymers are taken up by the microbes and are assimilated in them.
- Finally, microbe mineralizes the assimilated plastic fragments into simple molecules and excretes them into the environment.

The enzymes secreted by the microbial communities hydrolyze the polymers into monomers, oligomers, and dimers that can go across the bacterial or fungal cell membrane and act as a carbon source and energy source. Finally, utilization of these compounds is completed to form end products like CO_2, H_2O, and CH_4 (Fig. 10.4). Though this process sounds simple and happens spontaneously, in natural environment, it takes very long durations, even hundreds of years, to get accomplished. Thus, investigation of efficient and effective microbial communities and their enzymes becomes essential.[65]

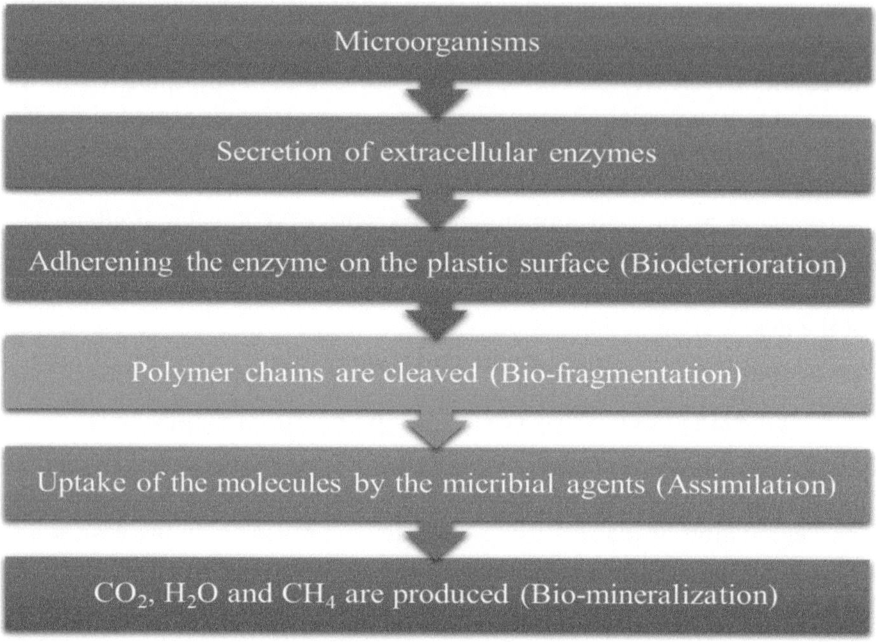

FIGURE 10.4 Mechanism of biodegradation of plastics.

Among all the polymers, polyethylene and polyethylene terephthalate are used extensively for most of the applications and comprise a major portion of municipal and industrial wastes.[51,63,65]

10.6 POLYETHYLENE AND ITS ENZYMATIC BIODEGRADATION

Polyethylene is the polymer of the linear hydrocarbon (C_nH_{2n}) and is widely used due to its exclusive mechanical properties. It has a high density in the range of 0.91–0.97 g/m^3.[2,31] The easy and low-cost production of polyethylene is the main reason for its versatile applications.[2] It is a highly inert material and recalcitrant to degradation or to be affected by any other factor. Therefore, its degradation methodologies are still an ongoing topic of research. The two premises given to sort out its accumulation problem are: Separation of polyethylene wastes and their recycling; and Replacement with biodegradable plastics.

However, due to processing and financial heaviness, both the strategies are facing a challenge. A spectrum of microbes possesses the enzymes such as oxidases and peroxidases capable of acting on polyethylene.

The molecular weight of the polyethylene to be degraded plays a significant role in its biodegradation. The polymers of low molecular weight (low-density polyethylene, LDPE) are easily utilized by the microbial agents, but only a little knowledge is available for the biodegradation of the polyethylene having a high molecular weight (high-density polyethylene, HDPE).[31] Thus, the high molecular weight of this polymer remains as havoc that it cannot directly enter the bacterial or fungal cells to be utilized or broken down by intracellular enzymes. Therefore for their degradation, specific strains of microbes that produce extracellular enzymes were selected based on the following three criteria:

- Analyzing the species of *Streptomyces* genera, which produce ligninolytic enzymes as lignin, is also considered one of the insoluble macromolecules as polyethylene.
- Collection of gram-positive bacteria residing in the area of higher alkanes as they can have the ability to use polyethylene as substrate.
- Investigation of the strains from the places where a heap of polyethylene is present over many years.

Growth of the strain utilized along with the changes on the surface following reduction of polymer weight of polyethylene can be considered its biodegradation. The strains of *Enterobacter, Aspergillus niger, Aspergillus flavous, Gliocladium virens, Penicillium pinophillum, Phanerochaete chrysosporium, Chladosporium chladosporioides, Rhodococcus rhodochroust, Nocordia asteroids, Anthrobacter paraffineus, Rhodococcus ruber, Brevibacillus borstilensis, Streptomyces setonii, Streptomyces badius, Mucor rouxii* have been reported to possess the enzymes capable of polyethylene degradation.[6,12,14,17,40,62].

10.7 POLYETHYLENE TEREPHTHALATE (PET) AND ITS BIODEGRADATION

It consists of monomers of terephthalic acid (TPA) and ethylene glycol bonded with ester bonds. It is widely used in the packaging industry for making single-use plastic bottles and is the most abundantly used polymer among all plastic polymers. Films of PET have exceptionally high resistance to chemicals, high elasticity and tensile strength, less weight, and are stable in the temperature range of −60°C to 220°C even. All these properties contribute to its high mechanical strength as equal as in concrete and make PET durable too.[32]

A large amount of waste is accumulated due to its immense use and improper waste management. The thermal and chemical means of recycling PET have been developed and are well accepted, but despite these recycling methods, which often lead to the release of toxic chemicals and need high infrastructure cost, an eco-amiable route for its recycling and degradation is necessary. As per the consideration of its biological degradation, less enough information is available. Few strains are known to have the capability to utilize it as a substrate for their growth. Examples include *Fusarium solani* and *Fusarium oxysporum* which are reported to grow on the PET yarns.[15]

The first enzyme reported for PET degradation in 2005 by Muller et al. was polyester hydrolase produced by *Thermobifida spp.* (TfH).[35,39] Apart from this, some other microbial species have been reported for the same, which are illustrated in Table 10.3.

TABLE 10.3 List of Microbial Species for PET Degradation.

Substrate	Enzyme	Source	Reference
Film of PET	Thh Est	*Thermobifida halotolerans*	[24]
Packaging film of PET	Cut190	*Sacchromonospora viridis*	[55]
PET	HiC	*Thermomyces insolens*	[43]
PET beverage bottles	TfH	*Thermobifida fusca*	[35]
PET bottle	Cut190	*Streptomyces spp.*	[13]
PET film nanoparticles	Tcur0390	*Thermomonospora curvata*	[60]
Transparent PET sheets	LC-Cutinase	*Nocordia spp.*	[48]

10.7.1 CUT 190 (CUTINASE LIKE) ENZYME

According to a research of Yoshiji et al., a mutant strain of *Saccharomyces cerevisiae* produces cutinase-like enzyme Cut190*, which is stable at high temperature and has high enzymatic activity. To characterize the functionality of this enzyme a component of PET, bis(2-hydroxyethyl) terephthalate (BHET), was used. They studied the PET degrading activity of Cut190 under the varying concentration of Ca^{2+} and pressure of 400 MPa. Their study outcomes showed that the application of high pressure along with the enzyme action could effectively degrade the PET. The increased pressure enhanced the PET degradation by 20% at around 60°C in the presence of a 2.5 mM concentration of Ca^{2+}.[18]

10.7.2 THH EST ENZYME

Thh Est enzyme is an esterase enzyme produced by *Thermobifida halotolerans*. This enzyme is a member of the family of serine hydrolases and homologous to the enzymes obtained from *Thermobifida alba* and *Thermobifida fusca.* Doris et al., in their scientific study on PET degradation, cloned the Thh gene and expressed it in *E. coli* and proved the efficiency of the respective enzyme, that it can hydrolyze the surface of PET at 50°C. The confirmation of degradation was obtained by analyzing its hydrolyses through matrix associated laser desorption or ionization time of flight (MALDI-TOF) mass spectrometry and detecting its monomeric components (terephthalic acid and mono (2-hydroxyethyl) terephthalate (MHET)) that were released after degradation.[59]

In another study, Yoshida et al. (2016) screened the natural communities of microbes and isolated Ideonella sakaiensis 201-F6 novel strain capable of utilizing PET as its carbon and energy source. This strain was found to produce two hydrolyzing enzymes, PETase and MHETase, which were apt to degrade PET and MHET respectively, and convert them to their monomeric forms, namely terephthalic acid and ethylene glycol.[66]

10.7.3 HIC

Ronkvist et al. showed a detailed study on the enzyme HiC produced from *Humicola insolens* (now *Thermomyces insolens*) which catalyzed the degradation of PET at 70°C under the incubation period of 96 h. The amorphous portion of PET (having crystallinity up to 7%) was degraded up to 97 ± 3% under the abovementioned physiochemical environment. The morphological changes over PET surface were analyzed before and after the enzymatic treatment by SEM (Scanning Electron Microscopy) imaging and DCS (Differential Scanning Calorimetry).[43]

10.7.4 TFH

Wei et al., 2017, studied the strain of *Thermobifida fusca* and proved that it was able to degrade the amorphous PET film at a high rate. They showed its catalytic activity against PET in the temperature range of 55–65°C resulting in the weight loss by ≤14% within 48 h of incubation.[61] Apart from these enzymes, few others, like a double mutant of Cut190 produced from the

Saccharomonospora viridis AHK190 was reported for degrading PET at 63°C, cutting off the weight by 13.5 ± 0.5% after 3 days of incubation.[28] Other enzymes TfCut2 and variants of TfCut2 produced from *Thermobifida fusca* KW3 were active in the temperature range of 60–80°C resulting in the weight loss of PET film by ≤25% and ≤45% respectively after 45 and 50 h of incubation, respectively.[61]

10.8 FACTORS INFLUENCING PLASTIC DERADATION

The process of biodegradation is regulated not only by the microbes or the enzymes utilized for degradation but also by the extraneous factors that influence the activity of these microbes and their respective enzymes. Adequate conditions of the environment are important for the survival and growth of microorganisms. Inadequate conditions of environmental temperature, pH, aeration, water activity, and redox can adversely affect the growth of effective microbe and hence the enzyme production and can lead to little degradation. The type and hydration of soil also affect the process if the degradation is occurring in the soil system.[32] The major factors that pose a significant effect on the biodegradation process of plastic have been briefly described below:

10.8.1 CONDITIONS (AEROBIC/ANAEROBIC)

Oxygen is utilized by microbes both as a substrate to conduct oxygenase enzyme-catalyzed degradative reaction and terminal acceptor of an electron in aerobic respiration. It is used for cleavage of ring, hydroxylation of aromatic cyclic compounds, and for aliphatic compound oxidation. The degradation of plastics by biotic agents is thus greatly influenced by the presence or absence of oxygen. The process is effective and faster in the presence of oxygen. In aerobic condition, the end product formed is carbon dioxide whereas in anaerobic conditions methane and carbon dioxide are produced which are harmful to the environment.[32]

10.8.2 CHEMICAL COMPOSITION OF POLYMERS

The chemical composition of polymer is the most significant factor that influences the degradation of plastic as different microbes produce diverse enzymes, and their specificity differs for different polymers. The monomeric units that compose the plastic play a critical role in deciding the route of degradation. As in different plastic types, the bonds between monomers differ and thus need specific mechanisms to be broken down. Also, different enzymes exhibit different degradation potentials depending on their affinity for the substrate polymer. Polyethylene terephthalate contains ester bonds between the monomers and is degraded by different esterase, lipase, cutinase, and hydroxylases to a different extent as the chemical composition of the polymer is not equally favorable to all these enzymes. Generally, starch-based polymers are easily degraded by microorganisms producing amylases, while polymers of olefins are challenging to get degraded due to their complex chemical compositions.[3]

10.8.3 PHYSIOLOGICAL PARAMETERS

As the growth of microorganisms and the activity of the enzymes are prominently affected by the temperature, pH, water availability, sunlight, moisture content, and culture environment, their degradation potentials also differ in different physiochemical conditions. Most fungal strains need acidic pH to grow, while bacterial strains require pH of around 6.8. In the same way, enzymes also perform depending on the conditions of pH and temperatures. Also, as plastic polymers are resilient in nature, a pH toward acidic range and high temperature can favor the degradation by weakening the polymer bonds.[65]

10.8.4 MICROBIAL GROWTH

Many microorganisms degrade the polymer effectively irrespective of their source. These microorganisms use these polymers as their carbon and energy source. The species of *Mucor* and *Aspergillus* genera have shown their potential to degrade the plastics present in landfills. Some other microbial types produce specific enzymes that are either intracellular or extracellular and degrade the polymers by breaking down the bonds responsible for polymerization into monomers without consuming the products for their nourishment.[41,65]

10.9 SUBSTITUTE FOR PLASTIC

10.9.1 BIODEGRADABLE PLASTICS

Bioplastics are plastics that are produced by utilizing biomass and can be classified based on their degradability into biodegradable bio-plastic and non-biodegradable bioplastic. The biodegradable plastic forms are fit for replacement of petroleum-based plastics as they can be naturally degraded by microbial action and forces of nature. The common route of degradation of this type of bioplastics involves the enzymatic activity of microorganisms in the presence of air, water, and light.

Raw materials like starch, cellulose, casein, protein, lipids are preferred for the synthesis of bioplastics.[10] Although bioplastic production is low-yielding and high priced, the advancements in metabolic and genetic engineering of microbial strains have improved the yield and production cost.[24,29]

Some examples of bioplastic include PLA (polylactic acid), PHB (poly hydroxybutyrate), and PHA (polyhydroxyalkanoate).[65] The biodegradability of bioplastics is also dependent upon the crystallinity of the plastic, temperature, and pH of the surrounding environment, the microbial population of the locale, the presence, and absence of additives, etc. In oil-based plastics, crude oil is the basic component from which this type of plastics is made. Thus, they require higher energy input for their degradation.[54]

10.9.1.1 PLANT-BASED BIOPLASTIC

For plant-based plastics, comprising material is taken from the plants such as starch of potato, corn, or soy. They are easily degraded by 60% within 180 days and require light, air, and moisture for biodegradation. Polymers made up of starch represent 21.3% of bioplastic as per stats of 2019. Naturally occurring starch cannot be directly used for plastic preparation because of its intermolecular forces, hydrophilicity, and processability, providing high glass transition temperature and low melting temperature. To overcome this, further processing and addition of plasticizers like urea, glycerol, glycerine, or sorbitol are required at high shear force and temperature to improve its thermoplastic and plasticity characteristics, thus producing thermoplastic starch. Cellulose is another plant polymer that has been extensively investigated for the development of bioplastic. It can also be converted into nanoforms and could be utilized for the reason.[4]

In a past research, bioplastic was made from pumpkin seeds and flesh and was patented. The glycerin and other chemicals from the pumpkin seeds were extracted and mixed with the starch of pumpkin flesh under certain experimental conditions resulting in bioplastic. This bioplastic had super tensile strength in comparison to conventional plastics and greater biodegradability.[4,67]

10.9.1.2 DEGRADABLE POLYETHYLENE

Degradable starch polyethylene is a degradable form of polyethylene. Different degradation techniques, including chemical, biological, and photodegradation of this plastic form, have been investigated. Prooxidants are added to enhance the degradability of this plastic by chemical means as these prooxidants could generate free radicals that react with the oxygen of polymer thereby weakening the polyethylene matrix. UV light can also oxidize and lead to the production of free radicals, thus degrading the polyethylene. Streptomyces sp. have been widely investigated for its biological degradation by the release of effective enzymes. A less toxic additive is used to make it degradable within 60 days to 5 years if exposed to light and moisture.[40]

To further enhance the degradability of polyethylene, many weak sites are inserted in the polymeric chain, such as carbonyl groups, olefinic bonds, pro-oxidants, or photoinitiators. The carbonyl groups are inserted in the polymeric chains to upsurge the degradation. These carbonyl groups are introduced at the alpha position of short branched chains and become a part of polyethylene.[44] Introduction of carbonyl group is also possible at side chain by copolymerizing vinyl ketone with ethylene monomers. The rate of degradation is also increased by the introduction of diene content in the polymer by co-polymerization. This makes the polymer highly unsaturated in nature and thus reduces its melting point. Ziegler-Natta, a catalyst for the synthesis of composition, is photodegradable and is made of polyethylene consisting of unsaturation. Moreover, hydrolyzable linkages when introduced can in turn increase the degradability of polyethylene. Addition of ester group in polyethylene by copolymerizing 2-methylene-1,3-dioxepane with ethylene has been evaluated to enhance the degradability. Tercopolymer made by introduction of carboxy-ester in polyethylene by utilizing copolymerization of carbon monoxide, 2-methylene-1,3-dioxepane and ethylene with the help of free radical initiator was made and found to have increased degradation rate. Pro-oxidants as metioned above are inserted during the polyethylene formulation to increase the oxidation when exposed to light. These pro-oxidants are made by using transition metals in the form of long-chain acetylacetonates or carboxylates. The composition is generally made by using metals like manganese, iron, cobalt, and nickel.[25,44] Apart from pro-oxidants, starch molecules also act as weaker sites that increase the ease of degradation.[44]

10.9.2 COMPOSTABLE BIOPLASTICS

These are the advanced version of bioplastics having the properties of plastic (tensile strength, resistance to impacts, oxygen barrier) but consist of compostable components such as starch, cellulose, soy resins, etc. and can be composted within 90–180 days. The thickness and comprising components of compostable plastics are the main factors that ascertain their degrading margin and time.[4,65] All compostable plastics are biodegradable plastics, but all biodegradable plastics are not compostable.

10.10 FUTURE PROSPECTS

Keeping concerns with the huge utilization of plastic, lack of efficient alternatives and the harmful effects of plastic accumulation on the environment and life forms pose a great demand for exhaustive research on the investigation of efficient eco-friendly approaches for plastic degradation. Physical and chemical degradation processes have their own drawbacks, including adverse effects on the environment. Microbial enzymes have proved themselves effective candidates for the purpose with least or no harm to the environment. Further studies to enhance the catalytic activity of the enzymes by enzyme engineering could be an effective approach for efficient biodegradation. Recombinant microbes containing combinations of genes of different enzymes could be made to improve the degradation further as they will target more than one plastic polymer simultaneously in natural environmental conditions. Thus, exhaustive research in this concern will guide the development of efficient, eco-friendly enzyme systems that can be implemented to deal with the exponentially increasing plastic waste in a feasible manner.

10.11 SUMMARY

Plastics have multipurpose advantages of applicability in many fields. The daily use, disposal, and improper management of plastic wastes lead to the accumulation of waste around the globe. The recalcitrant nature of this waste resists its degradation posing adverse effects on the environment. Though chemical and physical degradation approaches have been investigated, they have severe disadvantages related to the release of toxic chemicals and the need of large, sophisticated infrastructures, respectively. Among all the approaches of degradation, the biotic approach seems eco-friendly and economically feasible. Microbes mediate biological degradation by secreting various enzymes like cutinases, proteases, lipases, esterases, which are the main key to the microbial degradation of plastic wastes. Numerous researches are going on to investigate specific high-efficiency strains producing effective enzymes for the degradation of plastic. Several specific enzymes having high potential to degrade the resistant plastic polymers have been reported and are being modified to enhance their specificity and efficiency as they may take several days to years to fully degrade the trash in their typical form in natural conditions. The major advantages related to environmental safety and simple, effective degradation processes of enzymes prove them efficient to be further investigated for large-scale purposes, especially in natural conditions.

KEYWORDS

- **biodegradation**
- **bioplastic**
- **enzymes**
- **microbes**
- **plastic polymers**

REFERENCES

1. Alauddin, M.; Choudhury, I. A.; El Baradie, M. A.; Hashmi M. S. J. Plastics and Their Machining: A Review. *J. Mater Process. Tech.* **1995**, *51* (1–4), 40–46.

2. Arutchelvi, J.; Sudhakar, M.; Arkatkar, A.; Doble, M.; Bhaduri, S.; Uppara, P. V. Biodegradation of Polyethylene and Polypropylene. *Indian J. Biotechnol.* **2008,** *7* (1), 9–22.
3. Awasthi, S.; Srivastava, N.; Singh, T.; Tiwary, D.; Mishra, P. K. Biodegradation of Thermally Treated Low Density Polyethylene by Fungus Rhizopus oryzae NS 5. *3 Biotech.* **2017,** *7* (1), 1–8.
4. Barbari, M.; Conti, L.; Rossi, G.; Monti, M. Compostable Packaging Based on Fibrous Natural Materials and Bioplastics. In *Conference: 11th International AIIA Conference on Biosystems Engineering Addressing the Human Challenges of the 21st Century*, July 5–8, 2017 Bari, Italy.
5. Bassi, A. Biotechnology for the Management of Plastic Wastes. Chapter 13. In *Current Developments in Biotechnology and Bioengineering*; Wong, J. W. C., Tyagi, R. D., Pandey, A., Eds.; Elsevier B. V., 2017; pp 293–310.
6. Bonhomme, S.; Cuer, A.; Delort, A. M.; Lemaire, J.; Sancelme, M.; Scott, G. Environmental Biodegradation of Polyethylene. *Polym. Degrad. Stab.* **2003,** *81* (3), 441–452.
7. Brems, A.; Baeyens, J.; Dewil R. Recycling and Recovery of Post-Cet Alonsumer Plastic Solid Waste in a European Context. *Therm. Sci.* **2012,** *16* (3), 669–685.
8. Chavan, A. J.; Email, W. S. Use of Plastic Waste in Flexible Pavements Miss. *IJAIEM,* **2013,** *2* (4), 540–552.
9. Christenson, E. M.; Patel, S.; Anderson, J. M.; Hiltner, A. Enzymatic Degradation of Poly (ether urethane) and Poly (carbonate urethane) by Cholesterol Esterase. *Biomaterials* **2006,** *27* (21), 3920–3926.
10. Demirbas, A. Biodegradable Plastics from Renewable Resources, Energy Sources, Part A Recover. *Util. Environ. Eff.* **2007,** *29* (5), 419–424.
11. Dussud, C.; Hudec, C.; George, M.; Fabre, P.; Higgs, P.; Bruzaud, S.; Delort, A. M.; Eyheraguibel, B.; Meistertzheim, A. L.; Jacquin, J.; Cheng, J.; Callac, N.; Odobel, C.; Rabouille, S.; Ghiglione, J. F. Colonization of Non-Biodegradable and Biodegradable Plastics by Marine Microorganisms. *Front. Microbiol.* **2018,** *9*, 1–13.
12. El-Shafei, H. A.; Abd El-Nasser, N. H.; Kansoh A. L.; Ali, A. M. Biodegradation of Disposable Polyethylene by Fungi and Streptomyces species. *Polym. Degrad. Stab.* **1998,** *62* (2), 361–365.
13. Farzi, A.; Dehnad A.; Fotouhi A. F. Biodegradation of Polyethylene Terephthalate Waste Using Streptomyces Species and Kinetic Modeling of the Process. *Biocatal. Agric. Biotechnol.* **2019,** *17*, 25–31.
14. Gilan, I.; Hadar, Y.; Sivan, A. Colonization, Biofilm Formation and Biodegradation of Polyethylene by A Strain of Rhodococcus Ruber. *Appl. Microbiol. Biotechnol.* **2004,** *65* (1), 97–104.
15. Grover, A.; Chandra, S.; Paul Khurana, S.; Gupta, A.; Kumari, A. Polythene and Environment IPA-Under Creative Commons license 3.0 Polythene and environment. *Int. J. Environ. Sci.* **2015,** *5* (6), 1091–1105.
16. Gyung Yoon, M.; Jeong Jeon, H.; Nam Kim, M. Biodegradation of Polyethylene by a Soil Bacterium and AlkB Cloned Recombinant Cell. *J. Bioremed. Biodegrad.* **2012,** *3* (4), 1–8.
17. Hadad, D.; Geresh, S.; Sivan, A. Biodegradation of Polyethylene by the Thermophilic Bacterium Brevibacillus borstelensis. *J. Appl. Microbiol.* **2005,** *98* (5), 1093–1100.

18. Hantani, Y.; Imamura, H.; Yamamoto, T.; Senga, A.; Yamagami, Y. Functional Characterizations of Polyethylene Terephthalate-Degrading Cutinase-Like Enzyme Cut190 Mutants Using Bis-(2-hydroxyethyl) Terephthalate as the Model Substrate. *AIMS Biophys.* **2018,** *5* (4), 290–302.
19. Hiraga, K.; Taniguchi, I.; Yoshida, S.; Kimura, Y.; Oda, K. Biodegradation of waste PET: A Sustainable Solution for Dealing with Plastic Pollution. *EMBO Rep.* **2019,** *20* (11), 1–5.
20. Hirai, H. Degradation of Polyethylene and Nylon-66 by the Laccase-Mediator System. *J. Polym. Environ.* **2001,** *103*, 1–20.
21. Hopewell, J.; Dvorak, R.; Kosior, E. Plastics Recycling: Challenges and Opportunities. *Philos. Trans. R. Soc. B Biol. Sci.* **2009,** *364* (1526), 2115–2126.
22. Hoshino, A.; Isono, Y. Degradation of Aliphatic Polyester Films by Commercially Available Lipases with Special Reference to Rapid and Complete Degradation of Poly(L-lactide) Film by Lipase PL Derived from Alcaligenes sp. *Biodegradation* **2002,** *13* (2), 141–147.
23. Huang, Q.; Hiyama, M.; Kabe, T.; Kimura, S.; Iwata, T. Enzymatic Self-Biodegradation of Poly(l-lactic acid) Films by Embedded Heat-Treated and Immobilized Proteinase K. *Biomacromolecules* **2020,** *21* (8), 3301–3307.
24. Kale, G.; Kijchavengkul, T.; Auras, R.; Rubino, M.; Selke, S. E.; Singh, S. P. Compostability of Bioplastic Packaging Materials: An Overview. *Macromol. Biosci.* **2007,** *7* (3), 255–277.
25. Kale, S. K.; Deshmukh, A. G.; Dudhare, M. S.; Patil, V. B. Microbial Degradation of Plastic: A Review. *J. Biochem. Tech.* **2015,** *6* (2), 952–961.
26. Kaminsky, W.; Predel, M.; Sadiki, A. Feedstock Recycling of Polymers by Pyrolysis in a Fluidised Bed. *Polym. Degrad. Stab.* **2004,** *85*, 1045–1050.
27. Kawai, F.; Kawabata, T.; Oda M. Current Knowledge on Enzymatic PET Degradation and Its Possible Application to Waste Stream Management and Other Fields. *Appl. Microbiol. Biotechnol.* **2019,** *103* (11), 4253–4268.
28. Kawai, F.; Oda, M.; Tamashiro, T.; Waku, T.; Tanaka, N.; Yamamoto, M.; Mizushima, H.; Miyakawa, T.; Tanokura, M. A Novel Ca2+-Activated, Thermostabilized Polyesterase Capable of Hydrolyzing Polyethylene Terephthalate from Saccharomonospora Viridis AHK190. *Appl. Microbiol. Biotechnol.* **2014,** *98* (24), 10053–10064.
29. Khazir, S.; Shetty, S. Bio-Based Polymers in the World. *Int. J. Life Sci. Biotechnol. Pharma Res.* **2014,** *3* (2), 35–43.
30. Kosior, E.; Mitchell, J.; Crescenzi, I. Plastics Recycling. *Issues Environ. Sci. Technol.* **2019,** *2019* (47), 156–176.
31. Koutny, M.; Lemaire, J.; Delort, A. M. Biodegradation of Polyehthylene Films with Prooxidant Additives. *Chemosphere* **2006,** *64*, 1243–1252.
32. Lee, Z. H.; Paul, S. C.; Kong, S. Y.; Susilawati, S.; Yang, X. Modification of Waste Aggregate PET for Improving the Concrete Properties. *Adv. Civ. Eng.* **2019,** *2019* (6942052), 1–11.
33. Liu, M.; Zhang, T.; Long, L.; Zhang, R.; Ding, S. Efficient Enzymatic Degradation of Poly (ε-Caprolactone) by an Engineered Bifunctional Lipase-cutinase. *Polym. Degrad. Stab.* **2019,** *160*, 120–125.
34. Muhonja, C. N.; Magoma, G.; Imbuga, M.; Makonde, H. M. Molecular Characterization of Low-Density Polyethene (LDPE) Degrading Bacteria and Fungi from Dandora Dumpsite, Nairobi, Kenya. *Int. J. Microbiol.* **2018,** *2018* (4167845), 1–10.

35. Müller, R. J.; Schrader, H.; Profe, J.; Dresler, K.; Deckwer, W. D. Enzymatic Degradation of Poly(Ethylene Terephthalate): Rapid Hydrolyse Using a Hydrolase from T. fusca. *Macromol. Rapid Commun.* **2005,** *26* (17), 1400–1405.
36. Nimchua, T.; Eveleigh, D. E.; Sangwatanaroj, U.; Punnapayak, H. Screening of Tropical Fungi Producing Polyethylene Terephthalate-Hydrolyzing Enzyme for Fabric Modification. *J. Ind. Microbiol. Biotechnol.* **2008,** *35* (8), 843–850.
37. Nimchua, T.; Punnapayak, H.; Zimmermann, W. Comparison of the Hydrolysis of Polyethylene Terephthalate Fibers by a Hydrolase from Fusarium Oxysporum LCH I and Fusarium solani f. sp. pisi. *Biotechnol. J.* **2007,** *2* (3) 361–364.
38. Ojha, N.; Pradhan, N.; Singh, S.; Barla, A; Shrivastava, A.; Khatua, P.; Rai, V.; Bose, S. Evaluation of HDPE and LDPE Degradation by Fungus, Implemented by Statistical Optimization. *Sci. Rep.* **2017,** *7* (39515), 1–13.
39. Papadopoulou, A.; Hecht, K.; Buller, R.; Papadopoulou A. Enzymatic PET Degradation. *CHIMIA* **2019,** *73* (9), 743–749.
40. Pometto, A. L.; Lee, B.; Johnson, K. E. Production of an Extracellular Polyethylene-Degrading Enzyme(s) by Streptomyces species. *Appl. Environ. Microbiol.* **1992,** *58* (2), 731–733.
41. Pramila, R.; Ramesh, K. V. Biodegradation of Low Density Polyethylene (LDPE) by Fungi Isolated From Municipal Landfill Area. *J. Microbiol. Biotechnol. Res.* **2011,** *1* (4), 131–136.
42. Production of Plastics Worldwide from 1950 to 2019 (in Million Metric Tons). Statista, 2019. https://www.statista.com/statistics/282732/global-production-of-plastics-since-1950/ (accessed June 06, 2021).
43. Ronkvist, A. M.; Xie, W.; Lu, W.; Gross, R. A. Cutinase-Catalyzed Hydrolysis of Poly(ethylene terephthalate). *Macromolecules* **2009,** *42* (14), 5128–5138.
44. Roy, P. K.; Hakkarainen, M.; Varma, I. K.; Albertsson, A. C. Degradable Polyethylene: Fantasy or Reality. *Environ. Sci. Technol.* **2011,** *45* (10), 4217–4227.
45. Sack, U.; Hofrichter, M.; Fritsche, W. Degradation of Polycyclic Aromatic Hydrocarbons by Manganese Peroxidase of Nematoloma frowardii. *FEMS Microbiol. Lett.* **1997,** *152* (2), 227–234.
46. Santo, M.; Weitsman, R.; Sivan, A. The Role of the Copper-Binding Enzyme – Laccase in the Biodegradation of Polyethylene by the Actinomycete Rhodococcus Ruber. *Int. Biodeterior. Biodegrad.* **2013,** *84*, 204–210.
47. Shah, A. A.; Hasan, F.; Hameed, A.; Ahmed, S. Biological Degradation of Plastics: A Comprehensive Review. *Biotechnol. Adv.* **2008,** *26* (3), 246–265.
48. Sharon, C.; Sharon, M. Studies on Biodegradation of Polyethylene Terephthalate: A Synthetic Polymer. *J. Microbiol. Biotechnol.* **2012,** *2* (2), 248–257.
49. Singh, B.; Sharma, N. Mechanistic Implications of Plastic Degradation. *Polym. Degrad. Stab.* **2008,** *93* (3), 561–584.
50. Singh, P.; Sharma, V. P. Integrated Plastic Waste Management: Environmental and Improved Health Approaches. *Procedia Environ. Sci.* **2016,** *35*, 692–700.
51. Sivan, A. New Perspectives in Plastic Biodegradation. *Curr. Opin. Biotechnol.* **2011,** *22* (3), 422–426.
52. Subramanian, P. M. Plastics Recycling and Waste Management in the US. *Resour. Conserv. Recycl.* **2000,** *28* (3–4), 253–263.

53. Sumathi, T.; Viswanath, B.; Sri Lakshmi, A.; SaiGopal, D. V. Production of Laccase by Cochliobolus sp. Isolated from Plastic Dumped Soils and Their Ability to Degrade Low Molecular Weight PVC. *Biochem. Res. Int.* **2016,** *2016* (9519527), 1–10.
54. Swain, S. N.; Biswal, S. M.; Nanda, P. K.; Nayak, P. L. Biodegradable Soy-Based Plastics: Opportunities and Challenges. *J. Polym. Environ.* **2004,** *12* (1), 35–42.
55. Takuya, M.; Mizushima, H.; Ohtsuka, J.; Oda, M; Kawai, F; Tanokura M. Structural Basis for the Ca2+-Enhanced Thermostability and Activity of PET-Degrading Cutinase-Like Enzyme from Saccharomonospora Viridis AHK190. *Appl. Microbiol. Biotechnol.* **2015,** *99* (10), 4297–4307.
56. Thipse, S. S.; Sheng, C.; Booty, M. R.; Magee, R. S.; Dreizin, E. L. Synthetic Fuel for Imitation of Municipal Solid Waste in Experimental Studies of Waste Incineration. *Chemosphere* **2001,** *44* (5), 1071–1077.
57. Tokiwa, Y.; Jarerat, A. Biodegradation of Poly(L-lactide). *Biotechnol. Lett.* **2004,** *26* (10), 771–777.
58. Wang, H.; Zhang, Y.; Wang, C. Surface Modification and Selective Flotation of Waste Plastics for Effective Recycling: A Review. *Sep. Purif. Technol.* **2019,** *226*, 75–94.
59. Webb, H. K.; Arnott, J.; Crawford, R. J.; Ivanova E. P. Plastic Degradation and its Environmental Implications with Special Reference to Poly(ethylene terephthalate). *Polymers (Basel).* **2013,** *5* (1), 1–18.
60. Wei, R.; Oeser, T.; Then, J; Kühn, N.; Barth, M.; Schmidt, J.; Zimmermann, W. Functional Characterization and Structural Modeling of Synthetic Polyester-Degrading Hydrolases from Thermomonospora curvata. *AMB Express* **2014,** *4* (1), 1–10.
61. Wei, R.; Zimmermann, W. Biocatalysis as a Green Route for Recycling the Recalcitrant Plastic Polyethylene Terephthalate. *Microb. Biotechnol.* **2017,** *10* (6), 1302–1307.
62. Yamada-Onoder, K.; Mukumoto, H.; Katsuyaya, Y.; Saiganji, A.; Tani, Y. Degradation of Polyethylene by a Fungus, Penicillium simplicissimum YK. *Polym. Degrad. Stab.* **2001,** *72* (2), 323–327.
63. Yamashita, K.; Kikkawa, Y.; Kurokawa, K.; Doi, Y. Enzymatic Degradation of Poly(L-Lactide) Film by Proteinase K: Quartz Crystal Microbalance and Atomic Force Microscopy Study. *Biomacromolecules* **2005,** *6* (2), 850–857.
64. Yang, Z.; Peng, H.; Wang, W.; Liu, T. Crystallization Behavior of Poly(ε-Caprolactone)/ Layered Double Hydroxide Nanocomposites. *J. Appl. Polym. Sci.* **2010,** *116* (5), 2658–2667.
65. Yogalakshmi, K. N.; Singh, S. Plastic Waste: Environmental Hazards, Its Biodegradation and Challenges. Chapter 6. In *Bioremediation of Industrial Waste for Environmental Safety*; Saxena, G., Bharagava, R. N., Eds.; Vol. 1; Springer Singapore: Singapore; 2020; pp 1–35.
66. Yoshida, S.; Hiraga, K.; Takehana, T.; Taniguchi, I.; Yamaji, H.; Maeda, Y.; Toyohara, K.; Miyamoto, K.; Kimura, Y.; Oda, K. A Bacterium that Degrades and Assimilates Poly(Ethylene Terephthalate). *Science (New York, N. Y.)* **2016,** *351* (6278), 1196–1199.
67. Zheng, Y.; Yanful, E. K.; Bassi, A. S. Review of Plastic Waste Biodegradation. *Crit. Rev. Biotechnol.* **2005,** *25* (4), 243–250.

PART II

PLANTS AND BIOPRODUCTS FOR INACTIVATION OF ENZYMES

CHAPTER 11

NATURAL MEDICINAL PRODUCTS AS POTENTIAL ENZYME INHIBITORS

ANNIKA DURVE GUPTA[1] and SONALI ZANKAR PATIL[2]

ABSTRACT

One of the drug discovery and strategy inquiry efforts in research organizations throughout the world is the hunt for economical, authentic, high-quality, and easily available natural substance enzyme inhibitors. Many natural substances are enzyme inhibitors and discovering and developing them is an important area of pharmacology and biochemistry study. Efficacy and specificity, which show the intended absorption to block the enzyme, are typically mediated by pharmacological enzyme inhibitors. Enzyme inhibitors are intriguing since they can be used to treat a variety of medical disorders. A lack of enzymes causes a variety of disorders. Enzyme inhibitors have a significant position as new drugs because they have been used to treat a variety of physiological disorders. Secondary metabolites with bioactive properties can provide excellent pharmacophore patterns for drugs used to treat a wide range of diseases. Natural enzyme inhibitors play an important role in the regulation of a variety of metabolic processes. Many treatment techniques are being researched for serious health issues, with the main enzyme inhibitory hypothesis being one of the most commonly used. Therapeutically relevant enzyme inhibitors can be found in a wide range of medications used to treat a wide range of illnesses. The majority of inhibitors interact with specific enzymes by inhibiting their activity toward physiological natural substances while leaving other enzymes alone. Some secondary metabolites, for example, have intriguing anti-acetylcholinesterase (anti-AChE) activity,

Enzyme Inactivation in Food Processing. Technologies, Materials, and Applications. Megh R. Goyal, PhD, Junaid Ahmad Malik, PhD & Ravi Pandiselvam, PhD (Eds.)

which has been linked to Alzheimer's disease development (AD). This chapter discusses the various natural products obtained from various sources that act as enzyme inhibitors against various diseases and infections, as well as new drug discovery strategies.

11.1 INTRODUCTION

In the broadest sense, a natural product is anything produced by life, which includes bio-based materials (e.g., bioplastics, cornstarch), biotic materials (e.g., wood, silk), body liquids (e.g., milk, plant exudates), and other natural materials previously found in living creatures (e.g., soil, coal).[5]

In a wider sense, a natural product is any organic molecule generated by a living organism. Natural commodities have a significant structural variation as well as specific pharmacological or biological activity as a result of natural selection and evolutionary processes that have molded their worth over hundreds of thousands of years. In actuality, the structural variety of natural molecules considerably exceeds the capability of laboratory-based synthetic organic chemists. As a result, natural products have been utilized to cure diseases in both traditional and modern medicine. Natural chemicals are commonly utilized as beginning points for the creation of medications, followed by synthetic modifications to decrease adverse effects and increase bioavailability. Natural items, in fact, provide the inspiration for approximately half of all FDA-approved medications in the United States.[35]

In addition to medicine, natural goods and their derivatives, such as spices and herbs, antibacterial agents, and antioxidants, are frequently employed as food additives to maintain food freshness and durability. Natural organic items may be found in nearly every area of our life, from clothing to plastics and rubber, health, and beauty products, and even the energy we use to power our cars.

Natural products are categorized based on their biological function, metabolic activity, or origin. Primary and secondary metabolites are two common natural product categorization groupings. Primary metabolites are organic compounds that have an intrinsic function that is critical to the organism's existence (i.e., the organism would die without these metabolites). Main metabolites include the building components (nucleic acids, amino acids, carbohydrates, and fatty acids) needed to produce the primary macromolecules that support life (DNA, RNA, proteins, carbs, and lipids). Secondary metabolites, on the other hand, are organic compounds with an extrinsic function that affects species other than the producer. Secondary metabolites are not required for survival, but they do aid the organism's ability to compete in its environment. Primary and secondary metabolites are commonly used to characterize natural compounds, particularly in the field of organic chemistry. In the areas of medicinal chemistry and pharmacognosy, the study and application of natural products in medicine, a more conservative definition restricting natural products to secondary metabolites is frequently used.[35]

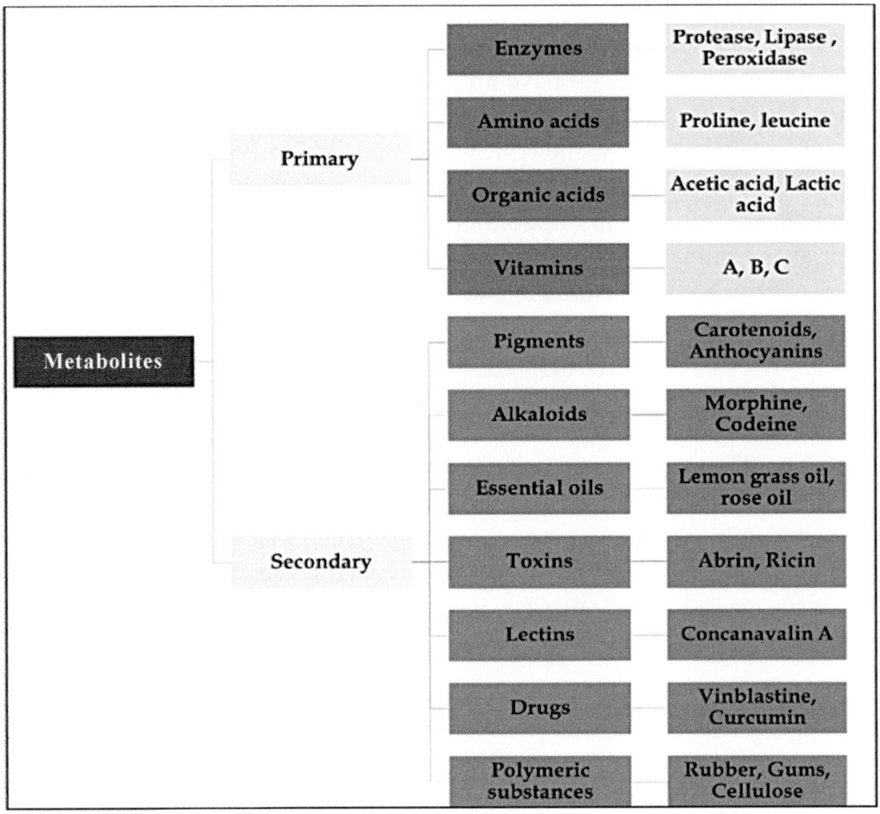

FIGURE 11.1 Primary and secondary metabolites.

Natural products can be extracted from microorganisms, plants, and animals' cells, tissues, and secretions. A crude (unfractionated) extract from any of these sources will contain a wide variety of structurally different and, in many cases, new chemical compounds. Scientists travel the world to gather samples for examination and assessment in drug discovery screens or bioassays since chemical variety in nature is based on biological diversity. Bioprospecting refers to the search of natural resources. Pharmacognosy, or the study of natural items having biological activity, provides the techniques for recognizing, selecting, and processing natural commodities intended for medicinal use. A natural extract generally possesses some biological activity that can be attributed to a single component or a set of related chemicals generated by the organism. These active molecules can be utilized directly in medication discovery and development or can be synthetically changed to increase biological characteristics or minimize adverse effects.[164]

Skin issues, gouty arthritis, hyperuricemia, diabetes, Alzheimer's disease, and other chronic illnesses have all become increasingly common in recent years. These diseases have been identified as a global health concern that affects many people throughout the world, and they are typically caused by the malfunction or overactivity of many metabolic enzymes involved in human metabolism.[133,134] Enzyme inhibitors are important medicines used to treat a variety of human diseases, including metabolic disorders, cardiovascular disease, neurological disorders, and cancer. As a result, inhibiting the catalytic activity of these therapeutically important enzymes has been identified as a critical discovery for treating the diseases described. As a result, using natural plant medicines for effective therapy is an important research field for treating many diseases while minimizing side effects. These enzymes and their inhibitors have been the subject of significant debate.[164] This chapter investigates various natural enzyme inhibitors derived from various sources (microbes, plants, and marine organisms) that have the potential to be used in the treatment of a variety of human diseases, including metabolic disorders, cardiovascular disease, neurological disorders, and cancer.

11.2 NATURAL PRODUCTS AS GOOD ENZYME INHIBITORS

Technological developments and the development of artificial intelligence in recent decades have led many academics to believe that it is possible to manufacture new medications entirely through in silico approaches. However, in the vast majority of situations, this method was futile. One of the primary reasons for this failure is that humans, in comparison to the natural scaffold molecules present in the world, have a limited ability to imagine new ones. Plants, algae, and microbes produce thousands of structurally unique compounds, the majority of which are rarely replicable in current organic chemistry laboratories. This evidence suggests that, despite significant technological progress, drug discovery will continue to be heavily reliant on the discovery of new natural bioactive molecules with drug-like activity in the coming years.

Natural products are abundantly available, and our forefathers have used their unique characteristics to heal ailments and injuries for ages. The potential of natural products as modulators of biological processes has been gradually recognized throughout evolution.[15] Natural products have been employed less frequently as a starting point for drug development by pharmaceutical companies in recent decades, due to the assumption that natural products are incompatible with drug discovery methodologies based on high throughput screening focused at molecular targets.[126] Furthermore, it was anticipated that combinatorial chemistry techniques could generate all of the chemical variation required for good lead discovery.

The findings of many huge combinatorial screening collections, on the other hand, have been rather disheartening, and it has already been established that variety within physiologically relevant "chemical space" is more essential than library size. To a degree, libraries of synthetic molecules have been created to replicate the chemical characteristics of natural substances.[73] Despite a lack of investment in natural goods as primary leads in drug development in recent decades, 34% of medications authorized by the US Food and Drug Administration (FDA) between 1981 and 2010 were natural compounds or directly derived from them.[127] Natural-source compounds have a high potential for drug development due to their extraordinary chemical and structural diversity. Approximately 40% of the chemical scaffolds present in natural products are missing from today's medicinal chemistry synthetic libraries. As a result, the utilization of nature inspired molecules is an excellent complement to synthetically manufactured molecules.[77,93]

One of the most important reasons for natural products' effectiveness as a source of bioactive molecules is their "drug likeness," which typically outperforms that of synthetic substances. Given their biosynthesis processes in live creatures, natural compounds have higher similarity and binding potential with biological structures, enhancing the likelihood of a successful interaction with various biological targets.[93]

One of the most notable characteristics of natural products is their three-dimensional conformation, which is due to their complex and unique structure, which is usually beyond the synthetic capability of medicinal chemistry. Natural products are frequently defined as having a "privileged" architecture that allows them to act as ligands for a wide range of enzymes and receptors. This phrase was coined by Evans in the late 1980s to refer to the benzodiazepine scaffold, which is distinguished by its capacity to bind not just to receptors in the central and peripheral nervous systems, but also to cholecystokinin receptors. In this way, and according to Evans's definition, a privileged structure displays affinity to several receptors/proteins.[17] During recent years various researchers around the world published an in-depth analysis that included a comprehensive list of favored scaffolds identified in both synthetic and natural medicines. Despite having the identical scaffold (6,6spiroacetal), spiketp, integramycin, and routiennocin have diverse bioactivities and are present in different species, demonstrating the evolution-driven proclivity for repetition once a good solution to a specific biochemical issue has been identified. This can also explain why macromolecular structures in biological systems do not form in a random fashion. As a result, 6,6 spiroacetal is a "privileged" scaffold found in a variety of natural products, with the potential to bind to diverse targets and exhibit distinct pharmacological effects.[17,47,203]

Given the huge diversity of chemicals that occupy the 'chemical space,' it is easy to infer that natural product span unique regions when compared to synthetic ones, having broader and more drug like characteristics. Rosen et al.[154] showed through computational screens that natural products cover portions of the chemical space that medicinal chemistry molecules do not cover, and as a result, these compounds may be beneficial for fresh leads.

Natural molecules differ from manufactured molecules in that they include less nitrogen, sulphur, or halogen atoms, are richer in oxygen, and have more hydrogen bond donors. Natural products' sterically complexity also plays a part in this equation, with these molecules displaying a greater number of rings and, overall, more chiral centers. Natural products are not created at random, but rather as a result of highly focused biosynthetic processes. As a result, these compounds are designed to interact with three dimensional and chiral chemical targets.[56,157]

Furthermore, the enzymes involved in biosynthesis are generally chiral in the sense that they generate a single isomer, which is not necessarily the case in medicinal chemistry, where racemic mixtures are commonly formed.[138] Enzymes involved in natural biosynthesis, as well as the molecular targets with which the natural product is intended to interact, are fundamentally three dimensional and chiral, much like human enzymes. As a result, there is a connection that helps explain why natural compounds can perform well as enzyme inhibitors.[138] The fact that the majority of natural products exhibit such properties indicates that they are the consequence of an evolutionary force that chooses molecules with a certain arrangement of atoms.[80,138] Furthermore, as compared to a randomly produced molecule, the likelihood of identifying a bioactive molecule in natural goods is significantly higher.[17] This is not surprising given that it results from nature's own high-throughput screening: not only are molecules prone to displaying the well-defined three-dimensional structure described above, but they are also produced to target well-conserved biological targets in a specific mechanism of action, which means that they are synthesized to exhibit some kind of activity toward a biological target.[4,203]

Another notable distinction between natural products and randomly generated molecules is the underlying synthesis method in both situations. Nature does not have a process like combinatorial chemistry, which may employ tens of thousands of distinct scaffolds as building pieces. Nature, in fact, has a limited number of building blocks available from a small number of biosynthetic pathways for biosynthesis, and as a result, any chemical novelty can only be achieved by branching out intermediates and creating a multitude of biosynthetic routes, ultimately generating different chemical entities. In combinatorial chemistry, a different method is used, in which the same series of reactions is frequently followed using different starting compounds. Nature's chemistry is fundamentally oxophilic, and as a result, it possesses enzymatic tools that conduct site selective CH activation to introduce oxygen atoms and distinguish between functional groups with varying degrees of oxidation.[24,72]

Natural products have coexisted with many species and habitats across time, going through the same iterative cycle of improvement and development as ever-changing biotic systems. Several natural substances and their derivatives function as enzyme inhibitors in a variety of therapeutic domains, including cancer, diabetes, hypertension, and infectious illnesses.[18] Teprotide, extracted from the venom of *Bothrops jararaca*, is a typical example. This chemical demonstrated long-lasting in vivo action against angiotensin-converting enzyme and was selected as a lead molecule for the development of angiotensin carboxypeptidase inhibitors.[97] Galanthamine, an anticholinesterase medication derived from the snowdrop plant *Galanthus nivalis* L., has been widely used to treat Alzheimer's disease.[137] Vialinin A, a pterphenyl chemical derived from the edible Chinese fungus *Thelephora vialis* plant, significantly suppresses the synthesis and release of tumor necrosis factor.[208]

Given the complicated structural diversity of these products, it is fair to doubt that these molecules would ever be viable as orally active medicines. Lipinski published a set of four criteria in 1997 that were shared by 90% of the more than 2000 medications and potential medicines in phase II clinical trials or later. These characteristics specify that a potential medication should contain less than five hydrogen donor groups, fewer than five hydrogen acceptor groups, a molecular weight less than 500 Da, and a log P less than 5. However, it is unmistakable that many physiologically active natural substances do not meet all of these criteria. As a result, Lipinski's criterion is violated by 12 of the 24 natural medicines licenced between 1970 and 2006.[138,143] In fact, only 51% of all FDA-approved small-molecule medicines were taken orally in 2006, therefore complying with this requirement.[210] Nonetheless, it is crucial to note that Lipinski's guidelines leave out several critical points. Natural goods, for example, typically benefit from transport mechanisms, making features such as molecular weight less significant in terms of intestine absorption.[56]

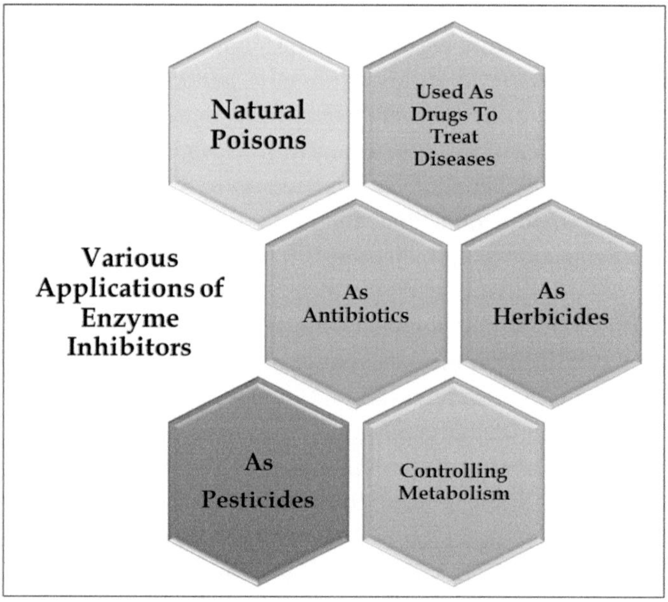

FIGURE 11.2 Applications of enzyme inhibitors.

Natural compounds generated from microbial sources are believed to be more chemically varied than those derived from plants. Microbial origin may be identified in roughly 53% of FDA-approved natural product-based medicines, primarily antiviral treatments.[144] Ara-A (also known as vidarabine) was one of the first antiviral nucleoside analogues identified, for example, in

Streptomyces antibioticus.[49] Several antiviral medicines based on microbial nucleosides were produced as a result.[170] Furthermore, some ansamycin-based antibiotics (such as rifamycin, 10) have demonstrated interesting antiviral effects against a wide range of infectious viruses.[45] Furthermore, the commonly used immunomodulatory medication mycophenolic acid,[11] which is produced from fungus, has demonstrated widespread antiviral action.[45]

11.3 SCREENING OF NATURAL PRODUCT EXTRACTS FOR POTENTIAL ENZYME INHIBITORS

Laboratory methods for screening enzyme inhibitors obtained from natural product extracts rely on various techniques such as spectroscopy, and enzyme activity is often measured using specially designed, labeled substrates that show a spectrometrically measurable signal as color reaction (absorbance) or fluorescence of a distinct wavelength after enzymatic cleavage.[149,59,94] To evaluate alpha-glucosidase activity, a chromogenic substrate p-nitrophenyl— d-glucopyranoside (pNPG) is often employed. pNPG, colorless compound with a glycosidically linked d-glucose residue and a p-nitrophenol moiety, is attacked at the glycosidic bond by α-glucosidase enzyme resulting in the release of p-nitrophenol (a yellow colored product), which is spectrophotometric measured at λ = 405 nm[14,89] (Fig. 11.3).

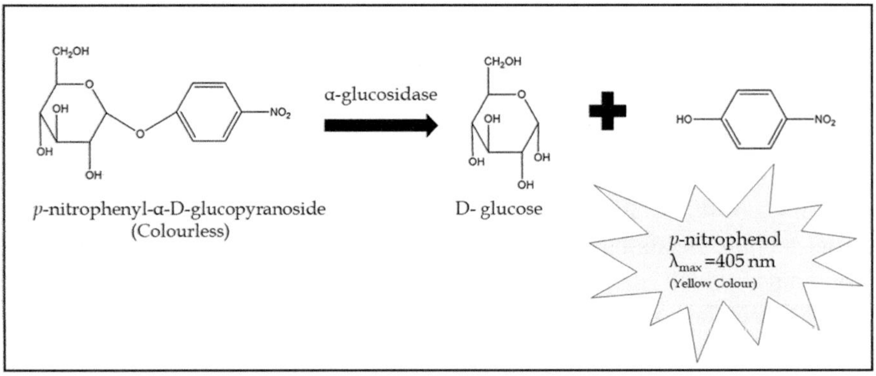

FIGURE 11.3 Enzymatic action of α-glucosidase results in the hydrolysis of colorless *p*-nitrophenyl-α-d-glucopyranoside to form a yellow-colored p-nitrophenol.[149]

Fluorescence based enzyme assays that employ fluorophore-linked substrates are also commercially available. The fluorophore is released by an enzymatic cleavage and subsequently fluoresces when stimulated at a specific wavelength.[14,89] The BODIPY ® FL-DQTM (boron-dipyrromethene fluorescent dye quenching) starch is widely used in fluorescence-based α-amylase tests. In spectrometry-based tests, turbidity, autofluorescence, and sample color induced by poor solubility can produce spectral interference.[175,189] As a result, while developing and carrying out enzyme-based tests with extracts of natural product, these variables must be taken into account (Fig. 11.4).

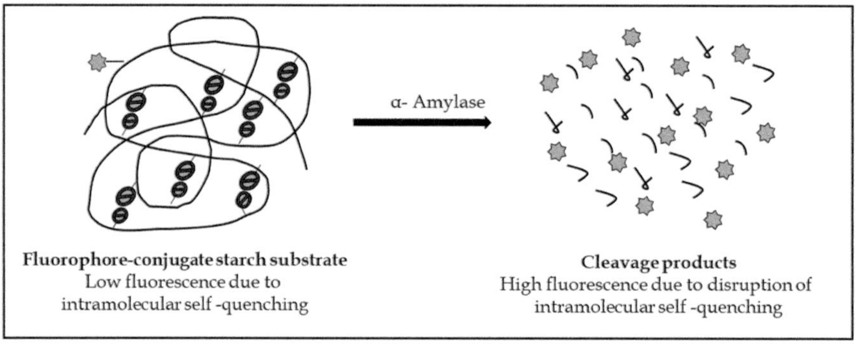

FIGURE 11.4 Representation of BODIPY®FL-DQ™ starch assay model.[185]

Plant extracts are typically vividly colored because they include naturally occurring, highly conjugated pigments. Absorbance values in spectrophotometric studies can be distorted by natural product coloring and plant extracts.[175] Carotenoids and Chlorophylls are the two primary photosynthetic plant pigments. Chlorophylls absorb a lot of light in the wavelength ranges 430–480 nm and 640–660 nm.[64] Carotenoids, which include β-carotene, phytoene, and lycopene, are a class of chemicals with a visible light absorption range of 400–530 nm.[65] The additional groups of pigments present in plants include betalains, which include betaxanthins and betacyanins,[6] and flavonoids, which include flavanols and anthocyanins.[58,115] All of these phytochemical groups can disrupt assays that employ wavelengths that overlap their absorption spectra (Fig. 11.5).

In addition to the pigments present naturally in plants, postharvest modifications can result in the creation of novel colored products. Polyphenol oxidases, for example, generate melanins and other dark pigments in harvested plant tissue exposed to oxygen. Enzymatic browning, commonly known as fruit browning, occurs often.[117,131,192] In sugarcane (*Saccharum* spp.) extracts, for example, the Maillard reaction and caramelization processes induce significant browning.[57] The colored by-products of such reactions can cause substantial hindrance in absorbance-based assays. For example, anthranilates, alkaloids, stilbenes, and coumarins fluoresce in the blue-violet range, that is, 400–520 nm; flavones and flavonoids fluoresce in the green-yellow range, that is, 520–590 nm; polycyclic aromatic quinones, tannins, and some alkaloids fluoresce in the orange range, that is, 635–590 nm; and porphyrins, chlorophyll in 600 to 730 nm.[40,43]

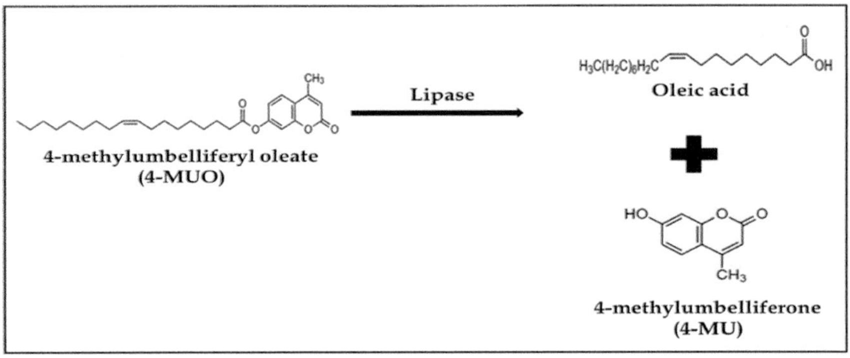

FIGURE 11.5 Hydrolysis of compound 4-MUO using enzyme lipase.[79]

Turbidity is produced by the presence of undissolved, suspended particles as a result of some extracts and compounds' failure to dissolve in the assay buffers, which can result in incorrect findings. Light experiences many scattering and absorption processes as it travels through a turbid medium. Photon absorption and scattering can potentially affect fluorescence measurements in a turbid media.[119]

In enzyme tests, the substrate can potentially be a source of error. Unstable substrates may progressively degrade to produce their product. Contamination of the substrate with the fluorogenic or chromogenic product creates a false signal and can result in an unrelated rise in fluorescence or absorbance.[51] Finally, assay interference from sample color, autofluorescence, and turbidity may contribute to measurement errors, lowering findings accuracy and reproducibility.[69,175] As a result, it is critical to reduce the impact of these interferences by blank-correcting raw data (RD) with appropriate sample and reagent blanks.

We have found natural compounds from medicinally significant plants that display anti-enzymatic action and have applications in treating diabetes, Alzheimer's disease, hypertension, and overcoming drug resistance difficulties in cancer treatment during the last several years.

11.4 VARIOUS NATURAL PRODUCTS USED AS ENZYME INHIBITORS

11.4.1 UREASE INHIBITORS

Urease (EC 3.5.1.5) is a crucial enzyme in the universal nitrogen cycle, which can exist in fungi, medicinal plants, and various bacteria. This type of hydrolase hastens urea hydrolysis by up to 100 times and transforms it to carbon dioxide and ammonia.[52,90,105,184] Meanwhile, this discovery in medicinal plants, including *Canavalia ensiformis*, a member of the Fabaceae family, has been thoroughly examined. This has changed the field of biochemistry understanding as this was the primary enzyme to be crystallized, studied,[38,181] as well as confirmed to be firmly dependent on nickel ions.[166]

Urease is the only hydrolytic enzyme that requires Ni^{2+} ions for catalytic activity.[52,105] In 1995, Jabri et al.[77] were able to fully study the three-dimensional structure of urease enzyme obtained from *Klebsiella aerogenes* using the technique of crystallography. Later, other structures for urease enzymes were found in *Bacillus pasteurii, Helicobacter pylori*, and most recently *C. ensiformis.*[67]

Urease enzyme, which raises the medium pH through NH_3 accretion in the presence of H. pylori, is an important medical urease.[184] Infections of the gastrointestinal tract or urine by ureolytic bacteria can cause various health issues in humans and other animals, which can include pyelonephritis, kidney stone formation, and, in the long run can result in hepatic coma and hepatic encephalopathy.[158] As a result, Gram-negative bacterium *H. pylori*, which flourishes in an acidic environment such as in the stomach (pH 2–4), has been linked to a number of major public health issues. *H. pylori* poisoning can cause stomach irritation as well as an increased risk of developing duodenal, gastric ulcers, gastric adenocarcinoma, and gastric lymphoma.[158] *H. pylori* can live in sick people's stomachs for the rest of their lives without causing illness. The widespread presence of *H. pylori* in the human population suggests that this bacterium has evolved methods to counteract host defences.[179] The main virulence component of this human disease is the urease enzyme, which is found in the cytoplasm or is superficially attached to *H. pylori.*[179] It is proposed that cytosolic ureases, which link to the surface of intact bacterial cells and are responsible for the hydrolysis of urea in human intestines at a concentration of 3 mM, are activated by pathogen cell lyses. The NH_3 produced raises the pH of the media, creating an ideal environment for the growth of *H. pylori.*[60] As seen in the last two decades, the advised first-line medication for *H. pylori* suppression included a combination of the antibiotics such as clarithromycin and amoxicillin, as well as omeprazole (proton pump inhibitor). The rise in *H. pylori* resistance to various antibiotics (particularly clarithromycin) complicates this treatment, which is an unappealing option in the modern day.[109]

Additional *H. pylori* infection-fighting strategies have emerged, including bismuth salts usage in conjunction with a proton pump cell inhibitor. In addition, inhibitors of urease could be used as active treatments to treat diseases caused by pathogenic bacteria that are dependent on urease. Though, commercially existing inhibitors of urease such as phosphorodiamidates, hydroxamic acid derivatives, and imidazoles are unstable and harmful, thwarting their use in clinical trials.[7] To enhance animal and human living quality, the primary search for new, recognized, innovative, and bioactive inhibitors of urease with reduced toxicity and improved stability is required.

11.4.2 XANTHINE OXIDASE

Gout is a serious public health problem. Gout, also referred to as hyperuricemia, affects 5–30% of the populace.[153] It appears to be spreading around the world and shows a significant risk factor for symptoms such as tophaceous gout, nephrolithiasis, and gouty nephropathy.[53,198] Hyperuricemia is caused by an excess or lack of uric acid excretion, which is aggravated by an intake of high dietary foods rich in nucleic acid iron, like leguminous pulses, meats, and marine food. Xanthine oxidase catalyzes the conversion of hypoxanthine and xanthine into uric acid uricosuric drugs, which promote uric acid elimination through the urinary tract, and inhibitors of xanthine oxidase, that halts the fatal step in uric acid biosynthesis, can reduce plasma uric acid concentrations and are commonly used to treat gout.[195]

Furthermore, xanthine oxidase functions as a significant organic oxygen source, releasing free radicals that contribute to the oxidative destruction of materials existing, resulting in a variety of extreme situations such as inflammation, cancer, hepatitis, ischemia reperfusion, and the elderly.[106,167] Allopurinol, the most commonly used inhibitor of xanthine oxidase in the therapy of gout, causes a slew of side effects, including allergic reactions, hepatitis, and nephropathy.[74] As a result, the hunt for new inhibitors of xanthine oxidase with increased medicinal potential and less side effects is centered not only on treating gout but also on addressing a number of other disorders associated with action of xanthine oxidase.

11.4.3 ANGIOTENSIN-CONVERTING ENZYME

Angiotensin-converting enzymes (ACE, peptidyldipeptide hydrolase, kininase II, EC 3.4.15.1) regulate blood pressure.[48] Angiotensin I-converting enzyme is a blood pressure regulator that catalyzes His-Leu release from carboxyl irredeemable angiotensin I, leading to the creation of angiotensin II, a powerful vasopressor octapeptide. A lack of the vasodilator bradykinin complicates angiotensin I-converting enzyme deficiency.[59]

Most, if not all, commercially marketed angiotensin I-converting enzyme inhibitors contain peptides derived from Brazilian viper *Bothrops jararaca* venom.[176] Furthermore, this animal base, microbes, and plants produce chemical compounds that inhibit angiotensin I-converting enzyme, which may be useful in the development of new angiotensin I-converting enzyme inhibitors. Angiotensin-converting enzyme inhibitors (ACEIs) prevent ACE from producing angiotensin II, lowering blood pressure and peripheral vascular tension. Coughing, taste differences, and skin rashes are common side effects of these synthetic medications.[5]

As a result, there has been a rise in interest in discovering food sources as angiotensin I-converting enzyme inhibitors for cost-effective and safe use. To effectively treat hypertension, more attractive angiotensin I-converting enzyme inhibitors have been devised and synthesized. Oral administration of these medications frequently results in unexpected adverse effects; dietary techniques may be a healthier way to control blood pressure. Furthermore, for a thorough evaluation of natural compounds, some trainings on specific plant species are required, where numerous types of angiotensin I-converting enzyme inhibitory molecules, such as xanthones, flavonoids, peptides, proanthocyanidins, and secoiridoids, must be recognized.[69]

11.4.4 α-AMYLASE

α- Amylase (EC 3.2.1.1) is a protein enzyme that hydrolyzes the bonds between big, *α*-bonded polysaccharides such as glycogen and starch, as well as elastic maltose and glucose.[156] This is the most prevalent amylase form found in other mammals and humans.[146,196] It is also found in seeds that cover starch as a reserve food material, and it is also secreted by certain fungi. Though amylase can be found in a variety of tissues, it is most typically found in saliva and pancreatic fluid, which each contain their own isoform of human α-amylase. They have the opposite effect on isoelectric focusing and can be distinguished in tests by consuming certain monoclonal antibodies.

Amylase is a salivary enzyme that catalyzes the breakdown of starch into maltose and dextrin. Ptyalin is another name for this type of amylase. It transforms huge, insoluble starch molecules into soluble starches, resulting in minor starches and, finally, maltose. Ptyalin operates on linear (1–4) glycosidic links, although partial hydrolysis requires a cleft product-specific enzyme. In the stomach, salivary amylase is deactivated by the gastric acid. In 20 min at 37°C, ptyalin was totally destroyed in gastric juice with a pH similar to 3.3–4. In contrast, 50% of activity of amylase is retained after 150 min of exposure to gastric juice at pH 4.3.[54] Starch, the substrate for ptyalin, and the result (glucose of short chains) can all work together to protect it against gastric acid inactivation. When ptyalin was put to buffer at pH 3.0, it was totally inactivated in 120 min; adding starch at 0.1% resulted in 10% residual activity while adding starch at 1.0% resulted in roughly 40% residual activity at 120 min.[155]

11.4.5 α-GLUCOSIDASE ENZYMES

Maltase, Maltase hydrolase, Glucosidosucrase, Glucoamylase, α-d-glucoside Glucohydrolases, Glucosidoinvertase, and α-1,4-glucosidase are glucosidases found at the brush end of the small intestine that act on (1–4) bonds.[120] α-glucosidase, on the other hand, decomposes starch and disaccharides into glucose. In the meantime, the maltase enzyme, which decomposes maltose, is virtually functionally equivalent. α-glucosidase's major purpose is to hydrolyze irreversible nonreducing (1–4) bound -glucose, releasing a single α-glucose molecule. α-Glucosidase is primarily a carbohydrate that hydrolyzes the release of α-glucose, which is distinct from β-glucose. β-Glucose remains unconfined through glucoamylase, a similar enzyme. The substrate molecule discrimination of α-glucosidase is superior to subsite attractions of enzyme active sites. A nucleophilic shift of an intermediate (oxocarbenium ion) is the key hypothesized mechanism.[110] *Rhodnius prolixus*, a blood-sucking bug, creates hemozoin after digesting the host's hemoglobin. The hemozoin synthesis is dependent on a substrate connected to the -glucosidase site.[182]

Carbonic anhydrase inhibition has been demonstrated by novel glycine and phenylalanine sulfonamide compounds.[50] The effects of thiamine, adrenaline, dopamine, and tyrosine on the glutathione reductase (GR) enzyme in bovine liver tissue have recently been established. Some natural amines are similarly affected by the GR in bovine liver.[169] In addition, the effects of different catecholamines on the characteristics of a purified carbonic anhydrase enzyme taken bovine kidney tissue are described. In the literature, some synthetic sulfonamide-containing compounds have also been found to have urease inhibitory potential.[194]

Many compounds are available which are obtained from natural sources showing effects against various diseases. Against HIV many compounds have been extracted from natural sources namely Tachyplesins (Peptide) obtained from Horseshoe crabs (*Tachypleus tridentatus* and *Limulus polyphemus*)[116]; Neamphamide A (Depsipeptide) produced by Sponge (*Neamphius huxleyi*)[134]; Halichlorine, Hymenialdisine, Clathsterol, Crambescidin 826, Dehydrofurodendin, Avarol are obtained from sponges.[90,185,149,22,26,100,151] 4-Acetoxy-plakinamine B (Alkaloid) obtained from Sponge (*Corticium* sp.) used as Acetylcholinesterase.[92] Callyspongynic acid (Polyacetylenic) obtained from Sponge (*Callyspongia)* is used against Diabetes.[121] Hymenialdisine (Alkaloid) obtained from Sponge (*Stylissa massa*) is used against Arthritis.[9] Norzoanthamine is obtained from Cnidarians that is used against Osteoporosis.[85] Eryloside F is obtained from *Erylus formosus* that is used for the treatment of Atherosclerosis.[178] 11-dehydrosinulariolide extracted from *S. flexibilis* is used against Parkinson's disease.[24]

11.5 NEW SOURCES AND NEW PRODUCTION METHODS

The development of new technologies and methods in recent years has significantly contributed to the beginning of a new era for natural products, which many regard as a "New Golden Age." Advances in metagenomics and microbial genomics, as well as advances in microbial culture, have opened up new avenues for understanding and manipulating biosynthetic pathways, thereby contributing to the rational discovery of natural products. Furthermore, advances in bioassays and high-throughput screening technology have had a significant impact in this field.[107,170] To cover the vast biodiversity of biological systems, interest in various sources of natural products has grown. Currently, there is a big concern about the study of species from several ecosystems, such as deep oceans (marine organisms) and extreme environments.[10] The marine environment, which covers 70% of the earth's surface, is a natural habitat for a significant number of species. Indeed, marine organisms such as tunicates, bryozoans, sponges, soft corals, fishes, sea hares, opisthobranch molluscs, nudibranchs, echinoderms, shells, prawns, sea slugs, and microorganisms are valuable resources for drug discovery due to their biodiversity.[41]

Secondary metabolites from these organisms have been studied for over 30 years, primarily by academic chemists who began isolating and elucidating novel compounds from these organisms in the 1970s. Since then, the molecules found in those species have received increased attention.[27] When compared to terrestrial natural products, marine natural products exhibit greater chemical novelty and a higher incidence of significant bioactivity (1% of marine samples tested for antitumor activity in a cytotoxicity screen performed by the National Cancer Institute show antitumor activity).[87,118] Furthermore, the distinct structure of some metabolites has resulted in the identification of novel mechanisms of action.

Ziconotide, for example, exerts analgesic action by reversibly blocking N-type voltage-sensitive calcium channels, thereby inhibiting the activity of pain sensing primary nociceptors.[112] This analgesic mechanism had not been reported prior to the discovery of this drug and is now a topic of interest for the improvement of other analgesic drugs.[108] The drug discovery potential of marine organisms resulted in the improvement of new classes of medicinal agents, some of which are already in use in clinics and others in clinical trials, primarily for their antimicrobial, antitumor, analgesic activity, and anti-inflammatory.

The FDA has already approved seven marine drugs (ziconotide, vidarabine, cytarabine, trabectedin, brentuximab vedotin, eribulin mesylate, and omega3 acid ethyl esters). Ziconotide, a peptide derived from a tropical marine cone snail, was the first drug to be certified for chronic pain treatment in spinal cord injury (2004). Trabectedin, a tetrahydroisoquinoline class alkaloid, was the first anticancer marine drug authorized in the European Union for the treatment of soft tissue sarcoma.[102] At the microbial level, there is growing interest in marine actinomycetes due to their diversity and demonstrated ability to produce novel bioactive compounds with specific structure variety and unique biological activities.[179] Salinoporamide A, a lactamlactone derived from *Salinispora tropica*, is a potent 20S proteasome inhibitor currently in phase I clinical trials for multiple myeloma treatment.[65] The difficulty in isolating and cultivating rare actinomycetes remains an impediment to their use as a natural products source. However, the continued advancement of improved cultivation methods and molecular skills promises to ensure that these limitations are overcome.[39,208]

Extreme-condition organisms found in the marine environment, on the other hand, have gotten a lot of interest. Palmerolide A is a powerful cytotoxic macrolide derived from the Antarctic circumpolar tunicate *Synoicum adareanum*, which has high anticancer action against melanoma cell lines.[39] Another marine organism that is gaining interest is cyanobacteria, which are photosynthetic bacteria that are known to be providers of a range of bioactive compounds. According to a survey of natural products with FDA approval and those in clinical trials, cyanobacteria are likely to be expected biosynthetic sources in 20% of these small molecules.[159] One of the biggest obstacles in natural product medication discovery, as previously indicated, is a shortage of supply. Despite this, significant breakthroughs in chemistry, fermentation, and biotechnology have found remedies.[10]

Total chemical synthesis advancements have greatly aided in overcoming this issue, allowing industrial scale production of these compounds for structure–activity relationship (SAR) studies, hit to lead optimization, and further industrial production.[113] Largazole, a histone deacetylase inhibitor isolated from the genus Symploc cyanobacterium, was chemically synthesized for the preparation of analogues and SAR studies, which revealed that the group is the pharmacophore responsible for antiproliferative activity.[204] Further examples of compounds first isolated from natural products and those now obtained by synthetic pathways are amphotericin B (an antibiotic isolated from *Streptomyces nodosus*), calicheamicin γ1 (an anticancer agent from *Micromonospora echinospora* ssp. *calichensi*), efrotomycin (an antibiotic from *Nocardia lactamdurans*), rapamycin (an immunosuppressant agent *Streptomyces hygroscopicus*), and taxol (an anticancer agent from *Taxus brevifolia*).[127]

Another key approach for addressing supply issues is the fermentation of a starting natural product or a major intermediary. Fermentation, on the other hand, cannot produce all natural products or beginning materials.[10,113] However, fermentation remains an important technique in the commercial manufacturing of many drugs: for example, trabectedin is made by fermenting the starting material cyanosafracin B from *Pseudomonas fluorescens*, followed by semisynthetic stages to obtain the finished medication.[33] Mixed fermentation has also been shown to be a useful strategy for accessing the metabolic potential of cultivatable bacteria, with the ability to improve yields of previously identified as well as undiscovered compounds.[143] Normal LC–MS analysis could not reveal emericellamides A and B in a pure culture of Emericella sp., but a higher enrichment was identified after inoculating the bacterial culture with the marine actinomycete *Salinispora arenicola.*[132]

Furthermore, improvements in biotechnology and metabolic engineering have had a substantial impact on the supply area, allowing for a better knowledge of the metabolic processes that result in secondary metabolite production. Knowing the biosynthetic pathway by which these metabolites are produced in vivo allows for an increase in their production rate in vitro by manipulating biosynthetic pathways and redirecting metabolic flows toward the desired product. Methods include boosting precursor supply, overexpressing or improving the efficiency of bottleneck enzymes, altering gene expression control, decreasing flow toward undesirable by-products or competing pathways, and reconstituting entire processes in a heterologous host.[144]

For example, by exchanging the non-ribosomal peptide synthase subunit and inactivating the tailoring enzyme (glutamic acid 3methyltransferase), genetic modifications of *Streptomyces roseosporus* for the daptomycin biosynthetic pathway provided sufficient amounts of compound for structure elucidation, in vitro screening, and further drug development.[110] Similarly, metagenomics approaches have been used to evaluate a wider range of bacterial synthetic capabilities, leading to the discovery of novel compounds. This strategy entails extracting the entire bacterial DNA from an environmental sample and cloning it in host organisms like Escherichia coli, which is then cultured and tested for the expression of bioactive metabolites.[66] Bryostatins, for example, are a class of protein kinase C modulators with significant central nervous system bioactivity. *Candidatus Endobugula sertula*, an uncultivated marine bacterial symbiont of the marine bryozoan *Bugula neritina*, contains these compounds. Following the discovery of bryostatin biosynthetic genes, the cultivation of this bacterial symbiont and heterologous expression of these genes was used to develop bryostatins and derived compounds.[191] These biotechnological methods are significant in drug development because they allow for the production of novel physiologically active compounds that are identical to the parent compound but have different chemical and occasionally biological features.[10] Weissman et al.[199] employed combinatorial biosynthesis to construct bacterial multienzyme polyketide synthases, resulting in the discovery of over 200 new polyketides as potential therapeutic candidates.

Because of advancements in purification and characterization methods, natural product screening became more compatible with the regular timeline of high-throughput screening operations.[72] In recent years, separation methods such as capillary electrophoresis (CE), counter current chromatography (CCC), high-performance liquid chromatography (HPLC), and supercritical fluid chromatography (SFC) have greatly advanced in terms of resolving power and efficiency.[10] Furthermore, the development of novel hyphenated spectroscopy methods such as HPLC-MS, HPLC-NMR, and HPLC-NMRMS allows for quick chemical identification and accelerates dereplication operations.[36] Table 11.1 contains a list of several medications that are presently being studied in clinical trials.

TABLE 11.1 Current Clinical Status of Various Drugs.[102]

Substance	Organism	Structure	Disease condition
Certified			
Ara-C , Cytarabine	Sponge	Nucleoside	Leukemia
Brentuximab vedotin	Shellfish/ cyanobacterium	ADC (MMAE)	Tumor, lymphoma
Ara-A, Vidarabine	Sponge	Nucleoside	Antiviral
Ziconotide	Snail	Peptide	Pain
Trabectedin	Urochordate	Alkaloid	Cancer
Phase III			
Plitidepsin	Urochordate	Depsipeptide	Tumor
Tetrodotoxin	Pufferfish	Alkaloid	Chronic aching
Soblidotin	Bacteria	Peptide	Tumor
Phase II			
Plinabulin	Fungus	Diketopiperazine	Tumor
Glembatumumab vedotin	Shellfish/ cyanobacterium	ADC (MMAE)	Breast tumor, melanoma
Elisidepsin	Shellfish	Depsipeptide	Tumor
PM1004	Nudibranch	Alkaloid	Tumor
TAsidotin, Synthadotin	Bacteria	Peptide	Tumor
Pseudopterosins	Coral (Soft)	Diterpene glycoside	Wound treatment
Phase I			
Pinatuzumab vedotin	Shellfish/ cyanobacterium	ADC (MMAE) N-HL,	CLL
Hemiasterlin	Sponge	Tripeptide	Tumor
HuMaxR-TF-ADC	Shellfish/ cyanobacterium	ADC (MMAE)	Tumor

In new decades, scientific advancements and the rise of artificial intelligence have led several researchers to consider that only in-silico methods can be used to create new medications.

However, in the vast majority of cases, this method proved ineffective. One of the primary reasons for this failure is that humans have a limited capability to understand novel scaffold molecules in contrast to the natural ones available in nature. Plants, algae, and microbes produce thousands of structurally distinct compounds, the vast majority of which are often intractable in modern organic chemistry laboratories. This study suggests that, despite significant technological progress, drug discovery will continue to rely heavily on the discovery of new natural bioactive compounds with drug-like activity in the coming years.

Nguyen et al.[126] studied the potential of certain *O. stamineus* pimaranediterpenes to inhibit PTP1B and promote glucose uptake in human adipocytes, indicating that such molecules might be helpful as lead compounds in the development of novel antihyperglycemic medicines. Alqahtani et al.[3] looked at the antidiabetic and antioxidant effects of plant extracts and purified components obtained from *Nuxia oppositifolia*, a plant whose aerial parts were utilized to make a well-known antidiabetic extract. The extract included two bioactive compounds, 3-oxolupenal and katononic acid, which were antioxidants and α-amylase and α -glucosidase inhibitors.

According to Ur-Rehman et al.,[192] the methanolic extract of *Boswellia elongata* resin contains various triterpene acids capable of inhibiting α-glucosidase, having IC50 values ranging from 9.9 to 56.8 M. The authors used kinetic and in silico docking techniques to investigate the mechanism of action. Dudoit et al.[42] investigated the biological properties of natural compounds obtained from Tannat grapes, determining the levels of stilbenes, anthocyanins, and flavan-3-ols as a ripening stage function, and discovered that these extracts had high in vitro inhibitory effects against alpha-glucosidase.

Kalhotra et al.[81] investigated the inhibitory activity of garlic-derived compounds on dipeptidyl peptidase-4, a serine protease that regulates glucose metabolism by catalyzing the degradation of glucagon-like peptides (GLP-1 and GLP-2) and the glucose-dependent insulin releasing polypeptide GIP. Tea catechins, according to Balestri et al.,[11] operate as differential inhibitors of aldose reductase, an enzyme implicated in the development of diabetic clinical consequences. Sadeer et al.[158] showed the biological characteristics of the phytochemical components of *Bruguiera gymnorhiza* (L.) Lam, a plant used to cure a variety of illnesses, including diabetes and certain related pathologies. Using two solvents, methanol and ethyl acetate, as well as other plant components, they extracted several extracts from the plant, notably leaves, roots, twigs, and fruits. After that, the extracts were evaluated for antioxidant and inhibitory action against a variety of enzymes, including α-amylase, α-glucosidase, tyrosinase, elastase, lipase, AChE, and BChE. The results showed that the extraction solvents employed had a substantial influence on the biological activities studied. Furthermore, the data given in this study offered a complete review of the pharmacological potential of *B. gymnorhiza*. Marin et al.[103] showed that naturally occurring drimane sesquiterpenoids such as isodrimeninol may be exploited as lead chemicals in the creation of novel antifungal medicines. Isodrimeninol, isolated from the bark of *Drimys winteri*, served as the starting material for the synthesis of four novel semisynthetic derivatives.

Coy-Barrera's[32] study sought to investigate the interactions of 58 natural 2-arylbenzofuran-containing compounds with cyclooxygenase (COX) enzymes. The author chose drugs with the lowest in vitro IC50 values and the most favorable binding energies for COX-2 inhibition using molecular dynamics simulations and kinetic analysis. The 3D-QSAR analysis revealed significant structural requirements for enzyme inhibition and optimizing ligand binding. Ly et al.[99] sought natural substances capable of inhibiting xylosyltransferase-I (XT-1), an enzyme participating in the onset of fibroproliferative disorders. The preliminary examination, which used a library of 96 natural compounds, found four compounds that had a strong inhibitory effect on XT-1. In silico docking was used to investigate the binding of such compounds with the enzyme, and their in vivo activities were proved by studying their impacts on human dermal fibroblasts.

Zazeri et al.[206] published the findings of a comprehensive research of the natural chemical piperine's interaction with interleukin-1 beta (IL-1) utilizing both computational and experimental molecular biophysical methods. The authors conducted thorough research using fluorescence spectroscopy to establish the affinity constant and the interaction mode of piperine. Furthermore, the authors used in-silico docking and molecular dynamic studies to investigate molecular interactions between the ligand and protein. The data gathered is valuable and can be used to improve future drug discovery investigations. Tisi et al.[187] gave a thorough analysis of natural compounds that are still identified and described as Ras protein family enzyme inhibitors. The activity of mutated/upregulated Ras proteins is vital for several types of cancer cells to survive. As a result, Ras inhibitors are a potent tool in the fight against tumor growth. According to the authors, the natural world is teeming with chemicals that can interfere with the synthesis, processing, activity, and signaling of Ras proteins. According to this study, natural compounds are the best source of inspiration for the development of novel molecules capable of suppressing the activity of Ras proteins or, at the very least, altering their function.

According to WHO official updates, the pandemic of a novel coronavirus (SARS-CoV-2) that began in Wuhan, China, in the end of 2019, has resulted in more than 9.3 million illnesses and more than 478,000 deaths across 216 countries that lack developed antiviral medicines or viable vaccinations. However, repurposing previously used drugs has resulted in some documented clinical improvements.[29] At this moment, global attempts are being done to describe molecular targets that will be critical in the development of anti-coronavirus therapeutics (Table 11.2). Ivermectin, a semisynthetic pentacyclic lactone produced from bacteria *Streptomyces avermitilis*, was discovered to be an efficient in vitro SARS-CoV 2 replication inhibition.[1,19] Sayed et al.[164] began simulated testing of a huge collection of microbial natural products with the goal of identifying a possible therapeutic candidate against SARS-CoV-2.

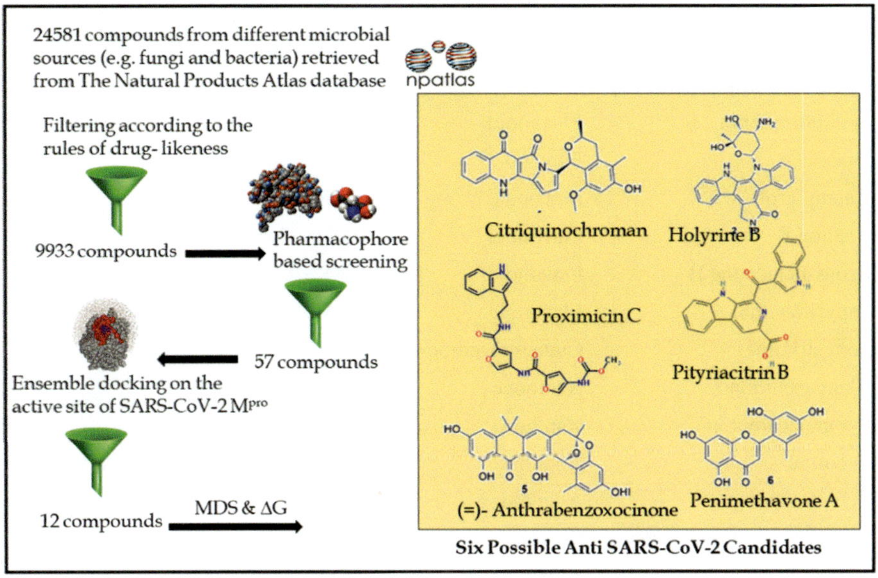

FIGURE 11.6 Applied strategy for selecting drugs against SARS-CoV-2,

TABLE 11.2 List of Potential Drugs Against Coronavirus Type Target.

Compound	Chemical group	Coronavirus type targeted	References
Glycyrrhizin	Saponin	SARS-CoV	[30]
Saikosaponin B2	Saponin	SARS-CoV	[25]
Saikosaponin A	Saponin	SARS-CoV	
Tetra-O-galloyl-β-D-glucose	Polyphenol	SARS-CoV	[203]
Luteolin	Flavonoid	SARS-CoV	
Sinigrin	Polyphenol	SARS-CoV	[98]
β-Sitosterol	Phytosterol	SARS-CoV	
Hesperetin	Flavonoid	SARS-CoV	
Amentoflavone	Flavonoid	SARS-CoV	[156]
Luteolin	Flavonoid	SARS-CoV	
Quercetin	Flavonoid	SARS-CoV	
Apigenin	Flavonoid	SARS-CoV	
Isobavachalcone	Flavonoid	SARS-CoV	[83]
Psoralidin	Flavonoid	SARS-CoV	
Tomentin A	Flavonoid	SARS-CoV	[28]
Tomentin B	Flavonoid	SARS-CoV	
Tomentin E	Flavonoid	SARS-CoV	
3′-O-Methyldiplacol	Flavonoid	SARS-CoV	
Isoliquiritigenin	Flavonoid	SARS-CoV	[136]
Quercetin	Flavonoid	SARS-CoV	
Kaempferol	Flavonoid	SARS-CoV	
Kazinol F	Flavonoid	SARS-CoV	
Broussochalcone B	Flavonoid	SARS-CoV	
Papyriflavonol A	Flavonoid	SARS-CoV	
Terrestrimine	Cinnamic amide	SARS-CoV	[176]
Blancoxanthone	Xanthone	HCoV 229E	[170]
Pyranojacareubin	Xanthone	HCoV 229E	
Lycorine	Alkaloid	SARS-CoV	[97]
Tylophorine	Alkaloid	TGEV	[201]
7-Methoxycryptopleurine	Alkaloid	TGEV	
Jubanine G	Alkaloid	PEDV	[82]
Jubanine H	Alkaloid	PEDV	
Nummularine B	Alkaloid	PEDV	

TABLE 11.2 *(Continued)*

Compound	Chemical group	Coronavirus type targeted	References
Tingenone	Triterpene	SARS-CoV	[156]
Iguesterin	Triterpene	SARS-CoV	
Pristimererin	Triterpene	SARS-CoV	
Dihydrotanshinone I	Diterpene	SARS-CoV	[137]
Cryptotanshinone	Diterpene	SARS-CoV	
Tanshinone IIA	Diterpene	SARS-CoV	
Schimperinone	Triterpene	PEDV	
Xanthoangelol	Chalcone	SARS-CoV	[138]
Hirsutenone	Diarylheptanoid	SARS-CoV	[137]
Rubranoside	Diarylheptanoid	SARS-CoV	
Curcumin	Diarylheptanoid	SARS-CoV	
APA	Lectin	SARS-CoV	[44]
UDA	Lectin	SARS-CoV	

11.6 NEW STRATEGIES FOR DELIVERY

Despite the bioactivity of many natural compounds in in vitro assays, their transposition for in vivo use is frequently disappointing, owing to their slow absorption rate. In fact, the majority of these drugs have poor solubility, a slow dissolution rate, and are unstable at high pH, all of which reduce their bioavailability.[162] It is also worth noting that many natural compounds do not meet the structural requirements of Lipinski's rule, and some of them may not exhibit oral bioavailability, despite the fact that, as previously stated, this is not a necessary requirement. However, technological advances, particularly in the field of nanotechnology, have enabled drug pharmacokinetics and pharmacodynamics to be controlled through the development of drug delivery systems. In addition to increasing bioavailability, these new strategies may aid in the management of harmful side effects by directing the drug to its intended target.[35] In a drug delivery system, microencapsulation is a technique. A matrix encapsulates small particles, such as nanoparticles.

Nanoparticles are generally defined as particles smaller than 100 nm in size; however, some authors divide them into two distinct groups based on size: microparticles, microspheres, or microcapsules (1–800 m range) and nanoparticles, nanospheres, or nanocapsules (below 1 m).[198] Polymer nanoparticles (nanocapsules and nanospheres), solid lipid nanoparticles, crystal nanoparticles, liposomes, micelles, and dendrimers are the most common types of nanoparticles used for drug delivery.[198] Polymeric nanoparticles have been used to improve the bioavailability of molecules such as luteolin, epigallocatechin gallate, and tea polyphenols[61,161] by using polymers such as polyethylene glycol and polyvinyl alcohol.

Liposomes are nanoparticles made of phospholipid bilayers that are similar to cell membranes, with a hydrophilic tail that can transport more polar drugs and a hydrophobic tail that can deliver lipophilic drugs. The type of delivery system should be chosen based on the drug's physicochemical properties. One or more hydrophilic drugs can be found in the liposomal aqueous compartment, which is formed by the hydrophilic head groups of the phospholipids. Lipophilic drugs, on the other hand, are better suited for delivery within a micelle, in which the drug-containing compartment is formed by the lipophilic tail of the phospholipid.[118] Although bergamot essential oil has anticancer properties, it lacks the acquired lipophilic properties required for solubilization. The encapsulation of this essential oil in liposomes increased the drug's solubility and resulted in increased cell death in vitro.[21]

Solid lipid nanoparticles have a solid hydrophobic core surrounded by phospholipids and are a good choice for hydrophobic drug delivery because they are more stable than liposomes and less toxic than polymeric nanoparticles.[95] Curcumin, a diarylheptanoid derived from turmeric, has piqued the interest of researchers as a potential treatment for a wide range of diseases, including cancer, inflammation, microbial infection, and angiogenesis. Poor bioavailability was a significant impediment to curcumin's therapeutic efficacy in clinical trials in this case. Orally administered liposomeen capsulated curcumin, on the other hand, has been shown to significantly increase this compound's bioavailability (up to 155 times at 1 mg/kg).[80,121,173,182] Solid dispersion, a molecular strategy that uses a hydrophilic inert carrier matrix to reduce aggregation size and increase dispersibility,[173] can also improve solubility.

Apigenin, a flavone derived from plants, has water solubility and fewer lipids. It was demonstrated that using solid dispersions of the compound with novel carbon nano powder could significantly improve its bioavailability: drug release profiles revealed that apigenin dissolution from the system improved by 275%.[37] The use of cyclodextrins is another method for increasing the water solubility of natural drugs and thus improving their bioavailability and stability. Cyclodextrins are a type of cyclic oligosaccharide with a lipophilic central cavity and hydrophilic outer surface that can be used to transport both lipophilic and hydrophilic drugs.[188] The supramolecular systems formed by encapsulation are dependent on the molecular interactions of both host and visitor particles.[142]

The physiological stability and insolubility issues of camptothecin, an anticancer agent, can be addressed by using loaded amphiphilic cyclodextrins, which may aid in its delivery.[31] Drug bioavailability can also be increased by targeting the drug to a specific target in a specific tissue/organ. Furthermore, once the drug is released in a specific area of the body, targeting strategies can reduce side effects, limiting toxicity to other organs. Targeting strategies can be divided into two categories.

The first is passive targeting, which does not rely on chemical/biological interaction and in its place relies on particles physical transport based on their natural properties (size, shape, and surface charge). The second type is active targeting, which involves attaching a ligand to the surface of a nanoparticle that is selectively recognized by receptors on the surface of cells. Active targeting is typically achieved by imbuing the nanoparticle with a protein, peptide, or antibody.[190] Passive targeting is a successful and less expensive method of cancer treatment that is widely used.

Numerous tumors have increased permeability due to faulty vasculature, which is important for the development of nanoparticles that can distinguish between tumor and healthy tissue. One example of this delivery system is encapsulated gambogic acid and vitamin E-containing telodendrimers for colon cancer treatment.[76] Because of their ability to deliver a cytotoxic compound to a specific growth cell while not affecting normal cells, antibody–drug conjugates are typical examples of active targeting and have been successfully used in therapeutics for hematological diseases and cancer.

A monoclonal antibody, a synthetic linker, and a powerful cytotoxic payload are the three main components of an antibody–drug conjugate.[62] There are many antibody–drug conjugates in use today, but only a few of them are natural product derivatives. Brentuximab vedotin is a chimeric antibody derived from dolastatin 10, an antitubulin agent found in *Symploca* sp., and is currently approved for Hodgkin's lymphoma treatment and anaplastic large cell lymphoma.[159] Trastuzumab DM1 is an antibody drug approved for metastatic breast cancer treatment. It is derived from maytansine, a powerful antimitotic compound obtained from the shrub *Maytenus ovatus*.[20]

Calicheamicin from *Micromonospora calichensis* is one of the most powerful DNA cleaving agents, and a calicheamicin1 derivative, Nacetylcalicheamicin, is now being conjugated to a monoclonal antibody. Anti-CD22 calicheamicin conjugate is currently being tested in phase 3 clinical trials for refractory or relapsed CD22 positive acute lymphoblastic leukemia.[84] The conjugation of nanoparticles with folic acid is also an intriguing approach to targeting cancer cells with overexpressed folic acid receptors. In this context, quercetin nanoparticles conjugated with folic acid have already been shown in some experiments to selectively target cancer cells in a folate-dependent manner.[46] Targeting can also be accomplished through the use of external forces. Magnetic fields have been widely considered for use in coordinating a conveyance framework.

Oncocalyxone A, an antitumor compound isolated from *Auxemma oncocalyx*, was given a new magnetic drug delivery nanosystem. Oleic acid coated iron oxide nanoparticles and oncocalyxone A were incorporated into the hydrophobic cores of copolymer micelles, allowing the system to be targeted to the tumor using a magnetic field.[13] The strategies discussed above are frequently solutions to the majority of the problems associated with the absence of bioavailability of natural products as drug leads. In many laboratories, more in-depth research on these approaches is being pursued with zeal. Novel drug delivery systems not only improve therapeutic value by increasing bioavailability and lowering toxicity, but also help to overcome noncompliance with the therapy by decreasing the number of times the drug is administered.

Badeggi et al.[8] investigated the antioxidant and antidiabetic activities of gold nanoparticles containing natural components obtained from *Leucosidea sericea*. Nanoparticles having IC50 values in the lower micromolar range and significant antioxidant activity when coupled with natural compounds can block both -glucosidase and -amylase. In conclusion, this work revealed that creating bioactive gold nanoparticles for natural chemical delivery is a reasonably easy process. Surprisingly, the authors suggested that such nanoparticles could be utilized to treat a wide range of diseases, including diabetes and inflammation.

11.7 DRUGS FOR NEW TARGETS

Given the foregoing, natural products have undoubtedly been a valuable source for the improvement and development of new drugs for a long time. However, their macromolecular targets remain largely unknown, hampering objective rational drug design and optimization methodologies.[151] With the advent of the sequenced human genome and new molecular biological approaches, rational molecular target-based screening emerged as a new perspective in drug discovery.[86] In this context, computational target prediction strategies based on chemical descriptors have been widely used in drug discovery to elucidate natural compound mechanisms of action and to enlighten researchers on their new potential therapeutic applications.[196] Indeed, there has been a shift in the use of many natural compounds, such that agents that were initially isolated and purified for potential therapeutic applications are now being studied for completely different targets. In this way, the previously applied concept of drug redirection in medicinal chemistry has been extended to natural products in the drug discovery field.[123]

Rapamycin, a microbial product, is a classic example of this crossover. Initially isolated as an antifungal compound produced by the bacterium *Streptomyces hygroscopicus*, it has been approved as an immunosuppressive drug and is also being tested (as a derivative) as an antineoplastic agent via specific inhibition of the mTOR protein kinase.[94,12] Other drugs, such as canagliflozin (a SGLT2 inhibitor) and darunavir (an HIV protease inhibitor), have been considered for the possibility of displaying crossover activity as renin inhibitors in addition to their classical approved targets, making them viable options for drug repurposing.[122] Many natural compounds target AMP-activated protein kinase, which is triggered by metformin (the main drug used to treat type 2 diabetes).

Salsalate (a salicylate prodrug) has been discussed as a nonsteroid anti-inflammatory drug that may also interfere with AMPK, improving metabolic parameters in subjects with insulin resistance. As a result, repurposing this drug could be effective in metabolic disorders treatment and even cancer, where AMPK activation is frequently sought.[70] In addition to its main target, Archazolid A, myxobacterium *Archangium gephyra* macrolide, has been pointed for other relevant macromolecular targets such as 5lipoxygenase and farnesoid X receptor, and it is now undergoing further studies for repurposing approaches.[151] This repurposing of authorized drugs strategy is a hopeful approach that could increase the capacity of natural products. In addition to financial benefits, repurposing has safety benefits. Existing drugs that are either approved or have been shown to be clinical trials safe, can take advantage of their lower development risk to potentially expand into new indications.

11.8 SUMMARY

Natural products have conventionally played an vital part in drug discovery and served as the foundation for the development of generic early drugs.

Progresses in methodologies, such as X-ray crystallography and NMR, as well as substitute drug discovery, primarily rational drug design and combinatorial chemistry, have placed a significant weight on natural products research in recent years, leading to a decline in interest among most pharmaceutical organizations, which have ended or significantly scaled down their operations based on these compounds.[18] Natural products frequently contain biologically significant molecular scaffolds and pharmacophore patterns that have evolved as preferred ligand–protein binding motifs. Nature has already chosen natural products as lead structures of biological significance in the nearly infinite universe of chemical space, and as a result, their chemical diversity and biochemical specificity must be explored. It is also worth noting that natural compounds can benefit from further improvements in target efficacy and selectivity to achieve ideal pharmacokinetic and pharmacodynamic properties.[153]

Recently and primarily as a result of advances in biotechnology and genomic approaches, there has been a revival of importance in organic/natural product research in a variety of therapeutic areas. Fundamental advances in purification and structure elucidation technologies have reduced the barriers to screening these molecules. The combination of these innovations, along with advances in genomics, metabolic engineering, and chemical synthesis, now provides an option for exploring the significant chemical diversity of natural products in the search for new drugs. These new approaches are expected to bring natural products back to the forefront of the drug development process. Furthermore, novel resources of natural products, like marine species, have been extensively investigated, and their contribution may be the most valuable asset in discovering novel structures with novel modes of action that can cover biologically relevant chemical space.

KEYWORDS

- **bioactive secondary metabolites**
- **enzyme deficiency**
- **human disease**
- **natural enzyme inhibitors**
- **natural product drugs**
- **natural products**
- **pharmacognosy**
- **primary metabolites**

REFERENCES

1. Abdelmohsen, U. R.; Bayer, K.; Hentschel, U. Diversity, Abundance and Natural Products of Marine Sponge-Associated Actinomycetes. *Nat. Prod. Rep.* **2014,** *31*, 381–399.
2. Algood, H. M. S.; Cover, T. L. *Helicobacter pylori* Persistence. An Overview of Interactions Between *H. pylori* and Host Immune Defenses. *Clin. Microbiol. Rev.* **2006,** *19* (4), 597–613.
3. Alqahtani, A. S.; Hidayathulla, S.; Rehman, M. T.; ElGamal, A. A.; Al-Massarani, S.; Razmovski-Naumovski, V.; Alqahtani, M. S.; El Dib, R. A.; AlAjmi, M. F. Alpha-Amylase and Alpha-Glucosidase Enzyme Inhibition and Antioxidant Potential of 3-Oxolupenal and Katononic Acid Isolated from Nuxia oppositifolia. *Biomolecules* **2019,** *10*, 61.
4. Anantharaman, V.; Aravind, L.; Koonin, E. V. Emergence of Diverse Biochemical Activities in Evolutionarily Conserved Structural Scaffolds of Proteins. *Curr. Opin. Chem. Biol.* **2003,** *7* (1), 12–20.
5. Atkinson, A. B.; Robertson, J. I. Captopril in the Treatment of Clinical Hypertension and Cardiac Failure. *Lancet* **1979,** *2* (8147), 836–839.
6. Azeredo, H. M. C. Betalains: Properties, Sources, Applications, and Stability—A Review. *Int. J. Food Sci. Technol.* **2009,** *44* (12), 2365–2376.
7. Azizian, H.; Nabati, F.; Sharifi, A.; Siavoshi, F.; Mahdayi, M.; Amanlou, M. Large-Scale Virtual Screening for the Identification of New *Helicobacter pylori* Urease Inhibitor Scaffolds. *J. Mol. Model* **2012,** *18*, 2917–2927.
8. Badeggi, U. M.; Ismail, E.; Adeloye, A. O.; Botha, S.; Badmus, J. A.; Marnewick, J. L.; Cupido, C. N.; Hussein, A. A. Green Synthesis of Gold Nanoparticles Capped with Procyanidins from Leucosidea sericea as Potential Antidiabetic and Antioxidant Agents. *Biomolecules* **2020,** *10*, 452.
9. Badger, A. M.; Cook, M. N.; Swift, B. A.; Newman-Tarr, T. M.; Gowen, M.; Lark, M. Inhibition of Interleukin-1-Induced Proteoglycan Degradation and Nitric Oxide Production in Bovine Articular Cartilage/Chondrocyte Cultures by the Natural Product, Hymenialdisine. *J. Pharmacol. Exp. Therap.* **1999,** *290* (2), 587–593.
10. Baker, D. D.; Chu, M.; Oza, U.; Rajgarhia, V. The Value of Natural Products to Future Pharmaceutical Discovery. *Nat. Prod. Rep.* **2007,** *24* (6), 1225–1244.
11. Balestri, F.; Poli, G.; Pineschi, C.; Moschini, R.; Cappiello, M.; Mura, U.; Tuccinardi, T.; Del Corso, A. Aldose Reductase Differential Inhibitors in Green Tea. *Biomolecules* **2020,** *10*, 1003.
12. Ballou, L. M.; Lin, R. Z. Rapamycin and mTOR Kinase Inhibitors. *J. Chem. Biol.* **2008,** *1* (1–4), 27–36.
13. Barreto, A. C. H.; Santiago, V. R.; Freire, R. M.; Mazzetto, S. E.; Denardin, J. C.; Mele, G. Magnetic Nanosystem for Cancer Therapy Using Oncocalyxone A, an Antitomour Secondary Metabolite Isolated from a Brazilian Plant. *Int. J. Mol. Sci.* **2013,** *14* (9), 18269–18274.
14. Beneyton, T.; Wijaya, I.P; Postros, P.; Najah, M.; Leblond, P.; Couvent, A. High-Throughput Screening of Filamentous Fungi Using Nanoliter-Range Droplet-Based Microfluidics. *Sci. Rep.* **2016,** *6*, 27223.
15. Beutler, J. A. Natural Products as a Foundation for Drug Discovery. *Curr. Protoc. Pharmacol.* **2009,** *46* (1), 9–11.

16. Boer, J. L.; Mulrooney, S. B.; Hausinger, R. P. Nickel-Dependent Metalloenzymes. *Arch Biochem. Biophys.* **2014,** *544*, 142–152.
17. Breinbauer, R.; Vetter, I. R.; Waldmann, H. From Protein Domains to Drug Candidates-Natural Products as Guiding Principles in the Design and Synthesis of Compound Libraries. *Angew. Chem. Int. Ed. Engl.* **2002,** *41* (16), 2879–2890.
18. Butler, M. S. Natural Products to Drugs: Natural Product-Derived Compounds in Clinical Trials. *Nat. Prod. Rep.* **2008,** *25* (3), 475–516.
19. Caly, L.; Druce, J. D.; Catton, M. G.; Jans, D. A.; Wagstaff, K. M. The FDA-Approved Drug Ivermectin Inhibits the Replication of SARS-CoV-2 In Vitro. *Antivir. Res.* **2020,** 104787.
20. Cassady, J. M.; Chan, K. K.; Floss, H. G.; Leistner, E. Recent Developments in the Maytansinoid Antitumor Agents. *Chem. Pharm. Bull.* **2004,** *52* (1), 1–26.
21. Celia, C.; Trapasso, E.; Locatelli, M.; Navarra, M.; Ventura, C. A.; Wolfram, J. Carafa, M.; Morittu, V. M.; Britti, D.; Di Marzio, L.; Paolino, D. Anticancer Activity of Liposomal Bergamot Essential Oil (BEO) on Human Neuroblastoma Cells. *Colloids Surf. B.* **2013,** *112*, 548–553.
22. Chang, L.; Whittaker, N. F.; Bewley, C. A. Crambescidin 826 and Dehydrocrambine A: New Polycyclic Guanidine Alkaloids from the Marine Sponge *Monanchora sp.* That Inhibit HIV-1 Fusion. *J. Nat. Products* **2003,** *66* (11), 1490–1494.
23. Chen, M. S.; White, M. C. A Predictably Selective Aliphatic C-H Oxidation Reaction for Complex Molecule Synthesis. *Science* **2007,** *318* (5851), 783–787.
24. Chen, W. F.; Chakraborty, C.; Sung C. S. Neuroprotection by Marine-Derived Compound, 11-Dehydrosinulariolide, in an In Vitro Parkinson's Model: A Promising Candidate for the Treatment of Parkinson's Disease. *Naunyn-Schmiedeberg's Arch. Pharmacol.* **2012,** *385* (3), 265–275.
25. Cheng, P. W.; Ng, L. T.; Chiang, L. C.; Lin, C. C. Antiviral Effects of Saikosaponins on Human Coronavirus 229E In Vitro. *Clin. Exp. Pharmacol. Physiol.* **2006,** *33*, 612–616.
26. Chill, L.; Rudi, A.; Aknin, M.; Loya, S.; Hizi, A.; Kashman, Y. New sesterterpenes from Madagascan *Lendenfeldia* Sponges. *Tetrahedron* **2004,** *60* (47), 10619–10626.
27. Chin, Y.-W.; Balunas, M. J.; Chai, H. B.; Kinghorn, A. D. Drug Discovery from Natural Sources. *AAPS J.* **2006,** *8* (2), E239–E253.
28. Cho, J. K.; Curtis-Long, M. J.; Lee, K. H.; Kim, D. W.; Ryu, H. W.; Yuk, H. J.; Park, K. H. Geranylated Flavonoids Displaying SARS-CoV Papain-Like Protease Inhibition from the Fruits of *Paulownia tomentosa*. *Bioorg. Med. Chem.* **2013,** *21*, 3051–3057.
29. Ciliberto, G.; Cardone, L. Boosting the Arsenal Against COVID-19 Through Computational Drug Repurposing. *Drug Discov. Today* **2020**, *25* (6), 946–948.
30. Cinatl, J.; Morgenstern, B. Bauer, G. Chandra, P.; Rabenau, H.; Doerr, H. W. Glycyrrhizin, an Active Component of Liquorice Roots, and Replication of SARS-Associated Coronavirus. *Lancett* **2003,** *361*, 2045–2046.
31. Cirpanli, Y.; Bilensoy, E.; Dogan, A. L.; Calis, S. Development of Polymeric and Cyclodextrin Nanoparticles for Camptothecin Delivery. *J. Controlled Release* **2010,** *148* (1), e21–e23.
32. Coy-Barrera, E. Discrimination of Naturally-Occurring 2-Arylbenzofurans as Cyclooxygenase-2 Inhibitors: Insights Into the Binding Mode and Enzymatic Inhibitory Activity. *Biomolecules* **2020,** *10*, 176.

33. Cuevas, C.; Perez, M.; Martin, M. J.; Chicharro, J. L.; Fernandez-Rivas, C.; Flores, M. Synthesis of Ecteinascidin ET-743 and Phthalascidin Pt-650 from Cyanosafracin B. *Org. Lett.* **2000,** *2* (16), 2545–2548.
34. *Cutler, S.; Cutler, H. G. Biologically Active Natural Products: Pharmaceuticals; CRC Press: Boca Raton, 2000; p 296.*
35. Devi, V. K.; Jain, N.; Valli, K. S. Importance of Novel Drug Delivery Systems in Herbal Medicines. *Pharmacogn. Rev.* **2010,** *4* (7), 27–31.
36. Dias, D. A.; Urban, S.; Roessner, U. A Historical Overview of Natural Products in Drug Discovery. *Metabolites* **2012,** *2* (2), 303.
37. Ding, S. M.; Zhang, Z. H.; Song, J.; Cheng, X. D.; Jiang, J.; Jia, X. B. Enhanced Bioavailability of Apigenin via Preparation of a Carbon Nanopowder Solid Dispersion. *Int. J. Nanomed.* **2014,** *9*, 2327–2333.
38. Dixon, N. E.; Gazzola, C.; Watters, J. J.; Blakeley, R. L.; Zerner, B. Jack Bean Urease (EC 3.5.1.5) [Letter]. A Metalloenzyme. A Simple Biological Role for Nickel? *J. Am. Chem. Soc.* **1975,** *97* (14), 4131–4133.
39. Diyabalanage, T.; Amsler, C. D.; McClintock, J. B.; Baker, B. J. Palmerolide A, a Cytotoxic Macrolide from the Antarctic Tunicate *Synoicum adareanum*. *J. Am. Chem. Soc.* **2006,** *128* (17), 5630–5631.
40. Donaldson, L. Autofluorescence in Plants. *Molecules* **2020,** *25* (10), 2393.
41. Donia, M.; Hamann, M. T. Marine Natural Products and Their Potential Applications as Anti-Infective Agents. *Lancet Infect. Dis.* **2003,** *3* (6), 338–348.
42. Dudoit, A.; Benbouguerra, N.; Richard, T.; Hornedo-Ortega, R.; Valls-Fonayet, J.; Coussot, G.; Saucier, C.-Glucosidase Inhibitory Activity of Tannat Grape Phenolic Extracts in Relation to Their Ripening Stages. *Biomolecules* **2020,** *10*, 1088.
43. Duval, R.; Duplais, C. Fluorescent Natural Products as Probes and Tracers in Biology. *Nat Prod Rep.* **2017,** *34* (2), 161–193.
44. E. Keyaerts, L.; Vijgen, C.; Pannecouque, E.; Van Damme, W.; Peumans, H.; Egberink, J.; Balzarini, M.; Van Ranst. Plant Lectins Are Potent Inhibitors of Coronaviruses by Interfering with Two Targets in the Viral Replication Cycle. *Antiviral Res.* **2007,** *75*, 179–187.
45. El Sayed, K. A. Natural Products as Antiviral Agents. In *Studies in Natural Products Chemistry*, Vol. 24; Elsevier: Amsterdam, The Netherlands, **2000**; pp 473–572.
46. El-Gogary, R. I.; Rubio, N.; Wang, J. T.; Al-Jamal, W. T.; Bourgognon, M.; Kafa, H. Polyethylene Glycol Conjugated Polymeric Nanocapsules for Targeted Delivery of Quercetin to Folate-Expressing Cancer Cells In Vitro and In Vivo. *ACS Nano.* **2014,** *8* (2), 1384–1401.
47. Evans, B. E.; Rittle, K. E.; Bock, M. G.; DiPardo, R. M.; Freidinger, R. M.; Whitter, W. L. Methods for Drug Discovery: Development of Potent, Selective, Orally Effective Cholecystokinin Antagonists. *J. Med. Chem.* **1988,** *31* (12), 2235–2246.
48. Fagugli, R. M.; Gentile, G.; Ferrara, G.; Brugnano, R. Acute Renal and Hepatic Failure Associated with Allopurinol Treatment. *Clin. Nephrol.* **2008,** *70*:523–6.
49. Farmer, P. B.; Suhadolnik, R. J. Nucleoside Antibiotics. Biosynthesis of Arabonofuranosyladenine by Streptomyces Antibioticus. *Biochemistry* **1972,** *11*, 911–916.
50. Fidan, I.; Salmas, R. E.; Arslan, M.; Senturk, M.; Durdagi, S.; Ekinci, D.; Senturk, E.; Cosgun, S.; Supuran, C. T. Carbonic Anhydrase Inhibitors: Design, Synthesis, Kinetic,

Docking and Molecular Dynamics Analysis of Novel Glycine and Phenylalanine Sulfonamide Derivatives. *Bioorg. Med. Chem.* **2015,** *23*, 7353–7358.

51. Földesi B. Guide to Enzyme Unit Definitions and Assay Design. 2019; p 14. https://www.biomol.com/resources/biomol-blog/guide-to-enzyme-unit-definitions-and-assay-design (accessed on Aug 25, 2021).
52. Follmer, C. Ureases as a Target for the Treatment of Gastric and Urinary Infections. *J. Clin. Pathol.* **2010,** *63*, 424–30.
53. Forbes, J. M.; Coughlan, M. T.; Cooper, M. E. Oxidative Stress as a Major Culprit in Kidney Disease in Diabetes. *Diabetes* **2008,** *57*, 1446–1454.
54. Fried, M.; Abramson, S.; Meyer, J. H. Passage of Salivary Amylase through the Stomach in Humans. *Dig. Dis. Sci.* **1987,** *32*, 1097–103.
55. Ganesan, A. The Impact of Natural Products Upon Modern Drug Discovery. *Curr. Opin. Chem. Biol.* **2008,** *12* (3), 306–317.
56. Geremias-Andrade, I. M.; Rocheto, A. C.; Gallo, F. A.; Petrus, R. R. The Shelf Life of Standardized Sugarcane Juice Stored Under Refrigeration. *Food Sci. Technol.* **2020,** *40* (1), 95–101.
57. Glover, B. J.; Martin, C. Anthocyanins. *Curr. Biol.* **2012,** *22* (5), R147–R150.
58. Goddard, J. P.; Reymond, J. L. Enzyme Assays for High-Throughput Screening. *Curr. Opin. Biotechnol.* **2004,** *15* (4), 314–322.
59. Gohlke, P.; Linz, W.; Scholkens, B. A.; Kuwer, I.; Bartenbach, S.; Schnell, A.; Unger, T. Angiotensin Converting Enzyme Inhibition Improves Cardiac Function. Role of Bradykinin. *Hypertension* **1994,** *23* (4), 411–418.
60. Graham, D. Y.; Fischbach, L. *Helicobacter pylori* Treatment in the Era of Increasing Antibiotic Resistance. *Gut* **2010,** *59*, 1143–1153.
61. Granja, A.; Pinheiro, M.; Reis, S. Epigallocatechin Gallate Nanodelivery Systems for Cancer Therapy. *Nutrients* **2016,** *8* (5), 310.
62. Gromek, S. M.; Balunas, M. J. Natural Products as Exquisitely Potent Cytotoxic Payloads for Antibody–Drug Conjugates. *Curr. Top. Med. Chem.* **2015,** *14* (24), 2822–2834.
63. Gross, J. Chlorophylls. Chapter 2; In: *Pigments in Vegetables: Chlorophylls and Carotenoids*; Springer Publishing: Boston, 1991; pp 3–74.
64. Gross, J. Carotenoids. Chapter 3; In: *Pigments in Vegetables: Chlorophylls and Carotenoids*; Springer Publishing: Boston, 1991; pp 75–278.
65. Gulder, T. A. M.; Moore, B. S. Salinosporamide Natural Products: Potent 20S Proteasome Inhibitors as Promising Cancer Chemotherapeutics. *Angew. Chem. Int. Ed. Engl.* **2010,** *49* (49), 9346–9367.
66. Gurgui, C.; Piel, J. Metagenomic Approaches to Identify and Isolate Bioactive Natural Products from Microbiota of Marine Sponges. *Methods Mol. Biol.* **2010,** *668*, 247–264.
67. Ha, N. C.; Oh, S. T.; Sung, J. Y.; Cha, K. A.; Lee, M. H.; Oh, B. H. Supramolecular Assembly and Acid Resistance of *Helicobacter pylori* Urease. *Nat. Struct. Biol.* **2001,** *8* (6), 505–509.
68. Hach. What Is the Difference Between a Reagent and a Sample Blank? 2019. https://support.hach.com/app/answers/answer_view/a_id/1000699/~/what-is-the-difference-between-a-reagent-blank-and-a-sample-blank%3F- (accessed on Aug 28, 2021).
69. Hansen, K.; Nyman, U.; Smitt, U. W.; Adseren, A.; Christensen, S. B.; Schwantner, Wagner, H. Angiotensin Converting Enzyme (ACE) Inhibitory Flavonoids from *Erythroxylum laurifolium*. *Phytomedicine* **1996,** *2*, 313–317.

70. Hardie, D. G. AMPK: A Target for Drugs and Natural Products with Effects on Both Diabetes and Cancer. *Diabetes* **2013,** *62* (7), 2164–2172.
71. Hartwig, J. F.; Cook, K. S.; Hapke, M.; Incarvito, C. D.; Fan, Y.; Webster, C. E.; Hall, M. B. Rhodium Boryl Complexes in the Catalytic, Terminal Functionalization of Alkanes. *J. Am. Chem. Soc.* **2005,** *127* (8), 2538–2552.
72. Harvey, A. L. Natural Products in Drug Discovery. *Drug Discov. Today* **2008,** *13* (19–20), 894–901.
73. Harvey, A. L.; Edrada-Ebel, R.; Quinn, R. J. The Re-Emergence of Natural Products for Drug Discovery in the Genomics Era. *Nat. Rev. Drug Discov.* **2015,** *14* (2), 111–129.
74. Hoey, B. M.; Butler, J.; Halliwell, B. On the Specificity of Allopurinol and Oxypurinol as Inhibitors of Xanthine Oxidase. A Pulse Radiolysis Determination of Rate Constants for Reaction of Allopurinol and Oxypurinol with Hydroxyl Radicals. *Free Radic. Res. Commun.* **1988,** *4*, 259–263.
75. Hong, J. Natural Product Diversity and Its Role in Chemical Biology and Drug Discovery. *Curr. Opin. Chem. Biol.* **2011,** *15* (3), 350–354.
76. Huang, W.; Wang, X.; Shi, C.; Guo, D.; Xu, G.; Wang, L. Bodman, A.; Luo, J. Fine-Tuning Vitamin E-Containing Telodendrimers for Efficient Delivery of Gambogic Acid in Colon Cancer Treatment. *Mol. Pharma.* **2015,** *12* (4), 1216–1229.
77. Jabri, E.; Carr, M. B.; Hausinger, R. P.; Karplus, P. A. The Crystal Structure of Urease from *Klebsiella aerogenes*. *Science* **1995,** *268*, 998–1004.
78. Jahn, M.; Zerr, A.; Fedorowicz, F. M.; Brigger, F.; Koulov, A.; Mahler, H. C. Measuring Lipolytic Activity to Support Process Improvements to Manage Lipase-Mediated Polysorbate Degradation. *Pharm Res.* **2020,** *37* (6), 118.
79. Jones, C. G.; Firn, R. D.; Malcolm, S. B. On the Evolution of Plant Secondary Chemical Diversity [and Discussion]. *Philos. Trans.: Biol. Sci.* **1991,** *333* (1267), 273–280.
80. Kakkar, V.; Singh, S.; Singla, D.; Kaur, I. P. Exploring Solid Lipid Nanoparticles to Enhance the Oral Bioavailability of Curcumin. *Mol. Nutr. Food Res.* **2011,** *55* (3), 495–503.
81. Kalhotra, P.; Chittepu, V. C. S. R.; Osorio-Revilla, G.; Gallardo-Velazquez, T. Phytochemicals in Garlic Extract Inhibit Therapeutic Enzyme DPP-4 and Induce Skeletal Muscle Cell Proliferation: A Possible Mechanism of Action to Benefit the Treatment of Diabetes Mellitus. *Biomolecules* **2020,** *10*, 305.
82. Kang, K. B.; Ming, G.; Kim, G. J.; Ha, T. K.; Choi, H.; Oh, W. K.; Sung, S. H. Jubanines F-J, Cyclopeptide Alkaloids from the Roots of *Ziziphus jujuba*. *Phytochemistry* **2015,** *119*, 90–95.
83. Kim, D. W.; Seo, K. H.; Curtis-Long, M. J.; Oh, K. Y.; Oh, J. W.; Cho, J. K.; Lee, K. H.; Park, K. H. Phenolic Phytochemical Displaying SARS-CoV Papain-Like Protease Inhibition from the Seeds of *Psoralea corylifolia*. *J. Enzyme Inhib. Med. Chem.* **2014,** *29*, 59–63.
84. Kim, E. G.; Kim, K. M. Strategies and Advancement in Antibody–Drug Conjugate Optimization for Targeted Cancer Therapeutics. *Biomol. Ther.* **2015,** *23* (6), 493–509.
85. Kinugawa, M.; Fukuzawa, S.; Tachibana, K. Skeletal Protein Protection: The Mode of Action of an Anti-Osteoporotic Marine Alkaloid, Norzoanthamine. *J. Bone Mineral Metabol.* **2009,** *27* (3), 303–314.
86. Knowles, J.; Gromo, G. A Guide to Drug Discovery: Target Selection in Drug Discovery. *Nat. Rev. Drug Discov.* **2003,** *2* (1), 63–69.

87. Kong, D. X.; Jiang, Y. Y.; Zhang, H. Y. Marine Natural Products as Sources of Novel Scaffolds: Achievement and Concern. *Drug Discov. Today* **2010,** *15* (21–22), 884–886.
88. Koyama, K.; Hirao, T.; Toriba, A.; Hayakawa, K. An Analytical Method for Measuring Alpha-Amylase Activity in Starch Containing Foods. *Biomed. Chromatogr.* **2013,** *27* (5), 583–588.
89. Krajewska, B.; Ureases, I. Functional, Catalytic and Kinetic Properties: A Review. *J. Mol. Catal. B Enzym.* **2009,** *59* (1), 9–21.
90. Kuramoto, M.; Tong, C.; Yamada, K.; Chiba, T.; Hayashi, Y.; Uemura, D. Halichlorine, an Inhibitor ofVCAM-1 Induction from the Marine Sponge *Halichondria okadai* Kadota. *Tetrahedron Lett.* **1996,** *37* (22), 3867–3870.
91. Lahlou, M. The Success of Natural Products in Drug Discovery. *Pharmacol. Pharm.* **2013,** *04* (03), 15.
92. Langjae, R.; Bussarawit, S.; Yuenyongsawad, S.; Ingkaninan, K.; Plubrukarn, A. Acetylcholinesterase-Inhibiting Steroidal Alkaloid from the Sponge *Corticium sp. Steroids* **2007,** *72* (9–10), 682–685.
93. Lankatillake, C.; Huynh, T.; Dias, D. A. Understanding Glycaemic Control and Current Approaches for Screening Antidiabetic Natural Products from Evidence-Based Medicinal Plants. *Plant Methods* **2019,** *15* (1), 105.
94. Law, B. K. Rapamycin: An Anti-Cancer Immunosuppressant? *Crit. Rev. Oncol. Hematol.* **2005,** *56* (1), 47–60.
95. Leonarduzzi, G.; Testa, G.; Sottero, B.; Gamba, P.; Poli, G. Design and Development of Nanovehicle-Based Delivery Systems for Preventive or Therapeutic Supplementation with Flavonoids. *Curr. Med. Chem.* **2010,** *17* (1), 74–95.
96. Lewis, R. J.; Garcia, M. L. Therapeutic Potential of Venom Peptides. *Nat. Rev. Drug Discov.* **2003,** *2* (10), 790–802.
97. Li, S. Y.; Chen, C.; Zhang, H. Q.; Guo, H. Y.; Wang, H.; Wang, L.; Zhang, X.; Hua, S. N.; Yu, J.; Xiao, P. G.; Li, R. S.; Tan, X. Identification of Natural Compounds with Antiviral Activities Against SARS-Associated Coronavirus. *Antiviral Res.* **2005,** *67*, 18–23.
98. Lin, C. W.; Tsai, F. J.; Tsai, C. H.; Lai, C. C.; Wan, L.; Ho, T. Y.; Hsieh, C. C.; Chao, P. D. Anti-SARS Coronavirus 3C-Like Protease Effects of *Isatis indigotica* Root and Plant-Derived Phenolic Compounds. *Antiviral Res.* **2005,** *68*, 36–42.
99. Ly, T. D.; Kleine, A.; Fischer, B.; Schmidt, V.; Hendig, D.; Kuhn, J.; Knabbe, C.; Faust, I. Identification of Putative Non-Substrate- Based XT-I Inhibitors by Natural Product Library Screening. *Biomolecules* **2020,** *10*, 1467.
100. M¨uller, W. E. G.; Schr¨oder, H. C. Cell Biological Aspects of HIV-1 Infection: Effect of the Anti-HIV-1 Agent Avarol. *Int. J. Sports Med.* **1991,** *12* (1), S43–S49.
101. Malfertheiner, P.; Megraud, F.; O'Morain, C.; Bazzoli, F.; El-Omar, E.; Graham, D. Current Concepts in the Management of *Helicobacter pylori* Infection: The Maastricht III Consensus Report. *Gut* **2007,** *56*, 772–781.
102. Malve, H. Exploring the Ocean for New Drug Developments: Marine Pharmacology. *J. Pharm. BioAllied Sci.* **2016,** *8* (2), 83–91.
103. Marin, V.; Iturra, A.; Opazo, A.; Schmidt, B.; Heydenreich, M.; Ortiz, L.; Jiménez, V. A.; Paz, C. Oxidation of Isodrimeninol with PCC Yields Drimane Derivatives with Activity Against Candida Yeast by Inhibition of Lanosterol 14-Alpha Demethylase. *Biomolecules* **2020,** *10*, 1101.
104. Maroney, M. J.; Ciurli, S. Nonredox Nickel Enzymes. *Chem. Rev.* **2014,** *114*, 4206–4028.

105. Matata, B. M.; Elahi, M. M. Sources of Reactive Oxidants Species in Biology and Disease. *Oxid. Stress* **2007,** *14*, 23–38.
106. McChesney, J. D.; Venkataraman, S. K.; Henri, J. T. Plant Natural Products: Back to the Future or Into Extinction? *Phytochemistry* **2007,** *68* (14), 2015–2022.
107. Malve, J. G. Targeting N-Type and T-Type Calcium Channels for the Treatment of Pain. *Drug Discov. Today* **2006,** *11* (5–6), 245–253.
108. Mégraud, F. The Challenge of *Helicobacter pylori* Resistance to Antibiotics: The Comeback of Bismuth-Based Quadruple Therapy. *Therap. Adv. Gastroenterol.* **2012,** *5* (2), 103–109.
109. Mehrani, H.; Storey, K. B. Characterization of Alpha-Glucosidases from Rainbow Trout Liver. *Arch. Biochem. Biophys.* **1993,** *306* (1), 188–94.
110. Miao, V.; Coeffet-Le Gal, M.-F.; Nguyen, K.; Brian, P.; Penn, J.; Whiting, A. Genetic Engineering in *Streptomyces roseosporus* to Produce Hybrid Lipopeptide Antibiotics. *Chem. Biol.* **2006,** *13* (3), 269–276.
111. Miljanich, G. P. Ziconotide: Neuronal Calcium Channel Blocker for Treating Severe Chronic Pain. *Curr. Med. Chem.* **2004,** *11* (23), 3029–3040.
112. Mlodzinska, E. Survey of Plant Pigments: Molecular and Environmental Determinants of Plant colors. *Acta Biol Crac Ser Bot.* **2009,** *51* (1), 7–16.
113. Montaser, R.; Luesch, H. Marine Natural Products: A New Wave of Drugs? *Future Med. Chem.* **2011,** *3* (12), 1475–1489.
114. Moon, K.; Kwon, E. B.; Lee, B.; Kim, C. Y. Recent Trends in Controlling the Enzymatic Browning of Fruit and Vegetable Products. *Molecules* **2020,** *25* (12), 2754.
115. Morimoto, M.; Mori, H.; Otake, T.; Ueba, N.; Kunita, N.; Niwa, M.; Murakami, T.; Iwanaga, S. Inhibitory Effect of tachyplesin I on the Proliferation of Human Immunodeficiency Virus In Vitro. *Chemotherapy* **1991,** *37* (3), 206–211.
116. Müller, M.G; Georgakoudi, I.; Zhang, Q.; Wu, J.; Feld, M. S. Intrinsic Fluorescence Spectroscopy in Turbid Media: Disentangling Effects of Scattering and Absorption. *Appl. Opt.* **2001,** *40* (25), 4633–4646.
117. Munro, M. H.; Blunt, J. W.; Dumdei, E. J.; Hickford, S. J.; Lill, R. E.; Li, S. The Discovery and Development of Marine Compounds with Pharmaceutical Potential. *J. Biotechnol.* **1999,** *70* (1–3), 15–25.
118. Muqbil, I.; Masood, A.; Sarkar, F. H.; Mohammad, R. M.; Azmi, A. S. Progress in Nanotechnology Based Approaches to Enhance the Potential of Chemopreventive Agents. *Cancers* **2011,** *3* (1), 428–445.
119. Mury, F. B.; da Silva, J. R.; Ferreira, L. S. Alpha-Glucosidase Promotes Hemozoin Formation in a Blood-Sucking Bug: An Evolutionary History. *PLoS One* **2009,** *4* (9), e6966.
120. Nakao, Y.; Uehara, T.; Matunaga, S.; Fusetani, N.; van Soest, R. W. M. Callyspongynic Acid, a Polyacetylenic Acid Which Inhibits α-Glucosidase, from the Marine Sponge *Callyspongia truncata*. *J. Nat. Prod.* **2002,** *65* (6), 922–924.
121. Narayanan, N. K.; Nargi, D.; Randolph, C.; Narayanan, B. A. Liposome Encapsulation of Curcumin and Resveratrol in Combination Reduces Prostate Cancer Incidence in PTEN Knockout Mice. *Int. J. Cancer* **2009,** *125* (1), 1–8.
122. Newman, D. J. Natural Products as Leads to Potential Drugs: An Old Process or the New Hope for Drug Discovery? *J. Med. Chem.* **2008,** *51* (9), 2589–2599
123. Newman, D. J.; Cragg, G. M. Natural Products as Sources of New Drugs Over the 30 Years from 1981 to 2010. *J. Nat. Prod.* **2012,** *75* (3), 311–335.

124. Newman, D. J.; Cragg, G. M. Natural Products as Sources of New Drugs from 1981 to 2014. *J. Nat. Prod.* **2016,** *79* (3), 629–661.
125. Newman, D. J.; Cragg, G. M.; Snader, K. M. The Influence of Natural Products Upon Drug Discovery. *Nat. Prod. Rep.* **2000,** *17* (3), 215–234.
126. Nguyen, P. H.; Tuan, H. N.; Hoang, D. T.; Vu, Q. T.; Pham, M. Q.; Tran, M. H.; To, D. C. Glucose Uptake Stimulatory and PTP1B Inhibitory Activities of Pimarane Diterpenes from Orthosiphon stamineus Benth. *Biomolecules* **2019,** *9*, 859.
127. Nicolaou, K. C.; Hale, C. R.; Nilewski, C.; Ioannidou, H. A. Constructing Molecular Complexity and Diversity: Total Synthesis of Natural Products of Biological and Medicinal Importance. *Chem. Soc. Rev.* **2012,** *41* (15), 5185–5238.
128. Nicolas, J. J.; Richard-Forget, F. C.; Goupy, P. M.; Amiot, M. J.; Aubert, S. Y. Enzymatic Browning Reactions in Apple and Apple Products. *Crit. Rev. Food Sci. Nutr.* **1994,** *34* (2), 109–157.
129. Nile, S. H.; Nile, A.; Gansukh, E.; Baskar, V.; Kai. G. Subcritical Water Extraction of Withanosides and Withanolides from Ashwagandha (Indian Ginseng) and Their Biological Activities. *Food Chem. Toxicol.* **2019,** *132*, 110659.
130. Nile, S. H.; Nile, A. S.; Liu, J.; Kim, D.H; Kai, G. Exploitation of Apple Pomace Towards Extraction of Triterpenic Acids, Antioxidant Potential, Cytotoxic Effects, and Inhibition of Clinically Important Enzymes. *Food Chem. Toxicol.* **2019,** *131*, 112459.
131. Niwa, H.; Watanabe, M.; Inagaki, H.; Yamada, K. Didemnilactones A and B and Neodidemnilactone Three New Fatty Acid Metabolites Isolated from the Tunicate *Didemnum moseleyi* (Herdman). *Tetrahedron* **1994,** *50* (25), 7385–7400.
132. Oh, D. C.; Kauffman, C. A.; Jensen, P. R.; Fenical, W. Induced Production of Emericellamides A and B from the Marine-Derived Fungus *Emericella* sp. in Competing Co-Culture. *J. Nat. Prod.* **2007,** *70* (4), 515–520.
133. Oku, N.; Gustafson, K. R.; Cartner L. K. Neamphamide A, a New HIV-Inhibitory Depsipeptide from the Papua New Guinea Marine Sponge *Neamphius huxleyi*. *J. Nat. Prod.* **2004,** *67* (8), 1407–1411.
134. Orhan, I. E.; Orhan, G.; Gurkas, E. An Overview on Natural Cholinesterase Inhibitors—A Multi-Targeted Drug Class—and Their Mass Production. *Mini-Rev. Med. Chem.* **2011,** *11* (10), 836–842.
135. Ortholand, J. Y.; Ganesan, A. Natural Products and Combinatorial Chemistry: Back to the Future. *Curr. Opin. Chem. Biol.* **2004,** *8* (3), 271–280.
136. Park, J. Y.; Jeong, H. J.; Kim, J. H.; Kim, Y. M.; Park, S. J.; Kim, D.; Park, K. H.; Lee, W. S.; Ryu, Y. B. Diarylheptanoids from Alnus Japonica Inhibit Papain-Like Protease of Severe Acute Respiratory Syndrome Coronavirus. *Biol. Pharm. Bull.* **2012,** *35*, 2036–2042.
137. Park, J. Y.; Kim, J. H.; Kim, Y. M.; Jeong, H. J.; Kim, D. W.; Park, K. H.; Kwon, H. J.; Park, S. J.; Lee, W. S.; Ryu, Y. B. Tanshinones as Selective and Slow-Binding Inhibitors for SARS-CoV Cysteine Proteases. *Bioorg. Med. Chem.* **2012,** *20*, 5928–5935.
138. Park, J. Y.; Ko, J. A.; Kim, D. W.; Kim, Y. M.; Kwon, H. J.; Jeong, H. J.; Kim, C. Y.; Park, K. H.; Lee, W. S.; Ryu, Y. B. Chalcones Isolated from Angelica Keiskei Inhibit Cysteine Proteases of SARS-CoV. *J. Enzyme Inhib. Med. Chem.* **2016**, *31*, 23–30.
139. Park, J. Y.; Yuk, H. J.; Ryu, H. W.; Lim, S. H.; Kim, K. S.; Park, K. H.; Ryu, Y. B.; Lee, W. S. Evaluation of Polyphenols from *Broussonetia papyrifera* as Coronavirus Protease Inhibitors. *J. Enzym. Inhib. Med. Chem.* **2017,** *32*, 504–515.

140. Pascolutti, M.; Quinn, R. J. Natural Products as Lead Structures: Chemical Transformations to Create Lead-Like Libraries. *Drug Discov. Today* **2014,** *19* (3), 215–221.
141. Patridge, E.; Gareiss, P.; Kinch, M. S.; Hoyer, D. An Analysis of FDA-Approved Drugs: Natural Products and Their Derivatives. *Drug Discov. Today* **2016,** *21*, 204–207.
142. Pereira, D. M.; Valentao, P.; Andrade, P. B. Nano- and Microdelivery Systems for Marine Bioactive Lipids. *Mar. Drugs* **2014,** *12* (12), 6014–6027.
143. Pettit, R. K. Mixed Fermentation for Natural Product Drug Discovery. *Appl. Microbiol. Biotechnol.* **2009,** *83* (1), 19–25.
144. Pickens, L. B.; Tang, Y.; Chooi, Y. H. Metabolic Engineering for the Production of Natural Products. *Annu. Rev. Chem. Biomol. Eng.* **2011,** *2*, 211–236.
145. Pugh, M. B., Ed. *Stedman's Medical Dictionary*, 27th ed.; Lippincott Williams & Wilkins: Baltimore, 2000; p 65.
146. Qurratul, A.; Ashiq, U.; Jamal, R. A.; Saleem, M.; Mahroof-Tahir, M. Alpha-Glucosidase and Carbonic Anhydrase Inhibition Studies of Pd (II)-Hydrazide Complexes. *Arab. J. Chem.* **2017,** *10* (4), 488–499.
147. Ramasubbu, N.; Paloth, V.; Luo, Y.; Brayer, G. D.; Levine, M. J. Structure of Human Salivary α-Amylase at 1.6 Å Resolution: Implications for Its Role in the Oral Cavity. *Acta Crystallogr. Sect. D.* **1996,** *52* (3), 435–446.
148. Rashid, M. A.; Gustafson, K. R.; Cartner, L. K.; Shigematsu, N.; Pannell, L. K.; Boyd, M. R. Microspinosamide, a New HIV Inhibitory Cyclic Depsipeptide from the Marine Sponge *Sidonops microspinosa. J. Nat. Prod.* **2001,** *64* (1), 117–121.
149. Rauf, A.; Patel, S. Pistagremic Acid as a Broad Spectrum Natural Inhibitor from *Pistacia integerrima* Stewart. *Nat. Prod. Res.* **2016,** *4*, 367–368.
150. Reddy, M. V. R.; Rao, M. R.; Rhodes D. Lamellarin α 20-Sulfate, an Inhibitor of HIV-1 Integrase Active Against HIV- 1 Virus in Cell Culture. *J. Medicinal Chem.* **1999,** *42* (11), 1901–1907.
151. Reker, D.; Perna, A. M.; Rodrigues, T.; Schneider, P.; Reutlinger, M.; Monch, B. Revealing the Macromolecular Targets of Complex Natural Products. *Nat. Chem.* **2014,** *6* (12), 1072–1078.
152. Riches, P. L.; Wright, A. F.; Ralston, S. H. Recent Insights Into the Pathogenesis of Hyperuricaemia and Gout. *Hum. Mol. Genet.* **2009,** *18*, R177–R184.
153. Rodrigues, T.; Reker, D.; Schneider, P.; Schneider, G. Counting on Natural Products for Drug Design. *Nat. Chem.* **2016,** *8* (6), 531–541.
154. Rosen, J.; Gottfries, J.; Muresan, S.; Backlund, A.; Oprea, T. I. Novel Chemical Space Exploration via Natural Products. *J. Med. Chem.* **2009,** *52* (7), 1953–1962.
155. Rosenblum, J. L.; Irwin, C. I.; Alpers, D. H. Starch and Glucose Oligosaccharides Protect Salivary-Type Amylase Activity at Acid pH. *Am. J. Physiol.-Gastrointestinal Liver Physiol.* **1988,** *254* (5), G775–G780.
156. Ryu, Y. B.; Jeong, H. J.; Kim, J. H.; Kim, Y. M.; Park, J. Y.; Kim, D.; Nguyen, T. T.; Park, S. J.; Chang, J. S.; Park, K. H.; Rho, M. C.; Lee, W. S. Biflavonoids from *Torreya nucifera* Displaying SARS-CoV 3clpro Inhibition. *Bioorg. Med. Chem.* **2010,** *18*, 7940–7947.
157. Sachs, G.; Weeks, D. L.; Wen, Yi.; Marcus, E. A.; Scott, D. R. Acid Acclimation by *Helicobacter pylori*. *Physiology* **2005,** *20*, 429–38.
158. Sadeer, N. B.; Sinan, K. I.; Cziáky, Z.; Jek″o, J.; Zengin, G.; Jeewon, R.; Abdallah, H. H.; Rengasamy, K. R. R.; Mahomoodally, M. F. Assessment of the Pharmacological

Properties and Phytochemical Profile of Bruguiera gymnorhiza (L.) Lam Using in Vitro Studies, in Silico Docking, and Multivariate Analysis. *Biomolecules* **2020,** *10*, 731.
159. Salvador-Reyes, L. A.; Luesch, H. Biological Targets and Mechanisms of Action of Natural Products from Marine Cyanobacteria. *Nat. Prod. Rep.* **2015,** *32* (3), 478–503.
160. Samuelson, G. Bohlin, L. *Drugs of Natural Origin: A Textbook of Pharmacognosy*; Swedish Pharmaceutical Press; Taylor and Francis Ltd.: Stockholm, Sweden, 2015; p 808.
161. Sanna, V.; Lubinu, G.; Madau, P.; Pala, N.; Nurra, S.; Mariani, A. Sechi, M. Polymeric Nanoparticles Encapsulating White Tea Extract for Nutraceutical Application. *J. Agric. Food Chem.* **2015,** *63* (7), 2026–2032.
162. Sansone, F.; Picerno, P.; Mencherini, T.; Villecco, F.; D'Ursi, A. M.; Aquino, R. P. Flavonoid Microparticles by Spray-Drying: Influence of Enhancers of the Dissolution Rate on Properties and Stability. *J. Food Eng.* **2011,** *103* (2), 188–196.
163. Sato, A.; Morishita, T.; Shiraki, T.; Yoshioka, S.; Horikoshi, H.; Kuwano, H.; Hanzawa, H.; Hata, T. *J. Org. Chem.* **1993,** *58*, 7632–7634.
164. Sayed A. M.; Alhadrami H. A.; El-Gendy A. O.; Shamikh Y. I.; Belbahri L.; Hassan, H. M.; Abdelmohsen U. R.; Rateb M. E. Microbial Natural Products as Potential Inibitors of SARS-CoV-2 Main Protease. *Microorganism* **2020,** *8*, 970.
165. Scapin, G. Structural Biology and Drug Discovery. *Curr. Pharm. Des.* **2006,** *12* (17), 2087–2097.
166. Schlesinger, N.; Dalbeth, N.; Perez-Ruiz, F. Gout—What Are the Treatment Options? *Expert Opin. Pharmacother.* **2009,** *10*, 1319–1328.
167. Seley-Radtke, K. L.; Yates, M. K. The Evolution of Nucleoside Analogue Antivirals: A Review for Chemists and Non-Chemists. Part 1: Early Structural Modifications to the Nucleoside Scaffold. *Antivir. Res.* **2018,** *154*, 66–86.
168. Sentuk, E.; Urcar, H.; Senturk, M.; Yildrim, S.; Gul, M. Bovine Liver Tissue On Glutathione Reductase Enzyme Determination of Effects of Thiamine, Tyrosine, Dopamine and Adrenaline. *Acta Physiol.* **2016,** *218*, 58.
169. Shen, B. A New Golden Age of Natural Products Drug Discovery. *Cell* **2015,** *163* (6), 1297–1300.
170. Shen, Y. C.; Wang, L. T.; Khalil, A. T.; Chiang, L. C.; Chem, P. W. Bioactive Pyranoxanthones from the Roots of *Calophyllum blancoi*. *Pharm. Bull.* **2005,** *53*, 244–247.
171. Shin, J.; Fenical, Fuscosides AD: Anti-Inflammatory Diterpenoid Glycosides of New Structural Classes from the Caribbean Gorgonian Eunicea fusca. *J. Org. Chem.* **1991,** *56* (9), 3153–3158.
172. Simeonov, A.; Davis, M. I. *Interference with Fluorescence and Absorbance, Assay Guidance Manual*; Eli Lilly & Company and the National Center for Advancing Translational Sciences: Bethesda, 2004; p 638.
173. Sinha, S.; Ali, M.; Baboota, S.; Ahuja, A.; Kumar, A.; Ali, J. Solid Dispersion as an Approach for Bioavailability Enhancement of Poorly Watersoluble Drug Ritonavir. *AAPS Pharm. Sci. Tech.* **2010,** *11* (2), 518–527.
174. Sirko, A.; Brodzik, R. Plant Ureases: Roles and Regulation. *Acta Biochim. Pol.* **2000,** *47*, 1189–1195.
175. Skeggs, L. T.; Kahn, J. R.; Shumway, N. P. The preparation and function of the hypertension converting enzyme. *J. Exp. Med.* **1956,** *103* (3), 295–299.

176. Song, Y. H.; Kim, D. W.; Curtis-Long, M. J.; Yuk, H. J.; Wang, Y.; Zhuang, N.; Lee, K. H.; Jeon, K. S.; Park, K. H. Papain-Like Protease (PLpro) Inhibitory Effects of Cinnamic Amides from *Tribulus terrestris* Fruits. *Biol. Pharm. Bull.* **2014,** *37*, 1021–1028.
177. Stead, P.; Hiscox, S.; Robinson P. S. Eryloside F, a Novel Penasterol Disaccharide Possessing Potent Thrombin Receptor Antagonist Activity. *Bioorg. Medicinal Chem. Lett.* **2000,** *10* (7), 661–664.
178. Stingl, K.; Altendorf, K.; Bakker, E. P. Acid Survival of *Helicobacter pylori*: How Does Urease Activity Trigger Cytoplasmic pH Homeostasis? *Trends Microbiol.* **2002,** *10* (2), 70–74.
179. Subramani, R.; Aalbersberg, W. Marine Actinomycetes: An Ongoing Source of Novel Bioactive Metabolites. *Microbiol. Res.* **2012,** *167* (10), 571–580.
180. Sumner, J. B. The Isolation and Crystallization of the Enzyme Urease. *J. Biol. Chem.* **1926,** *69*, 435–41.
181. Tadera, K.; Minami, Y.; Takamatsu, K.; Matsuoka, T. Inhibition of Alpha-Glucosidase and Alpha-Amylase by Flavonoids. *J. Nutr. Sci. Vitaminol.* **2006,** *52* (2), 149–153.
182. Takahashi, M.; Uechi, S.; Takara, K.; Asikin, Y.; Wada, K. Evaluation of an Oral Carrier System in Rats: Bioavailability and Antioxidant Properties of Liposome-Encapsulated Curcumin. *J. Agric. Food Chem.* **2009,** *57* (19), 9141–9146.
183. Takeuchi, T. On the Occurrence of Urease in Higher Plants. *J. Coll. Agric. Tokyo Imp. Univ.* **1909,** *1*, 1–14.
184. Tasdemir, D.; Mallon, R.; Greenstein, M. Aldisine Alkaloids from the Philippine Sponge *Stylissa massa* Are Potent Inhibitors of Mitogen-Activated Protein Kinase Kinase-1 (MEK-1). *J. Medicinal Chem.* **2002,** *45* (2), 529–532.
185. Thermofisher Scientific Inc. Detecting Peptidases and Proteases- Section 10.4.2020, 2020. https://www.thermofisher.com/in/en/home/references/molecular-probes-the-handbook/enzyme-substrates/detecting-peptidases-and-proteases.html (Accessed on Aug 28, 2021).
186. Thorne, N.; Auld, D. S.; Inglese, J. Apparent Activity in High-Throughput Screening: Origins of Compound-Dependent Assay Interference. *Curr. Opin. Chem. Biol.* **2010,** *14* (3), 315–324.
187. Tisi, R.; Gaponenko, V.; Vanoni, M.; Sacco, E. Natural Products Attenuating Biosynthesis, Processing, and Activity of Ras Oncoproteins: State of the Art and Future Perspectives. *Biomolecules* **2020,** *10*, 1535.
188. Tiwari, G.; Tiwari, R.; Rai, A. K. Cyclodextrins in Delivery Systems: Applications. *J. Pharm. BioAllied Sci.* **2010,** *2* (2), 72–79.
189. Toledo, L.; Aguirre, C. Enzymatic Browning in Avocado (Persea americana) Revisited: History, Advances, and Future Perspectives. *Crit. Rev. Food Sci. Nutr.* **2017,** *57* (18), 3860–3872.
190. Torchilin, V. P. Passive and Active Drug Targeting: Drug Delivery to Tumors as an Example. *Handb. Exp. Pharmacol.* **2010,** *197*, 3–53.
191. Trindade-Silva, A. E.; Lim-Fong, G. E.; Sharp, K. H.; Haygood, M. G. Bryostatins: Biological Context and Biotechnological Prospects. *Curr. Opin. Biotechnol.* **2010,** *21* (6), 834–842.
192. Ur-Rehman, N.; Halim, S. A.; Al-Azri, M.; Khan, M.; Khan, A.; Rafiq, K.; Al-Rawahi, A.; Csuk, R.; Al-Harrasi, A. Triterpenic Acids as Non-Competitive-Glucosidase Inhibitors from Boswellia elongata with Structure-Activity Relationship: In Vitro and In Silico Studies. *Biomolecules* **2020,** *10*, 751.

193. Urcar, H.; Senturk, E.; Senturk, M.; Gul, M.; Yildirim, S. Investigation of Effects of Some Catecholamines on the Activity of Carbonic Anhydrase Enzyme Purified from Bovine Kidney Tissue. 42st National Physiology Congress. *Acta Physiol.* **2016,** *218*, 57.
194. Vazquez-Mellado, J.; Alvarez Hernandez, E.; Burgos-Vargas, R. Primary Prevention in Rheumatology: The Importance of Hyperuricemia. *Best Pract. Res. Clin. Rheumatol.* **2004,** *18*, 111–24.
195. Voet, D.; Voet, J. G. Biochimie. (2e éd.). Bruxelles: De Boeck, 2005; p 1583.
196. Wassermann, A. M.; Lounkine, E.; Urban, L.; Whitebread, S.; Chen, S.; Hughes, K. A Screening Pattern Recognition Method Finds New and Divergent Targets for Drugs and Natural Products. *ACS Chem. Biol.* **2014,** *9* (7), 1622–1631.
197. Watanabe, S.; Kanellis, J.; Nakagawa, T.; Han, L.; Ohashi, R.; Lan, H.; Feng, L.; Johnson, R. J. Reducing Uric Acid as a Means to Prevent Cardiovascular and Renal Disease. *Expert Opin. Ther. Pat.* **2002,** *12*, 193–199.
198. Watkins, R.; Wu, L.; Zhang, C.; Davis, R. M.; Xu, B. Natural Product Based Nanomedicine: Recent Advances and Issues. *Int. J. Nanomed.* **2015,** *10*, 6055–6074.
199. Weissman, K. J.; Leadlay, P. F. Combinatorial Biosynthesis of Reduced Polyketides. *Nat. Rev. Microbiol.* **2005,** *3* (12), 925–936.
200. Welsch, M. E.; Snyder, S. A.; Stockwell, B. R. Privileged Scaffolds for Library Design and Drug Discovery. *Curr. Opin. Chem. Biol.* **2010,** *14* (3), 347–361.
201. Yang, C. W.; Lee, Y. Z.; Kang, I. J.; Barnard, D. L.; Jan, J. T.; Lin, D.; Huang, C. W.; Yeh, T. K.; Chao, Y. S.; Lee, S. J. Identification of Phenanthroindolizines and Phenanthroquinolizidines as Novel Potent Anti-Coronaviral Agents for Porcine Enteropathogenic Coronavirus Transmissible Gastroenteritis Virus and Human Severe Acute Respiratory Syndrome Coronavirus. *Antiviral Res.* **2010**, *88*, 160–168.
202. Yathindra, N. Structure of an Enzyme Revealed 80 Years After It Was Crystallized—Differential Functional Behaviour of Plant and Microbial Ureases Uncovered. *Curr. Sci.* **2010,** *99*, 566–568.
203. Yi, L.; Tsai, F. J.; Tsai, C. H.; Lai, C. C.; Wan, L.; Ho, T. Y. Anti-SARS Coronavirus 3C-Like Protease Effects of *Isatis indigotica* Root and Plant-Derived Phenolic Compounds. *J. Virol.* **2004,** *78*, 11334–11339.
204. Ying, Y.; Taori, K.; Kim, H.; Hong, J.; Luesch, H. Total Synthesis and Molecular Target of Largazole, a Histone Deacetylase Inhibitor. *J. Am. Chem. Soc.* **2008,** *130* (26), 8455–8459.
205. Yoshika, Y.; Namiki, D.; Makiuchi, M.; Sugaya, K.; Onose, J.; Ashida, H. Vialinin A and Thelephantin G, Potent Inhibitors of Tumor Necrosis Factor-Alpha Production, Inhibit Sentrin/SUMO-Specific Protease 1 Enzymatic Activity. *Bioorg. Med. Chem. Lett.* **2016,** *26* (17), 4237–4240.
206. Zazeri, G.; Povinelli, A. P. R.; de Freitas Lima, M.; Cornélio, M. L. The Cytokine IL-1_ and Piperine Complex Surveyed by Experimental and Computational Molecular Biophysics. *Biomolecules* **2020,** *10*, 1337.
207. Zhang, M. Q.; Wilkinson, B. Drug Discovery Beyond the 'Rule-of-Five'. *Curr. Opin. Biotechnol.* **2007,** *18* (6), 478–488.
208. Zotchev, S. B. Marine actinomycetes as an Emerging Resource for the Drug Development Pipelines. *J. Biotechnol.* **2012,** *158* (4), 168–175.

CHAPTER 12

ENZYMATIC INHIBITORS FROM MEDICINAL PLANTS

RAMU MEENATCHI, SAQIB HASSAN, KALPANA PACHILLU, SONIA BANSAL, NAMRATA DEKA, GEORGE SEGHAL KIRAN, and JOSEPH SELVIN

ABSTRACT

The natural world is one of the major sources of molecules with medicinal properties known to humankind. Natural products have long been utilized by humans to cure a myriad of ailments. Many diseases are still treated using plant or algal extracts in the developing world where modern medicine is still scarce. Each year hundreds of novel inhibitors from the natural world capable of targeting essential biological human proteins or enzymes are discovered. The chemical and structural diversity of these molecules interests the researchers for their use in drug discovery. Chemical synthesis of most of these molecules is difficult in standard organic chemistry laboratories. This suggests that despite significant technical advances, drug discovery will continue to rely heavily on the discovery of novel natural bioactive compounds with drug-like action in the coming years. Some of these compounds have been considered to be potential therapeutic molecules due to their high selectivity or potency, and have been used in clinical trials. Given the limited amount of investment in natural products as main leads in drug discovery during the last few decades, 34% of the medications recognized by the US Food and Drug Administration (FDA) between 1981 and 2010 were either natural products or direct extracts from them.

Enzyme Inactivation in Food Processing: Technologies, Materials, and Applications. Megh R. Goyal, PhD, Junaid Ahmad Malik, PhD & Ravi Pandiselvam, PhD (Eds.)

12.1 INTRODUCTION

Plants have thousands of years of history connected to the diversity of their natural products that are highly useful. These beneficial species have been termed "medicinal plants" because of their valuable compounds that aid in the therapeutic treatment of many diseases in the humans as well as in animals.[26] These compounds, commonly known as the natural products with medicinal properties isolated from the plants, are generally the secondary metabolites. The secondary metabolites of plants are produced in response to a stimulus from surroundings or as a defensive strategy against the predators. The evaluation of the natural products has positioned them as the most successful source of potential drug leads. The branch of study "Ethnopharmacology," with an objective involving the development of drugs for the treatment of patients and ultimately to validate the traditional use of medicinal plants, has been perceiving a decline because of introduction of the combinatorial chemistry.[63] The natural products have conversely been in continuous exploitation because of their outstanding chemical structures. At the same time, in finding the novel natural products from highly diverse natural environments, the percentage of secondary metabolites possessing biological activity counts to as low as 10%.[66] One of the significant challenges accountable to this lies in the question of what manner, means, or methods should be taken to approach this natural chemical diversity.[23]

World Health Organization (WHO) depicted a picture that estimates that around 80% of the World's population relies on medicinal plants and their natural products for healthcare.[64] The beliefs and trust of the population at various places of the world on traditional specialists, herbal medicines, and ayurvedic medicines still stand firm, besides the presence of allopathic medicine. Ayurvedic medicine includes the permutation of several constituents that are readily available from medicinal plants. However, the absence of pharmacological data related to the medicinal plants and their isolated natural products creates a challenge in gaining the required activity and potency from the isolated products. The primary role of the modern drug production procedure is to fill the gap mentioned and establish a way to gain the improvised sorts of the products with requisite activity.

Natural products have also been exploited as potential enzyme inhibitors. The enzyme inhibitors are bioactive secondary metabolites of the plants that bind with the enzyme and block its activity. This inhibitory process results in killing a pathogen or aids in improving various health-related issues.[72] The two labeling criteria, specificity, and effectiveness are considered in the medicinal enzyme inhibitors to assess the potentiality to inhibit the enzyme. Studies have shown that the inhibitors with greater specificity and potentiality prove to be a therapeutic drug with fewer side effects and low toxicity levels.

The increasing popularity and knowledge of the traditional medicines in the healthcare necessities have led to more and more exploration of medicinal plants, isolation of many natural products, and investigation of those products as potential medicines. One of the extensively used drugs extracted from the plants used in treating breast cancer is Taxol®. This drug is isolated from the bark of *Taxus brevifolia* and is known as paclitaxel (Taxol®).[80] Many other natural products have been explored for treating fever, chills, some of the autoimmune and inflammatory diseases, neurological disorders such as Alzheimer's disease, Parkinson's disease.

The ability of the extracts from the medicinal plants in inhibiting the replication of many viruses especially in HSV-2, HIV, HBV infections and also newly emerging viruses related to the poxvirus and SARS virus exhibit their antiviral potential.[51] The detection of phytochemicals such as alkaloids, phenolic acids, flavonoids, glycosides, saponins, polysaccharides, stilbenes, and tannins in the extracts from the plants has been explored for the natural products to be used as antidiabetic drugs.[75]

The discovery of new drugs consists of the initial step, the identification and isolation new enzyme inhibitors followed by other long procedures until the final approval of the drugs to be used by the population. In the past, the only way of discovering new drugs consisted of the trial-and-error method that moved toward combinatorial chemistry approach in the modern times where we explore the chemical structure and the molecular mechanisms. There are some drawbacks also associated with the use of natural products. Limited quantities of the beneficial natural compounds can be extracted from the natural sources as compared to the synthetic process. Moreover, with the increasing demand, synthetic process that is quick is preferred over the isolation of compounds from natural sources.

In this chapter, we have described the importance of plant- derived natural products as enzyme inhibitors, we have provided an overview of bioactive natural products derived from medicinal plants and the process of extracting these. We have also shed light on the application of these bioactive compounds in therapeutics, especially as anticancer, antiviral (including COVID19), and antidiabetic agents. Finally, we have discussed about the new approaches for identification of bioactive compounds from plants and also the challenges that are contributing to the decline of plant natural products as drug discovery source.

12.2 NATURAL PRODUCTS AS ENZYME INHIBITORS

Natural products are widespread and their distinctive properties have been used for healing ailments and injuries for centuries. Potential of natural products acting as the modulators of biological processes has been increasingly recognized throughout the evolution.[9] Natural products have been used less frequently as a starting point for drug development by pharmaceutical firms in recent decades, owing to a notion that natural products were incompatible with drug discovery techniques focused on high-throughput screening directed at molecular targets.[67] Additionally, it was assumed that combinatorial chemistry techniques would be capable of producing all of the diverse chemicals required for successful lead identification. To some extent synthetic molecules have been created to replicate the chemical characteristics of natural substances.[39] Despite a lack of investment in natural products as important leads in drug discovery in recent decades, the US Food and Drug Administration (FDA) certified 34% of medications between 1981 and 2010 that were natural products or directly derived from them.[68]

Compounds present in natural products have a lot of potential in drug discovery because of their chemical and structural diversity. Natural molecules are an excellent complement to synthetically generated ones.[42,53] The drug-likeness of natural bioactive compounds frequently exceeds that of synthetic drugs. This is one of the most important reasons for their effectiveness as a source of bioactive molecules. These natural molecules have a higher degree of resemblance and binding potential with biological structures, enhancing the chances of a successful interaction with a wide range of biological targets.[53] Natural products are known for their excellent three-dimensional conformation, which is due to their complex and unique structure, which is generally beyond the synthetic potential of medicinal chemistry.

Natural products are frequently described as having a "privileged" scaffold that allows them to bind to a wide range of enzymes and receptors. Evans coined this term in the late 1980s to describe the benzodiazepine scaffold, which is characterized by its capacity to bind not just to the central and peripheral nervous system receptors, but also to the cholecystokinin receptors. According to Evans, a privileged structure binds to many receptors/proteins.[13,29] These 'privileged' scaffolds are present in a variety of natural products and have the ability to bind to several targets, and lead to a multitude of pharmacological effects.[82] Given the vast diversity of compounds that make up the so-called chemical space, it's easy to assume that natural products, as opposed to synthetic ones, cover a larger range of characteristics and are more druglike. Rosén et al.[74] demonstrated through computational screening that natural products cover the chemical space that medicinal chemistry compounds do not. As a result, these compounds may be important for novel leads. The variation in chemical composition gives rise to distinction between natural products and other sources of molecules in terms of their capacity to show biological properties.[34]

Generally, the composition of natural molecules differs from synthetic ones in that they have less nitrogen, sulphur, or halogen atoms, have higher oxygen atoms, and possess more hydrogen bond donors. The sterical complexity is also a factor in this equation, since these natural compounds have higher number of rings and chiral centers. The obvious explanation for such chemical traits is that they are not produced randomly, but rather as the consequence of highly specialized biosynthetic processes. As a result, these compounds are known to interact with molecular targets that are three-dimensional and chiral in nature. Furthermore, the enzymes involved in biosynthesis are generally chiral, which leads to the formation of only one isomer, which is a characteristic not always observed in pharmaceutical chemistry, where racemic mixtures are common. The enzymes used for biosynthesis as well as the molecular targets with which the natural product is supposed to interact, are both three-dimensional and chiral by nature. Thus, it may be the reason for the natural products displaying effective results as enzyme inhibitors.[70] The fact that the most of natural products have these features indicates that they are the consequence of an evolutionary drive that favors molecules with a specific configuration.[70,58]

Furthermore, as compared to a randomly synthesized molecule, the likelihood of discovering a bioactive molecule in natural goods is significantly higher.[13] The underlying synthesis strategy in both natural products and randomly synthesized molecules is another significant difference. In contrast to combinatorial chemistry, which may use tens of thousands of distinct scaffolds as building components, nature lacks such a process. Nature only has a limited number of building blocks accessible from a limited number of biosynthetic pathways, any chemical innovation can only be attained by branching out intermediates and establishing a plethora of biosynthetic routes, ultimately resulting in distinct chemical entities. In combinatorial chemistry, a different method is used, in which the same series of reactions is frequently repeated with various starting molecules. In combinatorial chemistry, a different method is used, in which the same series of reactions is frequently repeated with various starting molecules. Nature's chemistry is basically oxophilic, and as a result, it possesses enzymatic tools that undergo site-selective C–H activation to introduce oxygen atoms and distinguish between functional groups with varying degrees of oxidation.[16,38]

Natural products have coexisted with diverse species and habitats over time, going through the same recurrent cycle of improvement and evolution with changing biotic systems. Several natural compounds or their derivatives are used as enzyme inhibitors in the treatment of cancer, diabetes, hypertension, and infectious illnesses.[15] Teprotide, extracted from the venom of the *Bothrops jararaca*, is a classic example. This compound was chosen as a lead compound for the development of angiotensin carboxypeptidase inhibitors because of its long-lasting in vivo activity against angiotensin converting enzyme.[55] Galanthamine, an anticholinesterase drug obtained from *Galanthus nivalis L.* (Snowdrop plant), is widely used to treat Alzheimer's disease.[28] Vialinin A, a pterphenyl chemical derived from the edible Chinese fungus *Thelephora vialis*, suppresses the synthesis and release of tumor necrosis factor-α.[83]

12.3 NATURAL PRODUCTS FROM PLANTS

Medicinal applications of plants have been extensively recognized for thousands of years. Over millions of years, they have developed and adapted to withstand bacteria, insects, fungi, and weather; they produce structurally diverse secondary metabolites. Their ethnopharmacological potentials have been exploited as a main source of drug discovery.[30,62] Traditional medicine (complementary or alternative herbal products) has prompted more research into medicinal plants as possible medications, leading to the isolation of many natural compounds that have gone on to become well-known pharmaceuticals. Plants have supplied more than 60% of anticancer medicines, either directly or indirectly, in the oncology sector. Natural products account for roughly half of all contemporary medicines.[10]

Antimicrobial resistance has resulted in the emergence of MDR in the last three decades, as a result of the recurrent and regular use of single drug for the same therapeutic target. As a result, it serves as a warning to pharmaceutical researchers to develop a new weapon against MDR and an opportunity to look for a new antibiotic spectrum to combat AMR.[57] Natural products might be a source of novel biologically active molecules, leading to the development of new therapeutics.[73] Secondary metabolites (small organic compounds) are produced by plants and are not necessary for regular growth or development, but are important for reproduction and defense against bacteria, fungi, viruses, vertebrates, and other organisms. These items have a lot of potential to be used as drugs.[11,61] The biochemistry of drugs derived from natural sources has made a significant contribution to public health and improved the development of affordable medicines throughout the world.[76] Chemical composition (containing nitrogen or not), chemical structure (e.g., having rings, including sugar), biosynthetic pathway (e.g., phenylpropanoid, which produce tannins), and solubility are used to classify them. They have been categorized in three main groups alkaloids, terpenes, and phenolics.[2,48] Antimicrobial properties of polyphenols have been fueled by high demand in the pharmaceutical and food industries to develop new food preservatives to avoid synthetic preservatives and novel therapies for the treatment of various microbial infections to combat microbial resistance against conventional antibiotics.

12.3.1 LANT-DERIVED ENZYME INHIBITORS AS ANTICANCER DRUGS

Enzyme inhibition plays a key role in drug discovery, especially in the discovery of cancer drugs. Numerous enzymes involved in cell cycle process are acting as enzyme targets for plant drug leads. For instance, the human lactate dehydrogenase subtype *A* protein expression in cancerous cells causes rapid tumor progression, metastatic condition, and also resistance against radiation and chemotherapy. *Acacia Arabica*, *Melaphis chinensis* gallnut, and other plant extracts showed inhibitory activity against LDH-A enzyme[25] aiding promising treatment against various types of cancers. Vinca alkaloids, Paclitaxel, docetaxel, Camptotecin, irinotecan, Podophyllotoxin, and analogues are potential anticancer drugs available in the commercial market with enzyme inhibition potential, especially the mechanism of action involves the decrease in expression of NFκB and COX-2 enzymes in cancerous cells. More detailed information on plant-derived enzyme inhibitors as anticancer drugs was listed in Table 12.1.

TABLE 12.1 Some Anticancer Drugs/Drug Leads from Plants as Enzyme Inhibitors with Pre-Clinical and Clinical Data.

S. No.	Name of the drug lead	Plant Origin	Type of cancer	Mechanism of action	*In vitro* study (cell lines) models	*In vivo* study (animal) models	Clinical trial -Stage	Status of drug-FDA Approved	References
1.	Vincristine, Vinblastine, Semisynthetic Analogues (vinflunine, vinorelbine and vindesine)	*Catharanthus roseus* G. Don.	leukemias, lymphomas, advanced testicular cancer, breast and lung cancers, Kaposi's sarcoma	Inhibition of the DNA synthesis, disrupt microtubules, apoptoptic cell death	Numerous cell lines	Rat, mice	Phase III (vinflunine)	FDA approved second line drug for advanced metastatic breast cancer and urothelial carcinoma	[20]
2.	Etoposide and Teniposide	*Podophyllum peltatum L.* and *Podophyllumemodii* Wallich	lymphomas, and bronchial and testicular cancers	inhibit topoisomerase II	mouse L1210 and human VA-13 and HT-29 cell lines	Rat, mice	Phase I and IIa (Etoposide)	FDA approved drug for small cell lung cancer and Refractory Testicular Tumors	[5]
3.	Paclitaxel and Docetaxel	*Taxus brevifolia Nutt.* and *Taxus baccata.*	breast, ovarian, and non-small cell lung cancer (NSCLC)	promote the polymerization of tubulin heterodimers to microtubules, suppressing dynamic changes in microtubules	Osteosarcoma and neuroblastoma cell lines	NOD/SCID xenograft and metastatic neuroblastoma mouse models	Phase II trial—breast cancer	FDA approved	[59]

TABLE 12.1 *(Continued)*

S. No.	Name of the drug lead	Plant Origin	Type of cancer	Mechanism of action	*In vitro* study (cell lines) models	*In vivo* study (animal) models	Clinical trial -Stage	Status of drug-FDA Approved	References
4.	Topotecan, irinotecan, belotecan	*Camptotheca acuminata Decne*	small cell lung cancer and pancreatic cancer.	inhibition of topoisomerase II	oral squamous cell carcinoma cell lines		Phase I	FDA approved	[21]
5.	Resveratrol	Grapes, Peanuts, Cranberries and Blueberries	Breast cancer, Osteosarcoma, Cervical cancer, Colorectal cancer,	Cell cycle inhibition, decreasing the expression of transcription factors (NF-κB)	colorectal (SW480 cell line), intestine (Caco2 cell line), liver (HepG2 hepatic derived cell line), etc.	rats and mice	Phase I (Colorectal Cancer)	not FDA approved	[77]
6.	**Curcumin**	*Curcuma longa*	Melanoma, Breast Cancer, Cervical Cancer, Colon Cancer,	inhibit the activity of NF-κB and COX-2, G2/M phase cell-cycle arrest	MDA-MB-231, MCF-7, BT-474, SK-BR-3-hr, Cervical Cancer, HeLa	Human cervical cancer HeLa cells xenograft model in nude mice	Phase II	FDA approved "Generally Recognized As Safe" (GRAS) food supplEment	ClinicalTrials.gov Identifier: NCT01042938
7.	Epigallocatechin-3-gallate	*Camelia sinensis* L.	pancreatic, breast, bladder and lung carcinomas	inhibition of the Wnt/β-catenin pathway	A549, SW480 cells	A549 cells xenograft model in BALB/c nude male mice	Phase I and Phase II clinical trials	'no questions' GRAS approval from FDA	ClinicalTrials.gov Identifier: NCT00917735

TABLE 12.1 *(Continued)*

S. No.	Name of the drug lead	Plant Origin	Type of cancer	Mechanism of action	*In vitro* study (cell lines) models	*In vivo* study (animal) models	Clinical trial -Stage	Status of drug-FDA Approved	References
8.	Quercetin	*Fruit and vegetables*	Prostate, Cervical Cancer, Colorectal Adenocarcinoma,	induce cell cycle arrest, apoptosis, necrosis, autophagy, tumor angiogenesis, metastasis inhibition	murine melanoma cells, Caco-2, HeLa cells	MIA PaCa-2 cells Orthotopic xenograft model in nude mice	Phase I and II	Not approved yet	ClinicalTrials.gov Identifier: NCT01538316
9.	Betulinic acid	*Spina date seed*	Melanoma, pancreatic cancer, Colon Cancer	promotes apoptosis, inhibiting the activation of mTOR	PANC-1 and SW1990	HCT116 cells xenograft model in mice	Phase I and II	FDA approved as topical treatment for metastatic melanoma	ClinicalTrials.gov Identifier: NCT00346502 (Suspended)
10.	Artemisinin	*Artemisia annua*	Breast, colon, cervical cancer	induction of caspase-dependent apoptosis	HCT116, MCF-7	CLY cells xenograft model in female athymic nude mice (Balb/c nu/nu)	Phase I	FDA approved as treatment for malaria; not approved yet for cancer treatment	ClinicalTrials.gov Identifier: NCT00764036

12.3.2 PLANT-DERIVED ENZYME INHIBITORS AS ANTIDIABETIC DRUGS

Worldwide people are mostly affected by various metabolic disorders irrespective of ages; the metabolic disorders include diabetes, obesity, cardiovascular diseases such as atherosclerosis, hypertension, renal disease, glaucoma, and so on. These metabolic disorders are composed of various metabolic enzymes as checkpoints that can act as target for drug leads. Some of the enzyme targets that are being utilized in the treatment of metabolic disorders are pancreatic lipase, 3-Hydroxy-3-methylglutaryl coenzyme A (HMG-CoA) reductase, Glucosidase, Carbonic anhydrase, Aldose reductase, xanthine oxidase, angiotensin-converting enzyme etc. Medicinal plants useful in treatment of metabolic disorders include *Allium sativum, Aloe vera, Andrographis paniculata, Berberis aristate, Bombax malabaricum, Camellia sinensis, Clitoria ternatea, Coffea arabica, Emblica officinallis, Curcuma longa, Glycyrrhiza glabra*, and so on.[65]

Pancreatic lipase catalyzes the conversion of triglycerides to monoacyl glycerol which is the storage lipid. To be an antiobese drug, several plant compounds including polyphenols, saponins, terpenes are acting as potent pancreatic lipase inhibitors. Inhibition of HMG-CoA reductase by plant metabolites can reduce the cholesterol biosynthesis that makes plant metabolite a good candidate for cardiovascular disease management.[60] Glucosidases catalyze the breakdown of glycosidic bonds between oligosaccharides and other glycoconjugates. α-Glucosidase enzyme inhibition reduces the digestion and absorption process of metabolism, thereby reduces the blood glucose level; thus, the plant metabolites targeting the inhibition of α-glucosidase are potential antidiabetic drugs (e.g., Metformin from plant *Galega officinalis*). Carbonic anhydrase inhibition by plant metabolites makes them the drug candidate for various properties including diuretics, antiglaucoma, antiobesity, antiepileptic agents. Plant-derived metabolites inhibiting aldose reductase activity act as good antidiabetic agents. Inhibition of xanthine oxidase by plant metabolites helps in the clinical management of gout disease. Angiotensin-converting enzyme inhibitors from plant source act as antihypertensive agents.

12.3.3 PLANT-DERIVED ENZYME INHIBITORS AS ANTIVIRAL DRUGS

Virus infections, notably infections caused by the human immunodeficiency virus types 1 (HIV-1) and 2 (HIV-2), as well as newly developing infectious viruses, have posed a threat to mankind globally. Medicinal plants have shown promise in the treatment of a number of viral diseases, and several of them exhibit broad-spectrum antiviral activity. In the past, due to (a) the highly infectious nature of viruses and (b) lack of adequate separation procedures for the identification of antiviral compounds from plants, investigations for checking the antiviral activity of natural compounds isolated from medicinal plants was limited. The development of vector-based methods, in which a noninfectious molecular clone of a virus could be utilized for antiviral screening, as well as advancements in separation technologies are backing the use of medicinal plants in modern drug discovery.

Cyanovirin N (CV-N), an 11-kDa protein derived from the cyanobacterium *Nostoc ellipsosporum*, has been identified as a possible antiviral agent. Apart from suppressing HIV-1 infection, the CV-N has broad-spectrum antiviral activity.[12] Various plants and their component parts (stems, leaves, roots, etc.) have been compared in their ability to block viral reverse transcriptase (RT) and integrase, the two key enzymes in HIV infection.[8] As RT regulates viral infectivity and replication, several studies have linked the ability of medicinal plants to inhibit RT activity to an indirect measure of anti-HIV effect.[32,50,82] Medicinal herbs have also been found to inhibit viral protease (PR), an enzyme required for the proteolytic processing of polyprotein precursors into proteins required for viral particle formation.[44,45] Natural RT and protease inhibitors found in a variety of medicinal plants also provide a reservoir of antiviral chemicals that are yet to be discovered.

Extracts from plans like *Podophyllum peltatum L, Phyllanthus urinaria* etc. have been used to treat human herpesviruses (HHV) infections.[7] Anti-influenza activity has been found in a wide range of phytochemicals, including alkaloids, flavonoids, glucosides, polyphenols, and saponins.[78] The methanolic extracts of *Acacia nilotica L. Willd ex Delile*, *Boswellia carterii, Embelia schimperi, Quercus infectoria, Trachyspermum ammi L.* and aqueous extracts of *Piper cubeba L., Q. infectoria,* and *Syzygium aromaticum L.*, have been found to have potential to inhibit HCV.[45] Traditional treatment for viral hepatitis has mostly concentrated on plants from the Euphorbiaceae family especially the *Phyllanthus* genus.

Virus infections are on the rise, posing a serious threat to humanity. Antiviral discovery programmes for such infections might include medicinal plants with wide antiviral properties. Glycyrrhizin, a bioactive component of liquorice (*Glycyrrhiza uralensis Fisch*), and lycorine, a bioactive component of *Lycoris radiata L.,* both of which were initially used for different purposes, exhibit significant anti-SARS-CoV activity.

12.3.3.1 PLANT-DERIVED ENZYME INHIBITORS IN COVID-19 RESEARCH

Carvacrol is monoterpene phenol found abundantly in essential oils of plants such as thyme and oregano which are being used as a food ingredient in Western countries.[46] Carcacrol specifically targets the enzyme activity of viral protease (Mpro in the nonstructural protein part of Sars-Cov 2) and Angiotensin-Converting Enzyme 2 (receptor for binding of virus onto the human respiratory tract).

12.4 BIOPROSPECTING OF BIOLOGICALLY ACTIVE COMPOUNDS FROM PLANTS

Bioprospecting is the search for small molecules, macromolecules, biochemical, and genetic information in natural sources that might be turned into economically viable products in the agricultural, aquaculture, bioremediation, cosmetics, nanotechnology, or pharmaceutical industries. The newfound technologies and methodologies of the recent past have brought on a novel and advanced phase of drug discovery. Historically, the era of plant-derived drug discovery expounded as the "New Golden Age" was preceded by the "Golden Age of Antibiotics." The worldwide incentives for the discovery of new antibiotics have acted as a catalyst for the pharmaceutical industries in initiating natural product discovery (NPD) programs. Development of new technologies, innovations in metagenomics and genomics have contributed in bioprospecting of natural bioactive compounds. Initially in the 1990s to early 2000s many of the renowned pharmaceutical industries withdrew the NPD programs with the introduction of automated high throughput screening (HTS) technology. The HTS approach accelerated the process pertaining to biological testing and facilitated the development of combinatorial chemistry as a better approach for producing "drug-like" compounds.

The HTS technology based on the combinatorial chemistry created large library of compounds. The large combinatorial libraries were then exploited for achieving the efficient "hits" and in the meantime classical natural product chemistry was swapped by the molecular target drug discovery approach.[69] The discovery process of drugs from natural products kept on advancing with time, technologies, and instruments aiding in rapid identification of novel compounds. The advanced knowledge made the synthetic chemists realize that the developed combinatorial libraries were lacking in many other natural products. The synthetic chemists then came up with the idea of synthesizing compounds that would resemble the natural products and termed the concept as diversity-oriented synthesis (DOS).

The foundation for the discovery of drugs initiates with exploration of novel bioactive natural products possessing potential biological activity. The next step comprises the dereplication process of identifying of known metabolites in extracts/mixtures. The importance of previously known compounds from the natural source has been considered important as it not only helps in the dereplication process but also aids in saving time and money. The decline in the isolation of the novel natural products during the 1970s was felt even though many environments were still to be explored for any novel products. The dereplication process considered to be the important identification process came again into the picture with the advent of new technologies such as HPLC-MS, HPLC-NMR, and HPLC-NMR-MS. These technological advancements accelerated the identification and dereplication process in the drug delivery process.

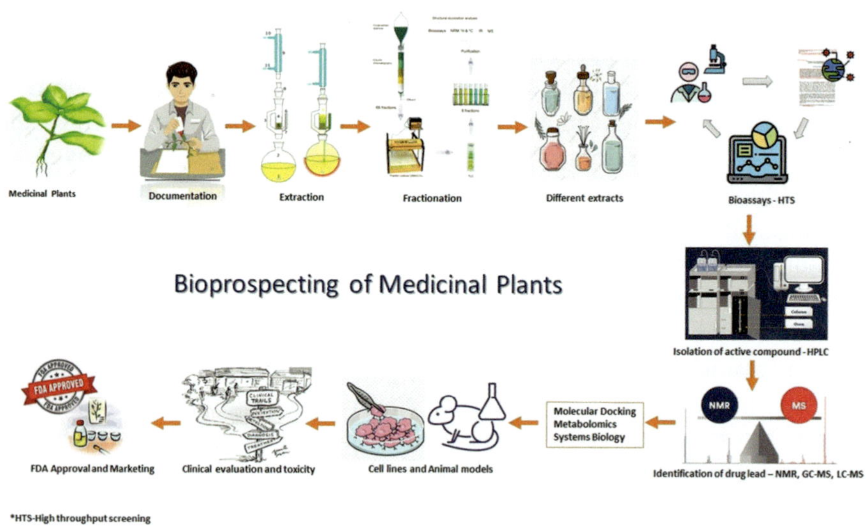

FIGURE 12.1 Bioprospecting of medicinal plants.

The emerging field of systems biology consists of various tools in post genomics era with the goal of differentiating and characterizing all genes as well as cell products in the most comprehensive manner. The metabolomics approach in the drug discovery system has been constructed for unbiased observations, development of highly reproducible data analytical tools, followed by data analysis to identify the correlation among all the available data. Mass spectrometry (MS), liquid chromatography (LC) and nuclear magnetic resonance spectroscopy (NMR) are the leading platforms for analytical procedures. The combination of metabolomics along with systems biology and other techniques have revolutionized the drug discovery process from natural sources (Fig. 12.1).

12.5 DESIGN AND DISCOVERY OF NOVEL PLANT-DERIVED ENZYME INHIBITORS

In recent years, selectively inhibiting the activity of certain enzymes responsible for a disease or a disorder is the strategy exploited by most drugs for the treatment of various clinical conditions such as cancer, diabetes, cardiovascular disease, neurological conditions, and so on. Enzyme inhibitors play a major role in promoting the quality of life of patients suffering from cancerous conditions.[52] Specifically, the enzyme human lactate dehydrogenase subtype-A (LDH-A), its protein expression in cancerous cells cause rapid tumor progression, metastatic condition, and also resistance against radiation and chemotherapy. Inhibition of LDH-A by using plant-derived bioactive compounds showed a promising effect to be a drug candidate.[25]

Conventional method to discover novel enzyme inhibitors is through trial-and-error method where various structurally different compounds from various medicinal plants will be screened to detect the inhibition of specific enzyme, lead compound will be subjected for identification. This conventional way of drug discovery is termed forward pharmacology. Recently, computational approach is also being followed for the process of drug discovery where three-dimensional structure of enzyme is screened for binding of various structurally determined ligand molecules (i.e., probable bioactive metabolite) and then the best binding molecules will be subjected to bioactivity screening and validation of compounds by clinical trials, which is the reverse pharmacology (Fig. 12.2).

Numerous plant-derived bioactive metabolites (viz., enzyme inhibitors) are still available to be explored. Practicing conventional methods for drug discovery from these numerous metabolites is time-consuming and tedious. Computational methods are high-throughput models for screening of vast metabolite composition. However, relying only on computational methods is not so advisable because computationally active compound may not show potentiality in the actual cellular environment. Hence, integration of computational methods and conventional process can ensure more probability for successful drug discovery. In other words, screening the compounds through computational methods and detecting their expected potential through bioassays is the best option to analyze a compound as a drug candidate.

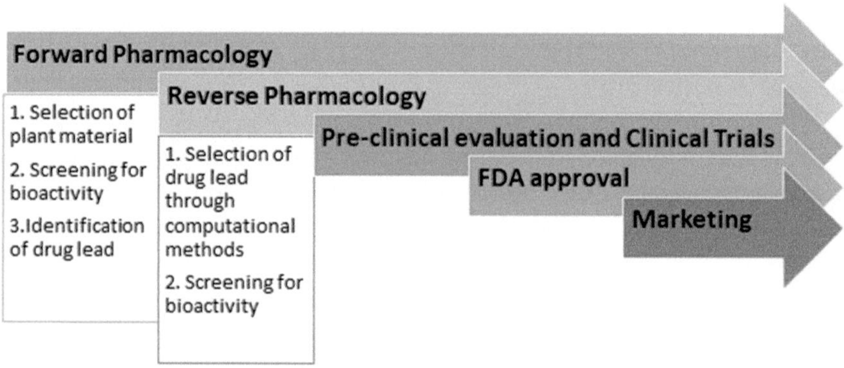

FIGURE 12.2 Recent advancements in the drug discovery of plant-derived enzyme inhibitors.

The process of drug discovery from plant-derived enzyme inhibitors involves the following steps:

(i) Selection of starting material
(ii) Bioassays for screening
(iii) Identification of target lead
(iv) Clinical evaluation
(v) Authority approval and marketing

12.5.1 SELECTION OF STARTING MATERIAL

Availability of plant-derived bioactive compounds is enormous and the process of selection of starting material itself is a tedious process. There are several strategies[3] followed by researchers which include:

a. Random screening—it is the process of selecting the starting material randomly based on the abundancy of plant material.
b. Ethnopharmacological screening involves the classical knowledge of plant materials where the medicinal information of the plants or

herbs is available in the literatures of Ayurveda, traditional medicinal systems, etc.

c. Chemosystematic or phylogenetic screening methods—this involves the combinatorial approach of phylogenetics and phytochemicals of plant material.
d. Ecological approach—this involves the use of traditional knowledge on animal behavior in eating or taking the plant material as self-medication by the animals which is referred to as the zoo-pharmacognosy.
e. Computational methods—compounds that accomplish efficiency grade in the *in silico* estimations can be used as starting materials for in vitro analysis to yield the drug lead.

These methods enable the designing of quantitative structure activity relationship (QSAR) models to give better results in the bioactivity screening.

1. Pharmacophore-based virtual screening involves three-dimensional analysis of physicochemical properties of enzyme-ligand binding.
2. Parallel virtual screening ("target fishing") method involves target identification if the bioactivity is already predicted.
3. Molecular docking accurately predicts the ligand-protein binding and estimates the strength of binding with docking scores

12.5.1.1 PREREQUISITES TO SELECT THE SCREENING METHODS

Prior to screening, there are several aspects to be specifically noted to focus an enzyme as a drug target[41] through which a drug candidate can be made successful in clinical trials. They are as follows:

(i) Physiological environment (environment mimicking the cells physiological condition for screening of bioactive compounds to inhibit target enzyme),
(ii) Enzyme binding cofactors or coenzymes,
(iii) Enzyme substrates (Substrates used for the enzyme inhibition assay should match the actual substrate found in the cells), and
(iv) Enzyme kinetics (rate of substrate binding and desorption plays a key role in kinetic selectivity of enzymes).[41]

The researchers have properly reviewed the role of mechanistic enzymology in the process of drug discovery and presented a detailed evaluation of enzyme kinetics for designing of enzyme inhibitors as drugs.

12.5.2 HIGH-THROUGHPUT SCREENING METHODS OR BIOASSAYS FOR SCREENING

Designing rapid screening models and developing high-throughput bioassays is considered to be a comparatively easier approach for the discovery of novel enzyme inhibitors. For example, Deiab et al.[25] has designed a high- throughput model for rapid screening of LDH-A inhibition assay via enzyme microarray setup to screen more than 900 compounds in a short time period. Lankatillake et al.[54] have designed more accurate, comprehensive, step-by-step procedures for high- throughput screening of various enzyme inhibition assays for α-glucosidase, α-amylase, and lipase enzymes in the 96-well setup. High-throughput screening of natural metabolites with diverse structural variability was reviewed efficiently by Henrich and Beutler.[40]

As most of the plant compounds are colored, there is possibility of difficulties that can be faced by researchers to screen for bioactivity through spectral analysis, especially detection of metabolites based on absorption at specific wavelength of light. Hence, high-throughput screening by biochemical assays involves increased steps of washing during the assays, screening various concentrations of compounds, specific selection of detecting agents-fluorescence, luminescence or color and involvement of additives to decrease the nonspecific binding of compounds that may otherwise interfere with screening process. The problem of nonspecificity can be rectified by detecting cytotoxic compounds before performing cell-based assays.

Other potential high- throughput screening methods include the capillary electrophoresis (multiplexed capillary electrophoresis equipment is used to separate enzymatic substrates from products), mass spectroscopy (direct quantification of bound enzyme-substrate complexes, indirect quantification of separated bound ligand by LC-MS, or through recognition of enzymatic substrates), and immobilization of protein receptors on fused silica columns. High-throughput screening of drug leads for antibacterial activity involves the use of antisense methods and genetic manipulation of yeasts can screen bioactivity of compounds and zebra fishes can be directly used to screen the potentiality of drug leads.[40]

Innovative extraction strategies for isolation of plant metabolites have revolutionized the process of drug discovery by high- throughput assays. This includes the involvement of green technological extraction methods such as super critical liquid extraction, pressurized fluid extraction, ultrasound assisted extraction, and microwave-assisted extraction strategies.[33]

12.5.3 IDENTIFICATION OF DRUG LEAD

Pure bioactive compounds are obtained by step-wise bioactivity-guided fractionation cycles. However, bioactivity-guided fractionation of drug leads is time-consuming and the yield of bioactive compounds reduces with each cycle of fractionation. Hence, an alternative method that is followed recently is the metabolomic approach. Plant metabolomics specifically targets the functionality of drug lead. Application of analytical techniques such as ultra-high-performance liquid chromatography (UHPLC), gas chromatography coupled with high resolution mass spectroscopy (HRMS; GC-MS), and nuclear magnetic resonance spectroscopy (NMR), LC-MS, tandem mass spectrometry (MS/MS), etc. In addition to these analytical methods, various in silico computational tools can be used for proper metabolite identification that includes METLIN,[35] CANPA[31] etc. Single-cell multiplexed activity metabolomics is the recent approach to profile the responses of single cell to bioactive compounds.[27]

There are several recent advancements in the process of drug discovery which includes the genome mining, heterologous gene expression, CRISPR (clustered regularly interspaced short palindromic repeats)-Cas9 technology, synthetic biology platform,[79] plant microbiome analysis, computer-aided drug designing (CADD), damage-associated molecular patterns (DAMPs) approach, etc. for the evaluation of various plant metabolites as enzyme inhibitors[4] (Fig. 12.3).

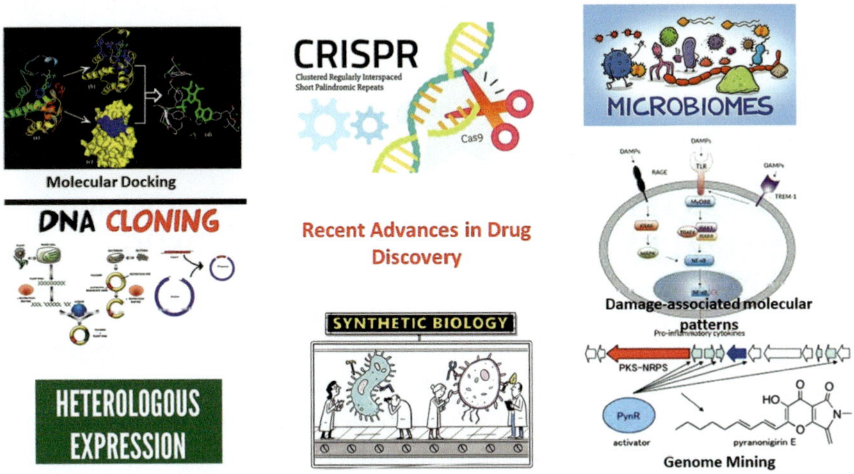

FIGURE 12.3 Recent advancement in the drug discovery of plant-derived enzyme inhibitors.

12.5.4 PRECLINICAL AND CLINICAL EVALUATION

Various plant-derived metabolites are being exploited in the drug development research. The drug development starts from bench to bed side process. Potential drug leads are evaluated pharmacologically in vitro (cell lines) and in vivo (animal models) before proceeding the development process to the clinical level (human trials). Toxicological, pharmacokinetic properties of the drug lead are evaluated extensively in the preclinical trials. Clinical trials in human subjects test the efficiency of the drug lead to treat the real-world diseased condition and also investigate the adverse side effects of using the drug lead. For any drug lead irrespective of the origin (either plant or microbe), the process of drug evaluation is same and it is mandatory to undergo the tedious process as guided by international agencies (e.g., FDA) to approve as a drug candidate before releasing into the commercial market.[1,17]

12.5.5 MARKETING

Several plant drugs were approved for clinical and dietary use by FDA. The approval of paclitaxel for the treatment of various cancer types has revolutionized the pharmaceutical market, where the market price of paclitaxel was raised to about $600,000 per kg in the commercial market as reported by Howat et al.[43] There is a market demand for plant-derived drugs, especially the anticancer drugs such as paclitaxel, etoposide, and vinblastine. To meet these demands, R &D units of many industries follow several approaches that include plant tissue culture, heterologous expression, hairy root cultures, metabolic engineering, organic synthesis, and semiorganic synthesis.[3,37]

12.6 OBSTACLES LEADING TO THE DECLINE OF THE PLANT-DERIVED NATURAL PRODUCTS AS A SOURCE OF DRUG DISCOVERY

Despite the tremendous opportunities these plant sources provide for drug discovery, it has been unable to rise to its potential. There is an imminent lack of knowledge base when it comes to identification of the correct plant variety for extraction of compounds, and this is compounded by the lack of specialists. Identification and subsequent documentation and preparation of the herbarium sample has thus far remained manual[24] and thus, lack of experts in this area creates a void that has become increasingly difficult to fill.[14] Identification of correct variety is also necessary so as to stop the damage to other indigenous plants that may resemble the desired plant variety.

Another factor that adds to the challenge is the quality and quantity of the drug obtained from the plant source. Often the amount of compound obtained from these natural sources is less and their extraction, purification, and processing take a lot of resources. This can become cost ineffective and the pharmaceutical companies thus look for alternatives in synthesizing the compounds directly rather than obtaining it from the natural sources.

On the other hand, large-scale harvesting of the natural source may lead to reduction in the plant habitat and pose serious risk to the stability of the resource itself. Ideally, obtaining the natural products from multiple sources is favorable for medicinal use, since that prevents over-exploitation of a particular plant variety. However, often times, once a plant source is identified to be of interest, it is difficult to contain its exploitation. The issue gets further aggravated by increased risk due to natural catastrophe and careless removal of the plant varieties due to lack of knowledge and proper regulations.[18] Organizations like WHO and European Medicines Agency (EMA) are constantly working toward development of guidelines to help with proper collection and harvesting of the medicinal plants.

Another complication in the extraction process is the quality of the compound that depends on the environment in which the plant was grown. This includes the presence of the microbes residing in the soil that interacts with the plants. The natural products obtained are, in reality, sometimes the result of these plant-microbe interactions and thus they can modulate the quality of the compound produced. Also, climate may play an important role and can affect the time window when the drug can be successfully obtained. Hence, climate, altitude, and soil composition are important factors to consider while choosing the source for drug extraction. Any compromise can lead to changes in the chemical composition and the quantity of the compound obtained.[47]

Further complications can arise if the compound is volatile or highly reactive. Most natural compounds have complex structural integrity and thus it becomes difficult to ascertain their large-scale synthesis and processing. The plant extract obtained may also contain additional substances that can interfere with the testing process and generate faulty results. Cytotoxic compounds that may be present in the plant extracts can reduce the efficiency of most cell-based assays[40] while presence of fluorescence quenchers can generate false-positive or false- negative results.[36] For instance, most plant extracts contain fatty acids and polyphenols. These compounds can interfere with a varied range of assays and therefore create difficulty in the extraction and purification process.[6] This also means that investigation of many products is simply overlooked because of lack of proper techniques required in sample preparations and reliable assays. To overcome this, development of sophisticated techniques is necessary.

Another major difficulty comes in establishing the therapeutic mechanism of these plant-derived drugs. The drug candidate may show excellent results; however, rigorous clinical trials are required before the drug can be commercialized.[19] Although the clinical trials are feasible, the process involving obtaining patents over the drug can be troublesome. Proper regulations are needed which will consider the welfare of indigenous interests while remaining cost effective for the pharmaceutical industries? The United Nation's Convention on Biological Diversity (CBD) aimed to provide guidelines and has helped in regulating and defining bioprospecting.[22] The guidelines, however, received major skepticism from both the sourcing countries and the pharmaceutical representatives. While some countries issued strict regulations to protect the indigenous plant varieties, many others were unable to establish appropriate institutions to carry out efficient legislation regarding permission of use. This further discouraged the pharmaceutical companies from utilizing these sources.[49]

The financial burden is worsened by the fact that there has been an increasing shift from small-molecule drug discovery to big-molecule-based drug discovery. This causes problems because further studies are required before a big-molecule-based drug can be approved for use. The cost of these additional studies remains even after expiration of patents and thus this becomes less lucrative for large-scale production. Because of the aforementioned challenges, the interest in plant-derived natural products has seen a steady decline. Thus, collective efforts are required to successfully revive the interest of big and medium pharmaceutical companies.

12.7 SUMMARY

Natural products from plants have always been an important source of drug discovery. There are numerous reasons to utilize plant-based medications, two of which may be particularly important. To begin with, plant-based natural compounds may offer additional modes of action, often encompassing a single medication to cure a particular condition. Second, plant-derived bioactive compounds can offer a high potential for generating biocompatible, cost-effective treatments, as well as accelerating the development of novel medications. There has been a growing interest in natural product research in several therapeutic areas recently, mainly due to the improvements in biotechnology and genomic techniques. The combination of these breakthroughs with developments in metabolic engineering, and chemical synthesis now provides a new way to investigate the chemical diversity of natural products in the search for novel medicines.

ACKNOWLEDGMENTS

The authors would like to thank Afrin Firdaus Idris, Research Scholar, TISS, for her help in formatting the references.

KEYWORDS

- **drug**
- **enzyme**
- **inhibitors**
- **medicinal**
- **natural**
- **plant**

REFERENCES

1. Ahn, K. The Worldwide Trend of Using Botanical Drugs and Strategies for Developing Global Drugs. *BMB Rep.* **2017,** *50* (3), 111.
2. Alamgir, A. N. M. Therapeutic Use of Medicinal Plants and Their Extracts; Vol. 1; Springer International Publishing: Berlin, Germany, 2017; pp 546.
3. Atanasov, A. G.; Waltenberger, B.; Pferschy-Wenzig, E. M.; Linder, T.; Wawrosch, C.; Uhrin, P, Temml, V., Wang, L., Schwaiger, S., Heiss, E. H., Rollinger, J. M. Discovery and Resupply of Pharmacologically Active Plant-Derived Natural Products: A Review. *Biotechnol. Adv.* **2015,** *33* (8), 1582–1614.
4. Atanasov, A. G.; Zotchev, S. B.; Dirsch, V. M.; Supuran, C. T. Natural Products in Drug Discovery: Advances and Opportunities. *Nat. Rev. Drug Dis.* **2021,** *20* (3), 200–216.
5. Atrafi, F.; Groen, H. J.; Byers, L. A.; Garralda, E.; Lolkema, M. P.; Sangha, R. S.; Calles, A. A Phase I Dose-Escalation Study of Veliparib Combined with Carboplatin

and Etoposide in Patients with Extensive-Stage Small Cell Lung Cancer and other Solid Tumors. *Clin. Cancer Res.* **2019,** *25* (2), 496–505.

6. Balunas, M. J.; Su, B.; Landini, S.; Brueggemeier, R. W.; Kinghorn, A. D. Interference by Naturally Occurring Fatty Acids in a Noncellular Enzyme-Based Aromatase Bioassay. *J. Nat. Prod.* **2006,** *69* (4), 700–703.
7. Bedows, E., Hatfield, G. M. An Investigation of the Antiviral Activity of *Podophyllum peltatum*. *J. Nat. Prod.* **1982,** *45* (6), 725–729.
8. Bessong, P. O.; Rojas, L. B.; Obi, L. C.; Tshisikawe, P. M.; Igunbor, E. O. Further Screening of Venda Medicinal Plants for Activity Against HIV Type 1 Reverse Transcriptase and Integrase. *Afr. J. Biotechnol.* 2006**,** 5 (6), 526–528.
9. Beutler John, A. Natural Products as a Foundation for Drug Discovery. *Curr. Protoc. Pharmacol.* **2009,** *46.*
10. Boucher, H. W.; Ambrose, P. G.; Chambers, H. F.; Ebright, R. H., Jezek, A.; Murray, B. E.; Newland, J. G.; Ostrowsky, B.; Rex, J. H. Infectious Diseases Society of America. White Paper: Developing Antimicrobial Drugs for Resistant Pathogens, Narrow-Spectrum Indications, and Unmet Needs. *J. Infect. Dis.* **2017,** *216* (2), 228–36.
11. Boy, H. I. A.; Rutilla, A. J. H.; Santos, K. A.; Ty, A. M. T., Alicia, I. Y.; Mahboob, T.; Tangpoong, J.; Nissapatorn, V. Recommended Medicinal Plants as Source of Natural Products: A Review. *Digit. Chin. Med.* **2018,** *1* (2), 131–142.
12. Boyd, M. R.; Gustafson, K. R.; McMahon, J. B.; Shoemaker, R. H.; O'Keefe, B. R.; Mori T. Discovery of Cyanovirin-N, a Novel Human Immunodeficiency Virus-Inactivating Protein that Binds Viral Surface Envelope Glycoprotein gp120: Potential Applications to Microbicide Development. *Antimicrob. Agents Chemother.* **1997,** *41* (7), 1521–1530.
13. Breinbauer, R.; Vetter, I. R.; Waldmann. H. From Protein Domains to Drug Candidates—Natural Products as Guiding Principles in the Design and Synthesis of Compound Libraries. *Angew. Chem. Int. Ed.* **2002,** *41* (16), 2878–2890.
14. Bucar F.; Wube, A.; Schmid, M. Natural Product Isolation—How to Get from Biological Material to Pure Compounds. *Nat. Prod. Rep.* **2013,** *30*, 525–545.
15. Butler, M. S. Natural Products to Drugs: Natural Product-Derived Compounds in Clinical Trials. *Nat. Prod. Rep.* **2008,** *25* (3), 475–516.
16. Chen, M. S.; White, M. C. A Predictably Selective Aliphatic C–H Oxidation Reaction for Complex Molecule Synthesis. *Science* **2007,** *318* (5851), 783–787.
17. Choudhari, A. S.; Mandave, P. C.; Deshpande, M.; Ranjekar, P.; Prakash, O. Phytochemicals in Cancer Treatment: From Preclinical Studies to Clinical Practice. *Front. Pharmacol.* **2020,** *10*, 1614.
18. Cordell, G. A. Sustainable Medicines and Global Health Care. *Planta Med.* **2011,** *77* (11), 1129–38.
19. Corson, T. W.; Crews, C. M. Molecular Understanding and Modern Application of Traditional Medicines: Triumphs and Trials. *Cell* **2007,** *130* (5), 769–774.
20. Cortes, J.; Perez-Garcia, J.; Levy, C.; Pardo, P. G.; Bourgeois, H.; Spazzapan, S.; Martínez-Jañez, N., Chao, T. C.; Espié, M.; Nabholtz, J. M.; Farré, X. G. Open-Label Randomised Phase III Trial of Vinflunine Versus an Alkylating Agent in Patients with Heavily Pretreated Metastatic Breast Cancer. *Ann. Oncol.* **2018,** *29* (4), 881–887.
21. Cragg, G. M.; Pezzuto, J. M. Natural Products as a Vital Source for the Discovery of Cancer Chemotherapeutic and Chemopreventive Agents. *Med. Princ. Pract.* **2016,** *25* (Suppl. 2), 41–59.

22. Cragg, G. M.; Katz, F.; Newman, D. J.; Rosenthal, J. The Impact of the United Nations Convention on Biological Diversity on Natural Products Research. *Na. Prod. Rep.* **2012,** *29* (12), 1407–1423.
23. Cragg, G. M.; Newman, D. J. Biodiversity: A Continuing Source of Novel Drug Leads. *Pure Appl. Chem.* **2005,** *77* (1), 7–24.
24. David B.; Wolfender, J. L.; Dias, D. A. The Pharmaceutical Industry and Natural Products: Historical Status and New Trends. *Phytochem. Rev.* **2015,** *14* (2), 299–315.
25. Deiab, S.; Mazzio, E.; Messeha, S.; Mack, N.; Soliman, K. F. High-Throughput Screening to Identify Plant Derived Human LDH-A Inhibitors. *Eur. J. Med. Plants* **2013,** *3* (4), 603–615.
26. Dias, D. A.; Urban, S.; Roessner, U. A Historical Overview of Natural Products in Drug Discovery. *Metabolites* **2012,** *2* (2), 303–36.
27. Earl, D. C.; Ferrell, P. B.; Leelatian, N.; Froese, J. T.; Reisman, B. J.; Irish, J. M.; Bachmann, B. O. Discovery of Human Cell Selective Effector Molecules Using Single Cell Multiplexed Activity Metabolomics. *Nat. Commun.* **2018,** *9* (1), 1–12.
28. Erdogan Orhan, I.; Orhan, G.; Gurkas, E. An Overview on Natural Cholinesterase Inhibitors-A Multi-Targeted Drug Class-and Their Mass Production. *Mini Rev. Med. Chem.* **2011,** *11* (10), 836–842.
29. Evans, B. E.; Rittle, K. E.; Bock, M. G.; DiPardo, R. M.; Freidinger, R. M.; Whitter, W. L.; Lundell, G. F.; Veber, D. F.; Anderson, P. S.; Chang, R. S.; Lotti, V. J. Methods for Drug Discovery: Development of Potent, Selective, Orally Effective Cholecystokinin Antagonists. *J. Med. Chem.* **1988,** *31* (12), 2235–2246.
30. Fellows, L.; Scofield, A. Intellectual Property Rights and Biodiversity Conservation—An Interdisciplinary Analysis of the Values of Medicinal Plants. In *Chemical Diversity in Plants*; Swanson, T., Ed.; Cambridge University Press: Cambridge, England, 1998; pp 19–44.
31. Fox Ramos, A. E.; Pavesi, C.; Litaudon, M.; Dumontet, V.; Poupon, E.; Champy, P.; Beniddir, M. A. CANPA: Computer-Assisted Natural Products Anticipation. *Anal. Chem.* **2019,** *91* (17), 11247–11252.
32. Fu, M.; Ng, T. B.; Jiang, Y.; Pi, Z. F.; Liu, Z. K.; Li, L.; Liu, F. *J. Pharm. Pharm.* **2006,** *58* (9), 1275–1280.
33. Gallego, R.; Montero, L.; Cifuentes, A.; Ibáñez, E.; Herrero, M. Green Extraction of Bioactive Compounds from Microalgae. *J. Anal. Test.* **2018,** *2* (2), 109–123.
34. Ganesan, A. The Impact of Natural Products upon Modern Drug Discovery. *Curr. Opin. Chem. Biol.* **2008,** *12* (3), 306–317.
35. Guijas, C.; Montenegro-Burke, J. R.; Domingo-Almenara, X.; Palermo, A.; Warth, B.; Hermann, G.; Koellensperger, G.; Huan, T.; Uritboonthai, W.; Aisporna, A. E.; Wolan, D. W. METLIN: A Technology Platform for Identifying Knowns and Unknowns. *Anal. Chem.* **2018,** *90* (5), 3156–3164.
36. Gul S, Gribbon P. Exemplification of the Challenges Associated with Utilizing Fluorescence Intensity Based Assays in Discovery. *Expert Opin. Drug Dis.* **2010,** *5* (7), 681–690.
37. Gutierrez-Valdes, N.; Häkkinen, S. T.; Lemasson, C.; Guillet, M.; Oksman-Caldentey, K. M.; Ritala, A.; Cardon, F. Hairy Root Cultures—A Versatile Tool with Multiple Applications. *Front. Plant Sci.* **2020,** *11*, 33.

38. Hartwig, J. F.; Cook, K. S.; Hapke, M.; Incarvito, C. D.; Fan, Y.; Webster, C. E.; Hall, M. B. Rhodium Boryl Complexes in the Catalytic, Terminal Functionalization of Alkanes. *J. Am. Chem. Soc.* **2005,** *127* (8), 2538–2552.
39. Harvey, A. L.; Edrada-Ebel, R.; Quinn, R. J. The Re-Emergence of Natural Products for Drug Discovery in the Genomics Era. *Nat. Rev. Drug Dis.* **2015,** *14* (2), 111–129.
40. Henrich, C. J.; Beutler, J. A. Matching the Power of High Throughput Screening to the Chemical Diversity of Natural Products. *Nat. Prod. Rep.* **2013,** *30* (10), 1284–1298.
41. Holdgate, G. A.; Meek, T. D.; Grimley, R. L. Mechanistic Enzymology in Drug Discovery: A Fresh Perspective. *Nat. Rev. Drug Dis.* **2018,** *17* (2), 115–132.
42. Hong, J. Role of Natural Product Diversity in Chemical Biology. *Curr. Opin. Chem. Biol.* **2011,** *15* (3), 350–354.
43. Howat, S.; Park, B.; Oh, I. S.; Jin, Y. W.; Lee, E. K.; Loake, G. J. Paclitaxel: Biosynthesis, Production and Future Prospects. *New Biotechnol.* **2014,** *31* (3), 242–245.
44. Huang, L., Chen, C. Molecular Targets of Anti-HIV-1 Triterpenes. *Curr. Drug Targets-Infect. Disord.* **2002,** *2* (1), 33–36.
45. Hussein, G., Miyashiro, H., Nakamura, N., Hattori, M., Kakiuchi, N., Shimotohno, K. Inhibitory Effects of Sudanese Medicinal Plant Extracts on Hepatitis C Virus (HCV) Protease. *Phytother. Res.: Int. J. Devot. Pharmacol. Toxicol. Eval. Nat. Prod. Deriv.* **2000,** *14* (7), 510–516.
46. Javed, H.; Meeran, M. F. N.; Jha, N. K.; Ojha, S. Carvacrol, a Plant Metabolite Targeting Viral Protease (Mpro) and ACE2 in Host Cells Can Be a Possible Candidate for COVID-19. *Front. Plant Sci.* **2020,** *11*. 56–79.
47. Jones, W. P.; Kinghorn, A. D. Extraction of Plant Secondary Metabolites. Chapter 13. In *Natural Products Isolation*; Sarker, S. D., Nahar, L., Eds.; Humana Press: New Jersey, USA, 2012; pp 323–351.
48. Kabera, J. N.; Semana, E.; Mussa, A. R.; He, X. Plant Secondary Metabolites: Biosynthesis, Classification, Function and Pharmacological Properties. *J. Pharm. Pharmacol.* **2014,** *2* (7), 377–392.
49. Kingston, D. G. Modern Natural Products Drug Discovery and Its Relevance to Biodiversity Conservation. *J. Nat. Prod.* **2011,** *74* (3), 496–511.
50. Kostova, I. Coumarins as Inhibitors of HIV Reverse Transcriptase. *Curr. HIV Res.* **2006,** *4* (3), 347–363.
51. Kotwal, G. J.; Kaczmarek, J. N.; Leivers, S.; Ghebremariam, Y. T.; Kulkarni, A. P.; Bauer, G.; De Beer, C.; Preiser, W.; Mohamed, A. R. Anti-HIV, Anti-Poxvirus, and Anti-SARS Activity of a Nontoxic, Acidic Plant Extract from the *Trifollium* Species Secomet-V/Anti-Vac Suggests That It Contains a Novel Broad-Spectrum Antiviral. *Ann. N. Y. Acad. Sci.* **2005,** *1056* (1), 293.
52. Kupchan, S. M. Novel Plant-Derived Tumor Inhibitors and Their Mechanisms of Action 1, 2. *Cancer Treat. Rep.* **1976,** *60* (8), 1115–1126.
53. Lahlou, M. The Success of Natural Products in Drug Discovery. *Pharmacol. Pharm.* **2013,** *4*, 17–31.
54. Lankatillake, C.; Luo, S.; Flavel, M.; Lenon, G. B.; Gill, H., Huynh, T., & Dias, D. A. Screening Natural Product Extracts for Potential Enzyme Inhibitors: Protocols, and the Standardization of the Usage of Blanks in α-Amylase, α-Glucosidase and Lipase Assays. *Plant Meth.* **2021,** *17* (1), 1–19.
55. Lewis, R. J.; Garcia, M. L. Therapeutic Potential of Venom Peptides. *Nat. Rev. Drug Dis.* **2003,** *2* (10), 790–802.

56. Li, S. Y.; Chen, C.; Zhang, H. Q.; Guo, H. Y.; Wang H, Wang, L.; Zhang, X.; Hua, S. N.; Yu, J.; Xiao, P. G.; Li, R. S. Identification of Natural Compounds with Antiviral Activities Against SARS-Associated Coronavirus. *Antivir. Res.* **2005,** *67* (1), 18–23.
57. Luepke, K. H.; Suda, K. J.; Boucher, H.; Russo, R. L.; Bonney, M. W.; Hunt, T. D.; Mohr, I. I. I. J. F. Past, Present, and Future of Antibacterial Economics: Increasing Bacterial Resistance, Limited Antibiotic Pipeline, and Societal Implications. *Pharmacother.: J. Hum. Pharmacol. Drug Ther.* **2017,** *37* (1), 71–84.
58. Malcolm, S. B.; Jones, C. G. On the Evolution of Plant Secondary Chemical Diversity: Discussion. *Philos. Trans. R. Soc. Lond. Ser. B* **1991,** *333* (1267), 279–280.
59. Martín, M.; Chacón, J. I.; Antón, A.; Plazaola, A.; García-Martínez, E.; Seguí, M. A.; Sánchez-Rovira P.; Palacios, J.; Calvo, L.; Esteban, C.; Espinosa, E. Neoadjuvant Therapy with Weekly Nanoparticle Albumin-Bound Paclitaxel for Luminal Early Breast Cancer Patients: Results from the NABRAX Study (GEICAM/2011-02), a Multicenter, Non-Randomized, Phase II Trial, with a Companion Biomarker Analysis. *Oncologist* **2017,** *22* (11), 1301.
60. Massy, Z. A.; Keane, W. F.; Kasiske, B. L. Inhibition of the Mevalonate Pathway: Benefits Beyond Cholesterol Reduction? *Lancet* **1996,** *347* (8994), 102–103.
61. Mawalagedera, S. M.; Callahan, D. L.; Gaskett, A. C.; Rønsted, N.; Symonds, M. R. Combining Evolutionary Inference and Metabolomics to Identify Plants with Medicinal Potential. *Front. Ecol. Evol.* **2019,** *7*, 267.
62. McRae, J.; Yang, Q.; Crawford, R.; Palombo, E. Review of the Methods Used for Isolating Pharmaceutical Lead Compounds from Traditional Medicinal Plants. *Environmentalist* **2007,** *27* (1), 165–174.
63. Mishra, B. B.; Tiwari, V. K. Natural Products: An Evolving Role in Future Drug Discovery. *Eur. J. Med. Chem.* **2011,** *46* (10), 4769–807.
64. Mukhtar, M.; Arshad, M.; Ahmad, M.; Pomerantz, R. J.; Wigdahl, B.; Parveen Z. Antiviral Potentials of Medicinal Plants. *Virus Res.* **2008,** *131* (2), 111–120.
65. Mukherjee, P. K.; Harwansh, R. K.; Bahadur, S.; Chanda, J.; Biswas, S.; Banerjee, S. Enzyme Inhibition Assay for Metabolic Disorders—Exploring Leads from Medicinal Plants. In *Animal Biotechnology*; Verma, A., Singh, A., Eds.; Academic Press: Cambridge, USA, 2020; pp 631–653.
66. Mushtaq, S.; Abbasi, B. H.; Uzair, B.; Abbasi, R. Natural Products as Reservoirs of Novel Therapeutic Agents. *EXCLI J.* **2018,** *17*, 420.
67. Newman, D. J. Natural Products as Leads to Potential Drugs: An Old Process or the New Hope for Drug Discovery? *J. Med. Chem.* **2008,** *51* (9), 2589–2599.
68. Newman, D. J.; Cragg, G. M. Natural Products as Sources of New Drugs Over the 30 Years from 1981 to 2010. *J. Nat. Prod.* **2012,** *75* (3), 311–335.
69. Ojima, I. Modern Natural Products Chemistry and Drug Discovery. *J. Med. Chem.* **2008,** *51* (9), 2587–2588.
70. Ortholand, J. Y.; Ganesan, A. Natural Products and Combinatorial Chemistry: Back to the Future. *Curr. Opin. Chem. Biol.* **2004,** *8* (3), 271–80.
71. Patwardhan, B.; Warude, D.; Pushpangadan, P.; Bhatt, N. Ayurveda and Traditional Chinese Medicine: A Comparative Overview. *Evid.-Based Complement. Altern. Med.* **2005,** *2* (4), 465–473.
72. Rauf, A.; Jehan. N. Natural Products as a Potential Enzyme Inhibitor from Medicinal Plants. In *Enzyme Inhibitors and Activators*; IntechOpen: Rijeka, Croatia, 2017; pp 165–177.

73. Rodrigues, T.; Reker, D.; Schneider, P.; Schneider, G. Counting on Natural Products for Drug Design. *Nat. Chem.* **2016,** *8* (6), 531–541.
74. Rosén, J.; Gottfries, J.; Muresan, S.; Backlund, A.; Oprea, T. I. Novel Chemical Space Exploration via Natural Products. *J. Med. Chem.* 2009, *52* (7), 1953–1962.
75. Salehi B, Ata A, Kumar NV, Sharopov F, Ramírez-Alarcón K, Ruiz-Ortega A.; Abdulmajid Ayatollahi S.; Valere Tsouh Fokou, P.; Kobarfard, F.; Amiruddin Zakaria, Z.; Iriti, M. Antidiabetic Potential of Medicinal Plants and TheirActive Components. *Biomolecules* **2019,** *9*, 551.
76. Soukup, R. W.; Soukup, K. The Series Progress in the Chemistry of Organic Natural Products: 75 Years of Service in the Development of Natural Product Chemistry. In *Progress in the Chemistry of Organic Natural Products 100*; Kinghorn, D. A., Falk, H., Kobayashi, J., Eds.; Springe International Publishing: Cham, Switzerland, 2015; pp 453–588.
77. Vervandier-Fasseur, D.; Latruffe, N. The Potential Use of Resveratrol for Cancer Prevention. *Molecules* **2019,** *24* (24), 4506.
78. Wang, X., Jia, W., Zhao, A., Wang, X. Anti-Influenza Agents from Plants and Traditional Chinese Medicine. *Phytother. Res.: Int. J. Devot. Pharmacol. Toxicol. Eval. Nat. Prod. Deriv.* **2006,** *20* (5), 335–341.
79. Wang, X.; Shang, X.; Huang, X. Next-Generation Pathogen Diagnosis with CRISPR/Cas-Based Detection Methods. *Emerg. Microb. Infect.* **2020,** *1*, 1682–1691.
80. Wani, M. C.; Horwitz, S. B. Nature as a Remarkable Chemist: A Personal Story of the Discovery and Development of Taxol®. *Anti-Cancer Drugs* **2014,** *25* (5), 482.
81. Welsch, M. E.; Snyder, S. A.; Stockwell, B. R. Privileged Scaffolds for Library Design and Drug Discovery. *Curr. Opin. Chem. Biol.* **2010,** *14* (3), 347–361.
82. Woradulayapinij, W., Soonthornchareonnon, N., Wiwat, C. In Vitro HIV Type 1 Reverse Transcriptase Inhibitory Activities of Thai Medicinal Plants and *Canna indica L. rhizomes*. *J. Ethnopharmacol.* **2005,** *101* (1–3), 84–89.
83. Yoshioka, Y.; Namiki, D.; Makiuchi, M.; Sugaya, K.; Onose, J. I.; Ashida, H.; Abe, N. Vialinin A and Thelephantin G, Potent Inhibitors of Tumor Necrosis Factor-A Production, Inhibit Sentrin/SUMO-Specific Protease 1 Enzymatic Activity. *Bioorganic Med. Chem. Lett.* **2016,** *26* (17), 4237–4240.

CHAPTER 13

MUSHROOMS AS A SOURCE OF ENZYME INHIBITORS

MALLIKA MAZUMDER and ANUP KUMAR SARKAR

ABSTRACT

Edible mushrooms and their derived natural products are widely used as potential enzyme inhibitors. Mushrooms are the unique group of fungi which provide multidirectional benefits to mankind. Mushrooms have long been used for both food and medicine. Many of them have a high nutritional property and a huge number of people from different regions of the world have consumed them in various ways. Almost all mushrooms act as reservoirs of novel bioactive secondary metabolite, and some of the compounds serve as the potential candidate for enzyme inhibition. Such chemicals are typically linked to changes in an enzyme's catalytic characteristics, which delay or halt catalysis. Fruiting bodies, mycelia, culture broth, submerged cultivation mycelia, and fermentation byproducts are all reservoirs of bioactive chemicals in mushrooms. The attention of researchers in recent decades has been focused on the comprehensive study of different enzyme inhibitors isolated from different mushroom species, which belong particularly to Ascomycetes and Basidiomycetes. In this chapter different aspects of mushroom-derived enzyme inhibitors are critically discussed. Exploration of the activity of these enzyme inhibitors is one of the vital trends for betterment of industrial and biotechnological applications.

Enzyme Inactivation in Food Processing: Technologies, Materials, and Applications. Megh R. Goyal, PhD, Junaid Ahmad Malik, PhD & Ravi Pandiselvam, PhD (Eds.)

13.1 INTRODUCTION

Enzyme inhibitors are the chemical substances that reduce the activity of enzymes in minute quantities in a specific chemical way. Use of enzyme inhibitors is recommended as a preferable management practice for the betterment of agriculture, environment, disease diagnosis, medical therapy, etc. Efficient use of enzyme inhibitors can reduce nutrient run off from the agricultural land and minimize the release of ammonia and other greenhouse gasses. Edible mushrooms and their derived natural products are many times found as potential enzyme inhibitors. Traditional healers and modern scientists have been studying mushrooms for decades to see if they might help with a variety of health issues, particularly chronic and degenerative diseases, such as obesity and cardiovascular ailments.[1] Though people have always been concerned about their health, some illnesses and diseases, such as diabetes, cardiovascular disease, and neuro-degenerative disorders, are unavoidable owing to poor dietary habits. In most cases, these diseases are executed by one or few enzymes. The consumption of mushrooms as food provides multiple pharmacological activities against such diseases and many other health ailments.[30,53] There is a valid demand to replace synthetic food additives with natural products based on the increasing evidence of the association between natural compound intake and health. Wild and cultivated mushrooms can produce several secondary metabolites, with the potential for major biological activity.[27,78] Mushrooms or macrofungi are utilized in traditional medicine in a variety of methods to address diverse conditions. The traditional use of mushrooms can be scientifically investigated using contemporary techniques and can reveal interesting substances that can be applied in the treatment of many diseases. The interference of additives produced from mushrooms with the organoleptic qualities and quality of food has been satisfactorily evaluated using modern methodologies.[74] Mushroom metabolites have several qualities that have been investigated for this substitution. Direct genetic manipulation with technique like CRISPR-Cas9 and gene recombination are currently viable strategies for scaling up metabolite production. Several mushroom metabolites have been shown to have pharmacological characteristics, as they exhibit modest to high anti-enzymatic activity against several therapeutically important enzymes. One of the main reasons why fungal products are successful as an enzyme inhibitor is their 'drug-like' nature that often goes beyond the activity of synthetic drugs.

This chapter critically examines various features of mushroom-derived enzyme inhibitors and their mode of action.

13.2 MUSHROOM: GENERAL INFORMATION AND UTILITY

Mushrooms are considered a special group of macrofungi, which grow above ground on soil or other food sources and produce spore-bearing fleshy fruiting bodies. Taxonomically, they belong to basidiomycotina and ascomycotina under kingdom Fungi. A mushroom is the fleshy, spore-bearing fruiting body of a fungus that is normally developed above ground on soil or on its food supply, and represents the reproductive structure of the fungi. Mushrooms lack the green pigment chlorophyll, which allows plants to use energy from sunlight to convert chemicals into food materials required for growth, a process known as photosynthesis. Instead, mushrooms create a diverse spectrum of extracellular enzymes. Ability of such enzyme to digest and retain secondary metabolites allows them to break down complex organic stuff into soluble molecules. Mushrooms can be divided into three categories based on their uses, edible (54%), medicinal (38%), and wild (8%).[64] Mushrooms can decompose organic matter and recycle nutrients back into the soil. Mushrooms can be used as food for a healthy state, medication for an ill condition, and dietary supplements for both healthy and sick states.[78] Many people throughout the globe use the fruiting body of mushrooms as a significant source of ingredients and nutrients in their cuisine. Mushrooms are a nutritious diet since they include carbs, fat, salt, and a variety of vital elements like selenium, potassium, riboflavin, niacin, vitamin D, proteins, and fiber.[80] Most of the edible mushrooms are characterized by cholesterol free and low content of calories.[77] Mushroom powder and its extract also have been increasingly incorporated in the formula because of their bioactive compounds.[45] Mushroom bioactive chemicals offer a great deal of potential for usage as natural performance enhancers for farm animals.[11] Medicinal mushrooms have been utilized to promote human health and strengthen the immune system in recent decades, and they have become important topics in medical and pharmaceutical study. Mushrooms have sparked a surge in popularity in recent years due to their ability to protect against a variety of ailments. People have long relied on products made from cultivated mushrooms to cure several ailments. There is a lot of scientific proof that mushrooms not only have a lot of nutrients, but they also have a lot of bioactive metabolites that have been shown to be powerful enzyme inhibitors. The diversity of macrofungi, that is, mushroom makes them a virtually limitless source of novel enzyme inhibitors with distinct properties.[67]

There are more than 2000 mushroom species in nature but around 25 of them are widely recognized as food and few are commercially grown such as *Agaricus bisporus*, *Lentinus edodes*, *Pleurotus* spp., and *Flammulina velutipes.*[80] Some other species of mushrooms are also possessing health benefits and cultivated in some parts of the world. In addition to these cultivated mushrooms, wild mushrooms are becoming increasingly important due to their nutritional, sensory, and, most importantly, pharmacological properties.[21] The growth of mushrooms and harvesting their metabolites is an important step in the development of biotechnological methods for obtaining enzyme inhibitor compounds. Submerged and solid-state fermentation are used to generate biomolecules like enzyme inhibitors from fungal sources, with the latter having an advantage over the former due to the adherence and anchoring support offered to the fungal mycelia, which improves fungal growth and biomolecule production.[71]

13.3 ENZYME INHIBITION AND ITS USES

Enzyme inhibitors are chemicals that bind to enzymes and reduce the activity of those enzymes. They adhere to enzyme active sites and reduce enzyme's affinity with substrates, preventing the formation of enzyme-substrate complexes. Enzymes are necessary for life processes of all organisms because they mediate biochemical processes such as metabolism, cellular signal transduction, cell cycling, and development. The malfunction, overexpression, or hyperactivation of the enzymes involved commonly leads in abnormalities in various biochemical systems, the underlying cause of which is frequently traceable to the dysfunction, overexpression, or hyperactivation of the enzymes involved. The identification of efficient enzyme inhibitors in human welfare practice has resulted from a better knowledge of enzyme kinetics at the molecular level. Enzyme inhibitors are the kinds of molecules that have pharmacological effects due to their suppressive activity on specific enzymes, allowing them to be used as drugs for a variety of human diseases, including communicable, metabolic, cardiovascular, neurological, and cancer diseases.[62] Some enzyme inhibitors are naturally present in the plants and animals whereas some are made synthetically. The presence of natural enzyme inhibitors governs the activity of the enzyme in plants and humans and ensures their intracellular and extracellular effects under physiological conditions. There are also intermediate products produced throughout some metabolic processes that are naturally occurring enzyme inhibitors.[8] Inhibitors made synthetically have been utilized in the past, but they almost always have negative effects. The toxicological impact of enzyme inhibition is dependent on two factors: intensity of inhibitor and post inhibition metabolism. Potency of inhibition is determined by some factors like degree and duration of enzyme inhibition, the nature of inhibition and the pharmacokinetics of the inhibitor, the biological importance of the enzyme and its ability to overcome inhibition by either synthesis of more enzyme or utilization of alternative biochemical pathways.[59] As a result, the demand for medications with no adverse effects derived from natural sources such as mushrooms is increasing. The inhibitions of enzymes are extensively studies because of their potential uses like;

1. Study related to enzyme inhibition enlightened different therapies against various enzymes that induce pathogenic states in plants, animals, humans and abnormalities in environment.
2. Enzyme inhibition is widely used to protect soil from pH elevation and loss of nutrient after use of chemical fertilizer through controlling of enzymatic hydrolysis.
3. The 3D structure and mode of action of many clinically important enzymes may be assessed using special enzyme inhibitors.[58]
4. To investigate novel or partially modified mechanisms of enzyme kinetics researchers of modern era prefer enzyme inhibitor based methodology.
5. The discovery of enzyme inhibitors has great importance on discovery of drug and drug designing research.[16,26]
6. The first findings indicate that using specific inhibitors to prevent and control enzymatic browning processes, and therefore increase the quality and nutritional content of foods, might be a viable option.

Umezawa[79] in Japan was the first to use the notion of screening for enzyme inhibitors. Numerous enzyme inhibitors have been developed in many laboratories of the world and are being used since then. Enzyme inhibitors derived from mushrooms are known for their "privileged" structure, which allows them to act as ligands for a wide range of enzymes and receptors.

13.4 ENZYME INHIBITORS FROM EDIBLE MUSHROOMS

Scientists and clinicians are continuously looking for a better, safer medicinal agent with fewer adverse effects. Mushrooms have been at the forefront of developing secondary metabolites and other inhibitors that stop enzymes. The finding of enzyme inhibitors from medicinal mushrooms may throw light on such kind of therapeutic agent. The clinical application of several mushroom-derived enzyme inhibitors results in the inactivation of several metabolic enzymes, lowering the risk associated with this metabolism. The effects of mushrooms and mushroom extracts on enzyme inhibition have been studied extensively. Extraction parameters (type of solvent, extraction temperature, and incubation time) and mushroom species may affect the quantity and quality of enzyme inhibitory activities in various mushroom extracts (strains, development stage, cultivation, and fruiting condition).[60] The proven pharmacological applications of such inhibitors involve in inactivation of several enzymes such as α-amylase, α-glucosidase, Angiotensin-converting enzyme, Aromatase, Myeloperoxidase, Protease, Tyrosinase, Cholinesterase, Urease, etc. In the traditional medicinal system, mushroom has played a major role in the treatment of several diseases and formulation made up of using extract of this group also reported to have numerous health benefits including tumor, cancer, viral diseases. Search for novel enzyme inhibitors from natural sources is always desired by the modern scientists. There is a great demand of α-amylase inhibitors due to their pharmacological efficacy to reduce the breakdown of carbohydrates. By inhibiting α-amylase, the breakdown of carbohydrates in the cell or any lumen can be slowed down. Partially inhibiting the activities of these the enzyme can be helpful in ameliorating diabetes condition.[6] Many researchers showed that extracts of different mushroom have α-amylase inhibitory activity. Bioactive compounds have been recognized and purified from different edible mushrooms, by various researchers, having alpha-amylase inhibitory activity.[77] Like α-amylase inhibitor there is also a great demand of **α**-Glucosidase. α-Glucosidase is a membrane-bound enzyme that normally resides on intestinal cells and catalyzes the final step of carbohydrate digestion to liberate free glucose, resulting in postprandial hyperglycemia. It is the leading cause of type 2 diabetes and affects approximately 2115 million people worldwide.[3] A huge number of antidiabetic mushrooms are available in nature and some of them are recommended as food such as *Agaricus* and *Pleurotus*. Like diabetes a huge number of people are suffering from blood pressure-related disorders. Angiotensin-converting enzyme (ACE), a Zn-metallopeptidase, plays a key role in the regulation of peripheral blood pressure mainly through the renin-angiotensin (RAS) and kallikrain-kinin systems (KKS).[43] It catalyzes the transformation of inactive Angiotensin-I to Angiotensin-II, which is a potent vasoconstrictor, enhances aldosterone secretion, subsequently promotes sodium and water re-absorption in the nephron, and therefore raises the arterial pressure by increasing intravascular fluid volume.[40] Comprehensive efforts of many researchers reveal that many bioactive compounds are actively inhibiting angiotensin-converting enzymes. Myeloperoxidase (MPO), a heme protein produced by activated leukocytes that contributes to host defence at inflammatory areas by generating halogenating chemicals such as hypochlorous acid (HOCl) and hypobromous acid (HOBr), is inhibited by certain mushroom metabolites.

Since ancient time some mushrooms have been recommended for the treatment of reproductive disorders and sexual hormone-related disease. Aromatase, also known as estrogen synthetase, is an enzyme that plays a crucial role in estrogen production. Aromatase inhibitors prevent postmenopausal women from producing estrogen. Aromatase inhibitors operate by inhibiting the enzyme aromatase, which converts androgen into oestrogen in minute concentrations in the body. This means that there is less oestrogen available to trigger hormone-receptor-positive breast cancer cell proliferation. Some metabolites of fungi effectively reduce the activity of some enzymes associated with neurodegenerative disorders. Cholinesterase is an enzyme that hydrolyzes acetylcholine in the synaptic gap and is important in the treatment of Alzheimer's disease. Cholinesterase inhibition may raise the level of acetylcholine in the synaptic cleft at the time, which might benefit Alzheimer's patients with cognitive impairment.[19] In animal tissues, there are two types of cholinesterases: acetylcholinesterase (AChE) and butyrylcholinesterase (BChE). AChE is a serine hydrolase mainly found at neuromuscular junctions and cholinergic brain synapses and is involved in the termination of impulse transmission in numerous cholinergic pathways in the central and peripheral nervous systems by rapid hydrolysis of the neurotransmitter acetylcholine. AChE activity has been related to Alzheimer's disease (AD), a progressive neurodegenerative illness of the brain that is one of the most common health issues in developed countries, with a high death rate owing to ageing populations.[55] Thus, AChE inhibitors have become one of the most extensively used treatments for Alzheimer's disease. BChE is known as the pseudo-cholinesterase and catalyzes the hydrolysis of many choline and non-choline esters and recently regarded as an important therapeutic target for the AD as AChE activity.[23]

Many metabolites produced from mushrooms are known to be protease inhibitors. Proteases are enzymes that catalyze the breakdown of peptide bonds. They are also known as peptidases or proteolytic enzymes. Excessive proteases can cause significant damage, which is regulated by enzyme breakdown caused by a change in pH, limiting proteolysis, or protease inhibitors. Protease inhibitors are common compounds found in all living things, and they can be derived naturally from bacteria, cyanobacteria, plants, animals, and more importantly from mushrooms. Proteins or peptides, polysaccharides, polyphenols, triterpenes, glycerolipids, and other nonprotein compounds with a lower molecular weight are all examples of protease inhibitors.[82] Mushroom protease inhibitors are different from those from other sources in terms of structure and inhibitory methods. Mushroom-derived protease inhibitors have been found to have a protective role in human under disease conditions. Mushrooms are a rich source of bioactive compounds that inhibit proteases, especially HIV protease and SARS-CoV-2 main protease.[61] Like proteases, tyrosinase is also a clinically significant enzyme. Tyrosinase (Tyr) is a family of orthologous enzymes found throughout nature that catalyzes the first two rate-limiting steps in the production of melanin.[48] It is a well-known copper-containing enzyme. With its monophenolase and diphenolase activity, tyrosinase plays an important role in melanin production.[5] Overproduction of melanin has been linked to a number of dermatological issues, including skin cancer. Natural products and particularly polyphenols are well-recognized as inhibitors of Tyr.[12,39] In general terms, inhibitors of tyrosinase are studied using monophenolic substrate like as tyrosine or diphenolic substrate like L-dopa and activity is evaluated on the basis of generation of dopachrome.[90] In broad sense different types of compounds such as specific tyrosinase inactivators, specific tyrosinase inhibitors, o-dopaquinone scavengers, alternative enzyme substrates, nonspecific enzyme inactivators, and denaturants are considered tyrosinase inhibitors, but, in reality, only the former two bind to the enzyme as genuine inhibitors and effectively stop it from working.[12] Specific tyrosinase inactivators are often called suicide inactivators or mechanism-based inhibitors. In layman terms, such compounds inhibit tyrosinase activity by inducing the enzyme catalyzing "suicide reaction." The enzymes bind to a "inhibitor" that has been reconfigured as a substrate. These inhibitors are then catalyzed by tyrosinase and create a covalent link with the enzyme, inactivating it irreversibly during the catalytic process.[90] Certain workers proposed that rearrangements in tertiary and quaternary structure of tyrosinase caused by the substratum and coordinated by solvent molecules could be the real explanation for the suicide inactivation. In recent years, a vast range of tyrosinase inhibitors ranging from mild to powerful have been found in various mushrooms, including phenolics, terpenes, steroids, chalcones, flavonoids, alkaloids, long-chain fatty acids, coumarins, sildenafil analogues, bipiperidines, biscoumarins, oxadiazole, tetraketones. In a study on *Hericium erinaceus* it was exhibited that Ascorbic acid, Caffeic acid, Chlorogenic acid, Cucumin, Pyridoxal, Quinic acid, Terephtalic acid, Vanillic acid etc. were effectively acted as inhibitors of tyrosinase.[38] Inhibitors of urease, a nickel-containing lyase enzyme that catalyzes the hydrolysis reaction to ammonia, carbon dioxide, or carbamate by a magnitude of 10^{14} faster than the spontaneous process, is found in certain mushrooms.[85] Urease is the most important enzyme in the global nitrogen cycle, which occurs in medicinal plants, fungi, and various microbes like bacteria, actinomycetes. In modern agriculture, excessive use of urea as fertilizer and subsequent enhancement of urease activity results in significant environmental and economic problems by releasing an extraordinarily large amount of ammonia into the environment.[14] Furthermore, it promotes the formation of peptic ulcers, pyelonephritis, kidney stones, hepatic encephalopathy, hepatic coma urolithiasis, urinary catheter encrustation, and similar disorders in humans.[33] Modern researchers are eager to focus on the finding of new urease inhibitors because it is well known that these compounds play an important role in the treatment of both plant and animal disorders. Urease catalyzes the hydrolysis of urea to produce CO_2 and NH_3 and its activity offers an environment for *Helicobacter pylori* colonization and neutralization of the acidity.[65,68] Inhibition of urease activity is therefore beneficial in the treatment of *H. pylori*-related disorders in people and animals, including stomach inflammation, ulcers, urolithiasis, pyelonephritis, hepatic encephalopathy, hepatic coma urolithiasis, and urinary catheter encrustation. Another advantage of urease inhibitors is that they prevent excessive ammonia loss to the atmosphere when added to urea or urea-containing fertilizers.[5]

TABLE 13.1 Enzyme Inhibitory Activities of Different Mushrooms.

Mushroom	Compound or Extract	Type of Enzyme Inhibition	References
Agaricus bisporus	Cell-free extracts from fruit bodies.	Urease inhibitor	[17]
	Mushroom extract	α-glucosidase inhibitory activities	[47]
	Mushroom extract	Inhibition of Lipase	[42]
	Peptides	Angiotensin I-converting enzyme inhibition	[41]
	Mushroom extract	Inhibition of trypsin	[81]

TABLE 13.1 *(Continued)*

Mushroom	Compound or Extract	Type of Enzyme Inhibition	References
Agaricus blazei	Mushroom extract	Inhibition of Lipase; Inhibition α-glucosidase	[73, 81]
Agrocybe cylindracea	Lovastatin	Inhibition of 3 hydroxy-3-methylglutaryl-CoA reductase	[13]
Amanita crocea	Mushroom extract	Strong inhibitor of Cholinesterase, moderate inhibitor of tyrosinase and α-amylase	[1]
Amanita ponderosa	Mushroom extract	Mild inhibition of lipase	[56]
Auricularia auricular	Mushroom extract	Mild inhibition of lipase	[56]
Boletus edulis	Mushroom extract	Inhibition of lipase	[56]
Catathelasma ventricosum	Aqueous extracts	Potent inhibition of α-glucosidase activity	[44]
Clitocybe maxima	Aqueous extracts	Potent inhibition of α-glucosidase activity	[44]
Clitocybe nebularis	Clitocypin	Inhibition of cysteine protease	[9]
		Strong inhibition of trypsin, weak inhibition of chymotrypsin	[4]
Collybia velutipes	Mushroom extract	Inhibition of lipase	[42]
Coprinopsis cinerea	Cospin	Strong inhibition of trypsin, weak inhibition of chymotrypsin	[54, 67]
Coprinus comatus	Methanolic extract	Inhibition of α-amylase and α-glucosidase	[72]
	Ergothioneine	Inhibition of Myeloperoxidase activity	[2]
Cordyceps militaris	Methanolic extract	Inhibition of α-glucosidase	[37,84]
Coriolus versicolor	Mushroom extract Mycellial extract	Strongest inhibitory effect on α-amylase activity and α-glucosidase	[73]
Craterellus cornucopioides	Aqueous extracts	Potent inhibition of α-glucosidase activity	[44]

TABLE 13.1 *(Continued)*

Mushroom	Compound or Extract	Type of Enzyme Inhibition	References
Cyclocybe aegerita	Methanolic extract; ceramide	Inhibition on cyclooxy-genase enzymes COX-1 and COX-2	[18, 87]
Cyclocybe cylindracea	Extract	Strong inhibition of Cholinesterase, Tyrosinase amylase and glucosidase activity	[1]
Daedalea dickinsii	Mycellial extract; Daedalin A	Inhibition of tyrosinase	[51]
Dictyophora indusiata	5-(Hydroxy methyl)-2-furfural	Anti-tyorsinase activity	[69]
Flammulina velutipes	Mushroom extract	α-glucosidase inhibitory activities	[47]
Fomitopsis pinicola	Mushroom extract	Inhibition of lipase	[42]
Ganoderma applanatum	Extract	Cholinesterase inhibition	[1, 88]
Ganoderma philippii	Hydroalcoholic extracts	Inhibition of α-amylase and α-amyloglucosidase	[32]
Ganoderma resinaceum	Mushroom extract	Cholinesterase inhibition	[1, 88]
Ganoderma lucidum	Mushroom extract	Lipase inhibition	[42]
Grifola frondosa	Mushroom extract	α-glucosidase inhibitory activities	[47]
	Mycellial extract	Strongest anti-lipase and anti-α-glucosidase activity	[73, 81]
	Mushroom extract	Mild inhibition of lipase	[56]
Hemileccinum depilatum	Mushroom extract	Tyrosinase inhibition	[1]
Hericium erinaceus	Mycellial extract	Lipase inhibition; Tyrosinase inhibition	[38, 81]
Hygrocybe acutoconica	Mushroom extract	Strong inhibition of tyrosinase, amylase and glucosidase activities	[1]
Hymenogaster aromaticus	Methanolic and Ethanolic extract	Inhibition of acetyl-cholinesterase, tyrosinase, α-amylase, and α-glucosidase	[1, 89]

TABLE 13.1 *(Continued)*

Mushroom	Compound or Extract	Type of Enzyme Inhibition	References
Hypsizygus marmoreus	Mushroom extract	α-glucosidase inhibitory activities	[47]
	Mushroom extract	Inhibition of lipase	[81]
Inonotus obliquus	Mushroom extract	Inhibition of lipase	[42]
	Methanolic extract	Inhibition of α-amylase and α-glucosidase	[72]
Laccaria amethystea	Aqueous extracts	Potent inhibition of α-glucosidase activity	[44]
Lactarius deliciosus	Mycellial extract	Acetylcholinesterase inhibition	[55]
	Aqueous and/or ethanol extract	Inhibitory effects on α-amylase and α-glucosidase	[86]
Lentinus edodes	Lentinus peptidase inhibitor	Strong inhibition of trypsin, weak inhibition of chymotrypsin	[54]
	Mushroom extract	Inhibition of lipase	[42]
Lenzites elegans	Hydroalcoholic extracts	Inhibition of α-amylase and α-amyloglucosidase	[32]
Lyophyllum shimeji	Mushroom extract	Inhibition of lipase	[81]
Macrolepiota procera	Macrocypins	Trypsin inhibition	[63]
Marasmius oreades	Mushroom extract	Mild inhibition of lipase	[56]
Morchella conica	Mushroom extract	Mild inhibition of lipase	[56]
	Methanolic extract	α-glucosidase inhibition	[72
Neoboletus erythropus	Extract	Inhibition of acetyl cholinesterase, tyrosinase, amylase and glucosidase	[1]
Phellinus linteus	Protocatechualdehyde, 5-hydroxymethyl-2-furaldehyde	Anti-tyrosinase activity	[31]
	Mushroom extract	Anti-lipase activity	[42]
Phlebopus portentosus	Mycellial extract	Anti-tyrosinase activity	[29]
Pleurotus cornucopiae	Oligopeptides of basidioma	Inhibition of angiotensin I-converting enzyme (ACE)	[20]

TABLE 13.1 *(Continued)*

Mushroom	Compound or Extract	Type of Enzyme Inhibition	References
Pleurotus eryngii	Mushroom extract	α-glucosidase inhibitory activities	[47]
	Mushroom extract	Anti-lipase activity	[81]
	Laccase and Pleureryn	HIV-1 reverse transcriptase	[22]
	Ergostane type sterols and Bisabolane-type sesquiterpenes	Inhibition of aromatase	[35, 36]
Pleurotus pulmonarius	Proteins	α-amylase inhibitory activity	[83]
Pleurotus ostreatus	Mushroom extract	α-glucosidase inhibitory activities	[47]
	Mushroom extract	Inhibition of Lipase	[42]
	Lovastatin	Anti-HMG-CoA reductase activity	[57]
	Crude cold-water extract	Angiotensin-I Converting Enzyme	[20]
	Lovastatin	α-amylase inhibitory activity	[75]
	Methanolic extract	Cyclindependent kinase inhibitor	[10]
Polyporus sulphureus	Mycellial extract	Potent inhibition of α-glucosidase activity	[55]
Ramaria aurea	Methanolic and Ethanolic extract	Inhibition of acetyl-cholinesterase, tyrosinase, α-amylase, and α-glucosidase	[1, 89]
Russula sanguinea	Methanolic and ethanolic extract	Inhibition of acetyl-cholinesterase, tyrosinase, α-amylase, and α-glucosidase	[1]
Rhizopogon luteolus	Methanolic and Ethanolic extract	Inhibition of acetyl-cholinesterase, tyrosinase, α-amylase, and α-glucosidase	[1, 89]
Rigidoporus ulmarius	Hydroalcoholic extracts	Inhibition of α-amylase and α-amyloglucosidase	[32]

TABLE 13.1 *(Continued)*

Mushroom	Compound or Extract	Type of Enzyme Inhibition	References
Rugiboletus extremiorientalis	Mycellial extract	Anti-tyrosinase activity	[29]
Russula emetica	Mycellial extract	Anti-tyrosinase activity	[29]
Russula extremiorientalis	Mycellial extract	higher α-glucosidase inhibitory activities	[29]
Sparassis crispa	Mushroom extract	Inhibition of lipase	[81]
Stropharia rugoso-annulata	Aqueous extracts	Potent inhibition of α-glucosidase activity	[44]
Tricholoma matsutakei	Mushroom extract	Inhibition of angiotensin-converting enzyme	[25]

There are numerous mushroom species that have yet to be investigated for their potential use in enzyme inhibition. These new mushroom species may have nutritional, clinical, and environmental effects that differ from those seen in previous species.

13.5 ACHIEVEMENTS AND FUTURE PERSPECTIVE OF TACTICS RELATED TO MUSHROOM-DERIVED ENZYME INHIBITION

Using chemistry as a research tool, mushroom-forming fungi have long been recognized as prolific makers of structurally varied and often bioactive secondary metabolites. The past decade has seen significant development in understanding natural product biosynthesis in mushrooms, owing to the collaborative efforts of investigators of chemical science, life science, medical science, and omics technology. Scientists and organizations have worked hard to improve the quality of farmed mushrooms, based on the therapeutic potential of mushroom-derived enzyme inhibitors. Comprehensive research also reveals a plethora of uses for such molecules in medicine, agriculture, and biotechnology. Many researchers have used virtual screening to look for possible enzyme inhibitors in low molecular weight mushroom compounds, and they have discovered some fascinating results. Most of the mushroom-derived inhibitors have gained attention because of their medicinal and therapeutic values. Tyrosinase inhibitors derived from mushroom can protect in fruits, beverages, vegetables, and edible mushrooms from enzymatic browning. Comprehensive investigations on various forms of tyrosinase inhibitors show that they might be used to treat aberrant melanin pigmentation.[46] The recommended treatments for Alzheimer's disease are based on inhibiting acetylcholinesterase, an enzyme that cleaves acetylcholine, allowing it to be elevated in the brain and facilitate neurotransmission.[52] The use of protease inhibitors derived from some mushrooms has insecticidal activities and is widely used insect pest management. Another advantage in protease inhibitors is that they provide cumulative protection from nematodes and harmful microorganisms in addition to the protection against herbivores. Many of the natural compounds derived from mushrooms have a direct impact on the inhibition of enzymes, whereas some of them have an indirect effect. Although quite a number of mushroom have been reported as a source of enzyme inhibitors, but extensive research on such topic may explore many facts regarding their mode of action and commercial application. The research on the chemistry and applications are growing exponentially owing to the uniqueness of such enzyme inhibitors.

13.6 LIMITATIONS OF MUSHROOM-DERIVED ENZYME INHIBITOR

Mushrooms' ability to produce possibly health-beneficial metabolite(s) is determined by their genome and gene expression. If mushrooms are being cultivated for medical purposes, it is essential that they are genetically identical to their wild counterparts.[76]

1. Mushrooms have been harvested wildly or are planted easily for use, while the few hard to grow species can be developed. Although there are over 3000 edible mushrooms, only about 3.3% of all edible species are produced in considerable quantities and are readily available for the purpose of sales.[34]
2. While mushrooms have enormous therapeutic potential, the thorough mechanisms involved in eliciting the majority of these pharmacological activities remain unclear and necessitate evidence-based insights. Hence studies and clinical trials must be conducted to lend scientific credibility to their attributes in providing health benefits by utilizing their active molecules.[7]
3. Natural product technologies and approaches have not yet evolved to the point where they can be employed effectively and efficiently in discovery of enzyme inhibitors and drugs from mushrooms and

similar organisms, resulting in a decline in interest in their development over the previous decades.

4. In some circumstances, when novel inhibitor molecules are discovered, structure elucidation can be problematic for some classes of natural compounds with high chemical complexity.
5. There have been numerous studies on enzyme inhibition conducted to date, but only a few of them have yielded promising results in field condition.
6. Intellectual property can also be a relevant issue that hinders further development of enzyme inhibitors from mushroom. With the growing awareness relating the economic aspect of natural products, many countries already have regulations in place to prevent or hinder sampling of natural products without prior authorization. Scientists and companies are demanding intellectual property rights over bioresources such as mushroom-derived enzyme inhibitors and the fast growth of patent applications for these products shows precisely this trend.

13.7 SUMMARY

Edible mushrooms are valuable for their therapeutic characteristics, as well as their presence of enzymes and bioactive chemicals, in addition to their culinary and nutritional values. Such mushrooms and their extracts are used by many people for their simple health necessities. Natural products obtained from mushrooms are widespread and our ancestors have explored their unique capabilities to treat ailments and injuries for centuries. Efforts to find small-molecule inhibitors of various enzymes for medicinal and aesthetic purposes are ongoing. Overall, it is thought that using a mushroom-derived enzyme inhibitor as a dietary supplement or as a coadjuvant in the treatment of pathogen-induced illnesses might give a wide range of advantages. There is still tremendous demand to investigate potential enzyme inhibitors in mushrooms, and there is also increasing use and application in agriculture and pharmaceutical industries of such inhibitors. Future study into the mechanisms of action of mushroom extracts will help us to better understand the intriguing activities and features of various enzyme inhibitors derived from mushrooms in the prevention and treatment of a variety of illnesses caused by enzyme dysfunction.

KEYWORDS

- **mushroom**
- **edible fungi**
- **enzyme inhibitor**
- **secondary metabolites**
- **bioactive compound**
- **human disorder**

REFERENCES

1. Alkan, S.; Uysal, A; Kasik, G; Vlaisavljevic, S; Berezni, S.; Zengin, G. Chemical Characterization, Antioxidant, Enzyme Inhibition and Antimutagenic Properties of Eight Mushroom Species: A Comparative Study. *J. Fungi* **2020,** *6* (3), 166.
2. Asahi, T.; Wu, X.; Shimoda, H.; Hisaka, S.; Harada, E.; Kanno, T.; Nakamura, Y.; Kato, Y.; Osawa, T.; A mushroom-Derived Amino Acid, Ergothioneine, Is a Potential Inhibitor of Inflammation-Related DNA Halogenations. *Biosci. Biotechnol. Biochem.* **2015,** *80* (2), 313–317.
3. Ata, A.; Naz, S.; Elias, E. M. Naturally Occurring Enzyme Inhibitors and Their Pharmaceutical Applications. *Pure Appl. Chem.* **2011,** *83*(9), 1741–1749.
4. Avanzo, P.; Sabotic, J.; Anzlovar, S.; Popovic, T.; Leonardi, A.; Pain, R. H.; Kos, J.; Brzin, J. Trypsin-Specific Inhibitors from the Basidiomycete *Clitocybe nebularis* with Regulatory and Defensive Functions. *Microbiology* **2009,** *155* (12), 3971–3981.
5. Ayvaz, M. C. Phenolic Profile of Three Wild Edible Mushroom Extracts from Ordu, Turkey and Their Antioxidant Properties, Enzyme Inhibitory Activities. *Br. Food J.* **2019,** *121*, (6), 1248–1260.
6. Bhattarcharjee, R.; Mitra, A.; Dey, B.; Pal, A. Exploration of Anti-Diabetic Potentials Amongst Marine Species—A Mini Review. *Indo Glob. J. Pharm. Sci.* **2014,** *4*(2), 65–73.
7. Bolaniran, T.; Jamiu, A. T.; Garuba, T.; Wudil, M. A.; Adeola, H. A.; Sabiu, S. An Appraisal of the Metabolites, Pharmacological and Biotechnological Significance of Edible Mushrooms. *Trans. R. Soc. S. Afr.* **2021,** *3*, 1–16.
8. Bjelakovic, G.; Stojanovic, I.; Bjelakovic, G. B.; Pavlovic, D.; Kocic, G.; Dakovic-Milic, A. Competitive Inhibitors of Enzymes and Their Therapeutic Application. *Facta Univ. Ser.: Med. Biol.* **2002,** *9* (3), 201–206.
9. Brzin, J.; Rogelj, B.; Popovic, T.; Strukelj, B.; Ritonja A. Clitocypin, a New Type of Cysteine Proteinase Inhibitor from Fruit Bodies of Mushroom *Clitocybe nebularis*. *J. Biol. Chem.* **2000,** *275*, 20104–20109.
10. Carrasco-Gonzalez, J. A.; Serna-Saldivar, S. O.; Gutierrez-Uribe, J. A. Nutritional Composition and Nutraceutical Properties of the *Pleurotus* Fruiting Bodies: Potential Use as Food Ingredient. *J. Food Compost. Anal.* **2017,** *58*, 69–81
11. Cateni, F.; Gargano, M. L.; Procida, G; et al. Mycochemicals in Wild and Cultivated Mushrooms: Nutrition and Health. *Phytochem. Rev.* **2021,** 1–45.

12. Chang, T. S. An Updated Review of Tyrosinase Inhibitors. *Int. J. Mol. Sci.* **2009,** *10* (6), 2440–2475.
13. Chen, S. Y.; Ho, K. J.; Hsieh, Y. J.; Wang, L. T.; Mau, J. L. Contents of Lovastatin, γ-Aminobutyric Acid and Ergothioneine in Mushroom Fruiting Bodies and Mycelia. *LWT—Food Sci. Technol.* **2012,** *47*, 274–278.
14. Chen, W.; Li, Y.; Cui, Y.; Zhang, X.; Zhu, H. L.; Zeng, Q. Synthesis, Molecular Docking and Biological Evaluation of Schiff Base Transition Metal Complexes as Potential Urease Inhibitors. *Eur. J. Med. Chem.* **2010,** *45*, 4473–4478.
15. Choi, J.; Lee, Y. M.; Jee, J. G. Thiopurine Drugs Repositioned as Tyrosinase Inhibitors. *Int. J. Mol. Sci.* **2018,** *19*, 77.
16. Copeland, R. A.; Harpel, M. R.; Tummino, P. J. Targeting Enzyme Inhibitors in Drug Discovery. *Expert. Opin. Ther. Targets* **2007,** *11*, 967–978.
17. De Windt, F. E.; Wagemaker, M. J. M.; Op den Camp, H. J. M.; van der Drift, C. Purine Degradation in the Edible Mushroom *Agaricus bisporus*. *Folia Microbiol.* **2002,** *47*, 672–676.
18. Diyabalanage, T.; Mulabagal, V.; Mills, G.; DeWitt, D. L.; Nair, M. G. Health Beneficial Qualities of the Edible Mushroom, *Agrocybe aegerita*. *Food Chem.* **2008,** *108*, 97–102.
19. Dorababu A. Critical Evaluation of Current Alzheimer's Drug Discovery (2018–19) & Futuristic Alzheimer Drug Model Approach. *Bioorg. Chem.* **2019,** *93*, 103299.
20. Ebigwai, J. K.; Edu, E. A.; Itam, E. H.; Mofunanya, A. J.; Activity of Crude Cold-Water Extract of the Culinary-Medicinal Oyster Mushroom, *Pleurotus ostreatus* (Jacq.: Fr.) P. Kumm. (higher Basidiomycetes), and timolol maleate on induced ocular hypertension. *Int. J. Med. Mushrooms* **2012,** *14*, 467–470.
21. Ergonul, P. G.; Akata, I.; Kalyoncu, F.; Ergonul, B. Fatty Acid Compositions of Six Wild Edible Mushroom Species. *Sci. World J.* **2013,** *2013*, 163964.
22. Fu, Z.; Liu, Y.; Zhang, Q. A Potent Pharmacological Mushroom: *Pleurotus eryngii*. *Fungal Genom. Biol.* **2016,** *6*(1), 1000139.
23. Gali, L.; Bedjou, F. Antioxidant and Anticholinesterase Effects of the Ethanol Extract, Ethanol Extract Fractions and Total Alkaloids from the Cultivated *Ruta chalepensis*. *S. Afr. J. Bot.* **2019,** *120*, 163–169.
24. Geng, X.; Tian, G.; Zhang, W.; Zhao, Y.; Zhao, L.; Ryu, M.; Wang, H.; Ng, T. B. Isolation of an Angiotensin I-Converting Enzyme Inhibitory Protein with Antihypertensive Effect in Spontaneously Hypertensive Rats from the Edible Wild Mushroom *Leucopaxillus tricolor*. *Molecules* **2015,** *20*, 10141–10153.
25. Geng, X.; Tian, G.; Zhang, W.; Zhao, Y; Zhao, L.; Wang, H.; Ng, T. B. A *Tricholoma matsutake* Peptide with Angiotensin Converting Enzyme Inhibitory and Antioxidative Activities and Antihypertensive Effects in Spontaneously Hypertensive Rats. *Sci. Rep.* **2016,** *6*, 24130.
26. Gregory, J. G Jr.; Ao, Z.; Michael, G. K.; Mei, Z.; Cyndi, R. M.; Erin, D.; Benjamin, T. B.; Matthew, T. G.; Krista, B. G.; Stephen, A. D.; Mark R. H.; Douglas, G. J. NADPH Oxidase-Dependent and Independent Mechanisms of Reported Inhibitors of Reactive Oxygen Generation. *J. Enzy. Inhib. Med. Chem.* **2013,** *28*, *1*, 95–104.
27. Gupta, S.; Summuna, B.; Gupta, M.; Annepu, S. K. Edible Mushrooms: Cultivation, Bioactive Molecules, and Health Benefits. In *Bioactive Molecules in Food, Reference Series in Phytochemistry*; Merillon, J. M., Ramawat, K. G., Eds.; Springer International Publishing: Basel, Switzerland, 2018; pp 23–42.

28. Hu, Y. H.; Zhuang, J. X.; Yu, F.; Cui, Y.; Yu, W.-W.; Yan, C. L.; Chen, Q. X. Inhibitory Effects of Cefotaxime on the Activity of Mushroom Tyrosinase. *J. Biosci. Bioeng.* **2016,** *121* (4), 385–389.
29. Kaewnarin, K.; Suwannarach, N.; Kumla, J.; Lumyong, S. Phenolic Profile of Various Wild Edible Mushroom Extracts from Thailand and Their Antioxidant Properties, Anti-Tyrosinase and Hyperglycaemic Inhibitory Activities. *J. Funct. Foods* **2016,** *27*, 352–364.
30. Kalaras, M. D.; Richie, J. P.; Calcagnotto, A.; Beelman, R. B. Mushrooms: A Rich Source of the Antioxidants Ergothioneine and Glutathione. *Food Chem.* **2017,** *233*, 429–433.
31. Kang, H. S., Choi, J. H., Cho, W. K.; Park, J. C.; Choi, J. S. A Sphingolipid and Tyrosinase Inhibitors from the Fruiting Body of *Phellinus linteus*. *Arch. Pharm. Res.* **2004,** *27* (7), 742–750.
32. Kaur, A.; Dhingra, G. S.; Shri, R. Evaluation of Carbohydrate Metabolism Inhibition by Some Species of Medicinal Mushrooms from India. *Int. J. Med. Mushrooms* **2015,** *17*(4), 351–360.
33. Khan, K. M.; Rahim, F.; Khan, A.; Shabeer, M.; Hussain, S.; Rehman, W.; Taha, M.; Khan, M.; Perveen, S.; Choudhary, M. I. Synthesis and Structure–Activity Relationship of Thiobarbituric Acid Derivatives as Potent Inhibitors of Urease. *Bioorg. Med. Chem.* **2014,** *22*, 4119–4123.
34. Khatun, S.; Islam, A.; Cakilcioglu U.; Chatterjee, N. C. Research on Mushroom as a Potential Source of Nutraceuticals: A Review on Indian Perspective. *Am. J. Exp. Agric.* **2012,** *2*(1), 47–73.
35. Kikuchi, T.; Kitaura, K.; Katsumoto, A.; Zhang, J.; Yamada, T.; Tanaka, R. Three Bisabolane-Type Sesquiterpenes from Edible Mushroom *Pleurotus eryngii*. *Fitoterapia* **2018,** *129*, 108–113.
36. Kikuchi, T.; Motoyashiki, N.; Yamada, T.; Shibatani, K.; Ninomiya, K.; Morikawa, T.; Tanaka, R. Ergostanetype Sterols from King Trumpet Mushroom (*Pleurotus eryngii*) and Their Inhibitory Effects on Aromatase. *Int. J. Mol. Sci.* **2017,** *18*, 2479–2489.
37. Kim, D. J.; Kang, Y. H.; Kim, K. K.; Kim, T. W.; Park, J. B.; Choe, M. Increased Glucose Metabolismand Alpha-Glucosidase Inhibition in *Cordyceps militaris* Water Extract-Treated HepG2 Cells. *Nutr. Res. Pract.* **2017,** *11*, 180–189.
38. Kim, S. Antioxidant Compounds for the Inhibition of Enzymatic Browning by Polyphenol Oxidases in the Fruiting Body Extract of the Edible Mushroom *Hericium erinaceus*. *Foods* **2020,** *9*, 951.
39. Kim, Y. J.; Uyama, H. Tyrosinase Inhibitors from Natural and Synthetic Sources: Structure, Inhibition Mechanism and Perspective for the Future. *Cell. Mol. Life Sci.* **2005,** *62* (15), 1707–1723.
40. Lafarga, T.; Hayes, M. Bioactive Peptides from Meat Muscle and By-Products: Generation, Functionality and Application as Functional Ingredients. *Meat Sci.* **2014,** *98*, 227–239.
41. Lau, C. C.; Abdullah, N.; Shuib, A. S.; Aminudin, N. Novel Angiotensin I-Converting Enzyme Inhibitory Peptides Derived from Edible Mushroom *Agaricus bisporus* (J. E. Lange) Imbach Identified by LC-MS/MS. *Food Chem.* **2014,** *148*, 396–401.
42. Lee, J. K.; Jang, J. H.; Lee, J. T.; Lee, J. S. Extraction and Characteristics of Antiobesity Lipase Inhibitor from *Phellinus linteus*. *Mycobiol.* **2010,** *38*(1), 52–57.
43. Li, G. H.; Le, G. W.; Shi, Y. H.; Shrestha, S. Angiotensin I-Converting Enzyme Inhibitory Peptides Derived from Food Proteins and Their Physiological and Pharmacological Effects. *Nutr. Res.* **2004,** *24*, 469–486.

44. Liu, Y. T.; Sun J.; Luo X. Y.; Rao S. Q.; Su Y. J.; Xu R. R.; Yang Y. J. Chemical Composition of Five Wild Edible Mushrooms Collected from Southwest China and Their Antihyperglycemic and Antioxidant Activity. *Food Chem. Toxicol.* **2012,** *50,* 1238–1244.
45. Lu, X.; Brennan, M. A.; Serventi, L.; Liu, J.; Guan, W.; Brennan, C. S. Addition of Mushroom Powder to Pasta Enhances the Antioxidant Content and Modulates the Predictive Glycaemic Response of Pasta. *Food Chem.* **2018,** *264*, 199–209.
46. Maghsoudi S, Adibi H, Hamzeh M, Ashrafi-Kooshka MR, Rezaei-Taviranic M, Khodarahmi R. Mushroom Tyrosinase Inhibition by Benzaldehyde Derivatives. *J. Rep. Pharm. Sci.* **2013,** *2* (2), 67–75.
47. Matsuur, H.; Asakawa, C.; Kurimoto, M.; Mizutani, J. Alpha-Glucosidase Inhibitor from the Seeds of Balsam Pear (*Momordica charantia*) and the Fruit Bodies of *Grifola frondosa*. *Biosci. Biotechnol. Biochem.* **2002,** *66* (7), 1576–1578.
48. Mayr, F.; Sturm, S.; Ganzera, M.; Waltenberger, B.; Martens, S.; Schwaiger, S.; Schuster, D.; Stuppner, H. Mushroom Tyrosinase-Based Enzyme Inhibition Assays Are Not Suitable for Bioactivity-Guided Fractionation of Extracts. *J. Nat. Prod.* **2019,** *82,* 136–147.
49. Mizutani, T.; Inatomi, S.; Inazu, A.; Kawahara, E. Hypolipidemic Effect of *Pleurotus eryngii* Extract in Fat-Loaded Mice. *J. Nutr. Sci. Vitaminol.* **2010,** *56*, 48–53.
50. Mojzych, M.; Tarasiuk, P.; Kotwica-Mojzych, K.; Rafiq, M.; Seo, S. Y.; Nicewicz, M.; Fornal, E. Synthesis of Chiral Pyrazolo[4,3-e][1,2,4]Triazine Sulfonamides with Tyrosinase and Urease Inhibitory Activity. *J. Enzy. Inhib. Med. Chem.* **2016,** *32* (1), 99–105.
51. Morimura, K.; Yamazaki, C.; Hattori, Y.; Makabe, H.; Kamo, T.; Hirota, M. A tyrosinase Inhibitor, Daedalin a, from Mycelial Culture of *Daedalea dickinsii*. *Biosci. Biotechnol. Biochem.* **2007,** *71*, 2837–2840.
52. Mukherjee, P. K.; Kumar, V.; Mal, M.; Houghton, P. J. Acetylcholinesterase Inhibitors from Plants. *Phytomedicine* **2007,** *14*, 289–300.
53. Muszynska, B.; Grzywacz-Kisielewska, A.; Kała, K.; Gdula-Argasińska, J. Anti-Inflammatory Properties of Edible Mushrooms: A Review. *Food Chem.* **2017,** *243*, 373–381.
54. Odani, S.; Tominaga, K.; Kondou, S.; Hori, H., Koide, T.; Hara, S.; Isemura, M.; Tsunasawa, S.; The Inhibitory Properties and Primary Structure of a Novel Serine Proteinase Inhibitor from the Fruiting Body of the Basidiomycete, *Lentinus edodes*. *Eur. J. Biochem.* **1999,** *262* (3), 915–923.
55. Orhan, I.; Ustun, O. Determination of Total Phenol Content, Antioxidant Activity and Acetylcholinesterase Inhibition in Selected Mushrooms from Turkey. *J. Food Compos. Anal.* **2011,** *24*, 386–390.
56. Palanisamy, M.; Gil-Ramirez, A.; Ruiz-Rodriguez, A.; Marin, F. R.; Reglero, G.; Soler-Rivas, C.; Testing Edible Mushrooms to Inhibit the Pancreatic Lipase Activity by an In Vitro Digestion Model. *Int. J. Food Sci. Technol.* **2012,** *47* (5), 1004–1010.
57. Patel, Y.; Naraian, R.; Singh, V. K. Medicinal Properties of *Pleurotus species* (Oyster Mushroom): A Review. *World J. Fungal Plant Biol.* **2012,** *3* (1), 1–12.
58. Paterson, R. R. M. Fungal Enzyme Inhibitors as Pharmaceuticals, Toxins and Scourge of PCR. *Curr. Enzy. Inhib.* **2008,** *4*, 46–59.
59. Pirmohamed, M.; Park, B. K. Adverse Drug Reactions: Role of Enzyme Inhibition and Induction. In *Variability in Human Drug Response: Proceedings of the Esteve*

Foundation Symposium VIII, Sitges, Spain, 7–10 October 1998; Elsevier Science & Technology: Amsterdam, Netherlands, 1999; pp 41–53.

60. Pongkunakorn, T.; Watcharachaisoponsiri, T.; Chupeerach, C.; On-nom, N.; Suttisansanee, U. Inhibitions of Key Enzymes Relevant to Obesity and Diabetes of Thai Local Mushroom Extracts. *Curr. Appl. Sci. Technol.* **2017,** *17* (2), 181–190.
61. Rangsinth,P.; Sillapachaiyaporn, C.; Nilkhet, S.; Tencomnao, T.; Ung, A. T.; Chuchawankul, S. Mushroom-Derived Bioactive Compounds Potentially Serve as the Inhibitors of SARS-CoV-2 Main Protease: An In Silico Approach. *J. Tradit. Complement. Med.* **2021,** *11* (2), 158–172.
62. Rengasamy, K. R. R.; Kulkarni, M. G.; Stirk,W. A.; Van Staden, J. Advances in Algal Drug Research with Emphasis on Enzyme Inhibitors. *Biotechnol. Adv.* **2014,** *32*, 1364–1381.
63. Renko, M.; Sabotic, J.; Mihelic, M.; Brzin, J.; Kos, J.; Turk D. Versatile Loops in Mycocypins Inhibit Three Protease Families. *J. Biol. Chem.* **2010,** *285* (1), 308–316.
64. Royse, D. J.; Baars, J.; Tan, Q. Current Overview of Mushroom Production in the World. In *Edible and Medicinal Mushrooms: Technology and Applications*; Zied, D. C., Pardo-Giminez, A., Eds.; John Wiley & Sons Ltd.: Hoboken, NJ, USA, 2017; pp 5–13.
65. Ruckriemen, J.; Klemm, O.; Henle, T. Manuka Honey (*Leptospermum scoparium*) Inhibits Jack Bean Urease Activity Due to Methylglyoxal and Dihydroxyacetone. *Food Chem.* **2017,** *230*, 540–546.
66. Sabotic, J.; Bleuler-Martinez, S.; Renko, M.; Avanzo Caglic, P.; Kallert, S.; Strukelj, B.; Turk, D.; Aebi, M.; Kos, J.; Kunzler, M. Structural Basis of Trypsin Inhibition and Entomotoxicity of Cospin, a Serine Protease Inhibitor Involved in Defence of *Coprinopsis cinerea* Fruiting Bodies. *J. Biol. Chem.* **2012,** *287* (6), 3898–3907.
67. Sabotic, J.; Trcek, T.; Popovic, T.; Brzin, J. Basidiomycetes Harbour a Hidden Treasure of Proteolytic Diversity. *J. Biotechnol.* **2007,** *128* (2), 297–307.
68. Sahin, H. Honey as an Apitherapic Product: Its Inhibitory Effect on Urease and Xanthine Oxidase. *J. Enzy. Inhib. Med. Chem.* **2016,** *31* (3), 490–494.
69. Sharma, V. K.; Choi, J.; Sharma, N.; Choi, M.; Seo, S. Y. In Vitro Anti-Tyrosinase Activity of 5-(Hydroxymethyl)-2-Furfural Isolated from *Dictyophora indusiata. Phytother. Res.* **2004,** *18* (10), 841–844.
70. Sima, V. H.; Patris, S.; Aydogmus, Z.; Sarakbi, A.; Sandulescu, R.; Kauffmann, J. M. Tyrosinase Immobilized Magnetic Nanobeads for the Amperometric Assay of Enzyme Inhibitors: Application to the Skin Whitening Agents. *Talanta* **2011,** *83* (3), 980–987.
71. Singhania, R. R.; Patel, A. K.; Soccol, C. R.; Pandey, A. Recent Advances in Solid-State Fermentation. *Biochem. Eng. J.* **2009,** *44*, 13–18.
72. Stojkovic, D.; Smiljkovic, M.; Ciric, A.; Glamoclija, J.; Van Griensven, L.; Ferreira, I. C. F. R.; Sokovic, M. An Insight into Antidiabetic Properties of Six Medicinal and Edible Mushrooms: Inhibition of α-Amylase and α-Glucosidase Linked to Type-2 Diabetes. *S Afr. J. Bot.* **2019,** *120*, 100–103.
73. Su, C. H.; Lai, M. N.; Ng, L. T. Inhibitory Effects of Medicinal Mushrooms on Alpha Amylase and Alpha-Glucosidase-Enzymes Related to Hyperglycemia. *Food Funct.* **2013,** *4* (4), 644–649.
74. Takahashi, J. A.; Barbosa, B. V. R.; Martins, B. d. A.; P. Guirlanda, C.; A. F. Moura, M. Use of the Versatility of Fungal Metabolism to Meet Modern Demands for Healthy Aging, Functional Foods, and Sustainability. *J. Fungi.* **2020,** *6*, 223.

75. Talkad, M. S.; Das, R. J.; Bhattacharjee, P.; Ghosh, S.; Shivajirao, U. P. Establishment of Enzyme Inhibitory Activities of Lovastatin, Isolated from *Pleurotus Ostreatus*. *Int. J. Appl. Sci. Biotechnol.* **2015,** *3*, 408–416.
76. Tamboli, E.; Bhatnagar, A.; Mishra, A. Effect of Natural Isolates from Mushroom mycelia on Alpha-Amylase Activity. *Int. J. Basic Appl. Biol.* **2017,** *4* (1), 12–14.
77. Tamboli, E.; Bhatnagar, A.; Mishra, A. Alpha-Amylase Inhibitors from Mycelium of an Oyster Mushroom. *Prep. Biochem. Biotechnol.* **2018,** *48* (8), 693–699.
78. Thakur, M. P. Advances in Mushroom Production: Key to Food, Nutritional and Employment Security: A Review. *Indian Phytopathol.* **2019,** *73*, 377–395.
79. Umezawa, H. Low-Molecular-Weight Enzyme Inhibitors of Microbial Origin. *Ann. Rev. Microbiol.* **1982,** *36* (1), 75–99.
80. Valverde, M. E.; Perez, T. H.; Lopez, O. P. Edible Mushrooms: Improving Human Health and Promoting Quality Life. *Int. J. Microbiol.* **2015,** *1*, 376387.
81. Vishvakarma, R.; Mishra, A. Production of a Protease Inhibitor from Edible Mushroom *Agaricus bisporus* and Its Statistical Optimization by Response Surface Methodology. *Prep. Biochem. Biotechnol.* **2017,** *47* (5), 450–457.
82. Vishvakarma, R.; Mishra, A. Protective Effect of a Protease Inhibitor from *Agaricus bisporus* on *Saccharomyces cerevisiae* Cells Against Oxidative Stress. *Prep. Biochem. Biotechnol.* **2019,** *49* (3), 244–254.
83. Wahab, N. A. A.; Abdullah, N.; Aminudin, N. Characterisation of Potential Antidiabetic-Related Proteins from *Pleurotus pulmonarius* (Fr.) Quel. (Grey Oyster Mushroom) by MALDITOF/ TOF Mass Spectrometry. *Biomed. Res. Int.* **2014,** 1–9.
84. Wu, L.; Sun, H.; Hao, Y.; Zheng, X.; g, Q.; Dai, hu, Z. SonS.; Z Chemical structure and inhibition on α-Glucosidase of the Polysaccharides from *Cordyceps militaris* with Different Developmental Stages. *Int. J. Biol. Macromol.* **2020,***148*, 722–736.
85. Xiao, Z. P.; Peng, Z. Y.; Dong, J. J., Deng, R. C.; Wang, X. D.; Ouyang, H.; Yang, P.; He, J.; Wang, Y. F.; Zhu, M.; Peng, X. C. Synthesis, Molecular Docking and Kinetic Properties of β-Hydroxy-β-Phenylpropionyl-Hydroxamic Acids as *Helicobacter pylori* urease inhibitors. *Eur. J. Med. Chem.* **2013,** *68*, 212–221.
86. Xu, Z.; Fu, L.; Feng, S.; Yuan, M.; Huang, Y.; Liao, J.; Zhou, L.; Yang, H.; Ding, C. Chemical Composition, Antioxidant and Anti-Hyperglycemic Activities of the Wild *Lactarius deliciosus* from China. *Molecules* **2019,** *24* (7), 1357.
87. Zhang, Y.; Mills, G. L.; Nair, M. G. Cyclooxygenase Inhibitory and Antioxidant Compounds from the Fruiting Body of an Edible Mushroom, *Agrocybe aegerita*. *Phytomedicine* **2003,** *10*, 386–390.
88. Zengin, G.; Sarikurkcu, C.; Gunes, E.; Uysal, A.; Ceylan, R.; Uysal, S.; Gungor, H.; Aktumsek, A. Two Ganoderma Species: Profiling of Phenolic Compounds by HPLC–DAD, Antioxidant, Antimicrobial and Inhibitory Activities on Key Enzymes Linked to Diabetes Mellitus, Alzheimer's Disease and Skin Disorders. *Food Funct.* **2015,** *6*, 2794–2802.
89. Zengin, G.; Uren, M. C.; Kocak, M. S., Gungor, H.; Locatelli, M.; Aktumsek, A.; Sarikurkcu, C., Antioxidant and Enzyme Inhibitory Activities of Extracts from Wild Mushroom Species from Turkey. *Int. J. Med. Mushrooms* **2017,** *19* (4), 327–336.
90. Zolghadri, S.; Bahrami, A.; Khan, M. T. H.; Munoz-Munoz, J.; Garcia-Molina, F.; Garcia-Canovas, F.; Saboury, A. A. A Comprehensive Review on Tyrosinase Inhibitors. *J. Enzy. Inhib. Med. Chem.* **2019,** *34* (1), 279–309.

CHAPTER 14

ENZYMATIC BROWNING—MECHANISM AND PREVENTION

SAQIB HASSAN[1], MUDASIR BASHIR MIR[2], AAMIR AFZAL MAGRAY[3], ANUSHARA PRABHAKARAN[1], NAMRATA DEKA[4], JOSEPH SELVIN[1], and GEORGE SEGHAL KIRAN[5]

ABSTRACT

Enzymatic browning is a phenomenon that commonly occurs in various fruits and vegetables when they are cut, peeled, exposed to air, undergoes mechanical damage during transportation, or is diseased. Oxidation of phenolic compounds present in fruits and vegetables takes place when these are exposed to air leading to the formation of brown melanins thus causing dark spots on fruits and vegetables. Enzymatic browning is caused by polyphenol oxidases (PPOs) (oxidoreductase enzymes) and peroxidases (POD) (oxidases), which are found in most fruits and vegetables, and in some seafoods. Dark coloration is mainly caused by the oxidation of phenolic compounds to quinones. PODs are thermostable enzymes belonging to a group of oxidases that cause oxidation of phenolic compounds and in this process, H_2O_2 acts as a catalyst. Enzymatic reactions change the color of fruits and vegetables, which is a serious issue during harvesting, transportation, storage, and processing. Undesirable changes in flavor, texture, color, and the nutritional quality of foods have been related to PODs and such changes make foods unacceptable to the consumers. Various physical and chemical methods are used for the prevention of enzymatic browning. These methods include thermal treatment, prevention of oxygen exposure, use of

Enzyme Inactivation in Food Processing. Technologies, Materials, and Applications. Megh R. Goyal, PhD, Junaid Ahmad Malik, PhD & Ravi Pandiselvam, PhD (Eds.)

low temperature, irradiation, acidification or reduction using antioxidants, chelating agents, or natural extracts.

14.1 INTRODUCTION

Enzymatic browning mostly occurs during storage and processing of fruit and vegetables. Color changes in fruit and vegetables can be caused by enzymatic reactions, which lower the quantity of bioavailable vitamins and antioxidants and also reduce the consumer attraction.[82] Browning reactions, on the other hand, maybe advantageous since they add color and taste to baked and fried foods, coffee, tea, chocolates, etc.[78] Enzymatic reactions cause discoloration as a result of the synthesis of new pigments or are involved in the breakdown of naturally occurring pigments.[1] Enzymatic browning of fruit and vegetables has been recognized as a serious issue that results in economic loss.[78] The oxidation process that causes enzymatic browning is another major cause of food spoilage following microbiological infection.

A vast variety of fruits and vegetables have been reported to show browning after experiencing a physiological or mechanical damage during harvesting or storage.[27] Brown discoloration in a wide range of fruits and vegetables reduces consumer acceptance and consequently has significant economic implications for the primary producer and the food processing industry.[29,45] As a result, preventing oxidation of fruits and vegetables during storage and processing becomes a top priority in the food industry. Various techniques for controlling enzymatic browning in fruits and vegetables include lowering the temperature, heating, packing, and coating, adding antioxidants, chelating and reducing chemicals that inhibit enzymatic action, restricting the substrates, and bleaching the pigments. The efficiency of antibrowning methods, on the other hand, is dependent on a number of parameters, including concentration, variety, and the interaction with the other factors such as pH and delivery methodology.

In this chapter, we discuss in detail the causes and mechanism of enzymatic browning. Control of enzymatic browning through various methods as well as the natural antibrowning agents has been delineated.

14.2 ENZYMATIC BROWNING IN FRUITS

Some of the fruits that are prone to enzymatic browning include:

14.2.1 BROWNING IN APPLES

Browning of harvested apples as well as the apple products is usually unpleasant and is linked to (i) physiological disorders like bitter pit, superficial senescent scalds, internal damage, watercore, core flush, and so on. (ii) Pitting, a flattened zone with brown flesh beneath, which is the most common defect; and (iii) improper preparation of slices and juices, causing undesirable browning. A number of postharvest physiological disorders affect apples, including superficial scald. It appears as a widespread browning of the skin that ranges from light to dark. Green-skinned apples and the un-blushed regions of red cultivars are more susceptible to superficial scald. Many factors, such as apple cultivar, growing temperatures, harvest maturity, postharvest chilling conditions, etc. influence the severity of the disorder.

Fruits that possess higher concentration of nitrogen and potassium but low in calcium are more prone to develop the condition. The conversion of alpha-farnesene to hazardous trienols has been linked to superficial scald and could be involved in the rupture of skin cells that leads to enzymatic phenol oxidation and the production of brown pigments. Chilling injury, senescent breakdown, and CO_2 injury under controlled atmosphere (CA) preservation can all cause internal browning in apples. Only a small percentage of apples appear to be prone to discoloration (browning) under CA storage, and it can range from a spot of brown on the flesh to nearly the whole flesh in extreme cases. The polyphenol oxidase (PPO) oxidation of apple phenolics is thought to be the cause of bruising and preparation-associated browning in apples.[51]

14.2.2 BROWNING IN AVOCADOS

During cold storage, brown discolorations on the skin and flesh of avocados can occur, most likely because of the enzymatic oxidation of phenolic compounds. During the climacteric rise, the fruit is particularly vulnerable, ethylene and a low calcium concentration increases vulnerability to chilling injury. Increases in PPO and guaiacol-POD activities have been reported during cold storage and shelf life at 20 °C, and have been linked to brown mesocarp discoloration, together with membrane permeability values.[22]

14.2.3 BROWNING IN OLIVES

Browning in bruised olives is thought to occur in two phases The activity of beta-glucosidases on the hydroxytyrosol glucoside and esterases on oleuropein results in the enzymatic release of the phenolic molecule hydroxytyrosol. PPO oxidizes hydroxytyrosol and verbascoside in the second step.[63]

14.2.4 BROWNING IN PEACH

Browning of the flesh in peach and pit cavity can occur in raw peach and nectarine, owing to freezing temperatures. Peach is more sensitive than nectarine in general, and late-season cultivars are the most susceptible. PPO activity and the concentration of phenolic compounds in peaches have been found to have risen up to the ripening stage, which coincided with the greatest degree of browning as indicated by absorbance measurements.

14.2.5 BROWNING IN PEARS

The pear, like the apple, is vulnerable to superficial scald, which is linked to higher levels of alpha-farnesene. Brown discolorations in pear have been identified, which may be linked to several metabolic processes. The discolorations can develop when pears are stored in hypoxic conditions with high CO_2 partial pressures, reduced level of oxidants, and due to the enzymatic oxidation of the major phenolic substrates like 5-caffeoylquinic acid (chlorogenic acid) and (-)-epicatechin.[3]

14.2.6 BROWNING IN TOMATO

In tomatoes, the enzymatic browning occurs comparatively at a moderate pace because of the presence of antioxidants like ascorbic acid and lycopene. PPO activity levels differ with cultivar, which can make certain cultivars more prone to enzymatic browning.[73] PPO expression can also be enhanced by higher watering levels.[7] Blossom end rot is a physiological disease in tomatoes that is connected to enzymatic browning. Symptoms can appear at any point during the fruit growth. A tiny, light brown mark appears at the flower end at first, then grows larger and darker as the fruit matures. The spot might be large enough to cover half of the fruit's surface, or it could be tiny and superficial. Caffeic and chlorogenic acids, as well as PPO and POD activities, have been reported to increase during calcium deficiency,[15] and there has been a significant link between blossom end rot and PPO activity.[11]

14.3 ENZYMATIC BROWNING IN VEGETABLES

14.3.1 BROWNING IN JICAMA (YAM BEAN)

External decay and a chill-induced dark staining of the flesh occur when jicama roots are stored for a few days at 10 °C or lower, independent of the cultivar. Increased concentrations of soluble phenolic components and phenylalanine ammonia lyase (PAL) activity have been linked to the discoloration, which has been related to enzyme-catalyzed phenolic oxidation. Damaged external tissues have been reported to have higher PPO and POD activities.[4] Enzymatic browning in yam tubers is particularly noticeable toward the stem end, where phenolic chemicals are abundant.

14.3.2 BROWNING IN LETTUCE

Enzymatic browning in raw lettuce can be caused by physiological effects or by physical damage. *Tipburn* (browning of leaf margins in the field due to poor calcium and other mineral distribution within the plant) is one of the physiological disorders' *russet spotting* (brown marks on the midrib caused by ethylene exposure during cold storage), *brown stains* (occur due to increased CO_2 levels, (around the midrib or darker edges), and *heart-leaf injury* (internal browning because of CO_2 exposure). Induction of the PAL enzyme has been linked to wound-related browning.[62]

14.3.3 BROWNING IN ONIONS

In onions, the outer scales of the bulb dry out and turn brown on aging. This process may involve the death of scale cells and disintegration of cellular compartmentalization may take place, with this there is release of some enzymes that can cause the primary phenolic compounds (quercetin glucosides) to become deglucosidated.[76] It has been observed that the quercetin generated by autoxidation yields a superoxide ion, which is subsequently converted to H_2O_2 and used by POD to oxidize quercetin to brown pigments.

14.3.4 ENZYMATIC BROWNING IN POTATOES

Prior to processing, that is during handling and cutting brown discoloration in raw potatoes has been associated with PPO activity. Potato cultivars that are genetically modified possessing reduced PPO activity exhibit a lower tendency to brown.[6] Enzymatic browning is not observed in the wild species *Solanum hjertingii*, which contains low concentration of certain PPO isoenzymes.[65] Chill storage correlated with lower ascorbate levels has been linked to greater browning potential in potatoes.[48]

14.4 CAUSES OF ENZYMATIC BROWNING

The phenomenon of enzymatic browning in fruits and vegetables is influenced by a number of factors such as PPO concentration, phenolic compounds content, presence of oxygen, pH, and temperature.[21] The role of PPO and concentration of phenolic compounds turns out to be predominant in enzymatic browning of fruits and vegetables.[82] PPO brings about the oxidation of phenolic compounds in fruits and vegetables leading to the formation of quinones, which are finally polymerized to melanin, a brown coloring pigment.

14.4.1 PHENOLIC COMPOUNDS

There is a wide variety of phenolic compounds present in fruits and vegetables.[67] Structurally, they are composed of aromatic rings with more than one hydroxyl groups,[43,66,68] and hence more often referred to as polyphenols. These compounds are secondary metabolites of plants produced through shikimate pathway by utilizing the intermediates of carbohydrate mechanism.[69] These chemical compounds differ mainly with respect to species, maturity stage, and environmental factors besides other factors. They are substrates for enzymes responsible for browning reactions. Catechin, chlorogenic acid, epicatechin are some of the phenolic compounds identified in apples.[58] Polyphenols also exhibit antioxidant potential besides affecting the color and flavor of fruits and vegetables. These polyphenols are oxidized by PPO enzymes resulting in the formation of quinones, which finally get converted brown colored pigments.

14.4.2 ENZYMES

Two enzymes have been identified, which are responsible for the synthesis of brown-colored pigments called as melanins[78,17] and these enzymes are PPO, and peroxidase. PPO is also known as catechol oxidase, diphenol oxidase, catecholase, phenolase, odiphenolase, cresolase, and tyrosinase. Copper is present at the active site of PPO which is vital for its activity. The molecular weight of this enzyme ranges from 57 to 62 kDa.[26] The oxidation of phenolic substrate is completed in two reactions. Phenolic substrate is hydroxylated at ortho-position in the first reaction. The second reaction is about the formation of o-benzoquinones and such compounds are formed by oxidation of diphenols. Oxygen in this case acts as a co-substrate for the reaction. PPO accomplishes the conversion of o-dihydroxy phenols to o-benzoquinones. It has also been reported that every PPO does not have the ability hydroxylate monophenols. The other enzymes that were found to have PPO activity are peroxidases (PODs). PODs are widely distributed in plant kingdom. However, their role in enzymatic browning reactions of vegetables and fruits is not clear. There are several reports suggesting that PPO increases the activity of peroxidase enzymes due to the production of hydrogen peroxide in PPO-catalyzed reactions.[60,74]

14.4.3 TEMPERATURE

Temperature significantly affects the rate of enzymatic browning reactions. Storing fruits and vegetables at low temperatures prevent their browning reactions. PPO stops activity at temperatures below 7 °C. However, the enzyme is not deactivated and resumes its activity after thawing of fruits and vegetables. Preservation of fruits and vegetables through refrigeration and chilling is therefore, preferred and practiced to prevent their spoilage. The sensitivity of PPOs to heat differs with their source. It has been reported that a temperature range of 70–95°C destroys the activity of PPO enzymes[14,50,54] and such an exposure even for shorter periods of time may completely inactivate the enzymes.

14.4.4 PH

pH plays a significant role in browning reactions of fruits and vegetables. It has been reported that PPO activity is inhibited in presence of acids.[69]

Enzymes responsible for browning work optimally at a pH range of 5.0–7.0. There is a drastic reduction in enzyme activity at pH values below 3.0. Fruits such as lemon, oranges, lime, etc. are highly acidic foods, and their high acid content prevents enzymatic browning in them.

14.5 MECHANISM OF ENZYMATIC BROWNING

The mechanism of enzymatic browning encompasses two important steps; first the oxidation of phenols (both mono and polyphenols) by the action of PPOs to ortho-quinones driven by oxygen as co-substrate,[80] followed by the nonenzymatic condensation or polymerization of these quinones to form polymers, which result into high molecular-weight, insoluble, complex, and dark-colored compounds known as melanins.[57,59,78] The active site of the enzyme is believed to be homologous to the oxygen-carrying nonheme pigment hemocyanin, consisting of two Cu (I) centers each coordinated in a pyramidal fashion by three imidazole moieties of the histidine residues, which are connected further to the glycoprotein chains.[25,35] The two Cu centers are not directly linked to each other and remain so until they undergo oxidative addition of O_2 molecule, which binds the two Cu atoms of the enzyme in μ-η^2:η^2 fashion.[24] The O_2 oxidizes each Cu (I) to Cu (II) while it gets reduced to peroxide ion (Fig. 14.1).

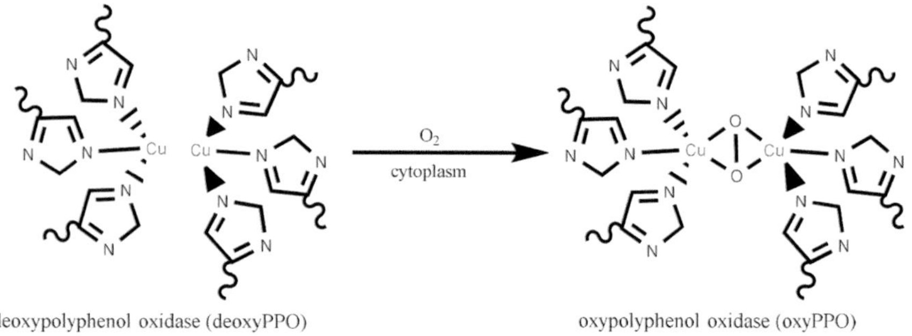

FIGURE 14.1 Binding site of the enzyme Phenol oxidase, before and after binding with the dioxygen molecule.

The proposed mechanism of action of PPO as in *Neurospora crassa*[35,70] appears to be in complete agreement with the most of polyphenyl oxidases. The mechanism involves two important reactions; one is dehydrogenation and other is hydroxylation (Fig. 14.2), taken care of by a common oxypolyphenol oxidase (oxyPPO) intermediate.

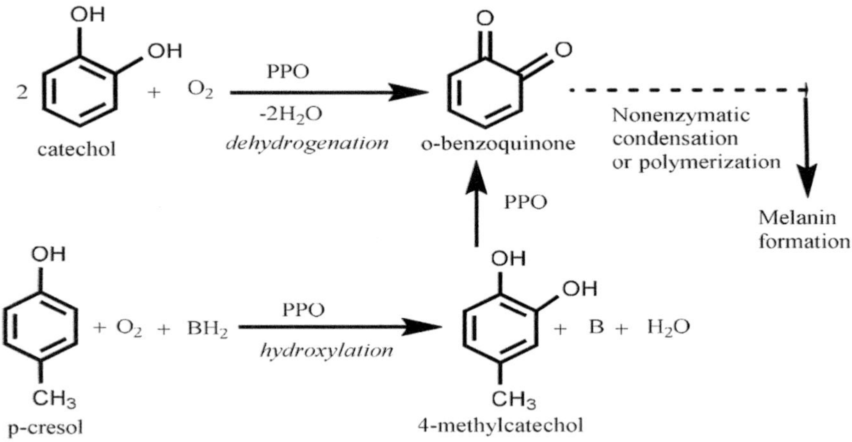

FIGURE 14.2 The representation of dehydrogenation and hydorxylation reactions as catalyzed by PPO.

Proposed pathway of the o-diphenol oxidation is shown in Figure 14.3. The catalytic cycle commences with the coordination of O_2 to both the two Cu (I) centers of deoxypolyphenol oxidase (deoxyPPO) to yield oxyPPO in which the O_2 has the characteristics of a peroxide, having a side-on interactin with both the metal centers and Cu centers being oxidized to Cu (II).[70] Next step involves the binding of substrate, that is, o-diphenol to the two Cu^{2+} centers of oxyPPO via two adjacent hydroxyl groups, to form the oxyPPO.diphenol complex. The o-diphenol is then further oxidized to o-benzoquinone rendering the enzyme as metPPO (metPPO or metpolyphenyl oxidase is the oxidized form of the enzyme in which the Cu metal centers exist in +2 oxidation states rather than +1). It is followed by the binding of another molecule of o-diphenol to metPPO, which is then oxidized to o-benzoquinone, reducing the enzyme back to deoxyPPO with Cu centers having +1 oxidation states.

The proposed pathway of hydroxylation of a monophenols by PPO as depicted from the in vitro studies involves BH_2, which is imperative for the conversion of metPPO to deoxyPPO. The resting state of the enzyme is believed to be >80% metPPO and <15% oxyPPO[79] which is ineffective until BH_2 is added which reduces metPPO to deoxyPPO which can then bind dioxygen to form oxyPPO. In the subsequent step, monophenol coordinates to one of the Cu^{2+} with the removal of one molecule of water. This follows with the change in the steric orientation of the bound substrate (monophenol) as well as the coordinated dioxygen molecule to make sure that the ortho position of the substrate falls within the bonding distance of the second Cu^{2+} center and consequently the ortho position is hydroxylated (not shown in the catalytic cycle here). So, the hydroxylated intermediate which is now a diphenol remains intact with the catalyst, still bonded to the two Cu centers and is transformed into o-benzoquinone in the next step. In the process, the enzyme is reduced to deoxyPPO and can again form oxyPPO without any need to transform into metPPO. The mechanism of monophenol hydroxylation followed by oxidation by the enzyme PPO is shown in Figure 14.4.

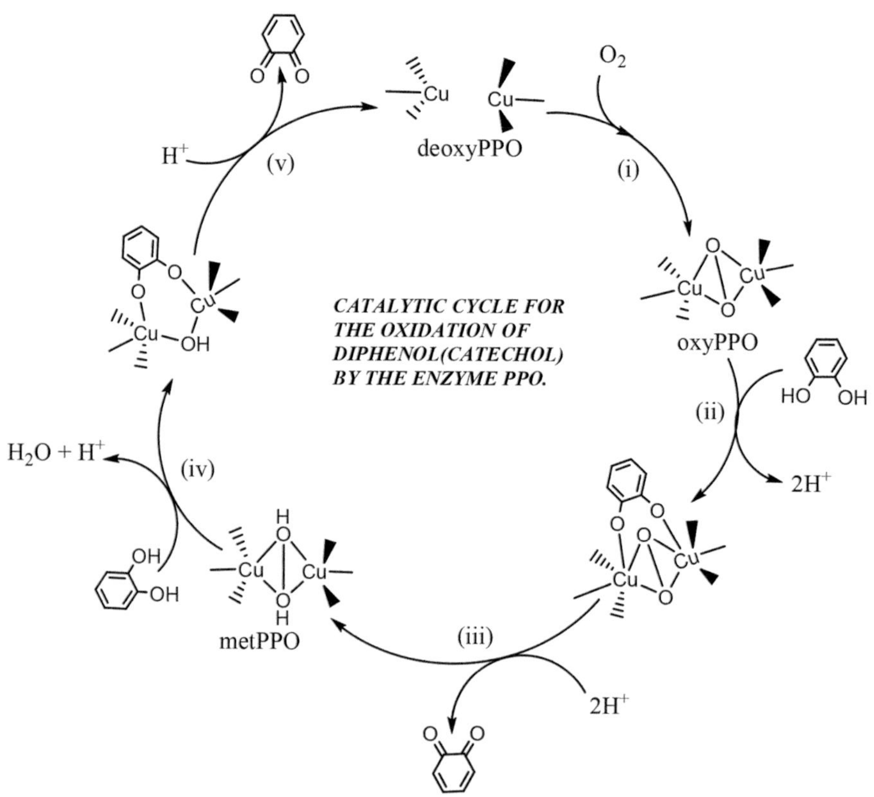

FIGURE 14.3 Catalytic cycle depicting the oxidation of o-diphenol (catechol) for *N. crassa* polyphenol oxidase.
Source: Adapted from Lerch [35] and Solomon et al. [70].

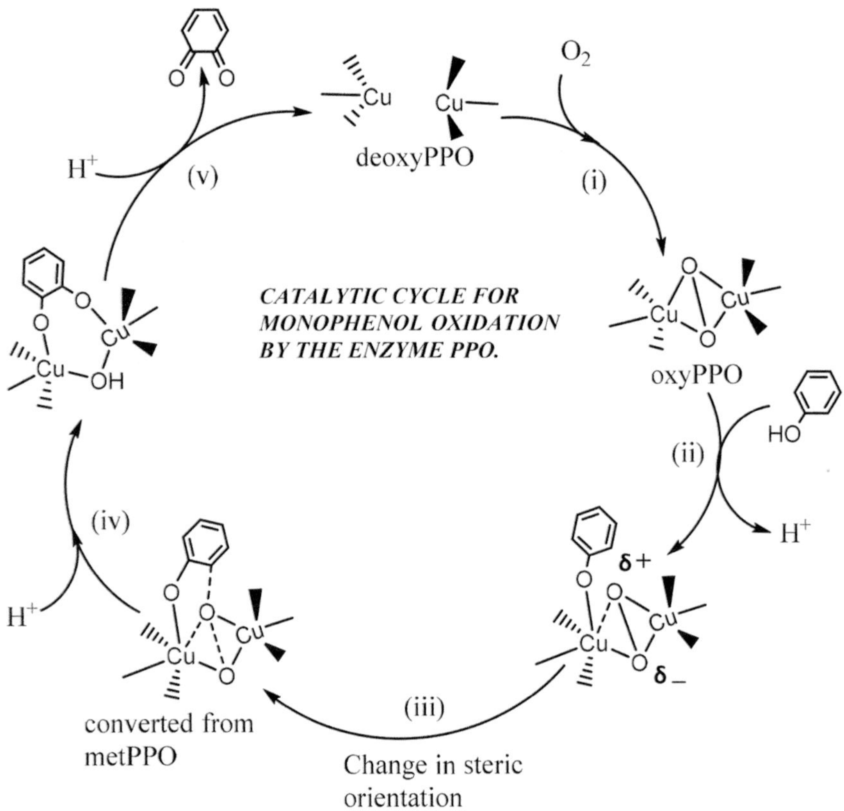

FIGURE 14.4 Catalytic cycle of monophenol hydroxylation followed by oxidation by the enzyme PPO.
Source: Adapted from Lerch [35] and Solomon et al. [70].

14.6 CONTROL OF ENZYMATIC BROWNING

14.6.1 TEMPERATURE CONTROLS

Biochemical reactions are regulated by temperature conditions that control the activity of the enzymes. Phenolic compounds react differently to different temperature conditions depending on the source but a range of 70–95°C may be responsible in destroying the enzymatic action of these compounds.[14,50,54] Heating by blanching and freezing at around 4 °C are commonly known temperature controls to halt enzymatic browning. Water blanching is known to be superior in halting browning to steam blanching methods.[64] But blanching is not preferred by consumers due to its loss in flavor and aroma of the fresh vegetables and fruits. For vegetables like tubers, microwave blanching or hot water microdrop sprays can be used as it gives good results for such vegetables.[72] When talking about freezing to slow enzymatic browning, the principle is that, the availability of water is decreased and hence it cannot take part in enzymatic reactions like polymerization and oxidation involving the phenolic compounds.[30] But in the case of freezing, freeze-thawing can cause more damage when the vegetable or fruit it thawed multiple times leading to increased browning in a short period of time.[27]

14.6.2 HIGH-TEMPERATURE HEAT SHOCK METHOD

The browning of fruits and vegetables can effectively control by the heat shock method also known as the high-temperature heat shock method. The compound known to cause browning—PPOs—are inhibited when the fruits and vegetables are washed in 45–70°C for a maximum of five minutes. In the observation by Loaiza-Velarde et al.[37] and Murata et al.[49] in fresh lettuce, heating in 50–60 °C preventing browning. It is known that vegetables like lettuce, celery, etc. have performed phenolic compounds which are accumulated after cutting.[23] Heat shock can redirect the synthesis of polyphenols by redirecting the synthesis of proteins from phenolic metabolism to site where there are safer heat-shock proteins.[44,62] Haque et al.[20] studied the response of PPO and POD in leaf of Pui vegetable (*Basella alba*) to low temperature treatment of 8 °C. The plants showed higher PPO and POD activities in the leaves indicating oxidative stress in low temperatures.

14.6.3 PRESSURE

In high pressures, some enzymes can be deactivated without affecting its nutrient value and flavor at about 3000–8000 bars. But the texture may be affected if vegetables are subjected to high pressures due to compression and decompression cycles breaking down tissue structures.[55,61] This method can be used to reduce enzymatic browning, but it is not a favorable method when considering fresh fruits, and leafy vegetables.

14.6.4 USE OF ANTIOXIDANTS AND REDUCING AGENTS

The early stages of browning can be suppressed by using antioxidants. The browning product can be inhibited by breakdown of the intermediate products involved in the process. But antioxidants act with correspondence to the different light conditions, pH, temperature, and atmospheric conditions.[36] Some examples of antioxidants are ascorbic acid, N-acetyl cysteine, hexylresorcinol, erythorbic acid, glutathione, and cysteine hydrochloride. On the other hand, reducing compounds decrease o-quinones back to their phenolic compounds by inactivating the PPO's.[5,53,69]

One of the widely used antibrowning agents is ascorbic acid commonly known as Vitamin C and is found widely in most citrus fruits. It is an essential nutrient to repair tissues and other biological functions.[5] It reduces enzymatic browning by reducing the oxidized substrates involved in browning of fresh foods.

14.6.5 CHELATING AGENTS

Due to the capacity of chelating agents to form complexes with Cu(II) present in PPO's, they can too be used to halt enzymatic browning. Metal ions such as Cu(II) and Fe(III) can be chelated by inactivation by the metabolite Kojic acid found in *Aspergillus* and *Penicillium.*[13] The action of Kojic acid in inhibiting enzymatic browning has been observed in apple slices. The use of chelating agents seems promising but it is not widely used due to difficulties related to high cost of extraction, and production. Carboxylic acids like tartaric acid, oxalic acid, lactic acid, malic acid, acetic acid, succinic acids too show antibrowning activities and have been observed in apple slices.[71]

Since a number of different methods can be used to control enzymatic browning, a combination of various methods seems useful to determine the best way for control. So, methods like temperature control and oxidizing agents can be clubbed together for treatment of particular fresh foods. It is imperative to figure out the best possible combination of agents that can be used that does not cause any harm to the produce or degrade flavor and texture. Additionally, consumers today are more inclined to choose products that have the least number of additives. Hence, it is beneficial to look into natural products that can be used to control enzymatic browning.

14.7 NATURAL ANTIBROWNING AGENTS

14.7.1 ONIONS

The flavonoids such as anthocyanins, quercetin, and some of the alk(en)yl cysteine sulphoxides (ACSOs) present in onions make it a choice of most of the health lovers.[19] Fresh onion extracts as well as heated onion extract are found to negatively influence the PPO activity in potatoes. Moreover, heated onion extract exhibited temperature-specific activity. The additional supply of glucose and glycine further enhanced the antibrowning property of onion extract against potato.[34] The impact of onion juice on browning of pear was also reported. Furthermore, the antibrowning effect was enhanced by increasing the temperature and time in this study also.[28] Onion extract prepared in water negatively influenced PPO and the heated juice further enhanced it. It is found that the temperature used to heat juice and the time duration used for heating is directly proportional to the inhibition of PPO.[28]

Additional studies emphasize the influence of onion extract on the browning, nutrient constitution, and antioxidant property of apple juice by comparing normal apple juice with heated one. Browning was decreased by the supply of heat even without the application of onion juice..[32] The properties of apple juice infused with fresh onion juice were found to be diminished browning, elevated total phenol concentration, total soluble solid, radical scavenging capacity, copper chelation, ferric reducing activities but there is no change in concentration of flavonoids.[32] Nevertheless, heated apple juice infused with onion extract exhibited elevated flavonoid concentration along with all other properties.[32] It is found that raising the temperature of onion extract improved its properties.

14.7.2 PINEAPPLE

Pineapple is one of the well-known fruits consumed all around the globe in raw form and in processed form. Among the products, pineapple juice is the favorite item because of its taste and smell.[12] Studies indicate that pineapple juice can also be used as an antibrowning agent.[42] Pineapple juice as well as ion-exchanged pine apple juice successfully inhibited the browning in raw and dried apple rings. Pineapple juice was made into varying fractions based on different parameters such as size, charge and at least 26% antibrowning activity of crude apple juice was exhibited by all the different fractions.[42]

The antibrowning activity of pineapple against banana was also addressed by studies. The pineapple juice was incubated with banana slices at 15°C for 3 days and the inhibitory effect was analogous to 8 Mm ascorbic acid. Moreover, citric acid along with malic acid are found to be the cause of inhibition of PPO in banana. The fractionation of pineapple juice confirmed that the most concentrated fraction inhibited PPO by 100%.[12]

14.7.3 LEMONS, GRAPES, AND WINE

Wheat pastry dough with the inclusion of 5 g of lemon juice inhibited browning in the dough due to the citric acid and ascorbic acid present in lemon juice. However, the luster of the dough diminishes with time due to the degradation of antibrowning agents.[9] Grape juice is composed of many polyphenols nevertheless; it exhibits less antibrowning property. White wine that is composed of alcohol, acid, and polyphenols also exhibited antibrowning activities in dough. When comparing antibrowning activity, lemon has the highest followed by wine and least for grape juice.[9] This higher antibrowning property of lemon accounts to its stronger chelating efficiency compared with others. The study also used mixture of lemon juice and white wine on dough, and the dough was prevented from browning for about four weeks.

14.7.4 DIETARY AND HERBAL COMPOUNDS

Xantone-related compound, Maclurin as well as swertiajaponin, a flavanoid incorporated in certain fruits and herbs have intense antioxidant properties and exhibit antibrowning activities in potato.[31,46,47] The antibrowning activity of Swertiajaponin was greater than ascorbic acid. Swertiajaponin being the competitive inhibitor of PPO and shows binding capacity with PPO. Moreover, it increases antioxidant property of potatoes when added to it.[47] It was evident from studies that Maclurin at a concentration of 10 μM decreased the generation of Reactive Oxygen Species (ROS) by 80% but vitamin C at same concentration could only reduce ROS generation by 43%. With the addition of Maclurin to the potato supernatant at a concentration of 1 and 10 μM, the ROS induced by SIN-1 and $ONOO^-$ was diminished. The ascorbic acid at 10 μM showed same effect on ROS and $ONOO^-$ as that of 1 μM of Maclurin.[46] Moreover, Maclurin retained antibrowning activity in potato supernatant for around 5 weeks when kept in 4°C.

Cyclodextrins are another category of antibrowning agent used in fruit juices.[41] They are cyclic oligosaccharides originated from starch comprising of 6, 7, or 8 glucose residues joined by α (1–4) glycosidic bonds.[75] These oligosaccharides chelates with PPO substrates so as to inhibit oxidation and formation of quinones followed by brown pigment development.[41] Maltosyl-α-Cyclodextrin was described as a "secondary antioxidant" since it inhibited browning in pear juice[38] and apple juice[39] along with enriching the antioxidant property of these juices. Different cyclodextrins were used to found out the extent of antibrowning effect on peach[40] and grape juices.[52]

Morus alba and *Ficus auriculate* plants from the family Moraceae are found to be remarkable tyrosinase inhibitors.[2,10] Flavones (30%), flavanones (14%), and 2-arylbenzofurans (10%) are the metabolites that help tyrosinase inhibition. The studies show that the activity of these compounds is similar to kojic acid.[10] The study by Paudel et al.[56] addressed the antibrowning and antioxidant properties of some 2-arylbenzofurans such as sanggenofuran A, mulberrofuran D2, mulberrofuran D, morusalfuran B, and mulberrofuran H extracted from root barks of *M. alba Linn*. Moreover, most of the compounds displayed DPPH radical scavenging property and their IC_{50} were found to be within 11.58–55.73 μM.[56] Among the compounds mulberrofuran H with IC_{50} of 11.58 ± 0.85 mM and morusalfuran B with IC_{50} of 12.99 ± 0.43 mM exhibited elevated antioxidant potential. However, mulberrofuran H showed highest browning inhibition with IC_{50}: 4.45 ± 0.55 μM for l-tyrosine and 19.70 ± 0.54 μM for l-DOPA. This is in analogy with kojic acid with an IC50 value of 4.49 ± 0.09 μM for l-tyrosine and 7.08 ± 0.57 μM for l-DOPA. All the other compounds showed average and changeable level of inhibition according to the substrate.[56] It has been reported that Cinnamon (bark oil) and ginger have antibrowning activity as they can inhibit the activity of the PPO enzyme.

14.7.5 OTHERS

Rosa roxburghii (RR) was found to be an efficient antibrowning agent of apple juice due to the presence of ascorbic acid and phenolic compounds.[81] The browning may be due to enzymatic activity of PPO or other chemical reactions without enzymes.[77] The antibrowning in *Eleocharis dulcis* (Burm. f.) Trin that is kept in storage was performed by 15 $\mu L\ L^{-1}$ H_2S treatment.[16] The elevated level of phenols and antioxidant activity prevented ROS accumulation and related oxidative damage. It was evident that the H_2S negatively regulated function of enzymes such as PAL, PPO, and peroxidase (POD).[16]

14.8 SUMMARY

The second most common cause of fruit and vegetable quality degradation is enzyme browning. PPO is found in a variety of fruits and vegetables. It causes enzymatic browning in fresh crops during processing such as slicing, and so on, which leads to cell damage. Refrigeration, freezing, blanching, and altering the product atmospheres are examples of physical methods for preventing enzymatic browning, whereas chemical methods included the use of agents such as sodium bisulfite, ascorbic acid, and chelating agents, as well as antioxidants and firmness agents, etc. However, after optimizing these treatments for various species and cultivars, they might be used in diverse combinations for greater efficacy. Undesirable enzymatic browning in food products is a major problem that must be investigated with the use of efficient enzyme inhibitors. Continuous research into better inhibitors derived from natural sources is required to ensure that they do not have any adverse side effects.

KEYWORDS

- **browning**
- **fruit**
- **melanin**
- **peroxidases**
- **polyphenols**
- **irradiation**

REFERENCES

1. Adams, J. B.; Brown, H. M. Discoloration in Raw and Processed Fruits and Vegetables. *Crit. Rev. Food Sci. Nutr.* **2007,** *47*, 319–333.
2. Adhikari, A.; Devkota, H. P.; Takano, A.; Masuda, K.; Nakane, T.; Basnet, P.; Skalko-Basnet, N. Screening of Nepalese Crude Drugs Traditionally Used to Treat

Hyperpigmentation: in Vitro Tyrosinase Inhibition. *Int. J. Cosmet. Sci.* **2008,** *30* (5), 353–360.

3. Amiot, M. J.; Tacchini, M.; Aubert, S. Y.; Oleszek, W. Influence of Cultivar, Maturity Stage, and Storage Conditions on Phenolic Composition and Enzymic Browning of Pear Fruits. *J. Agric. Food Chem.* **1995,** *43* (5), 1132–1137.
4. Aquino-Bolaños, E. N.; Mercado-Silva, E. Effects of Polyphenol Oxidase and Peroxidase Activity, Phenolics and Lignin Content on the Browning of Cut Jicama. *Postharvest Biol. Technol.* **2004,** *33* (3), 275–283.
5. Arias, E.; González, J.; Oria, R.; Lopez-Buesa, P. Ascorbic Acid and 4-Hexylresorcinol Effects on Pear PPO and PPO Catalyzed Browning Reaction. *J. Food Sci.* **2007,** *72* (8), C422–C429.
6. Bachem, C. W.; Speckmann, G. J.; van der Linde, P. C.; Verheggen, F. T.; Hunt, M. D.; Steffens, J. C.; Zabeau, M. Antisense Expression of Polyphenol Oxidase Genes Inhibits Enzymatic Browning in Potato Tubers. *Biotechnol.* **1994,** *12* (11), 1101–1105.
7. Barbagallo, R. N.; Chisari, M.; Branca, F.; Spagna, G. Pectin Methylesterase, Polyphenol Oxidase and Physicochemical Properties of Typical Long-Storage Cherry Tomatoes Cultivated Under Water Stress Regime. *J. Sci. Food Agric.* **2008,** *88*, 389–396.
8. Brandelli, A.; Lopes, C. H. Polyphenoloxidase Activity, Browning Potential and Phenolic Content of Peaches During Postharvest Ripening. *J. Food Biochem.* **2005,** *29* (6), 624–37.
9. Brütsch, L.; Rugiero, S.; Serrano, S. S.; Städeli, C.; Windhab, E. J.; Fischer, P.; Kuster, S. Targeted Inhibition of Enzymatic Browning in Wheat Pastry Dough. *J. Agric. Food Chem.* **2018,** *66* (46), 12353–12360.
10. Burlando, B.; Clericuzio, M.; Cornara; L. Moraceae Plants with Tyrosinase Inhibitory Activity: A Review. *Mini Rev. Med. Chem.* **2017,** *17* (2), 108–121.
11. Casado-Vela, J.; Sellés, S.; Bru, R. Purification and Kinetic Characterization of Polyphenol Oxidase from Tomato Fruits (*Lycopersicon esculentum cv. Muchamiel*). *J. Food Biochem.* **2005,** *29* (4), 381–401.
12. Chaisakdanugull, C.; Theerakulkait, C.; Wrolstad, R. E. Pineapple Juice and Its Fractions in Enzymatic Browning Inhibition of Banana, Musa (AAA Group) Gros Michel. *J. Agric. Food Chem.* **2007,** *55* (10), 4252–4257.
13. Chen, Y. H.; Lu, P. J.; Hulme, C.; Shaw, A. Y. Synthesis of Kojic Acid-Derived Copper-Chelating Apoptosis Inducing Agents. *Med. Chem. Res.* **2013,** *22* (2), 995–1003.
14. Chutintrasri, B.; Noomhorm, A. Thermal Inactivation of Polyphenoloxidase in Pineapple Puree. *LWT—Food Sci. Technol.* **2006,** *39* (5), 492–495.
15. Dekock, P. C.; Vaughan, D.; Hall, A.; Ord, B. G. Biochemical Studies on Blossom End Rot of Tomatoes. *Physiol. Plant.* **1980,** *48* (2), 312–6.
16. Dou, Y.; Chang, C.; Wang, J.; Cai, Z.; Zhang, W.; Du, H.; Zhu, L. Hydrogen Sulfide Inhibits Enzymatic Browning of Fresh-Cut Chinese Water Chestnuts. *Front. Nutr.* **2021,** *8*.
17. Espín, J. C.; García-Ruiz, P. A.; Tudela, J.; Varón, R.; García-Cánovas F. Monophenolase and Diphenolase Reaction Mechanisms of Apple and Pear Polyphenol Oxidases. *J. Agric. Food Chem.* **1998,** *46* (8), 2968–2975.
18. Ghidelli, C.; Rojas-Argudo, C.; Mateos, M.; Pérez-Gago, M. B. Effect of Antioxidants in Controlling Enzymatic Browning of Minimally Processed Persimmon 'Rojo Brillante'. *Postharvest Biol. Technol.* **2013,** 86, 487–493.

19. Griffiths, G.; Trueman, L.; Crowther, T.; Thomas, B.; Smith, B. Onions—A Global Benefit to Health. *Phytother. Res.* **2002,** *16* (7), 603–615.
20. Haque, M. S.; Islam, M. M.; Rakib, M. A.; Haque, M. A. A Regulatory Approach on Low Temperature Induced Enzymatic and Anti Oxidative Status in Leaf of Pui Vegetable (*Basella Alba*). *Saudi J. Biol. Sci.* **2014,** *21* (4), 366–373.
21. He, Q.; Luo, Y. Enzymatic Browning and Its Control in Fresh-Cut Produce. *Stewart Postharvest Rev.* **2007,** *3* (6), 1–7
22. Hershkovitz, V.; Saguy, S. I.; Pesis, E. Postharvest Application of 1-MCP to Improve the Quality of Various Avocado Cultivars. *Postharvest Biol. Technol.* **2005,** *37*, 252–264.
23. Hisaminato, H.; Murata, M.; Homma, S. Relationship Between the Enzymatic Browning and Phenylalanine Ammonia-Lyase Activity of Cut Lettuce, and the Prevention of Browning by Inhibitors of Polyphenol Biosynthesis. *Biosci. Biotechnol. Biochem.* **2001,** *65* (5), 1016–1021.
24. Holland, P. L. Metal–Dioxygen and Metal–Dinitrogen Complexes: Where Are the Electrons? *Dalton Trans.* **2010,** *39* (23), 5415–5425.
25. Huber, M.; Hintermann, G.; Lerch, K. Primary Structure of Tyrosinase from *Streptomyces Glaucescens. Biochemistry* **1985,** *24* (22), 6038–6044.
26. Hunt, M. D.; Eannetta, N. T.; Yu, H.; Newman, S. M.; Steffens, J. C. cDNA Cloning and Expression of Potato Polyphenol Oxidase. *Plant Mol. Biol.* **1993,** *21*(1), 59–68
27. Ioannou, I. Prevention of Enzymatic Browning in Fruit and Vegetables. *Eur. Sci. J.* **2013,** *9* (30), 310–341.
28. Kim, M. J.; Kim, C. Y.; Park, I. Prevention of Enzymatic Browning of Pear by Onion Extract. *Food Chem.* **2005,** *89* (2), 181–184.
29. Lattanzio, V.; Linsalata, V.; Palmieri, S.; Van Sumere, CF. The Beneficial Effect of Citric and Ascorbic Acid on the Phenolic Browning Reaction in Stored Artichoke (*Cynara scolymus L.*) Heads. *Food Chem.* **1989,** *33* (2), 93–106.
30. Lavelli, V.; Caronni, P. Polyphenol Oxidase Activity and Implications on the Quality of Intermediate Moisture and Dried Apples. *Eur. Food Res. Technol.* **2010,** *231* (1), 93–100.
31. Lee, B.; Moon, K. M.; Lee, B. S.; Yang, J. H.; Park, K. I.; Cho, W. K.; Ma, J. Y. Swertiajaponin Inhibits Skin Pigmentation by Dual Mechanisms to Suppress Tyrosinase. *Oncotarget* **2017,** *8* (56), 95530.a
32. Lee, B.; Seo, J. D.; Rhee, J. K.; Kim, C. Y. Heated Apple Juice Supplemented with Onion has Greatly Improved Nutritional Quality and Browning Index. *Food Chem.* **2016,** *201*, 315–319.
33. Lee, C. Y.; Whitaker, J. R. (Eds.) *Enzymatic Browning and Its Prevention.* American Chemical Society: Columbus, Ohio, 1995; pp 600.
34. Lee, M. K.; Kim, Y. M.; Kim, N. Y.; Kim, G. N.; Kim, S. H.; Bang, K. S.; Park, I. Prevention of Browning in Potato with a Heat-Treated Onion Extract. *Biosci. Biotechnol. Biochem.* **2002,** *66* (4), 856–858.
35. Lerch, K. Neurospora Tyrosinase: Structural, Spectroscopic and Catalytic Properties. *Mol. Cell. Biochem.* **1983,** *52* (2), 125–138.
36. Lindley, M. G. The Impact of Food Processing on Antioxidants in Vegetable Oils, Fruits and Vegetables. *Trends Food Sci. Technol.* **1998,** *9* (8–9), 336–340.
37. Loaiza-Velarde, J. G.; Tomás-Barberá, F. A.; Saltveit, M. E. Effect of Intensity and Duration of Heat-Shock Treatments on Wound-Induced Phenolic Metabolism in Iceberg Lettuce. *J. Am. Soc. Hortic. Sci.* **1997,** *122* (6), 873–877.

38. López-Nicolás, J. M.; García-Carmona, F. Use of Cyclodextrins as Secondary Antioxidants to Improve the Color of Fresh Pear Juice. *J. Agric. Food Chem.* **2007,** *55* (15), 6330–6338.
39. López-Nicolás, J. M.; Núñez-Delicado, E.; Sánchez-Ferrer, Á.; García-Carmona, F. Kinetic Model of Apple Juice Enzymatic Browning in the Presence of Cyclodextrins: The Use of Maltosyl-β-Cyclodextrin as Secondary Antioxidant. *Food Chem.* **2007,** *101* (3), 1164–1171.
40. López-Nicolás, J. M.; Pérez-López, A. J.; Carbonell-Barrachina, Á.; Garcia-Carmona, F. Use of Natural and Modified Cyclodextrins as Inhibiting Agents of Peach Juice Enzymatic Browning. *J. Agric. Food Chem.* **2007,** *55*(13), 5312–5319.
41. López-Nicolás, J. M.; Rodríguez-Bonilla, P.; García-Carmona, F. Cyclodextrins and Antioxidants. *Crit. Rev. Food Sci. Nutr.* **2014,** *54* (2), 251–276.
42. Lozano-De-Gonzalez, P. G.; Barrett, D. M.; Wrolstad, R. E.; Durst, R. W. Enzymatic Browning Inhibited in Fresh and Dried Apple Rings by Pineapple Juice. *J. Food Sci.* **1993,** *58*(2), 399–404.
43. Marshall, M. R.; Kim, J.; Wei, C. Enzymatic Browning in Fruits, Vegetables and Sea Foods. *FAO Rome* **2000,** *41*, 259–312.
44. Martin-Diana, A. B.; Rico, D.; Barry-Ryan, C.; Mulcahy, J.; Frias, J.; Henehan, G. T. Effect of Heat Shock on Browning-Related Enzymes in Minimally Processed Iceberg Lettuce and Crude Extracts. *Biosci. Biotechnol. Biochem.* **2005,** *69* (9), 1677–1685.
45. Mathew, A. G.; Parpia, H. A. Food Browning as a Polyphenol Reaction. In *Advances in Food Research*; Chichester, C. O., Ed.; Vol. 19; Academic Press: Cambridge, United States, 1971; pp 207–221.
46. Moon, K. M.; Kim, C. Y.; Ma, J. Y.; Lee, B. Xanthone-Related Compounds as an Anti-Browning and Antioxidant Food Additive. *Food Chem.* **2019,** *274*, 345–350.
47. Moon, K. M.; Lee, B.; Cho, W. K.; Lee, B. S.; Kim, C. Y.; Ma, J. Y. Swertiajaponin as an Anti-Browning and Antioxidant Flavonoid. *Food Chem.* **2018,** *252*, 207–214.
48. Munshi, C. B. *Effect of Mineral Nutrition, Sprout Inhibitors, and Naturally-Occurring Toxicants on Potato Quality*. MSc Thesis, Cornell University, New York, USA, 1994; pp 238.
49. Murata, M.; Tanaka, E.; Minoura, E.; Homma, S. Quality of Cut Lettuce Treated by Heat Shock: Prevention of Enzymatic Browning, Repression of Phenylalanine Ammonia-Lyase Activity, and Improvement on Sensory Evaluation during Storage. *Biosci. Biotechnol. Biochem.* **2004,** *68* (3), 501–507.
50. Ndiaye, C.; Xu, S. Y.; Wang, Z. Steam Blanching Effect on Polyphenoloxidase, Peroxidase and Colour of Mango (*Mangifera Indica L.*) Slices. *Food Chem.* **2009,** *113* (1), 92–95.
51. Nicolas, J. J.; Richard-Forget, F. C.; Goupy, P. M.; Amiot, M. J.; Aubert, S. Y. Enzymatic Browning Reactions in Apple and Apple Products. *Crit. Rev. Food Sci. Nutr.* **1994,** *34* (2), 109–157.
52. Núñez-Delicado, E.; Serrano-Megías, M.; Pérez-López, A. J.; López-Nicolás, J. M. Polyphenol Oxidase from Dominga Table Grape. *J. Agric. Food Chem.* **2005,** *53* (15), 6087–6093.
53. Oms-Oliu, G.; Aguiló-Aguayo, I.; Martín-Belloso, O. Inhibition of Browning on Fresh-Cut Pear Wedges by Natural Compounds. *J. Food Sci.* **2006,** *71* (3), S216–S224.

54. Özel, A.; Colak, A.; Arslan, O.; Yildirim, M. Purification and Characterization of a Polyphenol Oxidase from Boletus Erythropus and Investigation of Its Catalytic Efficiency in Selected Organic Solvents. *Food Chem.* **2010,** *119* (3), 1044–1049.
55. Palou, E.; Lopez-Malo, A.; Barbosa-Canovas, G. V.; Welti-Chanes, J. High Hydrostatic Pressure and Minimal Processing. *Minim. Process. Fruits Veg.* **2000,** 205–222.
56. Paudel, P.; Seong, S. H.; Wagle, A.; Min, B. S.; Jung, H. A.; Choi, J. S. Antioxidant and Anti-Browning Property of 2-Arylbenzofuran Derivatives from Morus Alba Linn Root bark. *Food Chem.* **2020,** *309*, 12573.
57. Peñalver, M. J.; Fenoll, L. G.; Rodríguez-López, J. N.; García-Ruiz, P. A.; García-Molina, F.; Varón, R.; et al. Reaction Mechanism to Explain the High Kinetic Autoactivation of Tyrosinase. *J. Mol. Catal. B: Enzy.* **2005,** *33* (1–2), 35–42.
58. Podsędek, A.; Wilska-Jeszka, J.; Anders, B.; Markowski, J. Compositional Characterisation of Some Apple Varieties. *Eur. Food Res. Technol.* **2000,** *210* (4), 268–272.
59. Queiroz, C.; Mendes Lopes, M. L.; Fialho, E.; Valente-Mesquita, V. L. Polyphenol Oxidase: Characteristics and Mechanisms of Browning Control. *Food Rev. Int.* **2008,** *24* (4), 361–375.
60. Richard-Forget, F. C.; Gauillard, F. A. Oxidation of Chlorogenic Acid, Catechins, and 4-Methylcatechol in Model Solutions by Combinations of Pear (*Pyrus communis cv.* Williams) Polyphenol Oxidase and Peroxidase: a Possible Involvement of Peroxidase in Enzymatic Browning. *J. Agric. Food Chem.* **1997,** *45* (7), 2472–2476
61. Rico, D.; Martin-Diana, A. B.; Barat, J. M.; Barry-Ryan, C. Extending and Measuring the Quality of Fresh-Cut Fruit and Vegetables: A Review. *Trends Food Sci. Technol.* **2007,** *18* (7), 373–386.
62. Saltveit, M. E. Wound Induced Changes in Phenolic Metabolism and Tissue Browning Are Altered by Heat Shock. *Postharvest Biol. Technol.* **2000,** *21* (1), 61–9.
63. Segovia-Bravo, K. A.; Jarén-Galán, M.; García-García, P.; Garrido-Fernández, A. Browning Reactions in Olives: Mechanism and Polyphenols Involved. *Food Chem.* **2009,** *114* (4), 1380–5.
64. Shivhare, U. S.; Gupta, M.; Basu, S.; Raghavan, G. S. V. Optimization of Blanching Process for Carrots. *J. Food Process Eng.* **2009,** *32* (4), 587–605.
65. Sim, S. K.; Ohmann, S. M.; Tong, C. B. Comparison of Polyphenol Oxidase in Tubers of *Solanum tuberosum* and the Non-Browning Tubers of *S. hjertingii*. *Am. Potato J.* **1997,** *74* (1), 1–3.
66. Singh, B.; Singh J. P.; Kaur, A.; Singh, N. Phenolic Composition and Antioxidant Potential of Grain Legume Seeds: A Review. *Food Res. Int.* **2017,** *101*, 1–16.
67. Singh, B.; Singh, J. P.; Kaur, A.; Singh, N. Bioactive Compounds in Banana and Their Associated Health Benefits—A Review. *Food Chem.* **2016,** *206*, 1–11.
68. Singh, B.; Singh, J. P.; Shevkani, K.; Singh, N.; Kaur, A. Bioactive Constituents in Pulses and Their Health Benefits. *J. Food Sci. Technol.* **2017,** *54* (4), 858–870.
69. Singh, B.; Suri, K.; Shevkani, K.; Kaur, A.; Kaur, A.; Singh, N. Enzymatic Browning of Fruit and Vegetables: A Review. *Enzy. Food Technol.* **2018,** 63–78.
70. Solomon, E. I.; Baldwin, M. J.; Lowery, M. D. Electronic Structures of Active Sites in Copper Proteins: Contributions to Reactivity. *Chem. Rev.* **1992,** *92* (4), 521–542.
71. Son, S. M.; Moon, K. D.; Lee, C. Y. Inhibitory Effects of Various Antibrowning Agents on Apple Slices. *Food Chem.* **2001,** *73* (1), 23–30.

72. Sotome, I.; Takenaka, M.; Koseki, S.; Ogasawara, Y.; Nadachi, Y.; Okadome, H.; Isobe, S. Blanching of Potato with Superheated Steam and Hot Water Spray. *LWT—Food Sci. Technol.* **2009,** *42* (6), 1035–1040.
73. Spagna, G.; Barbagallo, R. N.; Chisari, M.; Branca, F. Characterization of a Tomato Polyphenol Oxidase and Its Role in Browning and Lycopene Content. *J. Agric. Food Chem.* **2005,** *53* (6), 2032–2038.
74. Subramanian, N.; Venkatesh, P.; Ganguli, S.; Sinkar, V. P. Role of Polyphenol Oxidase and Peroxidase in The Generation of Black Tea Theaflavins. *J. Agric. Food Chem.* **1999,** *47* (7), 2571–2578.
75. Szente, L.; Szejtli, J. Cyclodextrins as Food Ingredients. *Trends Food Sci. Technol.* **2004,** *15* (3–4), 137–142.
76. Takahama, U. Oxidation of Vacuolar and Apoplastic Phenolic Substrates by Peroxidase: Physiological Significance of the Oxidation Reactions. *Phytochem. Rev.* **2004,** *3* (1), 207–219.
77. Walker, J. R. L. Enzymic Browning in Foods: Its Chemistry and Control. *Food Technol. NZ* **1977,** *12* (3), 19–21.
78. Whitaker, J. R.; Lee, C. Y. Recent Advances in Chemistry of Enzymatic Browning: An Overview. In *Enzymatic Browning and Its Prevention*; Lee, C. Y., Whitaker, J. R., Eds.; American Chemical Society: Washington, DC; 1995; pp 32–57.
79. Wilcox, D. E.; Porras, A. G.; Hwang, Y. T.; Lerch, K.; Winkler, M. E.; Solomon, E. I. Substrate Analog Binding to the Coupled Binuclear Copper Active Site in Tyrosinase. *J. Am. Chem. Soc.* **1985,** *107* (13), 4015–4027.
80. Yoruk, R.; Marshall, M. R. Physicochemical Properties and Function of Plant Polyphenol Oxidase: A Review. *J. Food Biochem.* **2003,** *27* (5), 361–422.
81. Yu, K.; Zhou, L.; Sun, Y.; Zeng, Z.; Chen, H.; Liu, J.; Liu, W. Anti-Browning Effect of *Rosa Roxburghii* on Apple Juice and Identification of Polyphenol Oxidase Inhibitors. *Food Chem.* **2021,** *359*, 129855.
82. Zawistowski, J.; Biliaderis, C. G.; Eskin, N. M. Polyphenol Oxidase. In *Oxidative Enzymes in Foods*; Robinson, D. S. R., Eskin, N. A. M., Eds. Elsevier: London, 1991; pp 217–273.

CHAPTER 15

ROLE OF METAL IONS AND ORGANIC COMPOUNDS ON LIGNOCELLULOLYTIC ENZYME ACTIVITIES

SAQIB HASSAN[1], AAMIR AFZAL MAGRAY[2], RAHIEL RASOOL[3], ANUSHARA PRABHAKARAN[1], HABIBA IBRAHIM[4], GEORGE SEGHAL KIRAN[5], and JOSEPH SELVIN[1]

ABSTRACT

Lignocellulolytic enzymes (LCEs) are crucial for the development of cost-effective, eco-friendly, and sustainable methods for pretreatment and degradation of lignocellulosic biomass, which results in the release of valuable end products such as enzymes, organic acids, other chemicals, feed, and biofuels. Enzymatic lignocellulose (LC) hydrolysis is important in microbially driven carbon cycling and energy conversion. LCEs have many applications, which include biofuel production, textile industries, waste product treatment, food and beverage industries, self-care products, and pharmaceutical industries. Cellulases (LC-active enzymes) are a diverse group of enzymes that are produced by fungus, bacteria, actinomycetes, etc. Several noncellulosic substances like metal ions and other organic compounds have been found to interfere with the activity of cellulases. Abiotic processes or microbial activities during natural cellulolysis may produce these substances and in case of industrial cellulolysis, they might arise from biomass feedstocks or upstream processing. For the development of biomass conversion technology and for understanding the microbiology of biodegradation of biomass, knowledge about cellulolysis-inhibiting enzymes/reactions is of great significance.

Enzyme Inactivation in Food Processing: Technologies, Materials, and Applications. Megh R. Goyal, PhD, Junaid Ahmad Malik, PhD & Ravi Pandiselvam, PhD (Eds.)

15.1 INTRODUCTION

Lignocellulolytic enzymes (LCEs) act as biocatalysts that breakdown lignin and other cellulosic materials into their constituents followed by hydrolysis to generate valuable products. Hydrolytic enzymes that breakdown resistive lignocellulose (LC)—a component of plant biomass are referred to as lignocellulases. Ligninolytic enzymes (peroxidases and oxidases) and hydrolytic enzymes (amylases, cellulases, chitinases, esterases hemicellulases, mannanases pectinases, and proteases) have been characterized under a diverse category of extracellular proteins.[81] Enzymes are macromolecules produced by living organisms that can aid in faster biological/biochemical reactions when put into a reaction. They help in the conversion of substrates into valuable end products by facilitating the reactions.[51,64] The six major classes of enzymes are hydrolases, isomerases, ligases, lyases oxidoreductases, and transferases.[64] The ligninolytic enzymes are classified as oxidoreductases, whereas the hydrolytic enzymes are classified as hydrolases.[118]

Fungi are well-known for degrading LC into its three main polymers. As the substrates are not soluble, hence, both bacterial and fungal breakdown must take place extracellularly.[96] Some lignocellulolytic fungi have been found to be capable of producing enzymes.[111] Producing lignocellulolytic bacterial enzymes has several advantages: bacteria grow rapidly than fungi, have multienzyme complexes with enhanced functionality and specificity, and can adapt to environmental stress conditions.[139] Consequently, bacteria are regarded as essential suppliers of critical enzymes and metabolites, and they are genetically very flexible. Many rare genera are still to be investigated and used for biotechnological applications. As a consequence, greater research into unique ecological habitats might lead to the identification of novel metabolite-producing communities that could be important contributors in developing green technology.[102,112] Extracellular enzymatic systems are known to exist in microorganisms in two different forms. They have a hydrolytic system that degrades hemicellulose (HC) and cellulose via production of hydrolases. They possess a lignin-depolymerizing oxidative and extracellular ligninolytic system. Thus, LCEs are classified as either ligninolytic or hydrolytic based on their enzymatic systems.

In this chapter, we provide a comprehensive knowledge about the LCEs along with the role of metal ions and organic compounds on LCEs. We also discuss the factors affecting enzyme adsorption on lignocellulolytic-based biomass and techniques used to evaluate the enzymatic adsorption onto lignocellulosic surfaces.

15.2 CELLULASES

LCEs are been widely employed in textile, detergent, and animal feed industry owing to their properties such as catalytic efficiency and thermos-stability and their use in ethanol production, bio-bleaching of Kraft pulp, etc.[20] Cellulases particularly are a group of enzymes commonly known as glycoside hydrolases (GHs) as they hydrolyze β-1,4 glucoside linkages in cellulose chains by acid-base hydrolysis. Cellulases are produced by microorganisms like bacteria, protozoa, especially filamentous fungi. These organisms secrete cellulose that are either bound or surface free in order to metabolize cellulose. Structurally they are made up of noncatalytic carbohydrate-binding molecules commonly known as CBMs, which are joined by a peptide linker molecule and these enzymes are also found to contain functionally known or unknown modules mostly CBMs that may be found at N or C terminus of the catalytic module.[146] The hydrolysis mechanism occurs by two catalytic residues of this particular enzyme, one being an acid donor and other one being a nucleophilic base.[38] In order to do retention this anomeric carbon C will undergo double displacement reaction by two step glycosylation/deglycosylation while the inversion will be done by single displacement nucleophilic hydrolysis.[135] So far 115 families of cellulases have been categorized on the basis of their amino acid chain similarities and crystal structure and nearly of them have been cloned and characterized.[113] Variety of topologies is being displayed by all of cellulases either in type of alpha helix or beta-barrel or alpha/beta complex.

15.2.1 TYPES OF CELLULASES

Cellulases have been grouped into three main types, which mediate complete hydrolysis:

(a) Endoglucanases, (b) Exoglucanases (including cellobiohydrolases), and (c) β-glucosidase.

15.2.1.1 ENDOGLUCANASES (EC 3.2.1.4)

Endoglucanases (EC 3.2.1.4) are also known as CMCases, they cleave β-1,4-bonds creating new ends. Organisms like bacteria, archea, fungi, etc. are its main producers belonging to families 5–9, 12, 44, 45, 48, 51, and 74. Fungal endoglucanases do not possess CBMs, while as the bacterial endoglucanases possess multiple CBMs as catalytic modules. For example, the catalytic modules bear active binding sites to cleave cellulose chain to produce glucose, soluble cellodextrins, or nonsoluble cellulose fragment. The structure of endoglucanase of *Thermobifi da fusca* exhibits a deep fissure at the active site, which is the main distinction to exoglucanases.[16,66,79]

15.2.1.2 EXOGLUCANASES (EC3.2.1.91)

Most widely studied exoglucanase is CBH (cellobiohydrolase). Exoglucanases act on the reducing or nonreducing ends of cellulose polysaccharide chains, thereby releasing either cellobiose or glucose as significant products. These can also be utilized to work upon microcrystalline cellulose, where they can peel off cellulose chains from its structure.[124] 2,Cellobiohydrolase 1 (CBH1) secreted by *Trichoderma reesei* accounts for ~60% of the total extracellular cellulases. CBHs are mainly produced many bacteria and fungi, and their catalytic modules belong to families 5, 6, 7, 9, 48, and 74 GHs Families 6 and 7 have aerobic fungal CBH while as families 6 and 48 produce bacterial CBH. On contrary family 48 is lined up with anaerobic fungal CBH, while as family 9 as well as 48 produce anaerobic bacterial CBH. The topology encompasses tunnel-shaped active site of exoglucanase, which enables the enzyme to hydrolyze cellulose as this tunnel covers the part of active site or the CBH entirely, a unique feature reported in family 7.[135] 48 exoglucanases so far reported are believed to play role in hydrolysis mediated by a bacterial Cellulase. This role is thought to be like to that of the *Trichoderma* CBHI (Cel7A).[124]

15.2.1.3 α-GLUCOSIDASE

They belong to families 1, 3, and 9 and they do not possess a CBM and can hydrolyze cellobiose and cellodextrins to glucose. They are unable to degrade insoluble cellulose but can actively degrade cellobiose. These are produced by Archea, Fungi, Bacteria, etc. belonging to the above-mentioned families. Mostly retention stereochemistry is possessed by families 1 and 3, while 9 show inversion stereochemistry. Their crystal structure has pocket-shaped cleft bearing active sites, which can bind to cellobiose and cellodextrin thereby clipping off glucose unit.[147]

15.2.2 METAL IONS ASSOCIATED WITH CELLULASE ACTIVITY

Metal ions due to their universal existence in microhabitat can be effectors and often serve as structural/functional cofactors [e.g., Ca(II), Zn(II), Fe(II), Mn(II), and Cu(II)] or inhibitors [e.g., Ag(I), Hg(II), and Pb(II)] of many enzymes, including cellulases. The metal ions that are associated with cellulase can either form protein complexes or act as electron donors or acceptors as Lewis acids. They promote inhibition by interacting with amine or sulfur groups. It has also been reported that ionic charge, ionic radius, and size have effect on activity and stability of enzyme. In general, large ionic radius does not have much effect on catalytic activity of amino acids, while smaller radius attracts charged amino acids that can damage their conformations also can lead to less activity. Different mono, di, and trivalent atoms have been reported to have inhibitory effects on cellulases secreted by different microorganisms. For example, endoglucanase produced by *Aspergillus fumigatus* has activating metal ions as Co^{2+} and Mg^{2+} while it gets inactivated by Pb^{2+}, Ni^{2+}, Cd^{2+}, Hg^{2+}, K^{+}, Mn^{2+}, Na^{+}, Cu^{2+}, Fe^{2+}, Fe^{3+}, Similarly, exoglucanase produced by *T. reesei* has activating metal ions such as Mn^{2+}, Ba^{2+}, and Ca^{2+}, while Hg^{2+} acts as its inactivator.[20]

Tejirian et al.[125] have reported cellulase characterization, in which divalent metal ions [e.g., Mg(II), Ca(II), etc.] as well as mono- or trivalent metal ions [e.g., Na(I), and Fe(III)] have been checked for their effect on cellulase activity, Hg(II) acts as a cellulase inhibitor which is attributed to its interaction with sulfur-containing amino acid residues. Amongst the divalent metal ions, only Cu(II) and Fe(II) have been reported to have robust inhibition on cellulose-hydrolyzing reactions of CBH, endo-β-glucanase (EG), or β-glucosidase (BG).

Tejirian et al.[125] have tested 1,10-phenanthroline and 2,2′-bipyridyl, which are chelating agents for the Fe(II) inhibition of enzymatic cellulolysis also considered as its potential mitigators. The chelators were found to have exhibit significant inhibition of hydrolysis. It was found that when chelating entities are present, the cellobiose titer in the hydrolysis products rose about twice than when it in the absent in the compounds thereby concluding that chelating inhibition restrained their use as mitigators against the enzymatic cellulolysis by Fe(II).

15.2.3 CHEMICAL AGENTS AND ORGANIC COMPOUNDS ASSOCIATED WITH CELLULASE ACTIVITIES

The inhibitory effect of various phenolic compounds on *T. reesei* was long back studied and it was observed that caffeic acid, vanillin, and catechol are cellulase inhibitors. Ximenes et al.[140] have reported that phenols like vanillin, hydrobenzoic acid, syringaldehyde, etc. showed inhibitory effect on cellulase activity rendering growth of *T. reesei* and also it precipitates the proteins and this inhibitory effect is likely to get amplified thereby reducing hydrolysis as reported by Kim et al.[56] and it was also revealed that the phenolic compounds can cause detrimental effect on cellulase activity. These compounds are reported to be potential inhibitors of endo/exoglucanases and β-glycosidases.

Gluconolactone as reported by Kou et al.[59] induces cgene expression in cellulolytic fungi *T. reesi*. It acts as β-glucosidase enzyme inhibitor by expressing gene expression to that of cellobiose as it significantly induces transcription of gene cbh1, as well as bgl1 and cel1a which encodes β-glucosidase intracellularly and extracellularly thereby inhibiting its activity. Moreover, lactose and transglyoclating derivate of cellobiose, Saphrose can cause inhibition of cellulase.

Chelating agents like EDTA (ethylene diamine), DPPE(1,2-bis diphenylphosphino-ethylene), and ethylene glycol also can inhibit cellulase activity as reported by Tejirian et al.,[125] which can cause sequestering of metal ions. When chelating agents form complex with metal ions, the active site of enzyme then becomes accessible to interact with the substrate, which signifies the positive effects of these compounds on cellulase activities, while as the negative effect of chelating agents on enzymatic activity indicates that enzyme activities rely on the sequestered inorganic ions.[88]

The presence of lignin in LC has been reported as one of the main barriers in enzyme hydrolysis process as it binds to enzymes unproductively via inhibiting reaction between carbohydrates and enzymes. CBMS mostly undergo adsorption onto lignin via hydrophobic interactions. Cellulases show higher catalytic activity when CBMs are eliminated thereby decrement in nonproductive adsorption on lignin. By adding Tween 20, polyethylene glycol as they block the free surface or remove nonspecific binding, they have been observed to improve the hydrolysis mechanism. Additionally, adsorption of cellulase onto cellulose surface has been found to be a rate-determining step as they depict degradation rates.[63]

It has been found that the sugars which are released by enzymatic hydrolysis of LC can be degraded and transformed into inhibitory compounds. Acidic conditions are favorable for conversion of glucose, mannose, and galactose into furan aldehydes such as hydroxymethylfurfurals. Hydroxymethylfurfurals can then be transformed into levulinic and formic acids. Specific xyloglucan endo-β-glucanase inhibitor proteins have been found in the cell walls of tomato, tobacco, and wheat plants and these enzymes have inhibitory effect on endoglucanases that act on xyloglucan.

Cellulase enzyme activities are also affected by drugs, fungicides, antibiotics and disinfectants sugars (final product inhibition), proteins (produced for defense mechanism by plants), CBM-binding organic compounds, products from sugar and degradation of lignin (phenolic compounds) food additives, plant hormones (auxins, Indoleacetic acid [IAA]), and ionic solids (Sodium azide).[20]

15.3 HEMICELLULASES

The polysaccharide, HC attaches to major cell wall component cellulose with the help of weak interaction such as hydrogen bonding and other weak force. This polysaccharide is heterogenous, since it is composed of diverse monomers such as D-xylose, D-mannose, D-galactose, etc.[116] Hemicellulases are distinct group of enzymes that accomplish the task of degrading hemicellulosic materials. These enzymes can be classified into GHs and carbohydrate esterases (CEs). Some of the GH family members include Endo-α-1,5-arabinanase, α-Glucuronidase, Endo-β-1,4-mannanase, which hydrolyze glycosidic bonds and certain CE family members which are Acetyl xylan esterase and Acetyl mannan esterase which break ester linkages of acetate or ferulic acid side groups.[116]

Alpha L arabinofuranosidases (AFases) act on arabinose comprising sidechains of arabinananas or arabinxylans.[58] Moreover, these group of enzymes are classified into diverse families of GHs and some of them are GH 3, 10, and 62.[85] α-glucuronidases act on sidechains of methyl glucuronic acid specifically the α-1,2-glycosidic linkage.[110] They are classified as GH 4, 67, and 115 families.[38] Endoxylanases cleave β-1,4-linkages of xylanes and these enzymes are grouped into GH 5,7, 8, 43, 10, and 11.[10] Xylosidases use xylobiose as substrate and catalyze the liberation of β-D-xylopyranosyl residues and these enzymes are grouped into GH 1, 3, 30, 39, 43, 51, 52, 54,116, and 120.[142]

15.3.1 METAL IONS ASSOCIATED HEMICELLULASE ACTIVITY

Metal ions such as Hg^{2+}, Zn^{2+}, Cd^{2+}, and Co^{2+} are known to inhibit the activity of AFases.[74] But some of the AFases such as G10 family enzymes are stabilized and activated by metal ions. AFs from *Bacillus stearothermophilus* is highly inhibited by Hg^{+}. However, they are merely stimulated by Mn^{2+}, Ca^{2+}, Mg^{2+}, and Ba^{2+}.[32] AFs produced by *Streptococcus* sp PC22 have been found to be completely inhibited by Hg^{2+} and Cu^{2+}, whereas Mn^{2+}, Fe^{2+}, and Zn^{2+} partially inhibit this enzyme. Ca^{2+} and Mg^{2+} should no effect on the enzyme activity.[105] Fe^{3+} and Co^{2+} were found to stimulate AFs produced by *Arthrobacter* species moreover Hg^{2+} negatively affected the activity and activity was not affected by Ca^{2+}, Mg^{2+}, and Mn^{2+}.[53] AFs from *Bifidobacterium longum* have been found to be negatively influenced by Hg^{2+} and Cu^{2+}, whereas enzyme activity was unaffected by Mg^{2+}, Mn^{2+}, Ca^{2+}, Co^{2+}, or Na^{+}.[74] β-xylosidase and AFases from *Thermomonospora Fusca* BD25 were found to be inhibited by Hg^{+} but were not affected on treatment with urea, SDS, EDTA, Zn^{2+}, and Mg^{2+}.[131]

Metal ions are not necessary for the function of α-glucuronidases. Some of the metal ions including Ag^{2+}, Zn^{2+}, and Cd^{2+} have negative effects.[52,133] α-Glucuronidase from *B. stearothermophilus* T-6 has been observed to show mild inhibition to Ni^{2+}, Zn^{2+}, and K^{+}, whereas Ag^{+}, Hg^{2+}, and Cu^{2+} exhibited elevated inhibition.[145] Ag^{+}, Hg^{2+}, Pb^{2+}, Sn^{2+}, Fe^{2+}, Fe^{3+}, and Al^{3+} highly inhibit α-glucuronidase.[133] One of the glucuronidase, p-nitrophenyl α-D-glucuronide-hydrolyzing enzyme (PNP-GAase) isolated from snail acetone powder shows immense inhibitory activity upon adding of Hg^{2+}.[52]

Xylanases are inhibited by Hg^{2+}, Fe^{2+}, Co^{2+}, Mn^{2+}, Ag^{2+}, Cu^{2+}, and Pb^{2+}.[49] The reactive groups on enzymes may sometimes combine with metal ions and leads to inhibition.[37] In the case of xylanases from *A. giganteus*, Hg^{2+} and Cu^{2+} are strong inhibitors whereas Ca^{2+}, Ba^{2+}, and Zn^{2+} are weak inhibitors.[28] In *Bacillus halodurans* xylanases it has been observed that catalysis is negatively affected by Hg^{2+}, Cu^{2+}, and Pb^{2+}. The decreased activity of enzyme has been linked to the activity of cysteine thiol group autooxidation that may lead to the disulfide linkages within or between molecules as well as production of sulfenic acid.[61]

15.3.2 CHEMICAL MOIETIES AND CARBON COMPOUNDS ASSOCIATED WITH HEMICELLULASE ACTIVITY

AFs produced by *Streptococcus species* PC22 have been found to be inhibited by N-bromosuccinimide (NBS) indicating the presence of tryptophan among the reactive group of the enzyme whereas iodoacetamide showed no activity suggesting that cysteine residues are not significant.[105] AFs from *Bifidobacterium longum* were neither affected by EDTA, which helps for chelation nor by DDT and β mercaptoethanol which take part in reduction reaction.[74] The impact of ionic detergents such as sodium dodecyl sulfate (SDS) on AFs was determined and the results were found to be concentration dependent. Lower concentration of SDS tends to enhance enzyme activity but elevated levels of SDS concentration inhibited the enzyme. SDS was found to affect the three-dimensional protein structure and it is evident that higher concentration of SDS may denature the enzyme.

The highest catalysis of α glucuronidases was exhibited with the addition of 2-mercaptoethanol, dithiothreitol, L cysteine, and NAD^+.[123] α-glucuronidases were highly affected by SDS, monoiodoacetic acid, and N–bromosuccinimide, and elevated level of enzyme activity was induced by reduced form of glutathione.[133] PNP-GAase is extensively prevented by p-chloromercuribenzoic acid, SDS, and N-bromosuccinimide.[52] NAD^+ dependent enzymes exist in family 4 and NAD^+ plays structural, functional, and catalytic roles in these enzymes.[103]

Xylanase is generally resistant to denaturants such as guanidine hydrochloric acid.[61] Two of the protein inhibitors of this enzyme include Xylanase Protein Inhibitor and *Triticum aestivum* xylanase inhibitor. Moreover, these are found in wheat, rye, and barley that interfere with the industrial production from these sources. Xylanase Protein Inhibitor specifically inhibit families such as G10 and G11 fungal xylanases, whereas *T. aestivum* xylanase inhibitor is specific inhibitors of G11 bacterial and fungal xylanases. EDTA showed less inhibitory effect whereas dithiothreitol activated xylanases from *A. giganteus.*[28] Another inhibitor of xylanase isolated from wheat, Taumatin-like Xylanase Inhibitor negatively affected most of the G 11 family enzymes but did not influence the G 10 family microbial enzymes.[13,90] *Talaromyces thermophilus* xylanase exhibited resistance to SDS but EDTA and dithiothreitol negatively affected the enzyme activity implying that the disulfide bonds are necessary for the maintenance of working enzyme configuration.[72]

Certain modulators such as N-bromosuccinimide (N-BS), ethyl 3-(3-dimethyl aminopropyl) carbodiimide (EDAC), and indoacetate (IAA) negatively influenced *B. halodurans* xylanase activity. The effect of NBS on enzyme activity indicates the presence of tryptophan residue as part catalytic group whereas effect of EDAC and WRK implies the significance of carboxylic group in enzyme activity.[134] Vanillic acid, acetosyringone, syringic acid, and syringaldehyde are reported as potent inhibitors of xylanase.[24] *Thermobaccilus xylanilyticus* produce an endoxylanase, which is highly susceptible to noncompetitive multisite inhibition by phenolic compounds. Some of those phenolic compounds comprise of cinnamic acid, p-coumaricacid, and 3,4,5-trimethoxy-cinnamic acid.[11]

The inhibitors of β xylosidase belonging to GH 43 purified from *Selenomonas ruminantium* were classified based on the fundamental properties of enzyme. Cellulobiose, D-glucose, Maltose, D-xylose, and L-xylose were grouped as single binding inhibitors. Moreover, D-arabinose, L-arabinose, D-erythrose, and D-ribose were grouped as double binding inhibitors.

15.4 LIGNIN

Lignin is a heterogenous, aromatic polymer encasing the cellulose fibers in the LC component of cell walls in plant cells. It is a complex organic polymer composed of phenylpropanoid monolignols, that is, sinapyl, p-coumaryl, and coniferyl alcohol building blocks clubbed together by the oxidative-coupling of these monolignols via strong ether and ester linkages. There is no precise definition for lignin due to its nature as there are no regularly repeating building units and the compositions and morphologies of lignin vary tremendously as a function of their origin. After cellulose, it is unequivocally the lignin that constitutes the second most plentiful organic polymer on earth.[47,119] It is hydrophobic in nature and cements together the cellulose microfibrils, thus provides plants with increased compression strength and shields them from attack by microorganisms.

15.4.1 LIGNIN-DEGRADING ENZYMES

Lignin-degrading enzymes represent both bioorganic and bioinorganic glycoproteins and are sometimes categorized as either lignin-degrading auxiliary enzymes or ligninolytic enzymes (also referred as lignin modifying enzymes). The enzymes that come under the banner of lignin-degrading auxiliary enzymes are the ones that are unable to depolymerize lignin on their own, thus require the involvement of some supplementary enzyme to serve their purpose of complete lignin degradation, in some instances via the action of some proteins that utilize oxidative H_2O_2.[46] Such additional enzymes include cellobiose dehydrogenase, glyoxal oxidase, pyranose 2-oxidase glucose oxidase, and aryl alcohol oxidases While most important examples of ligninolytic enzymes are nonheme oxidase, for example laccase and heme-containing peroxidases including manganese peroxidase (MnP), lignin peroxidase (LiP), and versatile peroxidase.[89,138] Several other enzymes have been reported to augment the activity of ligninolytic enzymes in the process of degradation of environmental waste, which include the enzymes such as feruloyl esterase, quinone reductases, xylanase, catechol 2,3-dioxygenase, and others.[36,89,101] It is reported that many species of both fungi and bacteria are responsible for the degradation of lignocellulosic waste, but growth limitations in fungi restricted their large-scale application.

15.4.1.1 LACCASE: STRUCTURE AND PROPERTIES

Laccase due to its chemistry is regarded as an ideal green catalyst for its vast oxidizing property for a plethora of compounds and for liberating H_2O as the only by-product in the process.[2,31] The low redox potential of this metalloenzyme renders it inefficient for directly taking part in catalysis so the mediators have a role to play. A mediator (whether natural or synthetic) that plays an essential role in the action of the laccase enzyme via a second pathway of action is a chemical compound that is forced through continuous redox cycles by first being oxidized by the enzyme followed by the reduction by the substrate. Hydroxycinnamic acid is the organic compound that plays a crucial role as a mediator for laccase during lignin degradation.[89]

Laccase is an extracellular metalloenzyme categorized within the class of copper-containing extracellular oxidases that consisting of glycoproteins as monomers, dimers, and tetramers, predominantly found in various species of bacteria, fungi, and actinomycetes.[46] Laccase is comprised of more or less 500 amino acid residues grouped together with Greek letter β cask topology in three consecutive domains. Laccase consists of two disulfide bonds, which bridge together the first domain with second domain as well as the first domain with third domain.[9,77] It has been reported that some laccases comprise three disulfide bridges with first one inside the first domain, second one between first domain and the second domain, and the third one between second domain and the third domain.[9] In the active catalytic site of the laccase enzyme, the four copper atoms are grouped into two, based upon their comparative distance from each other, one as the mononuclear location and the second as the trinuclear location.[57,77,100] Out of the four copper centers, the mononuclear copper termed as Type 1 (T1) is paramagnetic with a strong UV-visible absorption at 600 nm (responsible for the blue color of this center), due to a covalent bond between copper and amino-acid residue cysteine present in the glycoprotein chain (Fig. 15.1).[77]

Depending upon the nature of coordination of the copper centers, the four copper atoms have been further categorized into three types; Type 1 (T1) Cu that analogous to other multicopper oxidases has a triangular planar coordination geometry with two histidine residues (His) and one cysteine residue (Cys) as equatorial ligands whereas some other multicopper oxidases are reported to also possess in addition the methionine residue as an axial ligand. Type 2 (T2) Cu consists of two histidine residues and a water molecule coordinated to it and Type 3 (T3) Cu consists of two copper centers bridged by a water molecule and three histidine molecules coordinated to each copper center (Fig. 15.1).[100] The bridging of the water molecule between the two copper centers is believed to take place when the enzyme exists in its oxidized form. It is worthwhile to note that the crystalline structures of laccase revealed that any one of the three copper centers in the trinuclear site of the enzyme may act as T2 Cu where the other two copper centers reduce and become EPR inactive (copper in +1 oxidation state is diamagnetic and thus EPR inactive), while the other one remains EPR active (copper in +2 oxidation state is paramagnetic and thus EPR active).[30]

15.4.1.2 MECHANISM OF ACTION OF LACCASE ENZYME

In the catalytic cycle of laccase, the first step involves the action of molecular oxygen to oxidize the substrate molecules, which in fact is one electron oxidation of four equivalent, suitable reducing substrates such as amines or phenols. Such substrate is oxidized to form organic radicals and two water molecules are released by the reduction of molecular oxygen. The active catalytic site of laccase consists of four-membered copper cluster (though not bonded directly to each other), where the coordination of oxygen and its reduction as well as water formation and release takes place. T1 Cu that sometimes referred to as "blue site" is believed to be involved actively in substrate oxidation. On the other hand, T2 and T3 Cu centers (which form a trinuclear sub-cluster) catalyze the reduction of molecular oxygen to water.[97] If the substrate possesses the proper redox potential, laccase directly oxidizes it to the corresponding radical (Fig. 15.2). However, when a direct oxidation in not possible due to redox potential incompatibility or steric hindrance, the "chemical mediators" are employed which act as redox intermediates in a nature mimicking fashion.[48,80] Consequently, the free radicals generated due to the oxidation of these mediators then interact with both bulky and high redox-potential substrates, which is otherwise impossible in the absence of such valuable molecules (Fig. 15.2).

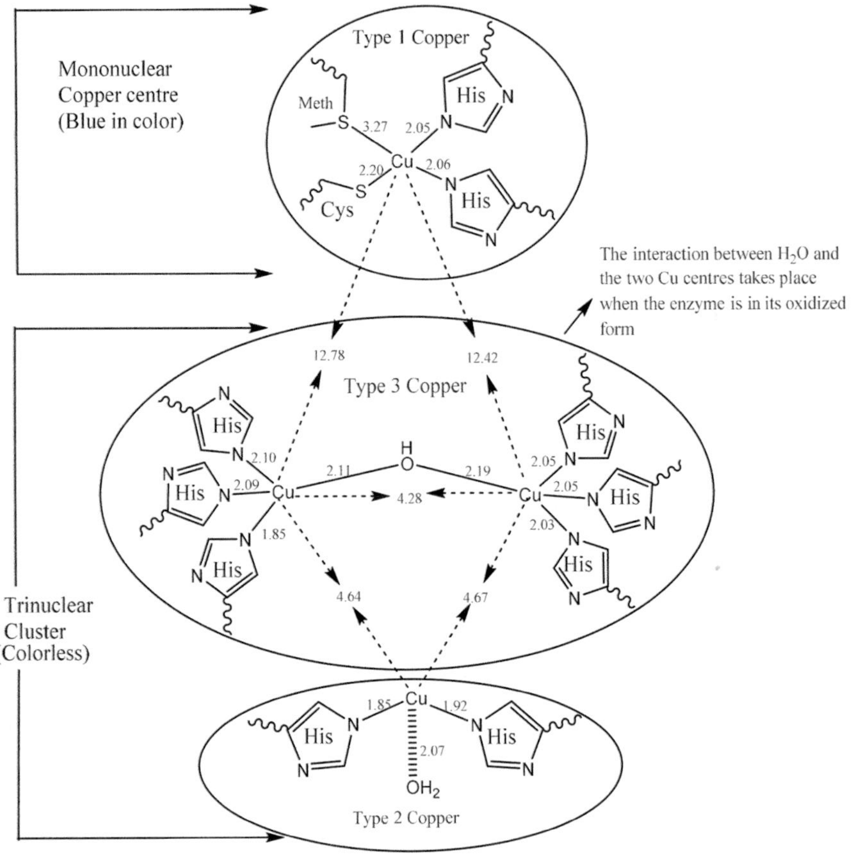

FIGURE 15.1 Depiction of the active site of laccase enzyme containing four copper centers forming two different sites-one as mononuclear site and other as trinuclear site.
Source: Adapted/modified from Ref. [27]

This redox system involving laccases with chemical mediators is termed as "laccase-mediator-system" and it is reported that such kind of redox system has laccase application in organic synthesis (for oxidizing activated hydroxyl groups to the corresponding carbonyls or acids) as well as in biotechnological and manufacturing processes.[39,76,93]

15.4.1.3 METAL IONS AND CHEMICAL AGENTS ASSOCIATED WITH THE ACTIVITY OF LACCASE ENZYME

Laccases are oxidases with multicopper centers meant to catalyze the one-electron oxidation during the depolymerization of lignin, conjoining the four-electron reduction of molecular oxygen (O_2) to water.[73] The groups of copper centers in the active site of the enzyme are T1 (mononuclear copper or blue copper, as it is paramagnetic), T2 (normal copper), and T3 (coupled binuclear copper). The former two are the primary electron acceptor centers, whereas the latter is reduced via intramolecular two-electron transfer from T1 and T2 copper sites in the process.[73,143]

Mostly, laccases do not require cofactors for their enzymatic action, still the addition of metal ions such as Cd^{2+}, Cu^{2+}, Mn^{2+}, Ni^{2+}, and Mo^{2+} increases their activity. While the ions including Ag^{2+}, Pb^{2+}, Zn^{2+}, and Hg^{2+} inhibit their activity. The compounds that are believed to inhibit the activity of laccases also include NaCl, H_2O_2, and NaN_3. The inhibition of 4-nitrophenol degradation by laccase was also observed on the addition of monovalent and divalent ions.[70]

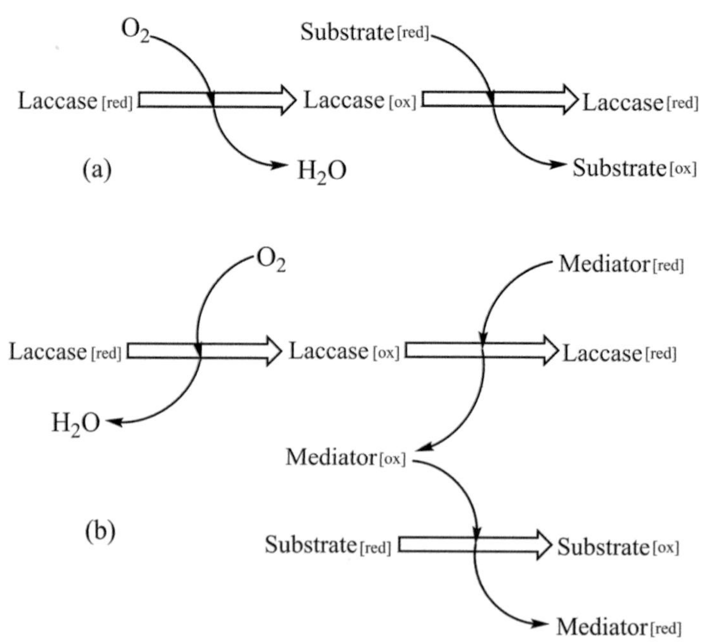

FIGURE 15.2 Redox cycles of substrate oxidation by laccase: (a) direct substrate oxidation; (b) substrate oxidation in the presence of a chemical mediator.

15.4.2 IGNIN PEROXIDASE (LIP)

LiP is a heme-containing metalloenzyme and like other peroxidases, it is one of the enzymes utilizing hydrogen peroxide (H_2O_2) for oxidative degradation of lignocellulosic waste.[25] It is a monomeric, Class II hemoprotein also called as "diaryl propane oxygenase."[99] LiP acts upon the nonphenolic components of lignin. Hydrogen peroxide gets reduced and acts as an electron acceptor during the oxidation process and the reaction looks like as follows:

$$2S + H_2O_2 + 2e^- \rightarrow 2e\text{–}2S_{ox} + 2H_2O \text{ 'S' represents the substrate}$$

LiP oxidizes the nonphenolic structures that constitute 90% of lignin due its high redox potential (700–1400 mV), which is optimal for the oxidation of such groups.[75] It plays a pivotal role in the enzymatic degradation of lignin due to its ability to oxidize and depolymerize numerous aromatic structures. It has been reported in various instances that LiP in addition to oxidizing phenolic structures, does oxidize a wide range of phenolic compounds as well.[5]

Firstly, LiP was discovered in the white-rot fungus; *Phanerochaete chrysosporium, where LiP was isolated from its extracellular medium.*[33,127] *Later on, various isoenzymes of LiP have been reported to be present in brown-rot fungi and some bacteria,*[14,91] *with bacterial LiP being least studied, thus demanding the need to search for new bacterial species, capable of producing highly active and versatile LiP.*

15.4.2.1 STRUCTURE AND MECHANISM OF LIP

LiP enzyme folds into a globular shape with dimensions as about 50 × 40 × 40 Å,[98] containing three short antiparallel β sheets, eight major α-helices, and minor helices.[15] The heme part of the protein segregates the proximal and the distal domains and remains fixed, yet accessible through two small channels in the protein. LiP weighs more or less 38–43 kDa, with an isoelectric point ranging from 3.3 to 4.7 and a pH of 3.0 when the substrate is vertaryl alcohol.[29] This low optimum pH of LiP is what sets it apart from various other peroxidases.

LiP is believed to be monomeric, heme-containing glycoprotein, consisting carbohydrates (four in number), water molecules (370 in number), two calcium ions, and about 343 amino acids residues.[15] Crystallographic studies of LiP revealed that it is larger in size with amino acid residues (about 343 in number), containing four disulfide bonds.[25] In comparison with cytochrome c peroxidase which is also a heme-containing peroxidase, rich in oxidizable amino acids, LiP consists of eight methionines, and just three tryptophans with no free cysteine and tyrosine being completely absent. The presence of tryptophan residues in the case of cytochrome c peroxidase and the replacement of these with phenylalanine residues at the contact point between the proximal and distal heme surfaces in LiP differentiate the two to a greater extent. Moreover, the aspartate (Asp-183) is linked to heme propionate via hydrogen bonding in LiP and disruption of this bond would destabilize the heme pocket and interestingly the observed low optimum pH of LiP is also attributed to the presence of the same bond. It is reported that the coordinate bond in the heme group of LiP existing between the nitrogen atom of the proximal histidine and the central iron is longer, thus slightly weaker, rendering the heme more electron deficient and consequently destabilizing the high oxidation states of iron which is more rational explanation for the higher redox potential of LiP.[15]

The catalytic cycle of LiP is illustrated in Figure 15.3. The catalytic cycle initiates with first reaction involving the formation of LiP-Compound-I (which is Fe^{4+}-oxo-porphyrin-radical intermediate) from the two-electron oxidation of the enzyme by the action of H_2O_2, which is imperative for the functioning of the LiP as a ligninolytic enzyme. The second step of the reaction follows with the conversion of LiP-Compound-I to yield the LiP-Compound-II, involving one-electron reduction of the former by substrate (S) such as nonphenolic aromatic structures of the lignin. In the third step, LiP-Compound-II receives a second electron from the reduced substrate returning the enzyme back to the resting (native) ferric oxidation state (Fe^{3+}) to terminate the cycle.[1]

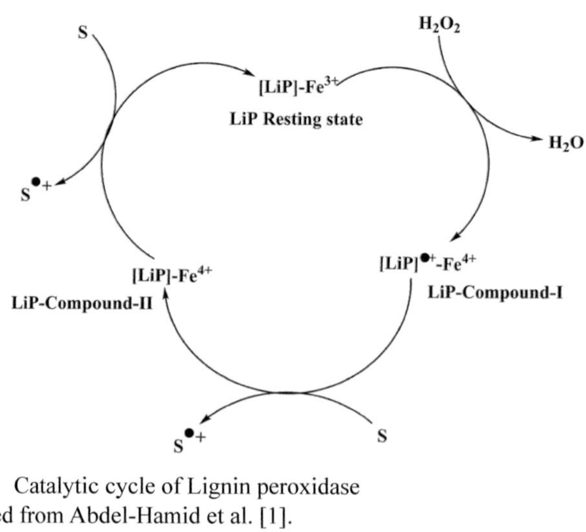

FIGURE 15.3 Catalytic cycle of Lignin peroxidase
Source: Adapted from Abdel-Hamid et al. [1].

15.4.2.2 METAL IONS AND CHEMICAL AGENTS ASSOCIATED WITH THE ACTIVITY OF LIP ENZYME

Since lignin-peroxidases are heme-containing peroxidases, so their active site consists of a porphyrin ring coordinated to Fe^{3+}. LiP in addition to the oxidation of a variety of phenolic compounds catalyzes the oxidative degradation (depolymerization) of nonphenolic structures of lignin and the compounds alike, driven by the action of H_2O_2.[19]

It is reported that metal ions including Fe^{2+}, Mn^{2+}, and Cu^{2+} if added to the aqueous reaction system of LiP, result in its enhanced catalytic activity, while the addition of Ag^{2+} ions to the same inhibits the activity.[4] The inhibition or denaturation of the enzyme results in the decreased activity of the LiP which is a function of the concentration of inhibitor molecules/ions in the aqueous reaction system. Moreover, different organic compounds including alcohols, esters, aldehydes, ketones, acids, amides, amines, ethers, cysteine sodium azide, TEMED, CTAB, EDTA, and H_2O_2 as well as some solvents that include acetonitrile, DMSO, and DMF are regarded as the potential inhibitors of LiP.

15.4.3 MANGANESE PEROXIDASE (MNP)

MnP also sometimes referred to as Manganese-dependent peroxidase was first discovered in *P. chrysosporium*.[62] So, in addition of being dependent on H_2O_2 for oxidizing stuff, this peroxidase also requires Mn^{2+} as a cofactor. MnP is regarded among other peroxidases as the commonly occurring peroxidase for lignin-degradation secreted by most white-rot fungi and others. MnP is better known for oxidizing the lignin phenolic structures. Structurally, MnP is a heme-containing metalloenzyme, which catalyzes the oxidation of one-electron donor cofactor Mn^{2+} ion to Mn^{3+} ion, which can then oxidize a wide range of phenolic substrates including the phenolic structures of lignin.[132] It is worthwhile to mention that there are reported instances where MnP also oxidizes otherwise nonoxidizable nonphenolic structures, but requires specific mediators to do that which include thiyl or lipid radicals.[1,106]

15.4.3.1 STRUCTURE AND MECHANISM OF ACTION OF MNPS

MnPs are heme-containing, Mn^{2+}-dependent glycoproteins having a molecular weight approximately ranging from 38 to 62.5 kDa, resembling 43%

with the structure of LiP containing ~350 amino acid residues. The structure of MnP consists of a heme group held in midst of two domains having five disulfide linkages with a minor helix and ten major helices. Since one of these five disulfide bonds exists at the manganese (Mn) bonding site, this is what distinguishes them from other peroxidases.[122]

The first step that commences the catalytic cycle of the enzyme is the formation of MnP-compound-I (Fe^{4+}-oxo-porphyrin-radical intermediate) via a two-electron transfer from the heme group to H_2O_2 which is in turn reduced to a molecule of water. MnP-Compound-I then subsequently induces the oxidation of monochelated Mn^{2+} ion to Mn^{3+} ion and itself is converted to MnP-Compound-II. The MnP-Compound-II that contains Iron in +4 oxidation state then gets reduced by one electron transfer from Mn^{2+} to yield Mn^{3+} and the native enzyme (MnP-Resting state), which has Iron in +3 oxidation state and can further be utilized to keep the catalytic cycle running.[1] Mn^{3+} ion can be stabilized by various small organic acid chelators including malonate, oxalate, glyoxylate, etc. and it is worthwhile to mention that the chelated Mn^{3+} possessing high redox potential and smaller size can then diffuse easily in the lignified cell wall. So, the chelated Mn^{3+} ions generated by MnP during the whole process act as low molecular mass, diffusible charge-transfer mediators, which oxidize various phenolic substrates including phenolic structures of lignin model compounds, simple phenols, amines, and dyes.[40,137] The release of Mn^{3+} ion that is responsible for the oxidation of substrate during lignin degradation would be impossible from the active site of the MnP enzyme, if not facilitated by these organic acid chelators. The whole catalytic cycle of MnP enzyme is depicted in (Fig. 15.4).

The catalytic cycle of MnP has a great resemblance with the catalytic cycles of other heme-containing peroxidases such as LiP. However, its uniqueness lies in its utilization of Mn^{2+} cofactor as the electron donor (Fig. 15.4). MnP in comparison with LiP is ineffective to be able to oxidize the more recalcitrant nonphenolic functional groups, but it is reported that the white-rot fungi, which produce only laccase and MnP (but not LiP), may cleave nonphenolic lignin substrates by producing certain mediators.[106]

15.4.3.2 METAL IONS AND CHEMICAL AGENTS ASSOCIATED WITH THE ACTIVITY OF MNP ENZYME

EDTA addition results in partial inhibition catalytic activity of MnP. EDTA being a chelating ligand coordinatively bonds with the metal ions present in the inorganic cofactors and prosthetic groups of the enzymes like MnP, LiP, etc. High concentrations of Fe^{2+} and Cu^{2+} ions are believed to enhance the catalytic activity of MnP.[3] The catalytic activity of MnP is completely inhibited by the addition of certain chemicals and metal ions, which include Ag^{+}, Pb^{2+}, Hg^{2+}, lactate, $CaCl_2$, TEMED, NaN_3, dithiothreitol, ascorbic acid, and β-mercaptoethanol.

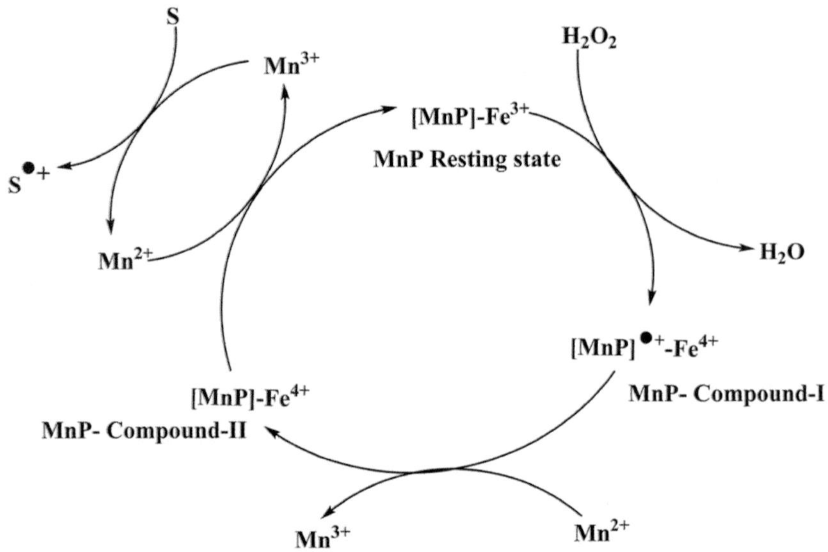

FIGURE 15.4 Catalytic cycle of Manganese peroxidase (MnP). (S represents Lignin or any phenolic substrate that is oxidized by MnP).
Source: Adapted from Abdel-Hamid et al. [1]

15.5 FACTORS INFLUENCING ENZYMATIC ADSORPTION IN LIGNOCELLULOSIC MATERIALS

Adsorption of enzymes on lignocellulosic materials is a sensitive and complex mechanism influenced by the properties of enzyme and the adsorbent substrate, as well as by the environmental dynamics, for example, the connected flowing states.[6] Knowledge about surface features-topography, roughness, substrate composition, and the interactions of enzymes with substrates are crucial for developing the understanding of adsorption process.

15.5.1 COMPOSITION OF VARIOUS LIGNOCELLULOSIC BIOMASS MATERIALS

Lignocellulosic material is one of the most commonly used raw materials for biofuel generation. The three major constituents found in lignocellulosic biomasses are cellulose, HC, and lignin. The interchain hydrogen bonding in the crystalline regions of cellulose account for its insolubility in most solvents. In addition, they make cellulose nonthermoplastic. Depending on the polymerization conditions, there are six possible configurations of glucose chains that are located in the elementary fibrils.[86] Native cellulose is a mix of two or more configurations.[86] The center of the primary fibrils of cellulose is water insoluble.[65,107] It is difficult to disrupt the crystalline arrangement of cellulose. Enzymatic action converts cellulose to cellobiose (a dimer of glucose), which is then broken down to glucose. The disruption of the linkages that hold the units together may cause cellulose to react or it may be due to the reaction of the common hydroxyl group. In most reactions, the hydroxyl groups arranged in the cellulose chain facilitate good enzyme binding/adsorption sites.[67]

HC is a combination of polysaccharides and has a molecular weight of around 30,000. Different sugars and uronic acids constitute HC. HC has D-glucosidic units linked by β-1,3 and β-1,4 bonds in its chemical structure. HC is a branching polysaccharide that attaches to the surface of each microfibril of cellulosic material.[42] By weight, cellulose makes up 40–50% of biomass, whereas HC constitutes 20–40%.[12,95,149] HC has five-carbon sugars, whereas cellulose contains six.

Lignin is an amorphous compound containing phenyl propanes [a 6-carbon ring connected to a 3-carbon chain Lignin strengthens cell architecture by binding polysaccharide fibers together]. Lignin includes variable quantities of methoxyl, hydroxyl, carbonyl, and carboxyl groups, depending on the origin and lignin separation technique used.[35]

15.5.2 PHYSICAL STRUCTURE OF BIOMASS

Physical structure of biomass is also an important factor for enzyme adsorption. If enzymes adhere to the surface, the interaction will most likely result in ineffective enzyme bonding. Aqueous medium is where most enzyme reactions take place. After soaking in aqueous media, substrate fibers become more visible, which aids in enzyme adsorption.

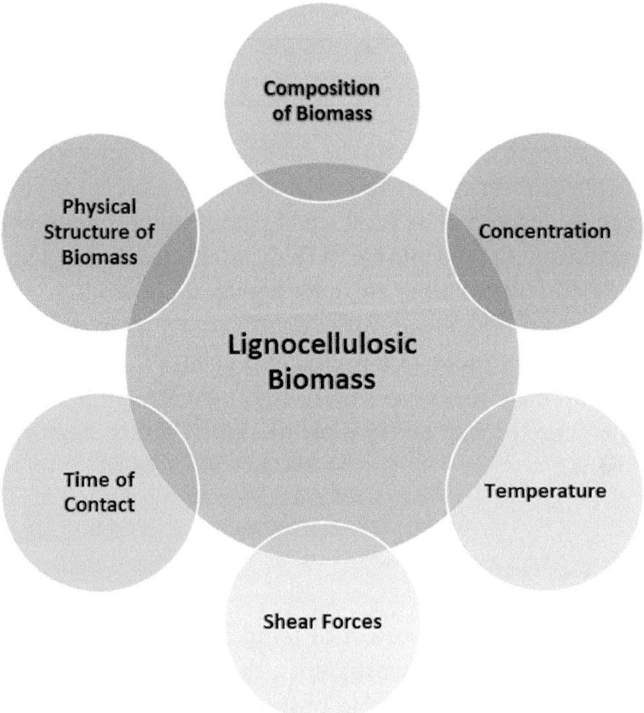

FIGURE 15.5 Factors affecting enzymatic adsorption on lignocellulosic biomass.

15.5.3 OTHER FACTORS AFFECTING ENZYME ADSORPTION

Adsorption of the enzymes on biomass-based substrates is influenced by a variety of variables, including related flow conditions, media temperature, enzyme contact time (incubation time), concentration of enzyme and substrate, and the existence of contaminants. The number of attached enzymes is determined by flow parameters, which have a significant impact on the adsorption process.[7,68]

15.5.3.1 SHEAR FORCES

Lower shear rates or less agitation are thought to result in greater adsorption. Agitation, shaking, and mixing generate shear forces. As a result, the process flow conditions should be carefully selected. Shaking, agitation, and altering rpm all these have an effect on adsorption of enzymes.[50,92,126] It is crucial to remember that enzymes which have been adsorbed will only convert the material to a product, whereas enzymes that have not been adsorbed will not be able to convert anything. As a result, adsorption is measured indirectly by conversion (hydrolysis).

The enzymes form a bond with the substrate once they come into contact. Consider the adsorption of a ligand molecule onto a receptor protein, where the number of adhesive bonds produced is determined by the charge densities of the ligand and the receptor.[21,109] To disintegrate a bond, a specific force is always needed. The number of links between enzymes and surface can be used to determine the shear stress necessary to separate enzymes from the substrate.[26,55] As the contact duration approaches its maximum value, the force required to remove enzymes increases, indicating that additional interconnections emerge progressively after the initial attachment of an enzyme with the substrate.

15.5.3.2 TEMPERATURE

The research findings on the effect of temperature on cellulase adsorption are debatable. According to investigators, the adsorption of cellulases onto biomass materials is an exothermic process controlled by enthalpy. It has been detected that when the temperature was raised, the concentration of cellulases adsorbed decreased,[43,47,54,78,87] on the other hand, some studies claimed that the adsorption of enzymes onto microcrystalline cellulose (MCC) was an endothermic reaction controlled by entropy, which increased with temperature. Tu et al.[130] studied action of cellulases on a softwood (lodgepole pine).

The soft wood is composed of 45–50% cellulose, 30–35% HC, 20–25% lignin, and 4–5% extractives. The adsorption experiment used lignin derived by steam exploding lodgepole pine (L-SELP) and lignin produced by ethanol-pretreated lodgepole pine (L-EPLP) and this experiment was carried out at a range of temperatures (4 °C, 25 °C, and 45 °C). At 45°C, maximum cellulase was adsorbed onto lignin than at 4 °C and 25 °C, indicating that cellulase adsorption onto L-SELP and L-EPLP is an endothermic process.[130,148] Since, there is a hydrophobic interaction between lignin and enzymes; this is true for the adsorption of cellulases onto lignin.[129] Entropy is the driving force for adsorption in the presence of hydrophobic interactions.[130,136] Lignin adsorbs more cellulases, and lignin adsorption increases as temperature rises.

15.5.3.3 TIME OF CONTACT

Singh et al.[117] established the adsorption pattern of cellulases onto microcrystalline cellulose (Avicel), revealing that most enzymes were adsorbed onto Avicel and pulp during the initial 10 min of contact at 30°C. According to Steiner et al.,[120] 50% adsorption of maximally adsorbed enzyme occurs within 1–2 min of contact with MCC; nevertheless, adsorption equilibrium took <30 min to achieve. The enzyme adsorption of pulp was better than that of Avicel. Within 10–20 min of contact, the majority of cellulase (approximately 18%) adsorbs into the xylan substrate. Jäger et al.[45] observed that cellulase adsorption on α-cellulose is fast, and cellobiohydrolase CBHI and endoglucanase EGI attain equilibrium in 20 and 30 min, respectively. Cellulases were adsorbed on spruce lignin (SP) and black cotton wood lignin by Pareek et al.,[94] and it was observed that more than 60% of the enzymes were adsorbed during the initial 30 min of contact time for SP and more than 2 h for black cotton wood lignin. An adequate amount of incubation time is needed to create dynamic adsorption equilibrium. Several researches have studied the adsorption of metal ions, dyes, and enzymes on a variety of lignocellulosic substrates at different temperatures. Their findings showed that increasing the temperature raised the equilibrium constant value while maintaining the time to achieve equilibrium constant.[18,23,44,54,78,115,121] Contact duration is critical for adsorption studies. Depending upon the enzyme–substrate system, contact time may vary also and the environment of the enzyme–substrate system (Fig. 15.5).

15.5.3.4 CONCENTRATION

Increase in the enzyme concentration increases adsorption for a certain amount of substrate before reaching the optimal value. The adsorption efficiency of a feedstock is determined by its resource and chemical composition, as well as the pretreatment procedures used. The effectiveness of enzymatic adsorption is dependent on the enzymes' accessibility to the cellulosic substrate; pretreatment methods play a crucial role. Baig[6] conducted a thorough investigation of the adsorption of cellulases on cellulose (Avicel PH 101) at room temperature, for eight different cellulase concentrations, and showed that adsorption increased with the increasing enzyme concentration until reaching an optimal value. Additional increase of the enzyme concentration does not make any difference. Adsorption on lignin takes more time as compared with cellulose. The repulsion between the amino acids like tyrosine and histamine from cellulases and the carbonyl from lignin may allow cellulase to undergo conformational changes in order to adsorb onto lignin, which might explain the delayed interaction. Thus, the presence of several positive and negative functional groups, lignin adsorption takes longer.

15.5.4 INTERACTIONS BETWEEN ENZYME AND LIGNOCELLULOSIC MATERIAL SURFACES AT THE MOLECULAR LEVEL

In the adsorption of enzyme onto lignin, three types of interactions are involved: (i) hydrophobic interactions[114,128]; (ii) electrostatic interactions[82,114]; and (iii) hydrogen-bonding.[104] Some hydrophobic groups have been isolated from lignin samples, including long-chain carboxylic acid, phenolic, and aliphatic hydroxyl groups.[8] At pH 5, generally the carboxylic acid groups found in lignin are deprotonated during the hydrolysis of lignocellulosic materials. Nakagame et al.[82] observed that at pH 4.8, isolated lignin possesses a negative charge, whereas *T. reesei* enzymes (Cel6A and Cel5A) have a positive charge, resulting in greater adsorption. It has been reported that repulsive electrostatic interactions between enzymes and lignin are enhanced at higher pH levels, resulting in decreased adsorption onto ligneous substrate.[69] Normally, enzyme reactions take place in aqueous media, this finding has practical implications in terms of adsorption and desorption.[6]

Noncovalent interactions between enzymes and the lignin surface are mediated by hydrophobic and electrostatic forces, with hydrogen bonding and dipolar interactions contributing a little.[60,71] Enzymatic cellulose disintegration is hampered on biomass surfaces over extended periods of time due to enzyme inaccessibility to cellulose induced by the gradual and regular buildup of lignin[22] The lignin-rich constituents of biomass adsorb enzymes more than nonlignin ones; this kind of adsorption is nonproductive since it terminates the hydrolysis of feedstock.[84,141] The size, concentration, and binding equilibrium of cellulose-binding modules (in cellulases) determine their penetrating ability. Cellulases contain 12 types of active amino acid residues. The amino acid residues on the surface of cellulases are charged. They possess amide and hydroxyl functions. For adsorption on cellulosic substrate, they probably interact with the C1 and C4 of the glucose molecule. The hydrogen bonding between cellulose and cellulase via hydroxyl groups on cellulose strengthens the adsorption. Enzyme modification is considered to improve substrate binding and provide protein stability at higher temperatures, implying a critical function.

15.5.5 TECHNIQUES USED TO EVALUATE THE ENZYMATIC ADSORPTION ONTO LIGNOCELLULOSIC SURFACES

Several techniques can be used to observe and evaluate the adsorbed enzymes, which include microscopy (from basic to advanced), radiolabeling, Spectrophotometry, and Coulter counter apparatus (Fig. 15.6).

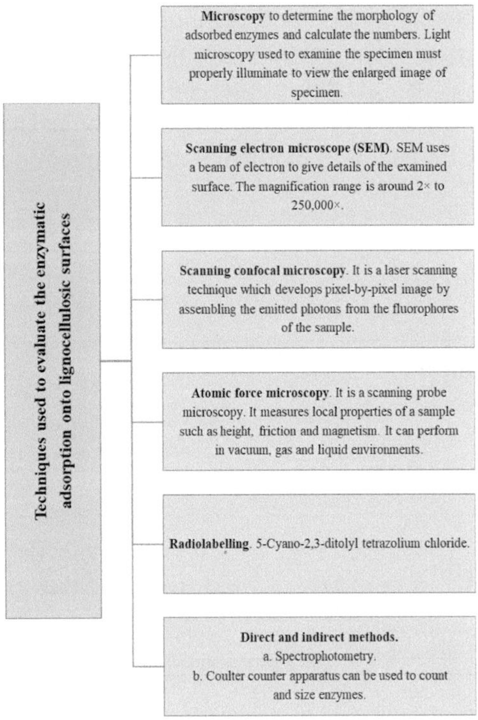

FIGURE 15.6 Various techniques used to evaluate the enzymatic adsorption onto lignocellulosic surfaces.

15.6 SUMMARY

Enzymes are important in many biological processes, which is why the role of LCEs in converting waste, especially the waste of lignocellulolytic origin, into valuable end products has gained a lot of attention at global level. As the availability of fossils is limited and these have detrimental effects on the environment, LC is an ecofriendly substitute. With the rising and impending food shortages, it is more important than ever to look into the waste stream source of lignocellulosic raw materials to ensure that food meant for human usage isn't diverted for energy production. Finding feasible, eco-friendly, and cost-effective pretreatment alternatives is critical to overcome the problems that come with using LC, and LCEs can play a key part. The discovery of new microbial-interactions might lead to the simultaneous synthesis of LCEs, which could assist overcome present cost and acceptability. More investigations are required to find unexplored microbial communities with potential for lignin degradation and cellulolysis. Once such new microbes are identified, further technical advancements can aid in replicating the optimum conditions so that these microbes can be cultivated in the lab, examined further, and may be used to biotechnological advances. As more information on LCEs becomes available, new applications are expected to emerge.

KEYWORDS

- **biodegradation**
- **enzymes**
- **cellulases**
- **inhibitors**
- **lignin**
- **lignocellulolytic**

REFERENCES

1. Abdel-Hamid, A. M.; Solbiati, J. O.; Cann, I. K. Insights into Lignin Degradation and Its Potential Industrial Applications. *Advances in Applied Microbiology*, **2013,** *82*, 1–28.
2. Alcalde, M.; Ferrer, M.; Plou, F. J.; Ballesteros, A. Environmental Biocatalysis: From Remediation with Enzymes to Novel Green Processes. *Trends Biotechnol.* **2006,** *24* (6), 281–287.
3. Asgher, M.; Iqbal, H. M. Characterization of a Novel Manganese Peroxidase Purified from Solid State Culture of *Trametes Versicolor* IBL-04. *BioResources* **2011,** *6* (4), 4317–4330.

4. Asgher, M.; Iqbal, H. M. N.; Irshad, M. Characterization of Purified and Xerogel Immobilized Novel Lignin Peroxidase Produced from *Trametes Versicolor* IBL-04 Using Solid State Medium of Corncobs. *BMC Biotechnol.* **2012,** *12* (1), 1–8.
5. Baciocchi, E.; Bietti, M.; Gerini, M. F.; Lanzalunga, O.; Mancinelli, S. Oxidation of Non-Phenolic β-O-Aryl-Lignin Model Dimers Catalysed by Lignin Peroxidase. Comparison with the Oxidation Induced by Potassium 12-Tungstocobalt (III) Ate. *J. Chem. Soc. Perkin Trans.* **2001,** *2* (9), 1506–1511.
6. Baig, K. S. *Strategic Adsorption/Desorption of Cellulases NS 50013 onto/from AVICEL PH 101 and Protobind 1000.* Doctoral Dissertation; Ryerson University, Toronto, Canada, 2016, pp 352.
7. Bekard, I. B.; Peter Asimakis, P.; Bertolini, J.; Dunstan, D. E. The Effects of Shear Flow on Protein Structure and Function. *Biopolymers* **2011,** *95* (11), 733–745.
8. Berlin, A.; Gilkes, N.; Kurabi, A.; Bura, R.; Tu, M.; Kilburn, D. et al. Weak Lignin-Binding Enzymes. *Appl. Biochem. Biotechnol.* **2005,** *121* (1), 163–170.
9. Bertrand, T.; Jolivalt, C.; Briozzo, P.; Caminade, E.; Joly, N.; Madzak, C.; Mougin, C. Crystal Structure of a Four-Copper Laccase Complexed with an Arylamine: Insights into Substrate Recognition and Correlation with Kinetics. *Biochemistry* **2002,** *41* (23), 7325–7333.
10. Bhardwaj, A.; Mahanta, P.; Ramakumar, S.; Ghosh, A.; Leelavathi, S.; Reddy, V. S. Emerging role of N- and C-Terminal Interactions in Stabilizing (β;/α) 8 Fold with Special Emphasis on Family 10 Xylanases. *Comput. Struct. Biotechnol. J.* **2012,** *2* (3), e201209014.
11. Boukari, I.; O'Donohue, M.; Rémond, C.; Chabbert, B. Probing a Family GH11 Endo-β-1, 4-Xylanase Inhibition Mechanism by Phenolic Compounds: Role of Functional Phenolic Groups. *J. Mol. Catal. B: Enzy.* **2011,** *72* (3–4), 130–138.
12. Brigham, J. S.; Adney, W. S.; Himmel, M. E. Hemicelluloses: Diversity and Applications. In *Handbook on Bioethanol: Production and Utilization*; Wyman, C. E., Ed.; Taylor and Francis: Washington, D.C., USA, 1996; pp 119–142.
13. Brutus, A.; Villard, C.; Durand, A.; Tahir, T.; Furniss.; C, Puigserver, A, et al. The Inhibition Specificity of Recombinant *Penicillium funiculosum* xylanase B Towards Wheat Proteinaceous Inhibitors. *Biochimica et Biophysica Acta (BBA)—Proteins Proteom.* **2004,** *1701* (1–2), 121–128.
14. Bugg, T. D.; Ahmad, M.; Hardiman, E. M.; Singh, R. The Emerging Role for Bacteria in Lignin Degradation and Bio-Product Formation. *Curr. Opin. Biotechnol.* **2011,** *22* (3), 394–400.
15. Choinowski, T.; Blodig, W.; Winterhalter, K. H.; Piontek, K. The Crystal Structure of Lignin Peroxidase at 1.70 Å Resolution Reveals a Hydroxy Group on the Cβ of Tryptophan 171: A Novel Radical Site Formed During the Redox Cycle. *J. Mol. Biol.* **1999,** *286* (3), 809–827.
16. Cohen, R.; Suzuki, M. R.; Hammel, K. E. Processive Endoglucanase Active in Crystalline Cellulose Hydrolysis by the Brown Rot Basidiomycete *Gloeophyllum trabeum. Appl. Environ. Microbiol.* **2005,** *71* (5), 2412–2417.
17. Creagh, A. L.; Ong, E.; Jervis, E.; Kilburn, D. G.; Haynes, C. A. Binding of the Cellulose-Binding Domain of Exoglucanase Cex from *Cellulomonas fimi* to Insoluble Microcrystalline Cellulose Is Entropically Driven. *Proc. Natl. Acad. Sci.* **1996,** *93* (22), 12229–12234.

18. Daoud, F. B.; Kaddour, S.; Sadoun, T. Adsorption of Cellulase *Aspergillus niger* on a Commercial Activated Carbon: Kinetics and Equilibrium Studies. *Coll. Surf. B: Biointerf.* **2010,** *75* (1), 93–99.
19. Dashtban, M.; Schraft, H.; Syed, T. A.; Qin, W. Fungal Biodegradation and Enzymatic Modification of Lignin. *Int. J. Biochem. Mol. Biol.* **2010,** *1* (1), 36.
20. de Cassia-Pereira, J.; Giese, E. C.; de Souza-Moretti, M. M.; dos Santos-Gomes, A. C.; Perrone, O. M.; Boscolo, M.; da Silva R.; Gomes, E.; Martins, D. A. Effect of Metal Ions, Chemical Agents and Organic Compounds on Lignocellulolytic Enzymes Activities. *Enzy. Inhib. Act.* **2017,** *29*, 139–164.
21. Ding, B.; Wang, X.; Yu, J.; Eds. *Electrospinning: Nanofabrication and Applications*. William Andrew: Norwich, England, 2019; pp 832.
22. Djajadi, D. T.; Jensen, M. M.; Oliveira, M.; Jensen, A.; Thygesen, L. G.; Pinelo, M.; et al. Lignin from Hydrothermally Pretreated Grass Biomass Retards Enzymatic Cellulose Degradation by Acting as a Physical Barrier Rather than by Inducing Nonproductive Adsorption of Enzymes. *Biotechnol. Biofuels*, **2018,** *11* (1), 1–13.
23. Du, R.; Su, R.; Zhang, M.; Qi, W.; He, Z. Cellulase Recycling After High-Solids Simultaneous Saccharification and Fermentation of Combined Pretreated Corncob. *Front. Energ. Res.* **2014,** *26*, 2–24.
24. Duarte, C. L.; Ribeiro, M. A.; Oikawa, H.; Mori, M. N. Study of Thermal Treatment Combined with Radiation on the Decomposition of Polysaccharides in Sugarcane Bagasse. *Radiat. Phys. Chem.* **2013,** *84*, 191–195.
25. Edwards, S. L.; Raag, R.; Wariishi, H.; Gold, M. H.; Poulos, T. L. Crystal Structure of Lignin Peroxidase. *Proc. Natl. Acad. Sci.* **1993,** *90* (2), 750–754.
26. Engler, A.J; Chan, M.; Boettiger, D.; Schwarzbauer, J. E. A Novel Mode of Cell Detachment from Fibrillar Fibronectin Matrix Under Shear. *J. Cell Sci.* **2009,** *122* (10), 1647–1653.
27. Enguita, F. J.; Martins, L. O.; Henriques, A. O.; Carrondo, M. A. Crystal Structure of a Bacterial Endospore Coat Component: A Laccase with Enhanced Thermostability Properties. *J. Biol. Chem.* **2003,** *278* (21), 19416–19425.
28. Fialho, M. B.; Carmona, E. C. Purification and Characterization of Xylanases from *Aspergillus giganteus*. *Folia Microbiol.* **2004,** *49* (1), 13–18.
29. Furukawa, T.; Bello, F. O.; Horsfall, L. Microbial Enzyme Systems for Lignin Degradation and Their Transcriptional Regulation. *Front. Biol.* **2014,** *9* (6), 448–471.
30. Garavaglia, S.; Cambria, M. T.; Miglio, M.; Ragusa, S.; Iacobazzi, V.; Palmieri, F.; Rizzi, M. The Structure of *Rigidoporus Lignosus* Laccase Containing a Full Complement of Copper Ions, Reveals an Asymmetrical Arrangement for the T3 Copper Pair. *J. Mol. Biol.* **2004,** *342* (5), 1519–1531.
31. Giardina, P.; Faraco, V.; Pezzella, C.; Piscitelli, A.; Vanhulle, S.; Sannia, G. Laccases: A Never-Ending Story. *Cell. Mol. Life Sci.* **2010,** *67* (3), 369–385.
32. Gilead, S.; Shoham, Y. Purification and Characterization of Alpha-L-Arabinofuranosidase from *Bacillus stearothermophilus* T-6. *Appl. Environ. Microbiol.* **1995,** *61* (1), 170–4.
33. Glenn, J. K.; Morgan, M. A.; Mayfield, M. B.; Kuwahara, M.; Gold, M. H. An Extracellular H_2O_2-Requiring Enzyme Preparation Involved in Lignin Biodegradation by the White Rot Basidiomycete *Phanerochaete Chrysosporium*. *Biochem. Biophys. Res. Commun.* **1983,** *114* (3), 1077–1083.

34. Glumoff, T.; Harvey, P.; Molinari, S.; Goble, M.; Frank, G.; Palmer, J. M.; et al. Lignin Peroxidase from *Phanerochaete Chrysosporium* Molecular and Kinetic Characterization of Isozymes. *Eur. J. Biochem.* **1990,** *187*, 515–520.
35. Gosselink, R. J.; Abächerli, A.; Semke, H.; Malherbe, R.; Käuper, P.; Nadif, A.; et al. Analytical Protocols for Characterisation of Sulphur-Free Lignin. *Ind. Crops Prod.* **2004,** *19* (3), 271–281.
36. Guillen, F.; Martinez, M. J.; Munoz, C.; Martinez, A. T. Quinone Redox Cycling in the Ligninolytic fungus *Pleurotus Eryngii* Leading to Extracellular Production of Superoxide Anion Radical. *Arch. Biochem. Biophys.* **1997,** *339* (1), 190–199.
37. Heinen, P. R.; Henn, C.; Peralta, R. M.; Bracht, A.; Simão, R. D.; da Conceição Silva, J. L. Xylanase from *Fusarium heterosporum*: Properties and Influence of Thiol Compounds on Xylanase Activity. *Afr. J. Biotechnol.* **2014,** *13* (9), 22–41.
38. Henrissat, B. A. Classification of Glycosyl Hydrolases Based on Amino Acid Sequence Similarities. *Biochem. J.* **1991,** *280* (2), 309–316.
39. Hilgers, R.; Vincken, J. P.; Gruppen, H.; Kabel, M. A. Laccase/Mediator Systems: Their Reactivity Toward Phenolic Lignin Structures. *ACS Sustain. Chem. Eng.* **2018,** *6* (2), 2037–2046.
40. Hofrichter, M. Lignin Conversion by Manganese Peroxidase (MnP). *Enzy. Microb. Technol.* **2002,** *30* (4), 454–466.
41. Hofrichter, M.; Ullrich, R.; Pecyna, M. J.; Liers, C., Lundell, T. New and Classic Families of Secreted Fungal Heme Peroxidases. *Appl. Microbiol. Biotechnol.* **2010,** *87* (3), 871–897.
42. Hopkins, W. G.; Huner, N. P. A. *Introduction to Plant Physiology*; 4th Ed.; Wiley Publishing: New York, UK, 2017; pp 542.
43. Hoshino, E.; Kanda, T.; Sasaki, Y.; Nisizawa, K. Adsorption Mode of Exo-and Endo-Cellulases from *Irpex lacteus* (*Polyporus tulipiferae*) on Cellulose with Different Crystallinities. *J. Biochem.* **1992,** *111* (5), 600–605.
44. Houston, R.; Labbé, N.; Hayes, D.; Daw, C. S.; Abdoulmoumine, N. Intermediate Temperature Water–Gas Shift Kinetics for Hydrogen Production. *React. Chem. Eng.* **2019,** *4* (10), 1814–1822.
45. Jäger, G.; Wu, Z.; Garschhammer, K.; Engel, P.; Klement T.; Rinaldi R.; et al. Practical Screening of Purified Cellobiohydrolases and Endoglucanases with α-Cellulose and Specification of Hydrodynamics. *Biotechnol. Biofuels* **2010,** *3* (1), 1–2.
46. Janusz, G.; Pawlik, A.; Sulej, J.; Burek, U. S.; Wilkołazka, A. J.; Paszczynski, A. Lignin Degradation: Microorganisms, Enzymes Involved, Genomes Analysis and Evolution. *FEMS Microbiol. Rev.* **2017,** *41*, 941–962.
47. Jaramillo, P. M. D.; Gomes, H. A. R.; Monclaro, A.; Silva, C. O. G.; Filho, E. X. F. Lignocellulose-Degrading Enzymes. Wiley Publishing: Hoboken, NJ, USA, 2015; pp 256.
48. Jin, X.; Yu, X.; Zhu, G.; Zheng, Z., Feng, F.; Zhang, Z. Conditions Optimizing and Application of Laccase-Mediator System (LMS) for the Laccase-Catalyzed Pesticide Degradation. *Sci. Rep.* **2016,** *6* (1), 1–7.
49. Juturu, V.; Wu, J. C. Microbial Xylanases: Engineering, Production and Industrial Applications. *Biotechnol. Adv.* **2012,** *30* (6), 1219–1227.
50. Kadić, A.; Palmqvist, B.; Lidén, G.; Effects of Agitation on Particle-Size Distribution and Enzymatic Hydrolysis of Pretreated Spruce and Giant Reed. *Biotechnol. Biofuels* **2014,** *7* (1), 1–10.

51. Karigar, C. S.; Rao, S. S. Role of Microbial Enzymes in the Bioremediation of Pollutants: A Review. *Enzy. Res.* **2011,** *2011.*
52. Kawabata, Y.; Ono, K.; Gama, Y.; Yoshida, S.; Kobayashi, H.; Kusakabe, I. Purification and Characterization of α-Glucuronidase from Snail Acetone Powder. *Biosci. Biotechnol. Biochem.* **1995,** *59* (6), 1086–1090.
53. Khandeparker, R.; Numan, M. T.; Mukherjee, B.; Satwekar, A.; Bhosle, N. B. Purification and Characterization of α-l-Arabinofuranosidase from *Arthrobacter* sp. MTCC 5214 in Solid-State Fermentation. *Process Biochem.* **2008,** *43* (7), 707–712.
54. Kim, D. W.; Hong, Y. G. Ionic Strength Effect on Adsorption of Cellobiohydrolases I and II on Microcrystalline Cellulose. *Biotechnol. Lett.* **2000,** *22* (16), 1337–1342.
55. Kim, H. W.; Han, S.; Kim, W.; Lim, J.; Kim, D. S. Modulating Wall Shear Stress Gradient via Equilateral Triangular Channel for In Situ Cellular Adhesion Assay. *Biomicrofluidics* **2016,** *10* (5), 054119.
56. Kim, Y.; Ximenes, E.; Mosier, N. S.; Ladisch, M. R. Soluble Inhibitors/Deactivators of Cellulase Enzymes from Lignocellulosic Biomass. *Enzy. Microb. Technol.* **2011,** *48* (4–5), 408–415.
57. Kokol, V.; Doliška, A.; Eichlerová, I.; Baldrian, P.; Nerud, F. Decolorization of Textile Dyes by Whole Cultures of *Ischnoderma Resinosum* and by Purified Laccase and Mn-Peroxidase. *Enzy. Microb. Technol.* **2007,** *40* (7), 1673–1677.
58. Koseki, T.; Okuda, M.; Sudoh, S.; Kizaki, Y.; Iwano, K.; Aramaki, I, et al. Role of Two α-L-Arabinofuranosidases in Arabinoxylan Degradation and Characteristics of the Encoding Genes from Shochu Koji Molds, *Aspergillus kawachii* and *Aspergillus awamori*. *J. Biosci. Bioeng.* **2003,** *96* (3), 232–241.
59. Kou, Y.; Xu, J.; Cao, Y.; Lv, X.; Zhao, G.; Chen, G.; Zhang, W. Gluconolactone Induces Cellulase Gene Expression in Cellulolytic Filamentous Fungus *Trichoderma reesei*. *RSC Adv.* **2014,** *4* (68), 36057–36063.
60. Kulkarni, C.; Bejagam, K. K.; Senanayak, S. P.; Narayan, K. S.; Balasubramanian, S.; George, S. J. Dipole-Moment-Driven Cooperative Supramolecular Polymerization. *J. Am. Chem. Soc.* **2015,** *137* (11), 3924–3932.
61. Kumar, V.; Satyanarayana, T. Biochemical and Thermodynamic Characteristics of Thermo-Alkali-Stable Xylanase from a Novel Polyextremophilic *Bacillus halodurans* TSEV1. *Extremophiles.* **2013,** *17* (5), 797–808.
62. Kuwahara, M.; Glenn, J. K.; Morgan, M. A.; Gold, M. H. Separation and Characterization of Two Extracelluar H_2O_2-Dependent Oxidases from Ligninolytic Cultures of *Phanerochaete Chrysosporium*. *FEBS Lett.* **1984,** *169* (2), 247–250.
63. Le Costaouëc, T.; Pakarinen, A.; Várnai, A.; Puranen, T.; Viikari, L. The Role of Carbohydrate Binding Module (CBM) at High Substrate Consistency: Comparison of *Trichoderma reesei* and *Thermoascus aurantiacus* Cel7A (CBHI) and Cel5A (EGII). *Bioresour. Technol.* **2013,** *143*, 196–203.
64. Lehninger, A. L.; Nelson, D. L.; Cox, M. C. Cell Biochemistry and Function. *Lehninger: Principles of Biochemistry*; 4th Ed., Vol. 23; Nelson, D. L., Cox, M. C., Eds.; W. H. Freeman & Co.: New York, NY, USA, 2004; p 1119.
65. Li, Y.; Wang, J.; Liu, X.; Zhang, S. Towards a Molecular Understanding of Cellulose Dissolution in Ionic Liquids: Anion/Cation Effect, Synergistic Mechanism and Physicochemical Aspects. *Chem. Sci.* **2018,** *9* (17), 4027–4043.
66. Li, Y.; Wilson, D. B. Chitin Binding by *Thermobifida fusca* Cellulase Catalytic Domains. *Biotechnol. Bioeng.* **2008,** *100* (4), 644–652.

67. Lindman, B.; Medronho, B.; Alves, L.; Costa, C.; Edlund, H.; Norgren, M. The Relevance of Structural Features of Cellulose and Its Interactions to Dissolution, Regeneration, Gelation and Plasticization Phenomena. *Phys. Chem. Chem. Phys.* **2017,** *19* (35), 23704–23718.
68. Lippok, S.; Radtke, M.; Obser, T.; Kleemeier, L.; Schneppenheim, R.; Budde, U, et al. Shear-Induced Unfolding and Enzymatic Cleavage of Full-Length VWF Multimers. *Biophys. J.* **2016,** *110* (3), 545–554.
69. Lou, H., Zhu, J. Y.; Lan, T. Q.; Lai, H.; Qiu, X. pH-Induced Lignin Surface Modification to Reduce Nonspecific Cellulase Binding and Enhance Enzymatic Saccharification of Lignocelluloses. *ChemSusChem* **2013,** *6*, 191–927.
70. Lu, C.; Cao, L.; Liu, R.; Lei, Y.; Ding, G. Effect of Common Metal Ions on the Rate of Degradation of 4-Nitrophenol by a Laccase-Cu^{2+} Synergistic System. *J. Environ. Manag.* **2012,** *113*, 1–6.
71. Lundell, S. J. *Quantum Mechanical Studies of N–H…N Hydrogen Bonding in Acetamide Derivatives and Amino Acids*. Master's Degree Thesis; Utah State University, Logan, USA, 2018; pp 129.
72. Maalej, I.; Belhaj, I.; Masmoudi, N. F.; Belghith, H. Highly Thermostable Xylanase of the Thermophilic Fungus *Talaromyces thermophilus*: Purification and Characterization. *Appl. Biochem. Biotechnol.* **2009,** *158* (1), 200–212.
73. Manavalan, T.; Manavalan, A.; Heese, K. Characterization of Lignocellulolytic Enzymes from White-Rot Fungi. *Curr. Microbiol.* **2015,** *70* (4), 485–498.
74. Margolles A, de los Reyes-Gavilan CG. Purification and Functional Characterization of a Novel α-l-Arabinofuranosidase from *Bifidobacterium longum* B667. *Appl. Environ. Microbiol.* **2003,** *69* (9), 5096–103.
75. Martínez, Á. T.; Speranza, M.; Ruiz-Dueñas, F. J.; Ferreira, P.; Camarero, S.; Guillén, F.; Andrade, J. C. D. R. Biodegradation of Lignocellulosics: Microbial, Chemical, and Enzymatic Aspects of the Fungal Attack of Lignin. *Int. Microbiol.* **2005,** *8*, 195–204.
76. Mate, D. M.; Alcalde, M. Laccase: A Multi-Purpose Biocatalyst at the Forefront of Biotechnology. *Microb. Biotechnol.* **2017,** *10* (6), 1457–1467.
77. Matera, I.; Gullotto, A.; Tilli, S.; Ferraroni, M.; Scozzafava, A.; Briganti, F. Crystal Structure of the Blue Multicopper Oxidase From the White-Rot Fungus *Trametes Trogii* Complexed With p-Toluate. *Inorganica Chimica Acta* **2008,** *361* (14–15), 4129–4137.
78. Medve, J.; Ståhlberg, J.; Tjerneld, F. Adsorption and Synergism of Cellobiohydrolase I and II of *Trichoderma reesei* During Hydrolysis of Microcrystalline Cellulose. *Biotechnol. Bioeng.* **1994,** *44* (9), 1064–1073.
79. Mejia-Castillo, T.; Hidalgo-Lara, M. E.; Brieba, L. G.; Ortega-Lopez, J. Purification, Characterization and Modular Organization of a Cellulose-Binding Protein, CBP105, a Processive β-1, 4-Endoglucanase from *Cellulomonas flavigena*. *Biotechnol. Lett.* **2008,** *30* (4), 681–687.
80. Morozova, O. V.; Shumakovich, G. P., Shleev, S. V.; Yaropolov, Y. I. Laccase-Mediator Systems and Their Applications: A Review. *Appl. Biochem. Microbiol.* **2007,** *43* (5), 523–535.
81. Mtui, G. Y. S. Lignocellulolytic Enzymes from Tropical Fungi: Types, Substrates and Applications. *Sci. Res. Essays* **2012,** *7*, 1544–1555.
82. Nakagame, S.; Chandra, R. P.; Kadla, J. F.; Saddler, J. N. The Isolation, Characterization and Effect of Lignin Isolated from Steam Pretreated Douglas-Fir on the Enzymatic Hydrolysis of Cellulose. *Bioresour. Technol.* **2011,** *102* (6), 4507–4517.

83. Niladevi, K. N. Ligninolytic Enzymes. In *Biotechnology for Agro-Industrial Residues Utilisation*; Springer: Dordrecht, 2009; pp 397–414.
84. Nitsos, C. K.; Lazaridis, P. A.; Mach-Aigner, A.; Matis, K. A.; Triantafyllidis, K. S. Enhancing Lignocellulosic Biomass Hydrolysis by Hydrothermal Pretreatment, Extraction of Surface Lignin, Wet Milling and Production of Cellulolytic Enzymes. *ChemSusChem* **2019,** *12* (6), 1179–1195.
85. Numan, M. T.; Bhosle, N. B. α-L-Arabinofuranosidases: The Potential Applications in Biotechnology. *J. Ind. Microbiol. Biotechnol.* **2006,** *33* (4), 247–260.
86. O'Sullivan, A. C. Cellulose: The Structure Slowly Unravels. *Cellulose* **1997,** *4*:173–207.
87. Ooshima, H.; Sakata, M.; Harano, Y. Adsorption of Cellulase from *Trichoderma viride* on Cellulose. *Biotechnol. Bioeng.* **1983,** *25* (12), 3103–3114.
88. Oviedo, C.; Rodríguez, J. EDTA: The Chelating Agent Under Environmental Scrutiny. *Quimica Nova* **2003,** *26*, 901–905.
89. Ozer, A.; Sal, F. A.; Belduz, A. O.; Kirci, H.; Canakci, S. Use of Feruloyl Esterase as Laccase-Mediator System in Paper Bleaching. *Appl. Biochem. Biotechnol.* **2019,** *190*, 1–11.
90. Paës, G.; Berrin, J. G.; Beaugrand, J. GH11 Xylanases: Structure/Function/Properties Relationships and Applications. *Biotechnol. Adv.* **2012,** *30* (3), 564–592.
91. Paliwal, R.; Rawat, A. P.; Rawat, M.; Rai, J. P. N. Bioligninolysis: Recent Updates for Biotechnological Solution. *Appl. Biochem. Biotechnol.* **2012,** *167* (7), 1865–1889.
92. Palmqvist, B.; Wiman, M.; Lidén, G. Effect of Mixing on Enzymatic Hydrolysis of Steam-Pretreated Spruce: A Quantitative Analysis of Conversion and Power Consumption. *Biotechnol. Biofuels* **2011,** *4* (1), 1–8.
93. Pankaj, K. C.; Shashi, L. B.; Manisha, S.; Sunil, K. S.; Rama, S. S. Y.; Sudha, Y. Fungal Laccases and Their Biotechnological Significances in the Current Perspective: A Review. *Curr. Org. Chem.* **2015,** *19* (19), 1916–1934.
94. Pareek, N.; Gillgren, T.; Jönsson, L. J. Adsorption of Proteins Involved in Hydrolysis of Lignocellulose on Lignins and Hemicelluloses. *Bioresour. Technol.* **2013,** *148*, 70–77.
95. Pauly, M.; Keegstra, K. Cell-Wall Carbohydrates and Their Modification as a Resource for Biofuels. *Plant J.* **2008,** *54* (4), 559–568.
96. Pérez, J.; Muñoz-Dorado, J.; de la Rubia, T.; Martínez. J. Biodegradation and Biological Treatments of Cellulose, Hemicellulose. *Int. J. Microbiol.* **2002,** *5*, 53–63.
97. Piontek, K.; Antorini, M.; Choinowski, T. Crystal Structure of a Laccase from the Fungus Trametes Versicolor at 1.90-Å Resolution Containing a Full Complement of Coppers. *J. Biol. Chem.* **2002,** *277* (40), 37663–37669.
98. Piontek, K.; Glumoff, T.; Winterhalter, K. Low pH Crystal Structure of Glycosylated Lignin Peroxidase from *Phanerochaete Chrysosporium* at 2.5 Å Resolution. *FEBS Lett.* **1993,** *315* (2), 119–124.
99. Piontek, K.; Smith, A. T.; Blodig, W. Lignin Peroxidase Structure and Function. *Biochem. Soc. Trans.* **2001,** *29* (2), 111–116.
100. Plácido, J.; Capareda, S. Ligninolytic Enzymes: A Biotechnological Alternative for Bioethanol Production. *Bioresour. Bioprocess.* **2015,** *2* (1), 1–12.
101. Pothiraj, C.; Kanmani, P.; Balaji, P. Bioconversion of Lignocellulose Materials. *Mycobiology* **2006,** *34* (4), 159–165.
102. Prakash, D.; Nawani, N. N.; Prakash, M.; Bodas, M.; Mandal, A.; Khetmalas, M, Kapadnis, B. Actinomycetes: A Repertory of Green Catalysts with a Potential Revenue Resource. *BioMed Res. Int.* **2013,** *2013*, 16–28.

103. Raasch, C.; Armbrecht, M.; Streit, W.; Höcker, B.; Sträter, N.; Liebl, W. Identification of Residues Important for NAD+ Binding by the *Thermotoga maritima* α-Glucosidase AglA, a Member of Glycoside Hydrolase Family 4. *FEBS Lett.* **2002,** *517* (1–3), 267–271.
104. Rahikainen, J. L.; Evans, J. D.; Mikander, S.; Kalliola, A.; Puranen, T.; Tamminen T.; Marjamaa, K.; Kruus, K. Cellulase–Lignin Interactions—The Role of Carbohydrate-Binding Module and pH in Non-Productive Binding. *Enzy. Microb. Technol.* **2013,** *53* (5), 315–321.
105. Raweesri, P.; Riangrungrojana, P.; Pinphanichakarn, P. α-L-Arabinofuranosidase from *Streptomyces* sp. PC22: Purification, Characterization and Its Synergistic Action with Xylanolytic Enzymes in the Degradation of Xylan and Agricultural Residues. *Bioresour. Technol.* **2008,** *99* (18), 8981–8986.
106. Reddy, G. V. B.; Sridhar, M., Gold, M. H. Cleavage of Nonphenolic β-1 Diarylpropane Lignin Model Dimers by Manganese Peroxidase from *Phanerochaete Chrysosporium*: Evidence for a Hydrogen Abstraction Mechanism. *Eur. J. Biochem.* **2003,** *270* (2), 284–292.
107. Reza, M.; Bertinetto, C.; Kesari, K. K.; Engelhardt Peter.; Ruokolainen, J.; Vuorinen, T. Cellulose Elementary Fibril Orientation in the Spruce S1–2 Transition Layer. *Sci. Rep.* **2019,** *9*, 3869.
108. Riva, S. Laccases: Blue Enzymes for Green Chemistry. *TRENDS Biotechnol.* **2006,** *24* (5), 219–226.
109. Robert P, Nicolas A, Aranda-Espinoza S, Bongrand P, Limozin L. Minimal Encounter Time and Separation Determine Ligand-Receptor Binding in Cell Adhesion. *Biophys. J.* **2011,** *100* (11), 2642–2651.
110. Roy, N.; Timell, T. E. The Acid Hydrolysis of Glycosides: IX. Hydrolysis of Two Aldotriouronic Acids Derived from a (4-O-Methylglucurono) Xylan. *Carbohydr. Res.* **1968,** *6* (4), 482–487.
111. Saadeddin, A. The Complexities of Hydrolytic Enzymes from the Termite Digestive System. *Crit. Rev. Biotechnol.* **2014,** *34* (2), 115–122.
112. Saini, A.; Aggarwal, N. K.; Sharma, A.; Yadav, A. Actinomycetes: A Source of Lignocellulolytic Enzymes. *Enzy. Res.* **2015,** *2015*, 69–78.
113. Schülein, M. Protein Engineering of Cellulases. *Biochimica et Biophysica Acta (BBA)—Protein Structure and Molecular Enzymology*, **2000,** *1543* (2), 239–252.
114. Scott, E. E.; Wolf, C. R.; Otyepka, M.; Humphreys, S. C.; Reed. J. R.; Henderson, C. J.; McLaughlin, L. A.; Paloncýová, M.; Navrátilová, V.; Berka, K.; Anzenbacher, P. The Role of Protein-Protein and Protein-Membrane Interactions on P450 Function. *Drug Metab. Dispos.* **2016,** *44* (4), 576–590.
115. Sethi, B.; Mishra, S.; Bisaria, V. S. Adsorption Characteristics of Cellulases from a Constitutive Mutant of *Trichoderma reesei*. *J. Ferment. Bioeng.* **1998,** *86* (2), 233–235.
116. Shallom, D.; Shoham, Y. Microbial Hemicellulases. *Curr. Opin. Microbiol.* **2003,** *6* (3), 219–28.
117. Singh, A.; Kumar, P. K.; Schügerl, K. Adsorption and Reuse of Cellulases During Saccharification of Cellulosic Materials. *J. Biotechnol.* **1991,** *18* (3), 205–12.
118. Singh, R. S.; Singh, T.; Pandey, A. Microbial Enzymes—An Overview. Chapter 1. In *Advances in Enzyme Technology*; Singh, R. S., Singhania, R. R., Pandey, A., Larroche, C., Eds.; Elsevier BV: Amsterdam, The Netherlands, 2019; pp 1–40.

119. Sokan-Adeaga, A. A.; Ana, G. R. E. E.; Sokan-Adeaga, M. A.; Sokan-Adeaga, E. D. Lignocelluloses: An Economical and Ecological Resource for Bio-Ethanol Production-A Review. *Int. J. Nat. Resour. Ecol. Manag.* **2016,** *1* (128), 144.
120. Steiner, W.; Sattler, W.; Esterbauer, H. Adsorption of *Trichoderma reesei* Cellulase on Cellulose: Experimental Data and Their Analysis by Different Equations. *Biotechnol. Bioeng.* **1988,** *32* (7), 853–865.
121. Sukumaran, R. K.; Abraham, A., Mathew, A. K. Enzymes for Bioenergy. In *Bioresources and Bioprocess in Biotechnology*; Abdulhameed, S., **Pradeep**, N. S., **Sugathan**, S., Eds.; Springer International Publishing: Gateway East, Singapore, 2017; pages 3–43.
122. Sundaramoorthy, M.; Kishi, K.; Gold, M. H.; Poulos, T. L. The Crystal Structure of Manganese Peroxidase from *Phanerochaete Chrysosporium* at 2.06-A Resolution. *J. Biol. Chem.* **1994,** *269* (52), 32759–32767.
123. Suresh, C.; Kitaoka, M.; Hayashi, K. A Thermostable Non-Xylanolytic α-Glucuronidase of *Thermotoga maritima* MSB8. *Biosci. Biotechnol. Biochem.* **2003,** *67* (11), 2359–2364.
124. Teeri, T. T. Crystalline Cellulose Degradation: New Insight into the Function of Cellobiohydrolases. *Trends Biotechnol.* **1997,** *15* (5) 160–167.
125. Tejirian, A.; Xu, F. Inhibition of Cellulase-Catalyzed Lignocellulosic Hydrolysis by Iron and Oxidative Metal Ions and Complexes. *Appl. Environ. Microbiol.* **2010,** *76* (23), 7673–7682.
126. Thomas, C. R.; Geer, D. Effects of Shear on Proteins in Solution. *Biotechnol. Lett.* **2011,** *33* (3), 443–456.
127. Tien, M.; Kirk, K. Lignin-Degrading Enzyme from the Hymenomycete *Phanerochaete chrysosporium*. *Science* **1983,** *221*, 661–663.
128. Tokunaga, Y.; Nagata, T.; Suetomi, T.; Oshiro, S.; Kondo, K.; Katahira, M.; Watanabe, T. NMR Analysis on Molecular Interaction of Lignin with Amino Acid Residues of Carbohydrate-Binding Module from *Trichoderma reesei* Cel7A. *Sci. Rep.* **2019,** *9* (1), 1–2.
129. Tu, M.; Chandra, R. P.; Saddler, J. N. Recycling Cellulases During the Hydrolysis of Steam Exploded and Ethanol Pretreated Lodgepole Pine. *Biotechnol. Progress* **2007,** *23* (5), 1130–1137.
130. Tu, M.; Pan, X.; Saddler, J. N. Adsorption of Cellulase on Cellulolytic Enzyme Lignin from Lodgepole Pine. *J. Agric. Food Chem.* **2009,** *57* (17), 7771–7778.
131. Tuncer, M. Characterization of\Beta-Xylosidase and\Alpha-L-Arabinofuranosidase Activities from *Thermomonospora Fusca* BD25. *Turk. J. Biol.* **2000,** 24 (4), 753–768.
132. Tuor, U.; Wariishi, H.; Schoemaker, H. E.; Gold, M. H. Oxidation of Phenolic Arylglycerol. Beta.-Aryl Ether Lignin Model Compounds by Manganese Peroxidase from *Phanerochaete Chrysosporium*: Oxidative Cleavage of an. Alpha.-Carbonyl Model Compound. *Biochemistry* **1992,** *31* (21), 4986–4995.
133. Uchida, H.; Nanri, T.; Kawabata, Y.; Kusakabe, I.; Murakami, K. Purification and Characterization of Intracellular α-Glucuronidase from *Aspergillus niger* 5–16. *Biosci. Biotechnol. Biochem.* **1992,** *56* (10), 1608–1615.
134. Verma, D.; Kawarabayasi, Y.; Miyazaki. K.; Satyanarayana, T. Cloning, Expression and Characteristics of a Novel Alkalistable and Thermostable Xylanase Encoding Gene (Mxyl) Retrieved from Compost-Soil Metagenome. *PLoS One* **2013,** *8* (1), e52459.
135. Vocadlo, D. J.; Davies, G. J. Mechanistic Insights into Glycosidase Chemistry. *Curr. Opin. Chem. Biol.* **2008,** *12* (5), 539–555.

136. Wang, D. W. *Optimizing Enzymatic Preparations of Mechanical Pulp Through the Characterization of New Laccases and Non-Productive Interactions Between Enzymes and Lignin*. Master's Degree Thesis, Department of Chemical Engineering and Applied Chemistry, University of Toronto, Canada, 2010; pp 429.
137. Wariishi, H.; Valli, K.; Gold, M. H. In Vitro Depolymerization of Lignin by Manganese Peroxidase of *Phanerochaete Chrysosporium*. *Biochem. Biophys. Res. Commun.* **1991,** *176* (1), 269–275.
138. Wong, W. S. D. Structure and Action Mechanism of Ligninolytic Enzymes. *Appl. Biochem. Biotechnol.* **2009,** *157*, 174–209.
139. Woo, H. L.; Hazen, T. C.; Simmons, B. A.; DeAngelis, K. M. Enzyme Activities of Aerobic Lignocellulolytic Bacteria Isolated from Wet Tropical Forest Soils. *Syst. Appl. Microbiol.* **2014,** *37*, 60–67.
140. Ximenes, E.; Kim. Y.; Mosier, N.; Dien, B.; Ladisch. M. Deactivation of Cellulases by Phenols. *Enzy. microb. technol.* **2011,** *48* (1), 54–60.
141. Yang M, Zhao W, Singh S, Simmons B, Cheng G. On the Solution Structure of Kraft Lignin in Ethylene Glycol and Its Implication for Nanoparticle Preparation. *Nanoscale Adv.* **2019,** *1* (1), 299–304.
142. Yang, X.; Shi, P.; Huang, H.; Luo, H.; Wang, Y.; Zhang, W.; et al. Two Xylose-Tolerant GH43 Bifunctional β-Xylosidase/α-Arabinosidases and One GH11 Xylanase from *Humicola insolens* and Their Synergy in the Degradation of Xylan. *Food Chem.* **2014,** *148*, 381–387.
143. Yaropolov, A. I.; Skorobogat'Ko, O. V.; Vartanov, S. S.; Varfolomeyev, S. D. Laccase. *Appl. Biochem. Biotechnol.* **1994,** *49* (3), 257–280.
144. Ying, W., Shi, Z.; Yang, H.; Xu, G.; Zheng, Z.; Yang, J. Effect of Alkaline Lignin Modification on Cellulase–Lignin Interactions and Enzymatic Saccharification Yield. *Biotechnol. Biofuels* **2018,** *11* (1), 1–3.
145. Zaide, G.; Shallom, D.; Shulami, S.; Zolotnitsky, G.; Golan, G.; Baasov, T.; et al. Biochemical Characterization and Identification of Catalytic Residues in α-Glucuronidase from *Bacillus stearothermophilus* T-6. *Eur. J. Biochem.* **2001,** *268* (10), 3006–3016.
146. Zhang, Y. H.; Himmel, M. E.; Mielenz, J. R. Outlook for Cellulase Improvement: Screening and Selection Strategies. *Biotechnol. Adv.* **2006,** *24* (*5*), 452–481.
147. Zhang, Y. P. Reviving the Carbohydrate Economy via Multi-Product Lignocellulose Biorefineries. *J. Ind. Microbiol. Biotechnol.* **2008,** *35* (5), 367–375.
148. Zheng, Y.; Zhang, S.; Miao, S.; Su, Z.; Wang, P. Temperature Sensitivity of Cellulase Adsorption on Lignin and Its Impact on Enzymatic Hydrolysis of Lignocellulosic Biomass. *J. Biotechnol.* **2013,** *166* (3), 135–143.
149. Zoghlami, A.; Paës, G. Lignocellulosic Biomass: Understanding Recalcitrance and Predicting Hydrolysis. *Front. Chem.* **2019,** *7*, 874.

CHAPTER 16

INSIGHT INTO CATALASE IMMOBILIZATION, APPLICABILITY, AND INACTIVATION MECHANISMS THROUGH PROCESS ENGINEERING STRATEGIES

ZEBA TABASSUM, ATIF KHURSHID WANI, and MADHURI GIRDHAR

ABSTRACT

Catalase, an antioxidant, is predominant in living organisms dwelling in oxygen-rich environments. From simple microorganisms to plants and animals, it is crucial in bestowing cell protection when exposed to oxidative stress. Hydrogen peroxide, a common subversive byproduct of metabolism, is acted upon by catalase catalyzing its decomposition process forming oxygen and water. Immobilization of catalase on suitable matrix has made it possible to reuse the enzyme to many folds, which has ensured its potential use in industries. Often catalase leads to cell membrane destruction in living system and overaction on hydrogen peroxide, which leads to lipid peroxidation making it unfriendly in both natural living systems and industries. Thus, careful regulation of catalase enzyme through engineering processes including inactivation mechanisms is necessary to understand in order to draw a balance in its use and minimize its harmful action. Since enzymes are active in complex configurations only, the change or unfolding of any domain results in decreasing enzyme activity or rendering them inactive. Thus, the fundamental focus in enzyme inactivation is active folded state, less folded

Enzyme Inactivation in Food Processing: Technologies, Materials, and Applications. Megh R. Goyal, PhD, Junaid Ahmad Malik, PhD & Ravi Pandiselvam, PhD (Eds.)

inactive state, and denatured state. In the case of catalase enzyme, the inactivation is attributed to medium components or the reactor. Pertaining to the hydrophobic interaction of catalase enzyme with nonpolar solvents causing fluctuation in its active structure is the main lead in its inactivation process. This chapter focuses on applications of immobilized catalase in dairy, food, textile, and pharmaceutical industries besides elucidating the role of suitable matrices. This chapter exemplifies information to understand and overcome limitations related to immobilization of catalase besides highlighting the importance of catalase inactivation and process engineering mechanisms to achieve its careful regulation.

16.1 INTRODUCTION TO CATALASE ENZYME

Catalase is a ubiquitous and essential enzyme present in almost all aerobic organisms, from bovines to cucumber, potato, banana peels, and bacteria like *Staphylococcus*, *Micrococus,* and many more. Because it eliminates the active form of oxygen, catalase is among the most important primary antioxidants in the defense system.[115] Catalase, an antioxidant agent that detoxifies hydrogen peroxide, plays a critical function during times of stress. Thénard in 1811 first discovered hydrogen peroxide and suspected a special chemical that degrade hydrogen peroxide.[22] After extensive work, Schonbein in 1863 proved the existence of that special suspected substance. Later, Loew[74] in 1900 named it as "Catalase."[48,74] Catalase is the first and earliest known antioxidant agent,[48] which is also one of the first isolated protein crystal.[37] In 1938, its molecular weight was calculated[112] and the amino acid sequence of the enzyme was worked out in 1969.[104] In 1981, finally 3-D structure of catalase was revealed by the efforts and discoveries of some scientists.[81] There are three different types of catalase: monofunctional catalase, bifunctional catalase peroxidases, and manganese catalases. All of these metalloenzymes developed into two distinct protein families: heme-containing (type I) and nonheme-containing (type II). Type I catalases possess an iron-porphyrin cofactor. Type II catalases, often known as "pseudo-catalases," feature a di-nuclear manganese active site and have lower specific activity than type I catalases that include heme. Molecular weight of catalase is 250 kDa (approx). The 3-D structure of catalase enzyme is tetrameric, consist of four identical tetrahedral subunits of polypeptide chains. Each polypeptide chain is composed of over 500 amino acids. Molecular weight of one subunit is 60 kDa. Each subunit contains a heme group and NADPH in its active center.

The heme group is responsible for reacting catalase with hydrogen peroxide. When hydrogen peroxide enters the active site, it interacts with the amino acid. Catalase decays hydrogen peroxide to oxygen and water. Catalytic power of catalase is very high, in each second it can convert thousands of hydrogen peroxide molecules in water and oxygen.[25] Main purpose of this chapter is to collect information about this enzyme, its several available sources, its limitation when native and its lots of possible applications in various industries. Main objective of this chapter is to analyze the availability and applicability of native and immobilized catalase enzyme.

$$2H_2O_2 \xrightarrow{\text{catalase}} 2H_2O_2$$

16.2 SOURCES OF CATALASE

Catalase is abundantly present in nearly all living organisms, fruits and vegetable, and microorganisms. There are some biologically active catalase enzymes extracted from sources discussed below.

16.2.1 ANIMAL SOURCES

Sumner and Dounce[111] first time extracted catalase from bovine liver and prepared catalase crystal. This procedure proved to be time consuming and failed for some others who tried this procedure. Later Tauber and Petit[117] in 1951 extracted and prepared catalase crystal by a comparative easy method within 48 h. Bovine liver catalase crystals were obtained by fractionation with acetone, dialysis, and re-crystallization from 50 mM phosphate buffer of pH 7.3.[117] Further in the year 1990 Chatterjee et al.[24] extracted catalase from goat liver and immobilized it on various solid supports.

16.2.2 PLANT SOURCES

Catalase has been isolated from cucumber (*Cucumis sativus) by Lamb et al.*[71] *in 1978. In 1986, Sakajo* and Asahi[99] extracted catalase from sweet potato roots. Catalase was extracted from *Zantedeschia aethiopica* leaves, which was stable between 0 and 50 °C and in pH 6–8.[119] Potato tubers (*Solanum tuberosum*) contain a huge amount of catalase enzyme. When extracted and purified, result showed remarkable similarity with bovine liver catalase.[12]

Maximum catalase-specific activity is obtained when it is extracted by chloroform. Purification of catalase was done from the marine red macro-alga (*Porphyra yezoensis).* Liquid nitrogen extraction from frozen and powdered materials proved more efficient than previous methods.[84] Catalase was successfully extracted by Usluoglu and Arabaci[122] from *Malva sylvestris* L. in 2012 and from red poppy in 2016.[8] Banana peels are also a good source of catalase enzyme as per the experiment carried by Farhan[43] in 2015. Some other examples are *Nicotiana tabacum,*[53] cotton,[88] sunflower,[40] and pumpkin.[126]

16.2.3 MICROBIAL SOURCE

Important microbial sources for catalase are *Aspergillus niger,*[68] *Micrococcus luteus,*[97] *and Enterococcus faecalis.*[47] *In 2002,* Amo *purified Pyrobaculum calidifontis* from hot springs, which proved to be a good source of catalase enzyme.[4] *Bacillus* sp. was first isolated from industrial effluents and then catalase was extracted from them.[64] There are some more examples of good sources of *Saccharomyces cerevisiae,*[77] *Caenorhabditis elegans* and *Neurospora crassa,*[102] *Micrococcus* sp., *E. coli,*[6] *Thermoascus aurantiacus,*[42] *Rhizobium radiobacte,*[85] *Staphylococcus aureus,*[82] and many more.

16.3 INDUSTRIAL APPLICATIONS OF CATALASE ENZYME

Sole function of catalase is to break down hydrogen peroxide; still it has diverse use as catalyst in industries like dairy, textile, food, pharmaceutical, cosmetic production, and also in making of some biosensors.

16.3.1 FOOD INDUSTRY

The use of catalase in food items has already been approved by the FDA and there are lots of patents that illustrate the use of catalase in the food industry.[75] Following are some of the examples, that is, hydrogen peroxide is used as bactericide to kill bacteria in food, further catalase is used to remove toxic hydrogen peroxide, from the food products.[10] Catalase is used in the production of variety of foods, for example sourdough, beer, wine, vinegar, etc. In the baking sector, catalase is used to eliminate glucose from egg whites before drying.[23] Researchers specified that by breaking down hydrogen peroxide, catalase provides oxygen for the oxidation of glucose. Catalase and glucose oxidase are frequently combined in the preservation of certain foods.[96] The combination of glucose oxidase and catalase is fruitful in the beverage industry also. It is used to eliminate oxygen from wine before bottling.[91] With other enzymes catalase can maintain sugar level and alcohol contain in wine.[98] Catalase with glucose-oxidase is employed in food products like mayonnaises for antioxidant effects.

16.3.2 DAIRY INDUSTRY

Catalase is widely utilized in the dairy industry to make cheese, yoghurt, as well as other dairy items.[96] In cheese products, catalase enzyme act as a bio-protective enzyme, enhance the flavor, shelf life, and safety of dairy products by converting the hydrogen peroxide to water and molecular oxygen.[110,114] Enzymes that are required for flavor development like Swiss cheese would be destroyed by the high heat of pasteurization. That's why in these types of cases, cold pasteurization is applied. Further in the process, hydrogen peroxide is used to eliminate pathogens and extend shelf life. Since hydrogen peroxide is a toxic substance, catalase eliminates all traces of hydrogen peroxide. The technique is already approved by FDA.[10] One of the most common illnesses in dairy cattle is infectious bovine mastitis. Catalase test for mastitis is used in checking milk quality.[10] This test is used to determine the quality of milk and can distinguish contaminated milk. Infected milk contains catalase, which disintegrates the substrate, hydrogen peroxide, into oxygen and water. The quantity of oxygen in the milk determines its quality.[69]

16.3.3 EGG PROCESSING

Eggs are good source of protein and its yolk contains fat too. Egg also contains significant amounts of iron, vitamins A and D, minerals, etc. Cracking eggs, sorting, combining, stabilizing, blending, pasteurizing, refrigerating or drying, and packaging are all examples of commercial egg processing. With rapid development in the modern biotechnology, there came high efficiency in industrial processes too. If hydrogen peroxide has been utilized in case of pasteurization of egg ingredients, catalase can be utilized to remove residual peroxide.[105]

16.3.4 FOOD PACKAGE

Food wrap or packaging increases the shelf life through inhibition from spoilage, oxidation, degradation, etc. Innovative food packaging attract consumer and help to go with market trends. One of such kind is active packaging, which covers different food items and drugs. In these cases, enzyme can be added directly to the product or can be incorporated into package structure through immobilization.

Nikola[90] in 2015 and Grigoras[51] in 2017, both mentioned Glucose oxidase-catalase-based food wrappers in their research. The enzyme combination Glucose Oxidase and Catalase was also employed to keep the fruit purees from browning. It eliminated 99% of the dissolved oxygen that causes browning. A very difficult step in food industries gets easy by using these enzymes together. Generally, chemical antioxidants are used, but that can be much harmful, whereas these catalase/glucose oxidases are natural antioxidants. They act as a suitable option to protect fruit purees from browning.

16.3.5 PHARMACEUTICALS

Catalase has been used in the pharmaceutical sector for a variety of redox reactions. Catalase is widely used in medicine in treatment of oxidative stress and in treatment of reverting malignant features.[10] Catalase has antioxidant property therefore used in antioxidant drugs. Such drugs are necessary for many diseases and especially for radiotherapy in different pathologies.[114] As mentioned above, active packaging is used in pharmaceuticals to protect drugs and medication through addition of enzyme directly or immobilization of enzymes. Such enzymes include catalase too.

16.3.6 TEXTILE INDUSTRY

Textile industries are one of the world's major businesses, involving the pretreatment, processing, and coloring of various fibers such as cotton, silk, wool, and synthetic fibers. Following treatment, huge volumes of water are utilized, resulting in the generation of a considerable number of effluents. Textile wastewater contains high suspended solids, chemical oxygen demand, color, acid, and other soluble substances.

In textile industry, catalase is used in bleaching process before dying. Natural fabrics (e.g., cotton) are generally bleached with hydrogen peroxide, and then removal of hydrogen peroxide is needed to color the cloth properly, which is done by using catalase. Use of catalase enzyme reduces water usage, time, and energy.[36] Textile industries release effluents that contain lots of toxic chemicals and ROS like hydrogen peroxide, which is a bleaching agent. Catalase not only breaks hydrogen peroxide, but also provides oxygen to the effluent water making it in better state.[5]

16.3.7 BIOREMEDIATION

Rapid urbanization and industrialization associated with environmental alteration, that is, pollution, which need waste management to reduce its intensity. Bioremediation is a waste management technique that uses enzymes like catalase to remediate polluted media like water and soil by changing environmental conditions to encourage the development of microorganisms that breakdown the target pollutants through their microbial activity. It is a cheap and environment friendly method to eliminate or reduce toxicity. Catalase is used in various facets of bioremediation, including as a hydrocarbon degradation indicator, the abolishment of reactive oxygen species such as hydrogen peroxide from industrial effluents, and the provision of oxygen in aerobic bioremediation processes where it is used to produce oxygen by the breakdown of hydrogen peroxide.[10,63] Crude oil is polluted with various toxic and cancerous substances that are hydrocarbon compounds. Catalase can act as an indicator of the hydrocarbon degradation in bioremediation.[41,66]

16.4 INACTIVATION OF CATALASE THROUGH PROCESS ENGINEERING STRATEGIES

Enzymes catalyze reactions and convert substrate into product. Catalase is one such enzyme and has several applications in various industrial bioprocesses. But its shelf life depends on its physical form. Being unstable and in liquid form, it usually has the shortest shelf life, powders and tablets are usually stable for short periods at room temperature, and under refrigeration for longer periods. Storage capacity of free catalase is very less and decreases its efficiency very rapidly. Stabilization of enzymes is one crucial problem in biocatalysis. Dissociation of subunits can inactivate enzymes. Enzymes can lose activity through their conformational alterations like denaturation or dissociation of its subunits, which is a crucial problem associated with process engineering. This mainly is caused because of various external factors. In such situations, it loses its activity and is not able to catalyze substrate into product, causing loss in industrial processes. Inactivation of catalase can happen due the medium components, protein, air bubbles, or any other substance in reactor. For example, NaCl[7] or ions like chloride ion, acetate anion, etc.[22,73] or by the substrate.[51,56,78,79]

Manufacturing enzymes with high specificity, low impurities, stable structure, and function is a very tough task and costly procedure.[31] Although, application of catalase in different industries is going on in full swing, enzymes have limitations in maintaining activities during catalytic processes, such as limitations related to pH sensitivity, temperature sensitivity, product purification, storage, reusability, and activity.[31,93] Activity of catalase is affected by various parameters, time and inhibitor, temperature and pH, storage and operational stability, etc. Additionally, it was observed that activity of free enzyme decreased with time[31] and remains unaffected even after increasing its concentration.[93]

Literature reviews suggest that native enzymes are very sensitive and their activity significantly differs with change in pH. Generally optimum pH of catalase found near pH 7.0. It was also found in various researches that catalases are almost inactive in highly acidic pH. Even some of them do not work in higher alkaline pH also. Like other enzymes, catalases also are sensitive to high temperature. They denatured with little increase in temperature, which is a big drawback for its use in industries. They show very less thermal stability due to disruption of intramolecular and intermolecular bonds.[31] Andrews and Martin[7] mentioned in their research that heat around 52–62 °C can decrease catalase activity, though does not complete inactivate, but activity reduces significantly.

Additional limitation includes, catalase, though being expensive, it lack reusability and is only capable of single-time use, which adds cost in industrial production. "Free" catalase enzymes are free to move everywhere. As a result, during its use in industrial process, it has high risk of either mixing with product, which increases product purification load, further increasing efforts and cost of production or there is a chance of washing out of enzyme with water or other liquid.

16.5 IMMOBILIZATION OF CATALASE

Immobilization means fixing, which restricts the free movement of an enzyme. Fixing of enzyme on the surface of any support and prevent to mix it with product or other things. Immobilization is popular among researchers for studies related to increase the applicability of enzymes in various fields. The first enzymatic immobilization was done in 1916, by adsorbing physically invertase on charcoal.[88] According to researchers, the bound and free enzymes are strikingly comparable, and the immobilized enzyme is stable and may be reused several times.[67] Immobilized enzymes, on the other hand, in their commercial applications have better thermal resilience, reliability, recuperation, renewability, high purity, and high product outputs.[122] It was found that free enzyme has 10 times lesser stability and lower yield than immobilized catalyst.[32] Because dissociation of subunits can neutralize enzymes, one of the key problems in biocatalysis is enzyme stabilization. Immobilization of enzymes can be a good response in this instance, as can the reusability of bio-derived catalysts, without which they would no longer be cost-effective.

16.6 TECHNIQUES OF IMMOBILIZATION

Catalase can be immobilized in physical as well as chemical methods. Physical methods include entrapment, adsorption, and encapsulation, whereas chemical methods are covalent attachment, cross-linking, ionic binding, and affinity binding.

16.6.1 ENTRAPMENT

Entrapment of enzymes can be done within gel by covalent or noncovalent bond. Other than gel, enzymes can also be entrapped into a film, fiber, and capsule.[54] Entrapment is a very easy, fast, and cheap method. Use of less quantity enzyme can result in higher entrapment.[118] Catalase that has been entrapped on an organic matrix such as cellulose acetate beads has a high mechanical stability and may be utilized under a variety of circumstances.[127] Another example is the immobilization of catalase by encasing it in alginate beads.[49]

16.6.2 ADSORPTION

Adsorption is also known as noncovalent binding.[54] Adsorption refers to adhesion of something to a surface. Weak electrostatic forces, such as Van der Waals forces, ionic bonding, and hydrogen bonding, are produced.[38] The method is simple and mild with a vast variety of carriers without any conformational changes. Supports already used in this procedure are like glass beads, charcoal or clay, etc. Water-insoluble transporters are used in the adsorption/carrier-binding technique. Ceramic, alumina, activated carbon, kaolinite, bentonite, porous glass, chitosan, dextran, gelatin, cellulose, and starch are some of the organic and inorganic matrices utilized.[38] This method is done by mixing the enzyme(s) and a support material with each other in adsorption properties, at optimum conditions for a time, after that immobilized enzyme is collected and washed to remove unbound enzymes. Adsorption is easy, cheap, fast method and little or no damage to enzyme/cells. But, sometime desorption happen due to changes in pH, temperature, or other physical factors like high flow and collisions of particles.[52]

16.6.3 ENCAPSULATION

Encapsulation is another type of immobilization technique in which enzymes are encapsulated or enclosed into a spherical capsule-like structure, which is having semipermeable porous membrane. Those capsules are generally made up of nylon or nitrocellulose or polystyrene, etc. A large surface area leads to higher catalytic efficiency. If the enzyme inside the capsule is stable then the method will be effective enough. But occasional inactivation of the enzyme has been reported in this method; despite of presence of high concentration enzyme.[38] Immobilization of catalase on activated carbon has improved stability with Gel encapsulation.[125]

16.6.4 COVALENT ATTACHMENT

Covalent attachment is a type of chemical method for enzyme immobilization. Covalent restricting strategy for immobilization is accomplished by binding of the enzyme with the matrix through the covalent linkage. Attached directly or indirectly with covalent bond to the groups like hydroxyl, amide, amino, carboxyl, etc. on the matrix. Matrix used typically in this techniques are polyacrylamide, porous glass or silica, agarose, Sepharose, methacrylic acid, styrene, etc.[38] When compared with soluble enzymes, catalase immobilized on green coconut fiber by covalent attachment demonstrated greater activity and enhanced heat stability.[31] This way of immobilization can promote higher resistance to harsh conditions, though sometimes it does not work every time and can leads to drastic changes.[38] Bovine liver Catalase was covalently attached onto amino acid-modified chitosan beads, after 100 reuse, it was still left with 40% of its high initial activity.[11] Immobilized catalase by covalent bonding on acetyl cellulose to create hydrogen peroxide sensors, applied in the determining azide in some fruit juices.[106]

16.6.5 CROSS-LINKING

Cross-linking immobilization of enzymes through cross-linking form a number of covalent bonds between enzyme and the matrix using reagents like glutardialdehyde, glutaraldehyde, glyoxal, disocyanates, hexamethylene diisocyanate, and toluene diisocyanate. For example, immobilization of Catalase on iron modifies collagen fiber through cross-linking of glutaraldehide.[26] However cross-linking can cause loss of enzyme or can have limitations caused by diffusion.[38]

16.6.6 IONIC BINDING

Interaction amongst enzyme molecules and a charged matrix is known as ionic binding. The larger the amount of enzyme attached to the matrix, the higher the surface charge density on the matrix. Ionic bond matrix for immobilization of catalase are polysaccharide derivatives, for example, chitosan, alumina, silicates, bentonite, silica gel, and polyvinylalcohol.[38]

16.6.7 AFFINITY BINDING

Affinity binding enzymes are attached to the career using specific ligands in this technique. Generally, ligands are naturally present on the enzyme, if not, then they are attached artificially. This mechanism is linked to minimal changes in enzyme conformation. The immobilized catalyst's stability and catalytic effectiveness are owing to the absence of active site involvement, and the presence of a high density of ligands results in a high immobilization efficiency. There are two option, either the support is bounded to an affinity ligand for target enzyme or the enzyme is joined with something that has affinity for the matrix.[38]

16.7 MATRIX OR SUPPORT FOR IMMOBILIZATION OF CATALASE

Several inorganic materials, using a variety of approaches, have the ability to immobilize enzymes. To be appropriate as solid matrices, such materials must possess the following characteristics: Under operating conditions, full insolubility in water, adequate mechanical strength, and chemical resistance are required.[131] The commercially available matrix or support for enzyme immobilization must be inexpensive, that is, cost-effective. Other requirements include inert/immobile, strong, stable, and the ability to improve enzyme specificity/activity, while reducing product inhibition, nonspecific binding via adsorption, and microbial contamination, among others.

TABLE 16.1 Various Matrix that Are Used to Immobilize Catalase.

Catalase matrix	Method of immobilization	Application and remark	Reference
Acetyl cellulose	Covalent bonding	Used to create H_2O_2 sensors and also biosensor which applied in the determining azide in some fruit juices.	[106]
Alginate Beads	Entrapment	Decomposed most of the substrate hydrogen peroxide in first 10 min. Gained a thermal stability due to having conformational stability by the immobilization.	[49]
Alumina	Covalent bonding and adsorption	Immobilized catalase showed higher stabilities at alkaline pH than the free enzyme. Another study states that specific activity of catalase was reduced to 0.5 fold after immobilization on alumina. Retained only 29% of native activity. Lost 55% activity after storing. As per another paper, catalase immobilized on γ-Al_2O_3 inactivated, unless treated with $FeCl_2$.	[24, 29, 39]
Amino acid modified chitosan beads	Covalent attachment	Maximal activity at pH 7 at 30°C and can retain 40% of its high initial activity after 100 times of reuse.	[11]

TABLE 16.1 *(Continued)*

Catalase matrix	Method of immobilization	Application and remark	Reference
Bentonite clay	Adsorption	Bentonite clay is inexpensive solid support, mud-like substance which derive from volcanic ash, has very high absorbent capacity, good to be used as enzyme carries. High chemical reactivity, low toxicity, and available functional groups allow the easy fixation of the enzyme.	[17, 20]
Bioskin	Adsorption	Half-life time of immobilized catalase is increased around 23.5 days.	[72]
$CaCO_3$ Nanoparticles	Adsorption	Hybrid epoxy-nano $CaCO_3$ matrix for catalase immobilization retains 92% of native activity. Km value for H_2O_2 was higher for immobilized enzyme than free enzyme. Vmax decreased. Thermal and storage stabilities improved. The half-life of enzyme increased five times when stored in 1 mM PPB, pH 7 at 5°C. The enzyme could be reused 30 times without any loss of its activity.	[56]
Calcined Montmorillonite	Adsorption	Immobilization resulted in increase of the enzyme's stability in high-temperature and high alkaline conditions as per needed in textile industries.	[113]
Calcium alginate gel	Adsorption	Max activity at pH7. Loss of activity of more than 6% with pH change. At higher temp., 4% gel achieves a more activity than 2% gel.	[78]
Carbon nanotubes	Adsorption	Adsorbed enzyme activities decreased in the order of Oxidized Single Walled Nano Tube > Single Wall Nano Tube > Multi Wall NanoTube	[129]

TABLE 16.1 *(Continued)*

Catalase matrix	Method of immobilization	Application and remark	Reference
Cellulose acetate beads	Entrapment	Organic matrix, highly resistant to mechanical stability, best results obtained with the beads activated with $Ce(SO_4)_2$. Used in Biosensor to detect H_2O_2 level and in decomposition of hydrogen peroxide in milk samples.	[127]
Cellulose Nanofibrous Membrane	Covalent bonding	Compared to free Catalase the half-life prolonged to around three times more, which indicates excellent storage stability. Retained residual activity around 66% of its initial activity upon 10 times reuse within 1 day.	[45]
PolyVinylAlcohol/ chitosan beads	Chemically crosslinked	PVA modified with APTES and chemically crosslinked beads with chitosan. Compared with control group, the activity of modified group increased about 1400 U/g support. The thermal stability of Catalase also enhanced after immobilization and also it retained more than 90% of its original activity after using it for nine times.	[128]
Chitosan	Covalently Attached	Though immobilized catalase in this case also showed inactivation due to temperature increase, immobilization yield of 95.91. Another study reveals that Km value of immobilized Catalase was higher than that of the free one which may be due to the decreased affinity of enzyme toward the substrate. Optimum temp. of immobilized catalase shifted to 40°C, by increasing 10°C. Storage stability was much better in both studies.	[8, 11, 65]
Chitosan–bentonite complex	Adsorption Crosslinking	Immobilization yield was 95.26. Storage stability showed 70% of its original activity after 60 days.	[65]

TABLE 16.1 *(Continued)*

Catalase matrix	Method of immobilization	Application and remark	Reference
Cibacron blue F3GA attached chitosan beads	Adsorption	Adsorption capacity of Chitosan beads is higher than that of Cibacron blue-Chitosan beads, but Cibacron blue-Chitosan bead-Catalase showed better activity. It was observed that with immobilization thermal, storage and operational stabilities of the enzyme were increased.	[18]
Collagen (Fe (III)) modified	Adsorption Cross-linking	Thermal stability of catalase was improved, kept 30% of relative activity after 5 h incubation at 75°C where free Catalase was completely inactive. As for the storage stability, the immobilized Catalase kept 88% of relative activity after stored at RT for 12 days while the free one was completely inactive under the same conditions. Immobilized Catalase preserved 57% of relative activity after being reused 26 times.	[26]
Cu(II) adsorbed chitosan beads (Glutaraldehyde pretreated)	Adsorption	Operational, thermal and storage stabilities of the Cu-Chitosan-Catalase were higher than that of Chitosan-Catalase.	[19]
Cu(II) chelated PolyVinylAlcohol nanofibrous membrane	Adsorption	High affinity for the matrix. The immobilized catalases showed better resistance to pH and temp. inactivation compared to the free form, and also showed higher thermal and storage stabilities.	[44]
Dextran	Covalent attachment	Got 100% activity retention. Attached carbohydrate stabilizes the catalase against disruption by heat.	[76]
Egg Shell	Adsorption	Simple low cost hydrogen peroxide biosensor fabricated successfully.	[24, 27]

TABLE 16.1 *(Continued)*

Catalase matrix	Method of immobilization	Application and remark	Reference
Egg-shell membrane	Covalent bonding	Results show a remarkable similarity between bound and soluble Catalase. In addition to that bound catalase is more stable and can be reused several times. Low cost hydrogen peroxide biosensor fabricated, through covalently immobilization. Shelf-life is long enough and stable. Successfully applied to the determine H_2O_2 concentration in some commercial samples.	[27]
Electrospun PolyVinylAlcohol/ PA6-Cu(II) nanofibrous membrane	Adsorption	Reusable support, helps in high producibility. Stability of immobilized catalase increased during storing and high temp.	[46]
$FeCl_2$-treated γ-Al_2O_3	Adsorption	Catalase/γ-Al_2O_3 is inactive. When treated with $FeCl_2$, it becomes active support.	[39]
Fullerenes C60	Covalent bonding	PZ crystal sensor created. Long lasting, after 95 days also it had 93% of its initial activity.	[28]
Gelatin	Encapsulation	Retain 65.69% of the native enzyme activity, stronger pH and temperature resistance. After 15 reuse, activated carbon with gelatin immobilized catalase still retained about 80% of its primary activity.	[24, 125]
Glassy Carbon electrode modified gold nanoparticles on a multiwalled carbon nanotubes/ chitosan film	Adsorption	Biosensor made to detect hydrogen peroxide, gave response in less than 10 s.	[87]
Glyoxal Agarose	Covalent bonding	Not good enough to involve all enzyme subunits. Washing makes strong decrease in the enzyme activity. Further cross-linking of previously immobilized enzyme with dextran-aldehyde allows the whole structure to fully stabilize.	[13, 50]

TABLE 16.1 *(Continued)*

Catalase matrix	Method of immobilization	Application and remark	Reference
Glutaraldehyde treated agarose	Crosslinking	All catalase activity disappeared. May be inactivation caused by even small amount of glutaraldehyde.	[13]
Glutaraldehyde treated alumina	Covalent bonding	The immobilized enzyme retained 90% of free catalase activity at 80°C and pH7. The half-life time of the immobilized catalase was increased to 2 h at pH12, and 60°C. Residual water from the washing process of H_2O_2-bleached fabrics was successfully treated with the immobilized catalyst and then reused for dyeing.	[121]
Glutaraldehyde-treated cellulose	Crosslinking	The activity of Catalase/glutaraldehyde–cellulose is notably increased after 6 months of storage, probably due to the formation of multipoint covalent enzyme–support bonds developed with time.	[39]
Glutaraldehyde-treated chitosan	Crosslinking	Catalase/glutaraldehyde–chitosan, a long-term interaction between catalase and NH2 groups of chitosan destroys the fundamental active structure of the enzyme.	[39]
Glutaraldehyde-treated organosilane	Cross-linking	Good capacity of catalase immobilization.	[2]
Gold and Gold nanoparticles	covalent immobilization ligand exchange	Active for 10 days when stored in room temperature.	[33, 92, 95]
Gold nanoparticle nanotubes and polythiophene hybrid	Adsorption	Innovative electrochemical hydrogen peroxide biosensor, have successfully synthesized. Fast response time of 7 s, good detection capacity for having excellent sensitivity. Good repeatability, high stability and acceptable reproducibility.	[86]

TABLE 16.1 *(Continued)*

Catalase matrix	Method of immobilization	Application and remark	Reference
Graphite	Adsorption	Immobilized catalase accelerate the electro oxidation of phenol. Graphite takes part in the electron transfer b/w the electrode and the substrate phenol, and accelerates its oxidation.	[57]
Hollow Silica Nanoparticles	Encapsulation	Increased resistance from the pH and thermally induced denaturation compared to free enzyme and resulted in enhanced enzyme properties. Beneficial in food industry, textile industry, biosensor, biomedical and various.	[109]
Kaolin	Adsorption	80% protein immobilized. At 70°C, immobilized Catalase was active four times higher than free Catalase. Immobilized Catalase was found to be stable in hexane at room temp. up to 12 days. Fully active even after washing.	[103]
Magnesium silicate	Covalently attached	Higher storage stabilities, operational stability of the catalase immobilized via glutaraldehyde than the catalase immobilized via glutaraldehyde + spacer.	[120]
Magnetic multiwalled carbon nanotubes	Adsorption	Have potential to save time and reduce the costs of various industrial enzyme applications (such as dyeing, cosmetics, leather production, paper production, etc.). Maintain structure and activity of the enzyme.	[60]
Mesoporous Silica	Encapsulation	These materials are designable for appropriate enzyme binding capacity and internal diffusivity.	[14, 61]
Mesoporous silica spheres	Encapsulation	Lifetime improved compared to their free forms.	[124]

TABLE 16.1 *(Continued)*

Catalase matrix	Method of immobilization	Application and remark	Reference
Microcapsule	Encapsulation	Mechanical stability is very high, long-term storage and operational stability as desired.	[108, 123]
Multi-walled carbon nanotubes–cysteine modified gold nanoparticles	Adsorption	Biosensor developed to determine hydrogen peroxide.	[55]
Multi-wall carbon nanotubes modified glassy carbon	Adsorption	MWCNT is a good carrier for immobilizing the bio-micromolecules. This may be useful in the biosensors and bio-fuel cells.	[101]
Nickel oxide nano-scale islands	Chemical adsorption by ionic bonding	The embedded catalase in NiO nanoparticles showed excellent catalytic activity toward hydrogen peroxide reduction. Excellent stability, long-term life and short preparations times are advantages	[100]
Nylon 6	Adsorption	Pre-activated nylon 6 proved to be very efficient matrix with high catalytic activity and higher half life. Though thermal stability of immobilized Catalase not good enough, showed inactivation due to temp. increase.	[94]
Polycaprolactone nanofibers	Encapsulation and Entrapment	Promising for biocatalyst immobilization due to their large surface area which is suitable in enzyme attachment, its stability, easy to separate, and fine porous structure.	[15]
Pillared Interlayer Clays	Adsorption	Stable to more alkaline pH., increasing the enzyme's temperature and pH stabilities so that can be used in textile industries.	[113]
Poly(isopropylacrylamide -co-hydroxyethylmethacrylate)	Entrapment	Optimum pH was the same for both free and immobilized enzyme. Operational, thermal and storage stabilities were found to increase, when immobilized. Maximum activity was achieved at 25°C.	[9]

TABLE 16.1 *(Continued)*

Catalase matrix	Method of immobilization	Application and remark	Reference
Polyacrylamide	Entrapment	Two optimum pH observed at 6.8 and 7.7 pH. Retained 57% of native catalase activity. After 15 days of immobilized catalase lost 27% of initial activity,	[24]
Polymeric hydrogels	Entrapped	pH sensitivity removed, but no improvement in the thermal stability.	[62]
Polyvinylpyrrolidone	Adsorption	Immobilization happens but enzyme inactivation observed due to increase in temperature.	[94]
Resin	Covalent bonding	The extent of crosslinking and the pH of the reaction medium have vital roles in the bound Catalase activity.	[70]
Sepiolite	Chemical adsorption by ionic bonds	Storage stability increased. Immobilized catalase lost 41.1% activity. Lost 9.8% of their initial activities after 10 times reuse.	[17]
Silica–alumina membrane	Encapsulation	Similar activity, with improved stability. Reused over 160 times.	[61]
Silica inverse opals	Sol Gel	Acidic and basic and thermal tolerance improved. Great potential for production of gluconic acid, biosensors, enzyme bioreactors, etc.	[130]
Slica sol–gel and cysteine modified gold electrode	Adsorption	H_2O_2 biosensor prepared. Activity increased with the addition of Al^{3+}.	[35]
Silicate	Adsorption	Maximum rate of the enzyme-catalyzed reaction was practically not affected by the immobilization matrix.	[58]
Siliceous mesocellular foam	Adsorptions	Thermal and storage stabilities improved due to the shield of the mesopores of MCF. High Catalase activity achieved.	[83]
Silk	Adsorption	Immobilization happens but enzyme inactivation observed due to increase in temperature.	[94]

TABLE 16.1 *(Continued)*

Catalase matrix	Method of immobilization	Application and remark	Reference
Silver nanoparticles	ligand exchange	Loses activity, but not much. Have much potential for further work.	[95]
Starch based polymers, that is, Starch Cellulose Acetate and Starch Poly Crapo Lactone	Covalent binding and Adsorption	Biodegradable material, sufficient enough as a matrix to stabilize Catalase and achieve higher half-life than free enzyme. Activity recovery of immobilized catalyst in this case is very much dependent on pH, though maintaining stability efficiently.	[30]
Titanate nanotubes	Covalent attachment	Relative activity of immobilized enzyme is about 60% of that of free enzyme. The immobilized catalase showed enhanced storage and recycling stability.	[1, 80]
Walnut	Adsorption	By activating carbon of walnut, and immobilizing them through physical adsorption, make the catalyst more stable.	[3]
Wool	Adsorption	Immobilization happens but enzyme inactivation observed due to increase in temperature.	[94]

16.8 FUTURE PERSPECTIVE

The advancement in enzyme engineering made during the last few years is the amalgam of genetic engineering and process engineering. Several enzyme characteristics are being altered through both genetic engineering and process engineering strategies to improve the yield by modulating enzyme kinetics. There is a growing demand for microorganisms and the enzyme-derivatives including catalases to be cloned and processed for safe and high production. Besides, diversifying the enzyme activities through process engineering the directed evolution approach can be used to engineer enzymes for wide range of applications. This requires the structural information and a comprehensive understanding of recombination mechanisms and protein mutation responses. The strategies that drive toward the random mutagenesis of a particular site of an enzyme, or which can introduce structural or sequence changes thereby paving way to accomplish more advanced enzyme engineering goals, need to be explored in detail. The systematic studies are necessary for process engineering to inactivate harmful or lethal enzymes with minimal impact on its positive aspects.

16.9 SUMMARY

After reviewing papers related to catalase enzyme immobilization, it can be concluded that catalase in most cases immobilized through adsorption and covalent binding. Whereas immobilization methods like crosslinking, entrapping, encapsulation, etc. are comparatively less in use. Nontoxic, biocompatible, resistant to microbial assault, readily accessible, structurally stiff, durable throughout exploitation, hydrophilic, with strong binding capacity, large surface area, and safe to use are all desirable characteristics for a catalase immobilization matrix. Catalase's industrial usage is improved by immobilization. Immobilized catalase enzyme on a good matrix increases thermal stability, whereas free enzymes easily denatures with little increasing of temp. Immobilized catalase proved to have more storage stability than free enzyme. In some cases, immobilized enzyme showed more activity than free enzyme. With some good matrix immobilized catalase showed very high reusability compared to native enzyme. Immobilization can overcome problems related to high production cost by increasing reusability of immobilized enzymes. Dissociation of subunits can inactivate the enzymes which affects stability of enzyme. Immobilizing Catalase on a ideal matrix helps to increase stability of catalase. Immobilized catalase is stronger pH and temperature resistance. Catalase enzyme has been studied since almost 100 years and research is still going on. The researchers are intending to find novel materials that are appropriate for immobilizing catalase. Immobilized catalase should be utilized more widely, thus research should be concentrated on overcoming the limits of immobilization methods.

KEYWORDS

- **catalase**
- **inactivation**
- **immobilization**
- **matrix**
- **adsorption**

REFERENCES

1. Ai, Q.; Yang, D.; Li, Y.; Shi, J. Highly Efficient Covalent Immobilization of Catalase on Titanate Nanotubes. *Biochem. Eng. J.* **2014,** *83*, 8–15.
2. Airoldi, C.; MonteiroJunior, O. A. C. Copper Adsorption and Enzyme Immobilization on Organosilane-Glutaraldehyde Hybrids as Support. *Polym. Bull.* **2003,** *50* (1), 61–68.
3. Alkan, S.; Gur, A.; Ertan, M.; Savran, A.; Gur, T.; Genel, Y. Immobilization of Catalase via Adsorption into Natural and Modified Active Carbon Obtained from Walnut in Various Methods. *Afr. J. Biotechnol.* **2009,** *8* (11), 2631–2635.
4. Amo T.; Atomi, H.; Imanaka, T. Unique Presence of a Manganese Catalase in a Hyperthermophilic Archaeon, Pyrobaculum Calidifontis VA1. *J. bacteriol.* **2002,** *184* (12), 3305–3312.
5. Amorim, A. M.; Gasques, M. D. G.; Andreaus, J.; Scharf, M. The Application of Catalase for the Elimination of Hydrogen Peroxide Residues After Bleaching of Cotton Fabrics. *Anais da Academia Brasileira de Ciencias* **2002,** *74* (3), 433–436.
6. Anca, G. G. Catalase Immobilization—A Review. *Biochem. Eng. J.* **2016,** *117*, 1–20.
7. Andrews, G. P.; Martin S. E. Heat Inactivation of Catalase from *Staphylococcus aureus* MF-31. *Appl. Environ. Microbiol.* **1979,** *37* (6), 1180–1185.
8. Arabaci, G.; Usluoglu, A. Immobilization Of Catalase From Red Poppy on Chitosan Beads. *J. Multidiscip. Eng. Sci. Stud. (JMESS)* **2016,** 2(1), 236–240.
9. Arica, M. Y.; Oktem, H.; Öktem, Z.; Tuncel S. Immobilization of Catalase in Poly(Isopropylacrylamide-Co-Hydroxyethylmethacrylate) Thermally Reversible Hydrogels. *Polym. Int.* **1999,** *48* (9), 879–884.
10. Arya S. K.; Kaushal, J.; Mehandia, S.; Singh, G.; Raina, A. Catalase Enzyme: Application in Bioremediation and Food Industry. *Biocatal. Agric. Biotechnol.* **2018,** *16*, 192–199.
11. Basak, E.; Aydemir, T. Immobilization of Catalase on Chitosan and Amino Acid-Modified Chitosan Beads. *Artif. cells Nanomed. Biotechnol.* **2012,** *41* (4),26–275.
12. Beaumont, F.; Jouvec, H. M.; Gagnon, J.; Gaillard, J.; Pelmont, J. Purification and Properties of a Catalase. From Potato Tubers (Solanum tuberosum). *Plant Sci.* **1990,** *72* (1), 19–26.
13. Betancor, L.; Hidalgo, A.; Fernández-Lorente, G.; Mateo, C.; Fernández-Lafuente, R.; Guisan, J. M. Preparation of a Stable Biocatalyst of Bovine Liver Catalase Using Immobilization and Postimmobilization Techniques. *Biotechnol. Progress* **2003,** *19* (3), 763–767.
14. Bolivar, J. M.; Gascon, V.; Marquez-Alvarez, C.; Blanco, R. M.; Nidetzky, B. Oriented Coimmobilization of Oxidase and Catalase on Tailor-Made Ordered Mesoporous Silica. *Langmuir: ACS J. Surf. Coll.* **2017,** *33* (20), 5065–5076.
15. Canbolat, M. F.; Savas H. B.; Gultekin F. Improved Catalytic Activity by Catalase Immobilization Using Gamma-Cyclodextrin and Electrospun PCL Nanofibers. *J. Appl. Polym. Sci.* **2016,** *134* (4), 44404.
16. Castellari, M.; Matricardi, L.; Arfelli, G.; Carpi, G.; Galassi, S. Effects of High Hydrostatic Pressure Processing and of Glucose Oxidase-Catalase Addition on the Color Stability and Sensorial Score of Grape Juice. *Food Sci. Technol. Int.* **2000,** *6* (1). 17–23.
17. Cengiz, S.; Cavas, L.; Yurdakoc, K. Bentonite and Sepiolite as Supporting Media: Immobilization of Catalase. *Appl. Clay Sci.* **2012,** *65–66*, 114–120.

18. Cetinus, S. A.; Oztop, H. N.; Saraydin, D. Immobilization of Catalase onto Chitosan and Cibacron Blue F3GA Attached Chitosan Beads. *Enzy. Microb. Technol.* **2007,** *41* (4), 447–454.
19. Cetinus, S. A.; Sahin, E.; Saraydin, D. Preparation of Cu(II) Adsorbed Chitosan Beads for Catalase Immobilization. *Food Chem.* **2009,** *114* (3), 962–969.
20. Ceylan, H.; Alkan, S.; Arsalan, O. Bentonite-Supported Catalase. *J. Serbian Chem. Soc.* **2005,** *70* (5), 721–726.
21. Chan, C.; Sepunaru, L.; Sokolov, S. V.; Katelhon, E.; Young, N. P.; Compton, R. G. Catalytic Activity of Catalase–Silica Nanoparticle Hybrids: From Ensemble to Individual Entity Activity. *Chem. Sci.* **2016,** *2017* (3), 2303–2308.
22. Chance, B. The Primary or Secondary Compounds of Catalase and Methyl or Ethyl Hydrogen Peroxide. *J. Biol. Chem.* **1949,** *180*, 947–959.
23. Chaplin, M. *Glucose Oxidase and Catalase in the Food Industry* from Enzyme Technology. 2014**.** http://www1.lsbu.ac.uk/water/enztech/god.html (accessed Sept 27, 2021).
24. Chatterjee, U.; Kumar, A.; Sanwal, G. G. Goat Liver Catalase Immobilized on Various Solid Supports. *J. Ferment. Bioeng.* **1990,** *70* (6)**,** 429–430.
25. Chelikani, P.; Fita, I.; Loewen, P. C. Diversity of Structures and Properties Among Catalases. *Cell. Mol. Life Sci.* **2004,** *61* (2), 192–208.
26. Chen, S.; Song, Na.; Liao, X.; Shi, B. Immobilization of Catalase on Fe (III) Modified Collagen Fiber. *Sheng Wu Gong Cheng Xue Bao—Chin. J. Biotechnol.* **2011,** *27* (7), 1076–1081.
27. Choi, M. M. F.; Yiu, T. P. Immobilization of Beef Liver Catalase on Eggshell Membrane for Fabrication of Hydrogen Peroxide Biosensor. *Enzy. Microb. Technol.* **2004,** *34* (1), 41–47.
28. Chuang, C. W.; Luo, L. Y. Detection of Hydrogen Peroxide and Glucose Based on Immobilized C60-Catalase Enzyme. *J. Chin. Chem. Soc.* **2009,** *56* (4), 771–777.
29. Costa, S. A.; Tzanov, T.; Paar A, Gudelj, M.; Gübitz, G. M.; Cavaco-Paulo, A. Immobilization of Catalases from Bacillus SF on Alumina for the Treatment of Textile Bleaching Effluents. *Enzy. Microb. Technol.* **2001,** *28* (9–10), 815–819.
30. Costa, S. A.; Reis, R. L. Immobilisation of Catalase on the Surface of Biodegradable Starch-Based Polymers as a Way to Change Its Surface Characteristics. *J. Mater. Sci. Mater. Med.* **2004,** *15* (4), 335–342.
31. Cristovao, R. O.; Silverio, S. C.; Tavares, A. P. M.; Brigida, A. I S.; Loureiro, J. M.; Boavemtura, R. A. R.; Macedo, E. A.; Coelho, M. A. Z. Green Coconut Fiber: A Novel Carrier for the Immobilization of Commercial Laccase by Covalent Attachment for Textile Dyes Decolourization. *World J. Microbiol. Biotechnol.* **2012,** *28* (9), 2827–38.
32. Cunha, A. G.; Fernández-Lorente, G.; Bevilaqua, J. V.; Destain, J.; Paiva, L. M. C.; Freire, D. M. G.; Fernández-Lafuente, R.; Guisán, J. M. Immobilization of Yarrowia Lipolytica Lipase—A Comparison of Stability of Physical Adsorption and Covalent Attachment Techniques. *Appl. Biochem. Biotechnol.* **2008,** *146* (1–3), 49–56.
33. Czechowska, E.; Ranoszek-Soliwoda, K.; Tomaszewska, E.; Pudlarz, A.; Celichowski G.; Gralak-Zwolenik, D.; Szemraj, J.; Grobelny, J. Comparison of the Antioxidant Activity of Catalase Immobilized on Gold Nanoparticles via Specific and Non-Specific Adsorption. *Coll. Surf. B Biointerf.* **2018,** *171*, 707–714.
34. Datta, S.; Christena, L. R.; Rajaram, Y. R. S. Enzyme Immobilization: An Overview on Techniques and Support Materials. *3 Biotech.* **2013,** *3*, 1–9.

35. Di, J.; Zhang, M.; Yao, K.; Bi, S. Direct Voltammetry of Catalase Immobilized on Silica Sol–Gel and Cysteine Modified Gold Electrode and Its Application. *Biosens. Bioelectron.* **2006,** *22* (2), 247–252.
36. Doshi, R.; Shelke, V. Enzyme in Textile Industry—An Environment Friendly. *Indian J. Fibre Text. Res.* **2001,** *26*, 202–205.
37. Dounce, A. L.; Sumner, J. B. Crystalline Catalase. *Science* **1937,** *85* (2206), 366–367.
38. Dwevedi, A. *Enzyme Immobilization: Advances in Industry, Agriculture, Medicine, and the Environment*; 1st Ed.; Springer International Publishing, 2016; p 132.
39. Eberhardt, A.; Pedroni, V.; Volpe, M.; Ferreira M. L. Immobilization of Catalase from Aspergillus Niger on Inorganic and Biopolymeric Supports for H_2O_2 Decomposition. *Appl. Catal. B Environ.* **2004,** *47* (3), 153–163.
40. Eising, R.; Trelease, R. N.; Ni, W. Biogenesis of Catalase in Glyoxysomes and Leaf-Type Peroxisomes of Sunflower Cotyledons. *Arch. Biochem. Biophys.* **1990,** *278* (1), 258–264.
41. Erdogan, E.; Karaca, A. Bioremediation of Crude Oil Polluted Soil. *Asian J. Biotechnol.* **2011,** *3* (3), 206–13.
42. Fang, F.; Li, Y.; Du, G. C.; Zhang, J.; Chen, J. Thermo-Alkali-Stable Catalase from Thermoascus Aurantiacus and Its Potential Use in Textile Bleaching Process. *Sheng wu gong Chang xue bao* **2004,** *20* (3), 423–428.
43. Farhan, L. O. Partial Purification and Characterization of Catalase from Banana Peels. *Baghdad Sci. J.* **2016,** *13* (2), 392–398.
44. Feng, Q.; Xia, X.; Wei, A.; Wang, X.; Wei, Q.; Huo, D. Preparation of Cu(II)-Chelated Poly(Vinyl Alcohol) Nanofibrous Membranes for Catalase Immobilization. *J. Appl. Polym. Sci.* **2011,** *120* (6), 3291–3296.
45. Feng, Q.; Hou, D.; Zhao, Y.; Xu, T.; Menkhaus, T. J.; Fong, H. Electrospun Regenerated Cellulose Nanofibrous Membranes Surface-Grafted with Polymer Chains/Brushes via the Atom Transfer Radical Polymerization Method for Catalase Immobilization. *Appl. Mater. Interf.* **2014,** *6* (23), 20958–20967.
46. Feng, Q.; Zhao, Y.; Wei, A.; Li, C.; Wei, Q.; Fong, H. Immobilization of Catalase on Electrospun PVA/PA6-Cu(II) Nanofibrous Membrane for the Development of Efficient and Reusable Enzyme Membrane Reactor. *Environ. Sci. Technol.* **2014,** *48* (17), 10390–10397.
47. Frankenberg, L.; Brugna, M.; Hederstedt, L. *Enterococcus faecalis* Heme-Dependent Catalase. *J. Bacteriol.* **2002,** *184* (22), 6351–6356.
48. Glorieux, C.; Calderon, P. B. Catalase, a Remarkable Enzyme: Targeting the Oldest Antioxidant Enzyme to Find a New Cancer Treatment Approach. *Biol. Chem.* **2017,** *398* (10), 1095–1108.
49. Görenek, G.; Akyilmaz, E.; Dinçkaya, E. Immobilization of Catalase by Entrapping in Alginate Beads and Catalase Biosensor Preparation for the Determination of Hydrogen Peroxide Decomposition. *Artif. Cells Blood Substit. Biotechnol.* **2004,** *32* (3), 453–461.
50. Grazu,V; Betancor, L.; Montes, T.; Lopez-Gallego, F.; Guisan, J. M.,; Fernandez-Lafuente, R. Glyoxylagarose as a New Chromatographic Matrix. *Enzy. Microb. Technol.* **2006,** *38* (7), 960–966.
51. Grigoras, A. G. Ctalase Immobilisation—A Review. *Biochem. Eng. J.* **2017,** *117* (B), 1–112.
52. Hassan, M. E., Tamer, T. M.; Omer, A, M. Methods of Enzyme Immobilization. *Int. J. Curr. Pharm. Rev. Res.* **2016,** *7* (6), 385–392.

53. Havir, E. A.; McaHale, N. A. Biochemical and Developmental Characterization of Multiple forms of Catalase in Tobacco Leaves. *Plant Phusiol.* **1987,** *84* (2), 450–455.
54. Homaei, A. Ravishankar, R. V.; Eds. *Advances in Food Biotechnology: Enzyme Immobilization and Its Application in the Food Industry*; John Wiley & Sons Ltd, 2015; pp 145–164.
55. Hong, J.; Yang, W.; Zhao, Y.; Xiao, B.; Gao, Y.; Yang, T.; Ghourchian, H.; Moosavi-Movahedi, Z.; Sheibani, N.; Li, J.; Moosavi-Movahedi, A. Catalase Immobilized on a Functionalized Multi-Walled Carbon Nanotubes–Gold Nanocomposite as a Highly Sensitive Bio-Sensing System for Detection of Hydrogen Peroxide. *Electrochimica Acta* **2013,** *89*, 317–325.
56. Hooda, V.; Preety. Immobilization and Kinetics of Catalase on Calcium Carbonate Nanoparticles Attached Epoxy Support. *Appl. Biochem. Biotechnol.* **2014,** *172* (1), 115–130.
57. Horozova, E.; Dimcheva, N.; Jordanova, Z. Electrooxidation of Phenol by Catalase Immobilized on Graphite Electrodes. *Bioelectrochemistry* **2001,** *53* (1), 11–16.
58. Horozova, E.; Dimcheva, N. Study of Catalase Immobilized on a Silicate Matrix for Non-Aqueous Biocatalysis. *Cent. Eur. J. Chem.* **2005,** *3* (2), 279–287.
59. Horst, M. F.; Rueda, E. H.; Ferreira, M. L. Activity of Magnetite-Immobilized Catalase in Hydrogen Peroxide Decomposition. *Enzy. Microb. Technol.* **2006,** *38* (7), 1005–1012.
60. Hossain-Ali, R. P., Nejadhosseinian, M.; Firouzi, M.; Masoum, S. Catalase Immobilization onto Magnetic Multiwalled Carbon Nanotubes: Optimization of Crucial Parameters Using Response Surface Methodology. *R. Soc. Chem.* **2019,** *43* (2), 593–600.
61. Itoh, T.; Ishii, R.; Matsuura, S.; Hamakawa, S.; Hanaoka, T.; Tsunoda, T.; Mizuguchi, J.; Mizukami, F. Catalase Encapsulated in Mesoporous Silica and Its Performance. *Biochem. Eng. J.* **2009,** *44* (2–3), 167–173.
62. Jiang, B.; Zhang, Y. Immobilization of Catalase on Crosslinked Polymeric Hydrogels—Effect of Anion on the Activity of Immobilized Enzyme. *Eur. Polym. J.* **1993,** *29* (9), 1251–1254.
63. Joergensen, R. G.; Schmaedeke, F.; Windhorst, K.; Meyer, B. Biomass and Activity of Microorganisms in a Fuel-Oil Contaminated Soil. *Soil Biol. Biochem.* **1995,** *27* (9), 1137–1143.
64. Kataria, M.; Saini, J.; Singh, M.; Kumar, K. Isolation of Catalase Producing Bacteria, Production of Catalase and Its Application to Degrade Hydrogen Peroxide from Effuelent. *Eur. J. Biotechnol. Biosci.* **2016,** *4* (6), 34–37.
65. Kaushal, J.; Seema; Singh, G.; Arya, S. K. Immobilization of Catalase onto Chitosan and Chitosan–Bentonite Complex: A Comparative Study. *Biotechnol. Rep.* **2018,** *18* (e00258), 1–7.
66. Karigar, C. S.; Rao, S. S. Role of Microbial Enzymes in the Bioremediation of Pollutants: A Review. *Enzy. Res.* **2011,** *2011*(805187), 1–11.
67. Kessi, E.; Arias, J. L. Using Natural Waste Material as a Matrix for the Immobilization of Enzymes: Chicken Eggshell Membrane Powder for β-Galactosidase Immobilization. *Appl. Biochem. Biotechnol.* **2019,** *187* (1), 101–115.
68. Kikuchi-Torii, K.; Hayashi, S.; Nakamoto, H.; Nakamura, S. Properties of Aspergillus Niger Catalase. *J. Biochem.* **1982,** *92* (5), 1449–1456.
69. Kilcawley, K. N.; Wilkinson, M. G.; Fox, P. F. Enzyme-Modified Cheese. *Int. Dairy J.* **1998,** *8* (1), 1–10.

70. Kumar, G. S. V.; Kumar, K. S. Immobilization of Catalase on a Novel Polymer Support, Crosslinked Polystyrene Ethylene Glycol Acrylate Resin: Role of the Macromolecular Matrix on Enzyme Activity. *Appl. Polym. Sci.* **2005,** *97*, 8–19.
71. Lamb, J. E.; Riezman, H.; Becker, W. M.; Leaver, C. J. Regulation of Glyoxysomal Enzymes During Germination of Cucumber: 2. Isolation and Immunological Detection of Isocitrate Lyase and L. *Plant Physiol. 62* (5), 754–760.
72. Legaz, M.; Solas, M. T.; Filho, L. C.; Vicenta, C. Quaternary Structure of Catalase Using a Bioskin-Immobilized Enzyme. *Biotechnol. Tech.* **1998,** *12* (16), 481–484.
73. Litchfield, W. J. Inactivation of Catalase by Chloride. *FEBS Lett.* **1977,** *83* (2) 281–284.
74. Loew, O. A New Enzyme of General Occurrence in Organisms. *Science* **1900,** *11* (279), 701–702.
75. Loncar, N.; Fraaije, M. W. Catalases Ass Biocatalysts in Technical Applications: Current State and Perspective. *Appl. Microbiol. Biotechnol.* **2015,** *99* (8), 3351–3357.
76. Marshall, J. J.; Rabinowitz, M. L. Stabilization of Catalase by Covalent Attachment to Dextran. *Biotechnol. Bioeng.* **1976,** *18* (9), 1325–1329.
77. Mate, M. J.; Zamocky, M.; Nykyri, L. M.; Herzog, C.; Alzari, P. M. Betzel, C.; Koller, F.; Fita I. Structure of Catalase-A from Saccharomyces Cerevisiae. *J. Mol. Biol.* **1999,** *286* (1), 135–149.
78. Milek, J.; Kwiatkowska-Marks, S.; Wójcik, M. Immobilization of Catalase from Aspergillus Niger in Calcium Alginate Gel. *CHEMIK* **2011,** *65* (4), 305–308.
79. Miłek, J. Estimation of the Kinetics Parameters for H_2O_2 Enzymatic Decomposition and for Catalase Deactivation. *Braz. J. Chem. Eng.* **2018,** *35* (3), 995–1004.
80. Muniyappan, S.; Solaiyammal, T.; Sudhakar, K.; Karthigeyan, A.; Murugakoothan, P. Conventional Hydrothermal Synthesis of Titanate Nanotubes: Systematic Discussions on Structural, Optical, Thermal and Morphological Properties. *Modern Electron. Mater.* **2017,** *3* (4), 174–178.
81. Murthy, M. R. N.; ReidIII, T. J.; Sicignano, A.; Tanaka, N.; Rossmann, M. G. Structure of Beef Liver Catalase. *J. Mol. Biol.* **1981,** *152* (2), 465–499.
82. Mustafa, H. S. I. Staphylococcus Aureus can Produce Catalase Enzyme when Adding to Human WBCs as a Source of H_2O_2 Productions in Human Plasma or Serum in the Laboratory. *Open J. Med. Microbiol.* **2014,** *4* (4), 249–251.
83. Na, W., Wei, Q., Sun, H.; Nie, Z. Catalase Immobilized on Siliceous Mesocellular foam with Controlled Window Size. *J. Porous Mater.* **2013,** *20*, 75–79.
84. Nakano, T.; Watanabe, M.; Sato, M.; Takeuchi, M. Characterization of Catalase from the Seaweed Porphyra Yezoensis. *Plant Sci.* **1995,** *104* (2), 127–133.
85. Nakayama, M.; Nakajima-Kambe, T.; Katayama, H.; Higuchi, K.; Kawasaki, Y.; Fuji, R. High Catalase Production by Rhizobium Radiobacter Strain 2–1. *J. Biosci. Bioeng.* **2008,** *106* (6), 554–558.
86. Nandini, S.; Nalini, S.; Sanetuntikul, J.; Shanmugam, S.; Niranjana, P.; Melo, J. S.; Suresh, G. S. Development of a Simple Bioelectrode for the Electrochemical Detection of Hydrogen Peroxide Using Pichia Pastoris Catalase Immobilized on Gold Nanoparticle Nanotubes and Polythiophene Hybrid. *Analyst* **2014,** *139* (22), 5800–5812.
87. Nashar, R. M. Flow injection Catalase Activity Measurement Based on Gold Nanoparticles/Carbon Nanotubes Modified Glassy Carbon Electrode. *Talanta* **2012,** *96*, 161–167.
88. Nelson, J. M.; Griffin, E. G. Adsorption of Invertase. *J. Am. Chem. Soc.*, **1916,** *38* (5), 1109–1115.

89. Ni, W.; Turley, R. B.; Trelease, R. N. Characterization of a cDNA Encoding Cottonseed Catalase. *Biochimica et Biophysica Acta (BBA)—Gene Struct. Expr.* **1990,** *1049* (2), 219–222.
90. Nikola. Catalases Ass Biocatalysts in Technical Applications: Current State and Perspective. *Appl. Microbiol. Biotechnol.* **2015,** *99* (8), 3351–3357.
91. Ough, C. S. Further Investigations with Glucose Oxidase-Catalase Enzyme Systems for Use with Wine. *Am. J. Enol. Viticult.* **1975,** *26*, 30–36.
92. Parker, M. C.; Patel, N.; Davies, M. C.; Roberts, C. J.; Tendler, S. J.; Williams P. M. A Novel Organic Solvent-Based Coupling Method for the Preparation of Covalently Immobilized Proteins on Gold. *Protein Sci.: A Publ. Protein Soc.* **1996,** *5* (11), 2329–2332.
93. Patel, H. H.; Kharat, A. R. Comparative Study of Free Enzyme and Immobilized Enzyme on BSA Digestion. *Int. J. Pharm. Sci. Res.* **2012,** *12* (9), 5015–5023.
94. Pifferi, P. G.; Bonora, V.; Spagna, G.; Tramontini, M. Immobilization of Catalase on Macromolecular Supports Activated with Acid Dyes. *Process Biochem.* **1993,** *28* (1), 29–38.
95. Pudlarz, A. M.; Czechowska, E.; Ranoszek-Soliwoda, K.; Tomaszewska, E.; Celichowski, G.; Grobelni, J.; Szemraj, J. Immobilization of Recombinant Human Catalase on Gold and Silver Nanoparticles. *Appl. Biochem. Biotechnol.* **2018,** *185* (3), 717–735.
96. Qureshi, M. A.; Khan, K. A.; Ahmed, P.; Urpit, S. Enzymes Used in Dairy Industries. *Int. J. Appl. Res.* **2015,** *1* (10), 523–527.
97. Raveendran, S.; Parameswaran, B.; Ummalyma, S. B.; Abraham, A.; Mathew, A. K.; Madhavan, A.; Rebello, S.; Pandey, A. Applications of Microbial Enzymes in Food Industry. *Food Technol. Biotechnol.* **2018,** *56* (1), 16–30.
98. Rocker, J.; Schmitt, M.; Pasch, L.; Ebert, K.; Grossmann, M. The Use of Glucose Oxidase and Catalase for the Enzymatic Reduction of the Potential Ethanol Content in Wine. *Food Chem.* **2016,** *210*, 660–670.
99. Sakajo, S.; Asahi, T. In Vitro Synthesis of Catalase Protein in Sweet Potato Root Microbodies. *Eisevier Sci. Publ. B. V. (Biomed. Div.)* **1986,** *205* (2), 337–340.
100. Salimi, A.; Sharifi, E.; Noorbakhsh, A.; Soltanian, S. Direct Electrochemistry and Electrocatalytic Activity of Catalase Immobilized onto Electrodeposited Nano-Scale Islands of Nickel Oxide. *Biophys. Chem.* **2007,** *125* (2–3), 540–548.
101. Salimi, A.; Noorbakhsh, A.; Ghadermarz, M. Direct Electrochemistry and Electrocatalytic Activity of Catalase Incorporated onto Multiwalled Carbon Nanotubes-Modified Glassy Carbon Electrode. *Anal. Biochem.* **2005,** *344* (1), 16–24.
102. Santoso, P.; Ambarsari, L.; Suryani; Yopi, Y. Purification and Characterization of Catalase from Indigenous Fungi of Neurospora crassa InaCC F226. *Int. J. Adv. Sci. Eng. Inf. Tachnol.* **2016,** *6* (4) 502–507.
103. Savran, A.; Alkan, S.; Demir, H.; Ceylan, H. Application of Natural Kaolin as Support for the Immobilization of Catalase from Bovine Liver. *Asian J. Chem.* **2006,** *18* (1), 413–418.
104. Schroeder, W. A.; Shelton, J. R.; Shelton, J. B.; Robberson, B.; Apell, G. The Amino Acid Sequence of Bovine Liver Catalase: A Preliminary Report. *Arch. Biochem. Biophys.* **1969,** *131* (2), 653–655.

105. Semih O. *Egg Processing with Catalase.* TRIPOD **2019.** http://prof.dr.semih.otles.tripod.com/enzymesused/ensymes/egg_processing_with_catalase.htm (accessed Aug 27, 2021).
106. Sezgintürk, M. K.; Goktug, T.; Dinckaya, E. A Biosensor Based on Catalase for Determination of Highly Toxic Chemical Azide in Fruit Juices. *Biosens. Bioelectron.* **2005,** *21* (4), 684–688.
107. Sharma, O.; Makkar, H. Egg Shell as a Carrier for Enzyme Immobilization. *Biotechnol. Bioeng.* **1983,** *25* (2), 595–597.
108. Shi, J.; Zhang, X.; Zhang, S.; Wang, X.; Jiang, Z. Incorporating Mobile Nanospheres in the Lumen of Hybrid Microcapsules for Enhanced Enzymatic Activity. *ACS Appl. Mat. Interf.* **2013,** *5* (21), 10433–10436.
109. Singh, H. P.; Gupta, N.; Sharma, R. K. Hollow Silica Nanoparticles as Support for Catalase Enzyme Immobilization. *Catal. Lett.* **2013,** *143*, 1304–1311.
110. Singh, R.; Kumar, M.; Mittal, A.; Mehta, P. K. Microbial Enzymes: Industrial Progress in 21st Century. *3 Biotech.* **2016,** *6* (2), 174.
111. Sumner, J. B.; Dounce, A. L. CATALASE. II. *J. Biol. Chem.* **1939,** *127* (2), 439–447.
112. Sumner, J. B.; Gralen, N. The Molecular Weight Of Crystalline Catalase. *Science (New York, N.Y.)* **1938,** *87* (2256), 284–284.
113. Suzuki, M.; Matsumoto, Y.; Sekikawa, T.; Iwahori, K. Adsorption and Enzymatic Activity of Immobilized Catalase on Calcined Montmorillonite and Pillared Interlayer Clays. *Jpn. J. Water Treat. Biol.* **2008,** *44* (3), 161–167.
114. Tehrani, H. S., Moosavi-Movahedi, A. A. Catalase and Its Mysteries. *Progress Biophys. Mol. Biol.* **2018,** *140*, 5–12.
115. Tamara, S. The Searching Of Active Catalase Producers Among The Microscopic Fungi. *Analele Universitatii din Oradea, Fascicula Biologie* **2014,** *XVIII* (2), 164–167.
116. Tarhan, L. Enzymatic Properties of Immobilized Catalase on Protein Coated Supports. *Biomed. Biochim. Acta* **1990,** *49* (5), 307–316.
117. Tauber, H.; Petit, E. L. Convenient Methods for Preparing Crystelline Catalase from Cow Liver. *J. Biol. Chem.* **1951,** *195* (2), 703–706.
118. Trevan, M. D. Enzyme Immobilization by Entrapment. *Methods Mol. Biol.* **1988,** *3*, 491–494.
119. Trindade, H.; Karmali, A.; Pais, M. S. One-Step Purification and Properties of Catalase from Leaves of Zantedeschia Aethiopica. *Biochimie* **1988,** *70* (12), 1759–1764.
120. Tukel, S. S.; Alptekin, O. Immobilization and Kinetics of Catalase Onto Magnesium Silicate. *Process Biochem.* **2004,** *39* (12), 2149–2155.
121. Tzanov, T.; Costa, S. A.; Carneiro, F.; Gübitz, G. M.; Cavaco-Paulo, A. Recycling Oftextile Bleaching Effluents for Dyeing Using Immobilized Catalase. *Biotechnol. Lett.* **2002,** *24* (3), 173–176.
122. Usluoglu, A.; Arabaci, G. Catalytic Properties and Immobilization Studies of Catalase from Malva Sylvestris L. *J. Chem.-Hindawi* **2012,** *2013*, 6.
123. Wang, X.; Jiang, Z,; Shi, J.; Liang, Y.; Zhang, C.; Wu, H. Metal–Organiccoordination-Enabled Layer-by-Layer Self-Assembly to Prepare Hybridmicrocapsules for Efficient Enzyme Immobilization. *ACS Appl. Mater. Interf.* **2012,** *4* (7), 3476–3483.
124. Wang, Y.; Caruso, F. Mesoporous Silica Spheres as Supports for Enzyme Immobilization and Encapsulation. *Chem. Mater.* **2005,** *17* (5), 953–961.

125. Wang, Y.; Guan, Y.; Yang, Y.; Yu, P.; Huang, Y. Enhancing the Stability of Immobilized Catalase on Activated Carbon with Gelatin Encapsulation. *J. Appl. Polym. Sci.* **2013,** *130* (3), 1498–1502.
126. Yamaguchi, J., Nishimura, M.; Akazawa, M. Purification and Characterization of Heme-Containing Low-Activity form of Catalase from Greening Pumpkin Cotyledons. *Eur. J. Biochem.* **1986,** 159 (2), 315–322.
127. Yildiz, H.; Akyilmaz, E.; Dinckaya, E. Catalase Immobilization in Cellulose Acetate Beads and Determination of Its Hydrogen Peroxide Decomposition Level by Using a Catalase Biosensor. *Artif. Cells Blood Substit. Immobil. Biotechnol.* **2009,** *32* (3), 443–452.
128. Zang, J.; Jia, S.; Liu, Y.; Wu, S. A Facile Method to Prepare Chemically Crosslinked and Efficient Polyvinyl Alcohol/Chitosan Beads for Catalase Immobilization. *Catal. Commun.* **2012,** *27*, 73–77.
129. Zhang, C.; Luo, S.; Chen, W. Activity of Catalase Adsorbed to Carbon Nanotubes: Effects of Carbon Nanotube Surface Properties. *Talanta* **2013,** *113*, 142–147.
130. Zhao, B.; Zhou, L.; Ma, L.; He, Y.; Gao, J.; Li, D.; Jiang, Y. Co-Immobilization of Glucose Oxidase and Catalase in Silica Inverse Opals for Glucose Removal from Commercial Isomaltooligosaccharide. *Int. J. Biol. Macromol.* **2018,** *107* (B), 2034–2043.
131. Zucca, P.; Sanjust, E. Inorganic Materials as Supports for Covalent Enzyme Immobilization: Methods and Mechanisms. *Molecules* **2014,** *19* (9), 14139–14194.

PART III

BIONANOMATERIALS FOR INACTIVATION OF ENZYMES

CHAPTER 17

NANOZYMES-BASED BIOSENSORS FOR SUSTAINABLE FOOD ENGINEERING: RECENT RESEARCH AND FUTURE PROSPECTS

VISHNU SHARMA, TARUN KUMAR KUMAWAT, ANJALI PANDIT, VARSHA KUMAWAT, RASHI GARG, AGRIMA BHATT, and MANISH BIYANI

ABSTRACT

Since the beginning of human history, people are consuming processed foods. Natural enzymes serve as important strong biocatalysts in the food processing industry, but they are limited by limitations such as exorbitant prices, unstable topologies, high susceptibility to external circumstances, low maintainability, etc. Natural enzymes are replaced with nanozymes, which are composed of nanomaterials and can catalyze biological processes. Nanozymes are more stable, cheaper, simpler to modify, and more efficient. Nanozymes exhibit excellent strength, are cost-effective, are faster and more reliable, and are intrinsically more productive. The nanozymes' enzyme mimicking strategies emit chromogenic signals that can be used to create inexpensive and simple biosensors. Nanozyme-based biosensors can detect hazardous toxins in food, including metals and pesticide residues, as well as biological pollutants like pathogens and biotoxins, and help prepare safe food. The current chapter details the evolution of nanozymes and some recent advancement in nanozymes-based biosensing, particularly in the detection of food toxins. The chapter also highlights the challenges and prospects of nanozymes to expand the frontiers in the food industry.

Enzyme Inactivation in Food Processing: Technologies, Materials, and Applications. Megh R. Goyal, PhD, Junaid Ahmad Malik, PhD & Ravi Pandiselvam, PhD (Eds.)

17.1 INTRODUCTION

Agriculture is vital to the growing economy, and it is also responsible for ensuring global food security. Agriculture is now confronted with a slew of problems, which include pollution from hazardous environmental contaminants like fertilizers and pesticides, and rising food demand due to a growing global population. As a result of these circumstances, the quality of food is deteriorating.[56] Food quality and safety have been a source of worry for consumers due to the difficulty of detecting particular food contaminants.[75] According to the report of the World Health Organization (2020) various food-related diseases due to microbial contamination, toxins, and other contaminants have affected around 600 million people.[64] Since these contaminants are present in trace amounts various analytical researches are going on to understand the sources of these contaminants and techniques to control them.[62] Among all the techniques, chromatography-based methods, electrophoresis, immunoassays, etc., are currently used to detect food-related contaminants, but due to high cost and time utility, these methods are not widely used.[81] It is essential to develop smart alternatives and more sustainable ways of food production and quality assurance in order to fight the continuing increase of the human population and the reckless depletion of the world's resources, which are both on the rise.[5]

Enzymes have long been recognized as critical components of food science. The enzymes have been discovered in nature thousands of years ago. But, the enzymes in their crystalline state were first isolated in 1926 by James B. Summer from the Jack bean.[80] Enzymes being a natural biocatalyst can be used to perform various biochemical reactions and has high catalytic efficiency, more specificity toward the substrate as well as great biocompatibility.[36] In ancient times also the enzymes were used in food processing, that is, to make fermented products like alcohol, cheese, and all the other foods that involve bacterial enzymes.[96] Since enzymes are more specific in catalyzing single substrate or breaking off small groups of compounds or any particular bonds present between molecules, they are found to be more beneficial than inorganic catalysts.[40] These enzymes are used in many food industries for several reasons, including the creation of various kinds of ingredients and the alteration of food textures (production of high fructose corn syrups, clarification of beverages, baking, meat tenderization, etc.).[87] Despite the fact that natural enzymes are extensively utilized in a variety of areas owing to their substrate selectivity and high catalytic activity, they are not well understood,[30,39] they frequently suffer from inherent flaws involving high preparation and filtration or purification costs, low mechanical stability, catalytic activity sensitivity to environmental conditions, and reusing and recycling difficulties.[2,66]

Nanotechnology is a field or technology that is based on the nanoscale.[33] Aspects of it that have potential uses in the food business include providing tools for pathogen detection, aiding in the transport of drugs and food to the target locations, and transferring bioactive substances to the target sites.[67] At the nanoscale level, food can change the safety levels, nutritional components, efficiency, and also the formation of new products or ingredients.[3,14] Bioprocessing, food analysis, detection, and quantification are possible uses for nanozymes in the food science field.[96] Recently, nanomaterial-based enzymes (also known as nanozymes) have piqued the attention of many researchers. They have distinct benefits such as high robustness and stability, as well as cheap manufacturing costs and ease of scaling up, which are essential as a replacement to natural enzymes and are urgently required.[50,73] Nanozymes-based biosensors can be utilized in different fields (food quality, evaluation of safety, monitoring of pollutants present in the environment).[85] This chapter discusses the nanozymes-based biosensors for sustainable food engineering applications.

17.2 NANOZYMES

Many artificial enzymes that mimic natural enzymes in structure and function have been developed recently. Nanozymes are nanomaterials having properties equal to natural enzymes[57] (Table 17.1). Nanozymes are a family of nanomaterials that may accelerate biological processes in the same manner as enzymes. Nanozymes have been utilized as small enzyme mimics for biosensing applications since 2007.[85] In nanozyme research, significant advances have been achieved, thanks to precise control and remarkable nanomaterial creation.[91] Enzyme-mimicking catalytic nanomaterials (nanozymes) have grown rapidly in recent years.[48] There are many benefits to utilizing nanozymes derived from a variety of nanomaterials in food-based applications, including carbon, metals, metal oxides, and metal-organic frameworks.[64] Nanozymes are enzymes with sizes in the nanometer range.[17] The main findings of colorimetric, electrochemical, fluorescent, and chemiluminescent signals are used to create nanozyme-based detection methods.[37]

Nanozymes have many advantages over biological enzymes, including strong activity, ease of manufacturing, and cheap cost, making them ideal for biochemical analysis[63] (Fig. 17.1). The critical similarity between natural enzymes and synthetic nanozymes is that enzyme-catalyzed reactions are formally comparable to those mediated by surfaces.[89] This hypothesis states that an enzyme binds to a substrate and forms an enzyme–substrate complex. Novel platinum group element nanozymes have developed recently.[31]

TABLE 17.1 Advantage and Disadvantage of Natural Enzymes and Nanozymes.

Enzymes	Advantages	Disadvantages
Natural Enzymes	Biocompatibility	High cost
	Good catalytic activity	Less stable
	Biocatalysis to a wide range	Difficulty in storage
	High substrate selectivity	Tedious separation and purification
Nanozymes	Highly stable	Low substrate selectivity
	Low cost	Ambiguous mechanism
	High catalytic activity	Potential nanotoxicity
	Long term storage	Less range of nanozymes

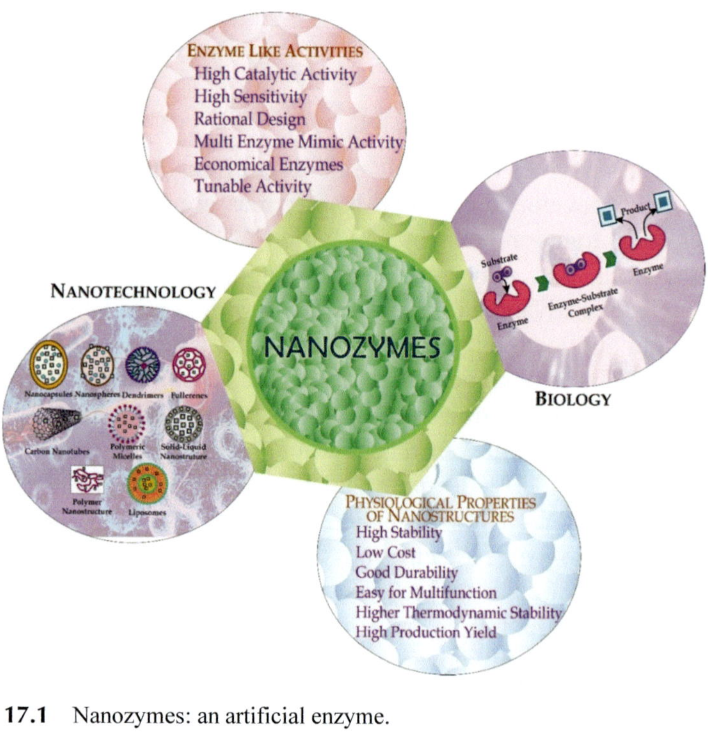

FIGURE 17.1 Nanozymes: an artificial enzyme.

Enzyme-simulating nanomaterials may substitute complicated natural enzymes and catalyze enzymatic processes in many hard-hitting applications, with the apparent benefit of high activity, stability, and efficiency.[94] The remarkable catalytic activity, high atomic utilization, and great selectivity of single-atom nanozymes have sparked interest in the area of nanozyme development.[20] Metal-organic framework nanozymes are gaining popularity as a new kind of nanozyme.[63] Its inorganic–organic hybrid structure may be engineered to possess various mimic enzyme functions.[84]

Huang et al.[38] have classified nanozymes into two families: Oxidoreductases and Hydrolases (Fig. 17.2).

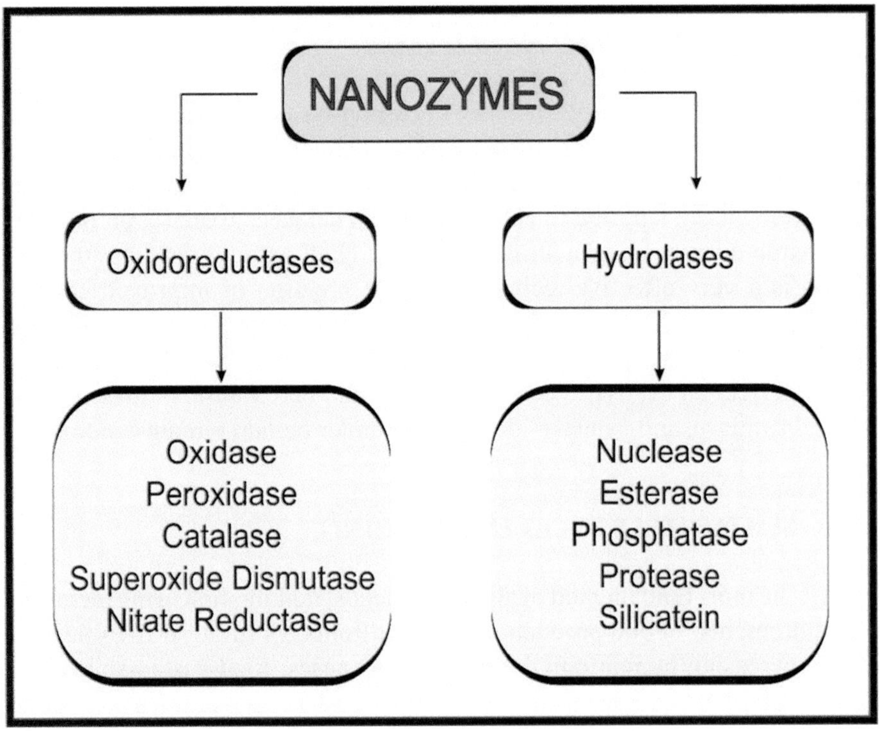

FIGURE 17.2 Classification of nanozymes.

Nanozymes may be classified into two groups depending on their function, namely, antioxidants (remove free radicals) and pro-oxidants (create free radicals).[76] Iron and carbon-based nanozymes have been studied the most since their discovery. This is due to their exceptional biocompatibility and great promise for applications in the biomedical field.

17.2.1 IRON-BASED NANOZYMES

The high ferrous and ferric iron content of iron oxide nanoparticles (Fe_3O_4NPs) gives them peroxidase-like activity. Iron active sites inside the heme group of natural enzyme Horseradish peroxide also have this activity. To enhance the efficacy of the Fe_3O_4 nanozyme, histidine residues were attached to the surface of Fe_3O_4NPs to imitate the iron active center architecture in Horseradish peroxide. The added histidine residues made it much easier for H_2O_2 substrates to find their way toward their catalytic site, increasing substrate binding capacity and catalytic performance.[21]

17.2.2 CARBON-BASED NANOZYMES

The presence of enzyme-like activity in carbon-based nanomaterials has also been reported. Carbon nanotubes, carbon nanodots,[34,92] graphene oxide,[79] fullerene,[35] carbon nitride,[49] and other carbon nanomaterials,[86] for example, have been widely reported to contain inherent catalase, oxidase, peroxidase, superoxide dismutase, or hydrolase activity. Heteroatomic loading to nanocarbon is a very effective technique for the creation of improved carbon-based nanozymes.[44] It was discovered that, after nitrogen was introduced into the carbon structure, the resulting carbon nanozymes not only had a positive effect on catalytic activity, but they had peroxidase, oxidase, superoxide dismutase, and catalase-like activity under certain circumstances.[22]

17.3 SUSTAINABLE FOOD ENGINEERING

Food is the most fundamental of all human needs, ranking first in the hierarchy of requirements.[42] Food production now confronts enormous difficulties due to the increasing human population. In most cases, food sustainability is a joint effort that includes both government and nongovernmental support for technical advancement.[61] Food security requires a reliable food supply. Risks from unpredictable supply chain disruptions impact food markets both locally and internationally.[58]

Approximately one-third of the food produced for human consumption is wasted. These losses affect the whole food value chain because they deplete natural resources including fresh water, energy, and soil.[53] Agriculture, food production, distribution, retail, catering, and consumer handling are all part of this enormous network. Food supply chain management includes agriculture and food logistics.[47] Sustainable food production is critical for human and environmental health, especially given the world's projected population growth of at least 25% by 2050.[72]

It is the goal of an innovative sustainable food engineering system to solve both food quality and food safety problems by quickly evaluating and diagnosing pathogens and chemical contamination in food.[27] More sustainable food production is enabled by nanotechnology, which creates sensors for observing physical, chemical, or organic properties and develops pathogen control technologies to improve food safety while reducing food waste.[70]

17.4 BIOSENSORS

Food safety and quality are two of the most important issues in today's food industry. The increasing complexity of the food chain requires the use of powerful analytical techniques to validate food safety and quality systems. In order to be effective, these techniques must have high sensitivity, selectivity, fast deployment, and automatic screening.[16,55] One of the most promising areas of study in analytical biology nowadays is the creation of novel biosensors. Among all biosensor types, enzyme-based biosensors offer an intriguing property: they exhibit intrinsic inhibition due to the creation of the enzyme-substrate complex.[46] Biosensors are an example of a novel approach to solving long-standing yet critical issues in a quality-conscious society.[12]

Biosensors are used to analyze food in three ways[71]:

(a) Food safety,
(b) Food quality,
(c) Food authenticity.

Many food industries utilize biosensors to monitor quality and safety.[54] Food safety screening is concerned with the identification of potentially harmful pollutants in food products.[15] Biosensors are used to detect a biological response on transducer surfaces. Enzymes, antibodies, cells, and microbes are fundamental detection elements.[11,32] The basic principle on which these biosensors work depends on the physicochemical changes that occur by the release of electrons, ions, gases, and heat by the biological compounds present in the test sample.[46,83] They are subsequently processed, amplified, and converted to digital signals that include information on the concentration of any particular component in the sample.[77]

The demand for fast and reliable detection of pathogenic microbes, insecticides, and toxins in food is growing,[15] and biosensor development has advanced, allowing for quantitative detection and screening to be carried out. The creation of gene-based sensors, which recognize and couple with DNA as the recognition or coupling entity, is the next step in the design of biosensors. Single-sensor platforms are being utilized in the agri-food sector to detect a broad variety of viruses and pollutants.[8,19]

17.5 APPLICATION OF NANOZYMES-BASED BIOSENSOR

Food safety has long been considered a major danger to human health throughout the globe, especially in the developing countries. Numerous methods for detecting biological and chemical pollutants in food have been developed to address food safety concerns.[51] Nanotechnology is transforming food processing and quality control.[6]

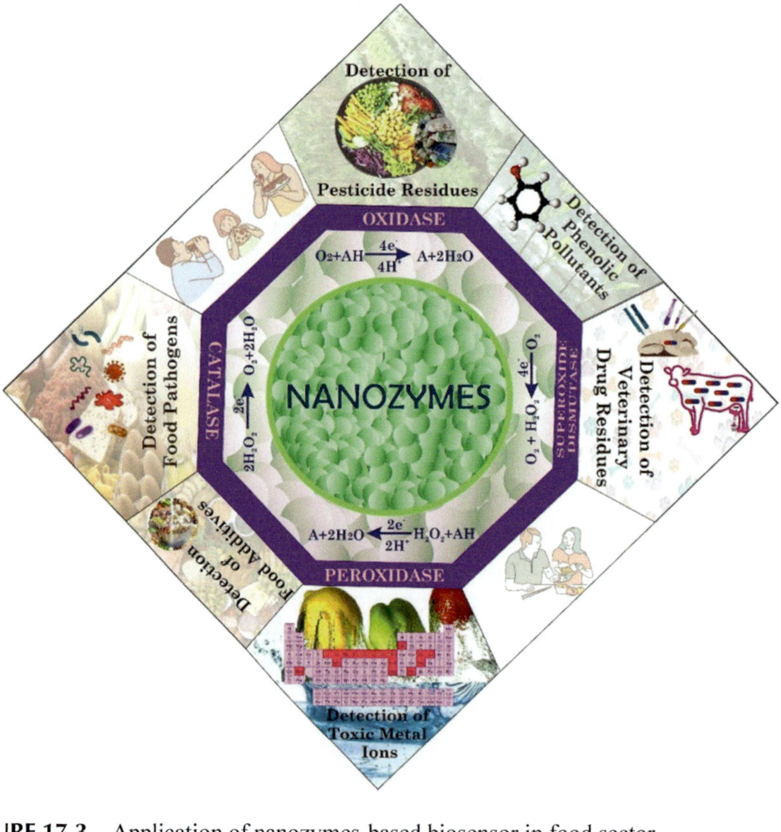

FIGURE 17.3 Application of nanozymes-based biosensor in food sector.

Nanozymes are nanomaterials that are capable of catalyzing biochemical processes in a variety of biosensing applications. A new class of nanozyme-based biosensors can detect ions, pesticide residues, and infections, which may affect food quality and safety[85] (Fig. 17.3).

17.5.1 INTERVENTION IN FOODBORNE PATHOGENS DETECTION

Microorganisms such as pathogens, viruses, and bacteria may be present in food and can impair the nutritional content of the meal partially or completely. Food-borne infections pose a significant threat to the food supply's safety. Pathogens are capable of producing serious diseases that account for 40% of the world's 50 million deaths per year.[45] The vast majority of illnesses caused by pathogenic microorganisms are caused by pathogens that are transferred via food,[7] which is a fact that should not be overlooked. The nutritional value, health, and general quality of life of food are all dependent on its quality. Chemical and microbiological changes that occur throughout the manufacturing and storage processes have an impact on food quality.[25] Food screening should be done at every stage of the supply chain due to the growing incidence of foodborne infections. Expensive analytical equipment and time-consuming sample preparation are used to detect food contaminants.[82] Foodborne pathogen detection often necessitates procedures that are simple in comparison to conventional methods that do not conform with regular food technology analysis.[4]

The food industry's most important criterion is that the food that is sold on the market be completely safe and of very high quality.[45] It has also been shown that microorganisms play a beneficial function in the creation of food, while some of them have been found to be associated with food degradation and the generation of toxins that may cause serious illnesses.[68] Microbial contamination in food is a significant contributor to large food-borne epidemics, which have had a significant effect on human health. Among the most frequent microorganisms known to cause food-borne epidemics are *Escherichia coli*, *Campylobacter*, *Salmonella*, *Shigella*, *Staphylococcus*, and *Bacillus cereus*.[6]

Biosensors can overcome these obstacles by providing accurate, fast, selective, qualitative, and quantitative analyte detection. These technologies may be used in food processing facilities and food safety labs to enable the fast identification of foodborne microorganisms and therefore improve food safety.[1] Incorporating nanomaterials into a biosensor may improve its performance.[65] In a number of fields, including food technology, MNPs-based sensors have showed significant potential.[69]

17.5.2 DETECTION OF PESTICIDES IN FOOD

Pesticides, particularly organophosphate pesticides, are a common pollutant found in fruits and vegetables that may be detrimental to human health even in small amounts.[37] Pesticides, which are extensively employed to control pests and regulate plant development, have presented a risk to human health. Several methods have been suggested to identify and monitor these impacts. Conventional techniques for pesticide residue analysis are not applicable in resource-constrained regions owing to their high cost, can't be used on-site, require technical knowledge, and take a long time to run, complexity, and reliance on costly equipment.[97] Over the past decade, researchers have conducted significant research into the nanozyme activity, which has helped them to achieve their goal.

On a worldwide scale, infectious illnesses are the main cause of mortality in people. Infections with pathogens may decrease food production and jeopardize biodiversity.[60] Due to pesticides' high toxicity and extensive usage, the discovery of sensitive, portable, and easy-to-use sensors is essential (Table 17.2). The nanozymes-based detection technology thus provides an excellent analytical method, since it's user-friendly, cost-effective, and simple to use.[52] A new fourth-generation biosensor for pesticide detection was developed by Snejdarkova et al.[78]

TABLE 17.2 Detection of Pesticides from Nanozyme Sensors.

Name of Pesticide	Food Sample	Nanozyme	Reference
Organophosphorus pesticides	Chinese cabbage	γ-MnOOH nanowires	[37]
Diazinon	Fruit Juice and Water	Fe_3O_4NPs@ZIF-8	[10]
Parathion	Water, pear, cabbage, and rice	AuNP/PtNP	[18]
Methyl paraoxon	*Semen armeniacae amarum, Nelumbinis semen, Dioscoreae rhizoma*	Nanoceria	[90]
Methyl parathion	Pear	Silver NPs	[13]

Pesticide detection optical molecular imprinted polymers sensors are increasingly used in food systems, particularly fluorescence, colorimetric, chemiluminescence, electrochemiluminescence, surface resonance plasm, and Raman surface spectroscopy sensors.[23] Weerathunge et al.[88] demonstrated acetamiprid sensor-regulated reversible Au nanozyme activity. Their 10-min test coupled the AuNP's intrinsic peroxidase activity with the acetamiprid-specific S-18 aptamer's high affinity and specificity. For the detection of organophosphate pesticides, Khairy et al.[43] investigated modified NPs (Nickel Oxide Nanoplatelets) (Parathion). Organophosphorus pesticides and nerve poisons may be quickly detected using Fe_3O_4 (MNP), a butxidase mimic, according to Liang et al.[48] The test comprises MNPs, AChE, and Choline Oxidase.

17.5.3 DETECTION OF HEAVY METALS IN FOOD

Heavy metals may accumulate in biological food chains and pose serious hazards to human health even when present in trace amounts.[9] The importance of food safety as a major global public health issue is being more recognized. Effective detection techniques are critical for ensuring food safety.[94] Many techniques are conducted to evaluate heavy metals at such low quantities, for example, Atomic Absorption Spectrometry and Inductively Coupled Plasma Mass Spectrometry. However, these techniques are time intensive, need sophisticated equipment, and require highly trained individuals. Thus, simple, fast, accurate, and cheap techniques for detecting heavy metals are desired.[41]

In several food science fields, enzymes have been and will continue to play a vital role for decades. The attention of food scientists has been lately drawn by nanozymes that combine nanomaterials with enzyme catalytic activity.[95] The detection technologies in nanotechnology are capable of improving food safety and quality, developing new approaches for fighting fraud, and providing important instruments in our bioterrorism armory.[93] In many applications, nanozymes are considered to be a better option than conventional enzymes.[26] Unique physical/chemical characteristics and improved environmental sensing capacity for different heavy metals have recently been discovered. The enormous potential of nanosensor-based biosensors for heavy metal ions as next generation and located sensing devices for food samples was studied.[74]

Mercury is harmful to human health, and nanobiosensors have shown promising uses for detecting Hg^{2+} ions. Recently, nanozymes, such as AuNPs and PtNPs, have been utilized to quantitatively detect Hg^{2+} ions. It would also be important to create some viable, practicable, and highly sensitive platforms for determining Hg^{2+} levels as low as nmol L^{-1} or pmol L^{-1}.[29] The gold nanozyme-based paper chip (AuNZ-PAD) was created by Han et al.[28] for detection of Hg^{2+}.

17.6 CHALLENGES AND FUTURE PERSPECTIVES OF NANOZYMES IN FOOD ENGINEERING

Enzymes have many applications in food processing but the major drawback that limits the catalytic functions of the enzymes and their utility in the food industries is its protein nature that causes structural configurations. In addition to these various other drawbacks include high cost, unstable structure, extreme sensitivity toward the external conditions, not being reusable, etc. The determination of nanozymes has been able to form a connection between the nanomaterials and the natural enzymes. Although nanozymes offer many benefits over natural enzymes, they are still limited. The first is that nanozymes are less efficient than natural enzymes and other organic catalysts, but that high-efficiency nanozymes may be produced in the future. Second, they sometimes show restricted selectivity to their substrates, which eventually affect the product formation.

Due to the complex structure of nanozymes, it is difficult for them to find their active sites and determining their catalytic activity. Most nanozymes are not capable of recognizing specific macromolecules or organelles that natural enzymes possess, but if the nanozymes based biosensors used are combined with some biomolecules (antibody, natural enzyme, etc.) then their specificity towards the targets can be improved. Nanozymes have risen in popularity with the discovery of enzyme-like activity in some nanomaterials, and are now being investigated in biosensors. Recent advances suggest that using nanozymes for food analysis, detection, and quantification may enhance the already significant effect of enzymes in food bioprocessing.

17.7 SUMMARY

Food is complicated and the quality of the foodstuff is important consideration. The assurance of food safety necessitates the rapid and precise detection and identification of all pollutants, chemicals, and microorganisms that are hazardous to human health. In recent decades, as consumer demand for safe and healthy foodstuffs has increased, a variety of creative, multidisciplinary, and multidimensional ideas have arisen. Nanobiotechnology has made significant strides in the agri-food sector, thanks to the development of highly reliable enzymatic nanobiosensors. Nanozymes, which are synthetic enzymes, have lately emerged as a powerful alternative to natural enzymes. These are specialized kinds of nanomaterials that show enzyme-like properties and are employed as artificial enzymes in nanobiosensors to detect pathogens and other contaminants. Nanobiotechnological interventions in the food processing sector have enabled the detection of food-borne diseases, toxins, pesticides, and adulterants at the early stages of their development. Undoubtedly, there will be an increase in the number of researchers working on nanozymes-based biosensors in the food industry in the coming years.

ACKNOWLEDGMENT

The authors would like to convey gratitude and obligation to the Chairman of the Biyani Group of Colleges, Jaipur, India, for his guidance and support.

KEYWORDS

- **biosensors**
- **food**
- **food processing**
- **nanotechnology**
- **nanozymes**
- **sustainable**

REFERENCES

1. Abdelhaseib, M. U.; Singh, A. K.; Bailey, M.; Singh, M.; El-Khateib, T.; Bhunia, A. K. Fiber Optic and Light Scattering Sensors: Complimentary Approaches to Rapid Detection of *Salmonella enterica* in Food Samples. *Food Control* 2016, *61*, 135–145.
2. Abuchowski, A.; Kazo, G. M.; Verhoest, C. R., Jr., Van Es, T.; Kafkewitz, D.; Nucci, M. L.; Viau, A. T.; Davis, F. F. Cancer Therapy with Chemically Modified Enzymes.

I. Antitumor Properties of Polyethylene Glycol-Asparaginase Conjugates. *Cancer Biochem. Biophys.* 1984, *7* (2), 175–186.

3. Aguilera, J. M. Why Food Microstructure? *J. Food Eng.* 2005, *67* (1–2), 3–11.
4. Ali, A.; Shah, T.; Ullah, R.; Zhou, P.; Guo, M.; Ovais, M.; Tan, Z.; Rui, Y. Review on Recent Progress in Magnetic Nanoparticles: Synthesis, Characterization, and Diverse Applications. *Front. Chem.* 2021a, *9*.
5. Ali, S. S.; Al-Tohamy, R.; Koutra, E.; Moawad, M. S.; Kornaros, M.; Mustafa, A. M.; Mahmoud, Y. A. G.; Badr, A.; Osman, M. E. H.; Elsamahy, T.; Jiao, H.; Sun, J. Nanobiotechnological Advancements in Agriculture and Food Industry: Applications, Nanotoxicity, and Future Perspectives. *Sci. Total Environ.* 2021b, *792*, 148359.
6. Arora, K. Advances in Nano Based Biosensors for Food and Agriculture. In *Nanotechnology, Food Security and Water Treatment. Environmental Chemistry for a Sustainable World*; Gothandam, K. M., Ranjan, S., Dasgupta, N., Ramalingam, C., Lichtfouse, E., Eds.; Springer International Publishing: Basel, Switzerland; 2018; pp 1–52.
7. Arora, K.; Chand, S.; Malhotra, B. D. Recent Developments in Bio-Molecular Electronics Techniques for Food Pathogens. *Anal. Chim. Acta* 2006, *568* (1–2), 259–274.
8. Ayrapetyan, M.; Oliver, J. D. The Viable but Non-Culturable State and Its Relevance in Food Safety. *Curr. Opin. Food Sci.* 2016, *8*, 127–133.
9. Babamiri, B.; Salimi, A.; Hallaj, R. Switchable Electrochemiluminescence Aptasensor Coupled with Resonance Energy Transfer for Selective Attomolar Detection of Hg^{2+} via CdTe@CdS/Dendrimer Probe and Au Nanoparticle Quencher. *Biosens. Bioelectron.* 2018, *102*, 328–335.
10. Bagheri, N.; Khataee, A.; Hassanzadeh, J.; Habibi, B. Sensitive Biosensing of Organophosphate Pesticides Using Enzyme Mimics of Magnetic ZIF-8. *Spectrochimica Acta Part A: Mol. Biomol. Spectrosc.* 2019, *209*, 118–125.
11. Bahadır, E. B.; Sezgintürk, M. K. A Review on Impedimetric Biosensors. *Artif. Cells Nanomed. Biotechnol.* 2016, *44* (1), 248–262.
12. Bahadır, E. B.; Sezginturk, M. K. Applications of Commercial Biosensors in Clinical, Food, Environmental, and Biothreat/Biowarfare Analyses. *Anal. Biochem.* 2015, *478*, 107–120.
13. Bian, Y.; Li, C.; Li, H. Para-Sulfonatocalix[6]Arene-Modified Silver Nanoparticles Electrodeposited on Glassy Carbon Electrode: Preparation and Electrochemical Sensing of Methyl Parathion. *Talanta* 2010, *81* (3), 1028–1033.
14. Blundell, J. E.; Thurlby, P. L. Experimental Manipulations of Eating: Advances in Animal Models for Studying Anorectic Agents. *Pharmacol. Ther.* 1987, *34* (3), 349–401.
15. Bunney, J.; Williamson, S.; Atkin, D.; Jeanneret, M.; Cozzolino, D.; Chapman, J.; Power, A.; Chandra, S. The Use of Electrochemical Biosensors in Food Analysis. *Curr. Res. Nutr. Food Sci. J.* 2017, *5* (3), 183–195.
16. Campuzano, S.; Yáez-Sedeño, P.; Pingarrón, J. Electrochemical Affinity Biosensors in Food Safety. *Chemosensors* 2017, *5* (1), 8.
17. Chatterjee, B.; Das, S. J.; Anand, A.; Sharma, T. K. Nanozymes and Aptamer-Based Biosensing. *Mater. Sci. Energy Technol.* 2020, *3*, 127–135.
18. Chen, G.; Jin, M.; Yan, M.; Cui, X.; Wang, Y.; Zheng, W.; Qin, G.; Zhang, Y.; Li, M.; Liao, Y.; Zhang, X.; Yan, F.; Abd El-Aty, A. M.; Hacımüftüoğlu, A.; Wang, J. Colorimetric Bio-Barcode Immunoassay for Parathion Based on Amplification by Using Platinum Nanoparticles Acting as a Nanozyme. *Microchimica Acta* 2019, *186* (6), 339.

19. Chen, T.; Cheng, G.; Ahmed, S.; Wang, Y.; Wang, X.; Hao, H.; Yuan, Z. New Methodologies in Screening of Antibiotic Residues in Animal-Derived Foods: Biosensors. *Talanta* 2017, *175*, 435–442.
20. Chi, L.; Zhang, Y.; Hua, Y.; Xu, Q.; Lv, M.; Wang, H.; Xie, J.; Yang, S.; Yong, Y. Fe-Based Single-Atom Nanozyme with Superior Peroxidase-Mimicking Activity for Enhanced Ultrasensitive Biosensing. *J. Nanosci. Nanotechnol.* 2021, *21* (12), 6126–6134.
21. Fan, K.; Wang, H.; Xi, J.; Liu, Q.; Meng, X.; Duan, D.; Gao, L.; Yan, X. Optimization of Fe_3O_4 Nanozyme Activity Via Single Amino Acid Modification Mimicking an Enzyme Active Site. *Chem. Commun.* 2017, *53*(2), 424–427.
22. Fan, K.; Xi, J.; Fan, L.; Wang, P.; Zhu, C.; Tang, Y.; Xu, X.; Liang, M.; Jiang, B.; Yan, X.; et al. In Vivo Guiding Nitrogen-Doped Carbon Nanozyme for Tumor Catalytic Therapy. *Nat. Commun.* 2018, *9*(1), 1–11.
23. Fang, L.; Jia, M.; Zhao, H.; Kang, L.; Shi, L.; Zhou, L.; Kong, W. Molecularly Imprinted Polymer-Based Optical Sensors for Pesticides in Foods: Recent Advances and Future Trends. *Trends Food Sci. Technol.* 2021, *116*, 387–404.
24. Gao, L.; Zhuang, J.; Nie, L.; Zhang, J.; Zhang, Y.; Gu, N.; Wang, T.; Feng, J.; Yang, D.; Perrett, S.; et al. Intrinsic Peroxidase-Like Activity of Ferromagnetic Nanoparticles. *Nat. Nanotechnol.* 2007, *2*(9), 577–583.
25. Garcia Lozano, M.; Pena García, Y.; Silva Gonzalez, J. A.; Ochoa Banuelos, C. V.; Luevanos Escareno, M. P.; Balagurusamy, N. Biosensors for Food Quality and Safety Monitoring: Fundamentals and Applications. Chapter 20. In *Enzymes in Food Biotechnology*; Kuddus, M., Ed.; Elsevier Inc.: Amsterdam, The Netherlands, 2019; pp 691–709.
26. Gomaa, E. Z. Nanozymes: A Promising Horizon for Medical and Environmental Applications. *J. Clust. Sci.* 2021, *2*, 1–23.
27. Guillard, V.; Gaucel, S.; Fornaciari, C.; Angellier-Coussy, H.; Buche, P.; Gontard, N. The Next Generation of Sustainable Food Packaging to Preserve Our Environment in a Circular Economy Context. *Front. Nutr.* 2018, *5*, 121.
28. Han, K. N.; Choi, J.-S.; Kwon, J. Gold Nanozyme-Based Paper Chip for Colorimetric Detection of Mercury Ions. *Sci. Rep.* 2017, *7* (1), 2806.
29. Hasan, A.; Nanakali, N. M. Q.; Salihi, A.; Rasti, B.; Sharifi, M.; Attar, F.; Derakhshankhah, H.; Mustafa, I. A.; Abdulqadir, S. Z.; Falahati, M. Nanozyme-Based Sensing Platforms for Detection of Toxic Mercury Ions: An Alternative Approach to Conventional Methods. *Talanta* 2020, *215*, 120939.
30. Haseloff, J.; Gerlach, W. Simple RNA Enzymes with New and Highly Specific Endoribonuclease Activities. *Nature* 1988, *334*, 585–591.
31. He, S.-B.; Yang, L.; Lin, M.-T.; Balasubramanian, P.; Peng, H.-P.; Kuang, Y.; Deng, H.-H.; Chen, W. Platinum Group Element-Based Nanozymes for Biomedical Applications: An Overview. *Biomed. Mater.* 2020, *16*(3), 1–22.
32. Helali, S.; Martelet, C.; Abdelghani, A.; Maaref, M. A.; Jaffrezic-Renault, N. A Disposable Immunomagnetic Electrochemical Sensor Based on Functionalised Magnetic Beads on Gold Surface for the Detection of Atrazine. *Electrochim. Acta* 2006, *51* (24), 5182–5186.
33. Hsu, C. S.; Robinson, P. R. (Eds.) *Introduction to Petroleum Technology*; Springer Handbooks, 2017; pp 83.

34. Hu, Y.; Gao, X.; Zhu, Y.; Muhammad, F.; Tan, S.; Cao, W.; Lin, S.; Jin, Z.; Gao, X.; Wei, H. Nitrogen-Doped Carbon Nanomaterials as Highly Active and Specific Peroxidase Mimics. *Chem. Mater.* 2018, *30*(18), 6431–6439.
35. Hu, Z.; Zhang, C.; Huang, Y.; Sun, S.; Guan, W.; Yao, Y. Photodynamic Anticancer Activities of Water-Soluble C60 Derivatives and Their Biological Consequences in A Hela Cell Line. *Chem. Biol. Interact.* 2012, *195* (1), 86–94.
36. Huang, L.; Sun, D.; Pu, H.; Wei, Q. Development of Nanozymes for Food Quality and Safety Detection: Principles and Recent Applications. *Compr. Rev. Food Sci. Food Saf.* 2019a, *18* (5), 1496–1513.
37. Huang, L.; Sun, D.-W.; Pu, H.; Wei, Q.; Luo, L.; Wang, J. A Colorimetric Paper Sensor Based on the Domino Reaction of Acetylcholinesterase and Degradable γ-MnOOH Nanozyme for Sensitive Detection of Organophosphorus Pesticides. *Sens. Actuators B: Chem.* 2019b, *290*, 573–580.
38. Huang, Y.; Ren, J.; Qu, X. Nanozymes: Classification, Catalytic Mechanisms, Activity Regulation, And Applications. *Chem. Rev.* 2019c, *119*(6), 4357–4412.
39. Hubatsch, I.; Ridderström, M.; Mannervik, B. Human Glutathione Transferase A4–4: An Alpha Class Enzyme with High Catalytic Efficiency in The Conjugation of 4-Hydroxynonenal and Other Genotoxic Products of Lipid Peroxidation. *Biochem. J.* 1998, *330* (1), 175–179.
40. James, J.; Simpson, B. K.; Marshall, M. R. Application of Enzymes in Food Processing. *Critical Rev. Food Sci. Nutr.* 1996, *36* (5), 437–463.
41. Kalyani, N.; Chatterjee, B.; Sharma, T. K. Aptamer Mediated Sensing of Environmental Pollutants Utilizing Peroxidase Mimic Activity of NanoZymes. In *Nanozymes for Environmental Engineering. Environmental Chemistry for a Sustainable World*; Daima, H. K., Lichtfouse, E., Eds.; Springer International Publishing: Basel, Switzerland, 2021; pp 111–143.
42. Kesavan, P. C.; Swaminathan, M. S. Modern Technologies for Sustainable Food and Nutrition Security. *Curr. Sci.* 2018, *115* (10), 1876.
43. Khairy, M.; Ayoub, H. A.; Banks, C. E. Non-Enzymatic Electrochemical Platform for Parathion Pesticide Sensing Based on Nanometer-Sized Nickel Oxide Modified Screen-Printed Electrodes. *Food Chem.* 2018, *255*, 104–111.
44. Kotov, N. Inorganic Nanoparticles as Protein Mimics. *Science* 2010, *330* (6001), 188–189.
45. Kumar, H.; Neelam. Enzyme-Based Electrochemical Biosensors for Food Safety: A Review. *Nanobiosensors Dis. Diag.* 2016, *2016* (5), 29–39.
46. Kurbanoglu, S.; Erkmen, C.; Uslu, B. Frontiers in Electrochemical Enzyme Based Biosensors for Food and Drug Analysis. *TrAC Trends Analy. Chem.* 2020, *124*, 115809.
47. Lazaridesa, H. N. Food Processing Technology in a Sustainable Food Supply Chain. *Procedia Food Sci.* 2011, *1*, 1918–1923.
48. Liang, M.; Yan, X. Nanozymes: From New Concepts, Mechanisms, and Standards to Applications. *Acc. Chem. Res.* 2019, *52*(8), 2190–2200.
49. Lin, T.; Zhong, L.; Wang, J.; Guo, L.; Wu, H.; Guo, Q.; Fu, F.; Chen, G. Graphite-Like Carbon Nitrides as Peroxidase Mimetics and Their Applications to Glucose Detection. *Biosens. Bioelectron.* 2014, *59*, 89–93.
50. Liu, B.; Liu, J. Surface Modification of Nanozymes. *Nano Res.* 2017, *10* (4), 1125–1148.
51. Lv, M.; Liu, Y.; Geng, J.; Kou, X.; Xin, Z.; Yang, D. Engineering Nanomaterials-Based Biosensors for Food Safety Detection. *Biosens. Bioelectron.* 2018, *106*, 122–128.

52. Mahmoudpour, M.; Karimzadeh, Z.; Ebrahimi, G.; Hasanzadeh, M.; Ezzati Nazhad Dolatabadi, J. Synergizing Functional Nanomaterials with Aptamers Based on Electrochemical Strategies for Pesticide Detection: Current Status and Perspectives. *Crit. Rev. Anal. Chem.* 2021, 1–28.
53. Massad-Ivanir, N.; Shtenberg, G.; Raz, N.; Gazenbeek, C.; Budding, D.; Bos, M. P.; Segal, E. Porous Silicon-Based Biosensors: Towards Real-Time Optical Detection of Target Bacteria in the Food Industry. *Sci. Rep.* 2016, *6* (1), 38099.
54. Mehrotra, P. Biosensors and Their Applications—A Review. *J. Oral Biol. Craniofac. Res.* 2016, *6* (2), 153–159.
55. Mishra, G.; Barfidokht, A.; Tehrani, F.; Mishra, R. Food Safety Analysis Using Electrochemical Biosensors. *Foods*, 2018, *7* (9), 141.
56. Mittal, D.; Kaur, G.; Singh, P.; Yadav, K.; Ali, S. A. Nanoparticle-Based Sustainable Agriculture and Food Science: Recent Advances and Future Outlook. *Front. Nanotechnol.* 2020, *2*, 10.
57. Munir, S.; Shah, A. A.; Rahman, H.; Bilal, M.; Rajoka, M. S. R.; Khan, A. A.; Khurshid, M. Nanozymes for Medical Biotechnology and Its Potential Applications in Biosensing and Nanotherapeutics. *Biotechnol. Lett.* 2020, *42* (3), 357–373.
58. Namany, S.; Govindan, R.; Alfagih, L.; McKay, G.; Al-Ansari, T. Sustainable Food Security Decision-Making: An Agent-Based Modelling Approach. *J. Clean. Prod.* 2020, *255*, 120296.
59. Naveen Prasad, S.; Bansal, V.; Ramanathan, R. Nanozyme-Based Sensors for Pesticide Detection. *Environ. Chem. Sustain. World* 2021, 145–175.
60. Neethirajan, S.; Ragavan, K. V.; Weng, X. Agro-Defense: Biosensors for Food from Healthy Crops and Animals. *Trends Food Sci. Technol.* 2018, *73*, 25–44.
61. Neethirajan, S.; Ragavan, V.; Weng, X.; Chand, R. Biosensors for Sustainable Food Engineering: Challenges and Perspectives. *Biosensors* 2018, *8* (1), 23.
62. Nerin, C.; Aznar, M.; Carrizo, D. Food Contamination during Food Process. *Trends Food Sci. Technol.* 2016, *48*, 63–68.
63. Niu, X.; Li, X.; Lyu, Z.; Pan, J.; Ding, S.; Ruan, X.; Zhu, W.; Du, D.; Lin, Y. Metal–Organic Framework Based Nanozymes: Promising Materials for Biochemical Analysis. *Chem. Commun.* 2020, *56* (77), 11338–11353.
64. Payal, A.; Krishnamoorthy, S.; Elumalai, A.; Moses, J. A.; Anandharamakrishnan, C. A Review on Recent Developments and Applications of Nanozymes in Food Safety and Quality Analysis. *Food Anal. Meth.* 2021, *14* (8), 1537–1558.
65. Perez-Lopez, B.; Merkoci, A. Nanomaterials Based Biosensors for Food Analysis Applications. *Trends Food Sci. Technol.* 2011, *22* (11), 625–639.
66. Posorske, L. Industrial-Scale Application of Enzymes to the Fats and Oil Industry. *J. Am. Oil Chem. Soc.* 1984, *61* (11), 1758–1760.
67. Rashidi, L.; Khosravi-Darani, K. The Applications of Nanotechnology in Food Industry. *Crit. Rev. Food Sci. Nutr.* 2011, *51* (8), 723–730.
68. Rawat, S. Food Spoilage: Microorganisms and Their Prevention. *Asian J. Plant Sci. Research*, 2015, *5*(4), 47–56.
69. Rocha-Santos, T. A. P. Sensors and Biosensors Based on Magnetic Nanoparticles. *TrAC Trends Anal. Chem.* 2014, *62*, 28–36.
70. Rodrigues, S. M.; Demokritou, P.; Dokoozlian, N.; Hendren, C. O.; Karn, B.; Mauter, M. S.; Sadik, O. A.; Safarpour, M.; Unrine, J. M.; Viers, J.; Welle, P.; White, J. C.; Wiesner,

M. R.; Lowry, G. V. Nanotechnology for Sustainable Food Production: Promising Opportunities and Scientific Challenges. *Environ. Sci.: Nano* 2017, *4* (4), 767–781.
71. Rotariu, L.; Lagarde, F.; Jaffrezic-Renault, N.; Bala, C. Electrochemical Biosensors for Fast Detection of Food Contaminants—Trends and Perspective. *TrAC Trends Anal. Chem.* 2016, *79*, 80–87.
72. Schroeder, J. I.; Delhaize, E.; Frommer, W. B.; Guerinot, M. Lou; Harrison, M. J.; Herrera-Estrella, L.; Horie, T.; Kochian, L. V.; Munns, R.; Nishizawa, N. K.; Tsay, Y. F.; Sanders, D. Using Membrane Transporters to Improve Crops for Sustainable Food Production. *Nature* 2013, *497* (7447), 60–66.
73. Shin, H. Y.; Park, T. J.; Kim, M. Il. Recent Research Trends and Future Prospects in Nanozymes. *J. Nanomater.* 2015, *2015*, 1–11.
74. Singh, H.; Bamrah, A.; Bhardwaj, S. K.; Deep, A.; Khatri, M.; Brown, R. J. C.; Bhardwaj, N.; Kim, K.-H. Recent Advances in the Application of Noble Metal Nanoparticles in Colorimetric Sensors for Lead Ions. *Environ. Sci.: Nano* 2021a, *8* (4), 863–889.
75. Singh, K.; Solanki, P.; Malhotra, B. D.; Pandey, A.; Singh, Ravindra. Introduction to Nanomaterials: An Overview Toward Broad-Spectrum Applications. In *Nanomaterials in Bionanotechnology Fundamentals and Applications*; Singh, R. P., Singh, K., Eds.; CRC Press:Boca Raton, Florida, 2021; pp 388.
76. Singh, S. Nanomaterials Exhibiting Enzyme-Like Properties (Nanozymes): Current Advances and Future Perspectives. *Front. Chem.* 2019, *7*, 46.
77. Singh, S.; Kumar, V.; Dhanjal, D. S.; Datta, S.; Prasad, R.; Singh, J. Biological Biosensors for Monitoring and Diagnosis. In *Microbial Biotechnology: Basic Research and Applications. Environmental and Microbial Biotechnology*; Singh, J., Vyas, A., Wang, S., Prasad, R., Eds.; Springer International Publishing: Basel, Switzerland, 2020; pp 317–335.
78. Snejdarkova, M.; Svobodova, L.; Nikolelis, D. P.; Wang, J.; Hianik, T. Acetylcholine Biosensor Based on Dendrimer Layers for Pesticides Detection. *Electroanalysis* 2003, *15* (14), 1185–1191.
79. Song, Y.; Qu, K.; Zhao, C.; Ren, J.; Qu, X. Graphene Oxide: Intrinsic Peroxidase Catalytic Activity and Its Application to Glucose Detection. *Adv. Mater.* 2010, *22* (19), 2206–2210.
80. Sumner, J. B. The Isolation and Crystallization of the Enzyme Urease. *J. Biol. Chem.* 1926, *69* (2), 435–441.
81. Thanushree, M. P.; Vimala, B. S. K., Moses, J. A.; Anandharamakrishnan, C. Detection Techniques for Insect Infestation in Stored Grains. *Agric. Eng. Today* 2018, *42* (4), 48–56.
82. Turasan, H.; Kokini, J. Novel Nondestructive Biosensors for the Food Industry. *Ann. Rev. Food Sci. Technol.* 2021, *12* (1), 539–566.
83. Verma, N.; Hooda, V.; Gahlaut, A.; Gothwal, A.; Hooda, V. Enzymatic Biosensors for the Quantification of Biogenic Amines: A Literature Update. *Crit. Rev. Biotechnol.* 2020, *40* (1), 1–14.
84. Wang, F.; Chen, L.; Liu, D.; Ma, W.; Dramou, P.; He, H. Nanozymes Based on Metal-Organic Frameworks: Construction and Prospects. *TrAC Trends Anal. Chem.* 2020, *133*, 116080.
85. Wang, W.; Gunasekaran, S. Nanozymes-Based Biosensors for Food Quality and Safety. *TrAC Trends Anal. Chem.* 2020, *126*, 115841.

86. Wang, X.; Qu, K.; Xu, B.; Ren, J.; Qu, X. Multicolor Luminescent Carbon Nanoparticles: Synthesis, Supramolecular Assembly with Porphyrin, Intrinsic Peroxidase-Like Catalytic Activity and Applications. *Nano Res.* 2011, *4* (9), 908–920.
87. Wasserman, B. P.; Montville, T. J.; Korwek, E. L. Food Biotechnology: A Scientific Status Summary. *Food Technol. (Chicago)* 1988, *42* (1), 133–146.
88. Weerathunge, P.; Ramanathan, R.; Shukla, R.; Sharma, T. K.; Bansal, V. Aptamer-Controlled Reversible Inhibition of Gold Nanozyme Activity for Pesticide Sensing. *Anal. Chem.* 2014, *86* (24), 11937–11941.
89. Wei, H.; Wang, E. Nanomaterials with Enzyme-Like Characteristics (Nanozymes): Next-Generation Artificial Enzymes. *Chem. Soc. Rev.* 2013, *42*(14), 6060.
90. Wei, J.; Yang, L.; Luo, M.; Wang, Y.; Li, P. Nanozyme-Assisted Technique for Dual Mode Detection of Organophosphorus Pesticide. *Ecotoxicol. Environ. Saf.* 2019, *179*, 17–23.
91. Wu, J.; Li, S.; Wei, H. Multifunctional Nanozymes: Enzyme-like Catalytic Activity Combined with Magnetism and Surface Plasmon Resonance. *Nanoscale Horiz.* 2018, *3* (4), 367–382.
92. Xu, Z.; Lan, J.; Jin, J.; Dong, P.; Jiang, F.; Liu, Y. Highly Photoluminescent Nitrogen-Doped Carbon Nanodots and Their Protective Effects Against Oxidative Stress On Cells. *ACS Appl. Mater. Interf.* 2015, *7* (51), 28346–28352.
93. Yang, T.; Duncan, T. V. Challenges and Potential Solutions for Nanosensors Intended for Use with Foods. *Nat. Nanotechnol.* 2021, *16* (3), 251–265.
94. Zhang, D.; Du, P.; Chen, J.; Guo, H.; Lu, X. Pyrazolate-Based Porphyrinic Metal-Organic Frameworks as Catechol Oxidase Mimic Enzyme for Fluorescent and Colorimetric Dual-Mode Detection of Dopamine with High Sensitivity and Specificity. *Sens. Actuat. B: Chem.* 2021, *341*, 130000.
95. Zhang, X.; Wu, D.; Zhou, X.; Yu, Y.; Liu, J.; Hu, N.; Wang, H.; Li, G.; Wu, Y. Recent Progress in the Construction of Nanozyme-Based Biosensors and Their Applications to Food Safety Assay. *TrAC Trends Anal. Chem.* 2019, *121*, 115668.
96. Zhang, Y.; Rui, X.; Simpson, B. K. Trends in Nanozymes Development Versus Traditional Enzymes in Food Science. *Curr. Opin. Food Sci.* 2021, *37*, 10–16.
97. Zhu, Y.; Wu, J.; Han, L.; Wang, X.; Li, W.; Guo, H.; Wei, H. Nanozyme Sensor Arrays Based on Heteroatom-Doped Graphene for Detecting Pesticides. *Anal. Chem.* 2020, *92* (11), 7444–7452.

CHAPTER 18

NANOENCAPSULATED ANTIMICROBIAL CARRIERS IN PACKED FOOD

JYOTI SARWAN[1], RAKESH R. JADHAV[2], HEENU SHARMA[1], MARIYAM CHAUDHARY[1], NIKET RANA[1], SHAYANTI ROY[1], NAZIM UDDIN[1], and JAGADEESH CHANDRA BOSE[1]

ABSTRACT

For several years, it has been concluded that foodborne diseases are the main cause of numerous gastrointestinal issues. These foodborne diseases are also responsible to weaken the digestive system and reduce the immune power of an individual. For decades, a variety of chemical preservatives and antimicrobial agents have been used to extend the shelf life of food products, but rising demand for organic and chemical-free foods has led to the development of a new method of food packaging: the encapsulation of nanoparticles to reduce microbial spoilage and improve food quality. Therefore, in this chapter, different types of food encapsulation methods have been mentioned with their applications.

18.1 INTRODUCTION

Foodborne diseases are now an increasing threat to a healthy life; around 600 million cases have been registered which cannot be neglected. According to a report by WHO,[54] about 420,000 cases of foodborne diseases are found every year. On average, around 23 million European people fall sick annually

Enzyme Inactivation in Food Processing: Technologies, Materials, and Applications. Megh R. Goyal, PhD, Junaid Ahmad Malik, PhD & Ravi Pandiselvam, PhD (Eds.)

only via contaminated food, with illnesses starting from mild diarrhea, fever, mild discomfort in digestion, and dizziness to severe infection leading to death. An estimated about 5000 annual deaths are registered from all over the world. The main cause of such sickness is a microbial infection occurring due to microbial contamination of food. These microbes are not only responsible to cause certain illnesses in the human body but also decrease the shelf life and quality of the food.[9] Hence, an efficient method to ensure food safety, as well as quality, is required. Active packaging and nanotechnology are trending in the food industry to achieve nutrient-rich food with longer shelf life. Over the last 20 years, nanotechnology is emerging widely in the food domain particularly to design some nanocarriers or nanodelivery systems as antimicrobial agents or shields. The technology of nanoencapsulation proved its effectiveness in various domains of the food industry such as food packaging, preservation, formation of functional foods, etc. New developments and novel trends in food technology had enabled the use of many nanotechnology-based devices to be used in the food industry, promising advanced methods for food processing, and preservation, eliminating the chances of microbial contamination and increasing the durability and stability of food under extreme environmental conditions. Most of the nanoscale particles used in active nanofood packaging have an effective antimicrobial activity that can be used as coating agents to provide an antimicrobial shield from microbial contamination. Nanoencapsulation of certain bioactive compounds provides a higher ratio of surface area/volume; hence, increasing the solubility, bioavailability, and dispersion dose as compared to the standard devices used in packaging. Moreover, the market of nanotech appliances such as nanobiosensors to detect contamination, quality assurance sensors, nanocomposites to control diffusion, etc. are in increasing demand; therefore, making the technology more promising and highly demanding.

This chapter focuses on the various advancements of nanotechnology in the food packaging industry, and methods and brief applications of the same are discussed.

18.2 FOOD PACKAGING

It is a vital step in phrases of meal protection. The main purpose of meal packaging is to stop frequent spoilage, contamination, and growth sensitivity by sanctioning enzyme action and cutting back weight reduction. In beneficial ingredients, bioactive detail usually gets devalued and deactivated because of the antagonistic surroundings. Therefore, the time period of the merchandise is reduced. The employment of nanostructured and nanomodified materials can be a stepping stone in the direction of the extended shelf existence of food. Food packaging is advanced with the aid of the utilization of beneficial nanomaterials with upgrades in physical chemistry, similar to temperature and moisture balance, barrier homes, mechanical power, sturdiness, and adaptability (progressed packaging). Nanomaterials with energetic properties, such as ultraviolet (UV) defense, antimicrobial, active packaging, antioxidative nanosensors, and their sensitivity or intelligent functions toward the detection and identification of gases, minute organic molecules, lively stage, and product identity, could also be used to advance packaging (smart packaging). The entire development of biobased products, such as biodegradability, eco-friendliness, cheap cost, and biodegradability, may be carried out with the use of bionanomaterials (biobased full packaging). Plastic polymers that are generally non-degradable, featuring actual packaging can be efficiently replaced by means of biobased effective packaging, in addition to perishable or biocompatible packaging.

18.2.1 IMPROVED FOOD PACKAGING

The primary goal of advanced food packaging that incorporates advantageous nanoparticles into polymeric fabric is to improve the mechanical and physical properties of the food, such as fuel resistance, temperature and humidity resistance, mechanical sturdiness, and flexibility. Many nanocomposites or particles incorporated into polymers or nanostructures have already made their way into the manufacturing unit designed to meet the demands and needs of the various drinks and oils industries, where the quantity/proportion of nanoparticles such as clay nanocomposites is as high as 5% (w/w). These improve the packaging's barrier properties, resulting in up to a 90% reduction in gasoline and dioxide permeability. Scientists experimentally alleged that too much-loaded clay in ethylene–vinyl alcohol copolymer-based completely nanocomposite films causes a loss of tensile properties as well as optical transparency, due to the formation of clay agglomerates. By the incorporation of entirely 3% (w/w) clay, fuel (by 59%) and vapor barrier (via 90%) features of the nanocomposite films had been progressed in assessment with the material with montmorillonite clay nanocomposites or particles being delivered.

Researchers experimentally believed that the inclusion of 4% (w/w) of nanoclay gives rise to associate growth in oxygen-blocking characteristics of styrene (PS) by using 51%.

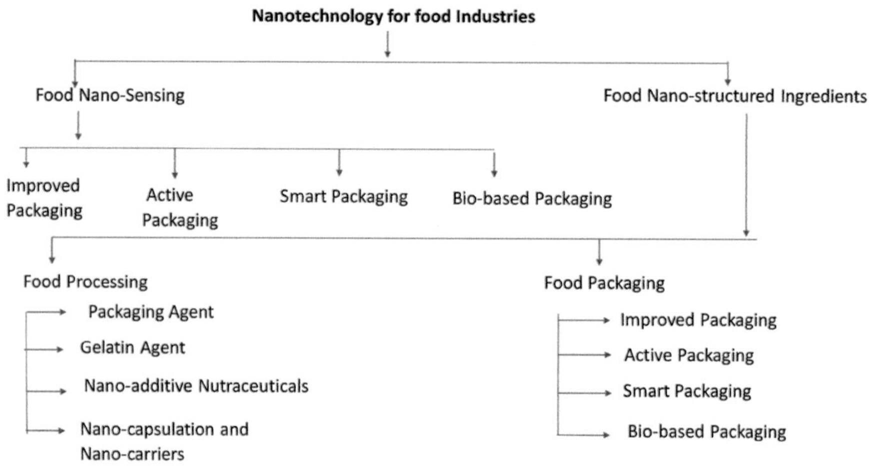

FIGURE 18.1 Representing the major role of nanoparticles in food industries.

18.3 NANOPARTICLES IN FOOD PACKAGING

A nanoparticle might be a tiny particle that tiers between 1 and 100 nm in length. Undetectable by the human eye, nanoparticles will exhibit significantly absolutely specific physical and chemical residences to their substantial material counterparts. Resulting from their very small size, nanoparticles have a certainly big expanse to quantitative relation while in comparison to bulk fabric, cherish powders, plate, and sheet. This feature permits nanoparticles to own surprising optical, physical, and chemical properties, as they are small enough to confine their electrons and create quantum effects that occur evidently, either due to the presence of materials of combustion reactions, or created purposefully through engineering to develop and perform a specialized character. Nanomaterials are used in a wide range of sectors, from mining and cosmetics to environmental maintenance and air purification, due to the ability of electrons to create materials in a very precise fashion to perform a specific function. Nanoclays in combination with golden nanoparticles have a potential future in the active packaging age, a new generation of food preservation based on mass transfer interactions between food and packaging systems. These nanocompounds are used in a variety of ways in the packaging of machine victimization.

Nanomaterials are best suited for the packaging of food as they have numerous advantages over normal packaging substances. As a result, nanocoating is a significant generation method for increasing food/meal packaging properties. Different forms of food coatings, such as thin layers/films, are commonly used to cover the food during production with a switching barrier. These can also be made from material that is suitable for human consumption. Edible food coatings are appropriate for direct application to the product, but non-edible coatings have the purpose of defensive instrumentation with or without being a part of meals. Fit for human consumption nanocoatings (5-nm thin coatings) are commonly used in the meat method exchange, agricultural industry (for the safety of fruit, greens products, and so on), cheese and bakeshop industry, and so on to provide flavor, color, enzymes, antioxidants, and browning resistant substances to products. Second, nanoclays as nanomaterials also are considerably used as well as studied for food packaging due to their mechanical, thermal, as well as barrier residences, and occasional rate. Many substances, such as natural polyelectrolytes (proteins, polysaccharides), charged lipids such as phospholipids, surfactants, and combination particles like micelles, vesicles, and droplets, are frequently used to form unique layers for the instruction of nanolaminates for food packaging.

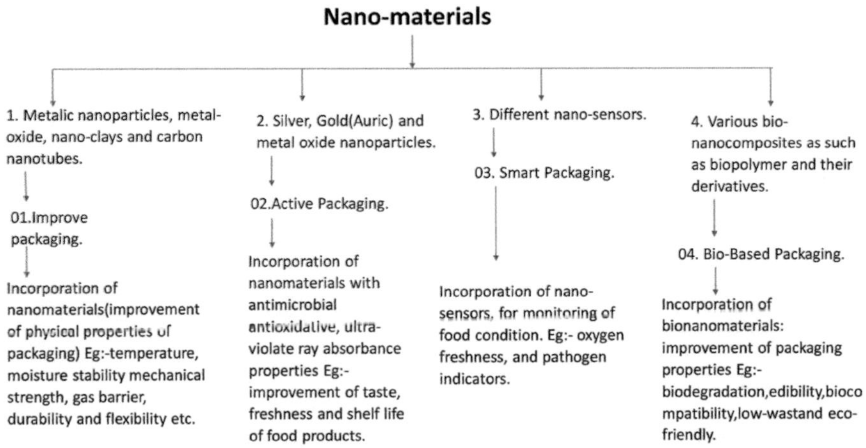

FIGURE 18.2 Role of nanoparticles in active food packaging.

Biopolymers for biodegradable film development for food packaging should be inexhaustible, inexpensive, and, if possible, sourced from waste. Polysaccharides such as cellulose, starch, chitosan, gelatin, alginate, carrageenan, pullulan, and kefiran are commonly used as ingredients for biodegradable coatings/biofilm arrangements. They can create masking films with exceptional boundary characteristics for use as a vehicle for O_2 and CO_2. Their tractable electrical values are similar to those seen in synthetic polymers.

For example, the tension advantages of the films/layers are dependent on a high amylose starch content and are similar to those seen in low-thickness polyethylene films. The employment of polysaccharides in conjunction with unique materials to work on the boundary and mechanical props of covering films for food use is frequently emphasized. To collect protein and polysaccharides-based films, the development of plasticizers (hydrophilic to increase and hydrophobic to decrease water fume porousness) is often required. On the mechanical properties of films, there are paintings of plasticizers. Plasticizer expansion allows for a reduction in pressure, hardness, thickness, and consistency of substances. In addition, the flexibility of the polymer chain as well as crack resistance may be improved. Showering, submersion, or scouring are all easy ways to apply consumable nanocoatings.

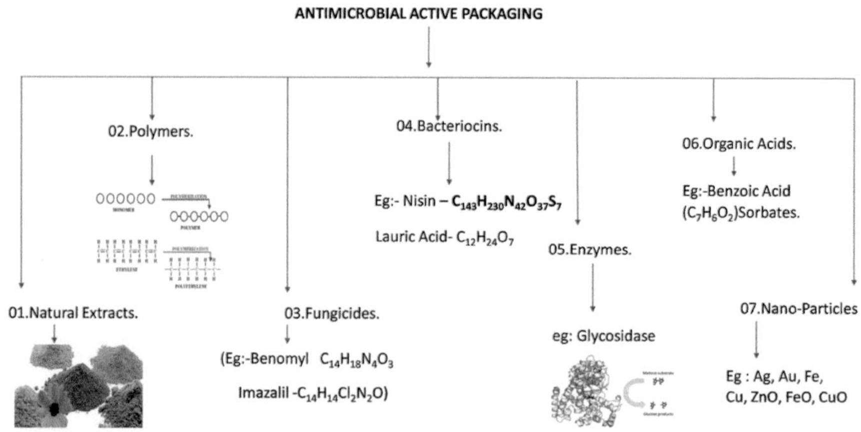

FIGURE 18.3 Various types of microbial-resistant packaging.

They usually contain chemicals that are not harmful to the environment, and they do not need to be worn out from food before use. By and large, those nanoprotecting films provide nanostructures in which common adhesives with antibacterial and mobile reinforcement actions might be fused to increase their high-quality impacts on the brand-new product first class. Nano-overlay films are typically made up of at least two layers of nanolayer film with physical/substance interlinking. The layer-by-layer assertion is the most well-known method for combining them. This approach equips floor coverings with one-of-a-kind nanolayers that act as interfacial layers and rely on proprietary nanomaterials.[27,58] To show and similarly create various layered homes, different adsorbing combinations such as conventional or biobased polyelectrolytes such as polysaccharides, proteins, colloidal debris such as vesicles, beads, micelles, and charged lipids (surfactants and phospholipids) can be used. Additional, unique dynamic mixes (e.g., antimicrobials, mobile reinforcements, anticooking experts, chemicals, flavors, heady fragrances, etc.) can be added to the movies to improve the quality and extend the time range of usage of packaged food devices.[22] As palatable nanoblanketed films/sheets, nano-overlaid coatings might also be made from eatable or biobased fixes. Because of the synergism present between the framework polymer and nanodust, as a small amount of nanofiller is joined within the polymer framework, polymer/earth nanocomposites also deal with the promising lessons in meal bundling due to the massive enlargement within the mechanical and boundary properties of these bundling. Food packaging materials made of dust nanocomposite have a longer lifespan, are lighter, more heat resistant, and are typically shatterproof.[49] The transportation of diffusing particles is obstructed by invulnerable particles/clays in a nanocomposite coated in clay. Interfacial areas/parts with different porousness characteristics compared to the essential polymer are generated as a result. As a result, the nonlinear route increases the mean fuel dissemination duration, extending the timeframe of the practical usefulness of quickly spoiled food kinds.[33,46]

18.3.1 METHODS OF NANOENCAPSULATION OF BIOCATALYTIC SUBSTANCES

Nanoencapsulation is an emerging revolutionary method in the food packaging industry, in which the nanosized particles of about 9–10-nm are produced using various approaches. The complete process of nanoencapsulation is divided into two parts. The initial part uses the antimicrobial compounds to pack the carriers and in the next stage, the size of these carriers is decreased to nanosizes (9–10 nm).[34] By doing so, the functional properties of the nanoframework such as adsorption, solvency, bioavailability, etc. improve because of the increased surface area of the carriers contrasted with their mass equivalents.[19] The selection of a suitable encapsulation technique is essential as the mechanism utilized for encapsulation influences the molecule size, size dissemination, surface region, shape, dissolvability, and embodiment proficiency. The decision of technique relies on the necessary size, physical–chemical properties, nature of the center, and wall material.[2]

In general, two approaches are used for the development of nanocapsules. These are:

18.3.1.1 TOP–DOWN APPROACH

In this, devices that permit size and shape alterations are utilized. Deformation and size reduction would help to achieve the desired use of the developed nanomaterials. It includes two methods, emulsification and emulsification–solvent evaporation.

18.3.1.1.1 Emulsification

Nanoemulsions are produced by encapsulating the bioactive components into an aqueous solution. These nanoemulsions are formed due to the colloidal dispersions between two immiscible liquids having droplet sizes between 50 and 1000 nm[38] and can be used directly in liquid form or are dried to powder form. The formation of emulsion beads requires high energy input, which can be achieved using chemical or mechanical means, as they produce a homogenous progression of littlest drop sizes with an adequate measure of energy in minimal time. Major methods used for emulsification are high shear stirring, homogenizers (high speed or pressure), microfluidizers, and/or ultrasonicators. The high kinetic strength possessed by these small emulsion beads assumes an indispensable part in the maintenance of the surface oil content of the product.[23] In microfluidization, up to 20,000 psi of pressing factor is required for the formation of emulsion droplets and is accounted as the best emulsification strategy with the highest encapsulation productivity. In high-pressure homogenizers, 100–2000 bar pressure is used to produce high shear pressure for breaking the particles into the nanometer size range. Additionally, homogenization can be performed both at high and/or low temperatures.[31,44] Moreover, ultrasonications and high shear mixing are likewise acceptable emulsification methods for the formation of nanodroplets with insignificant re-coalescence. All these methods of emulsification are proficient for the formation of nanoemulsions.

18.3.1.1.2 Emulsification–Solvent Evaporation

It is an advanced method of emulsification in which polymer solution is emulsified into an aqueous phase and the polymer–solvent is vaporized as nanospheres for polymer precipitation.[36] This technique is used in combination either with homogenization or ultrasonication to achieve a product of the required nanosize and better stability. Ethylcellulose, polycaprolactone (PCL), cellulose acetate phthalate and poly-(h-hydroxybutyrate), etc. are the commonly used polymers for emulsification- solvent evaporation techniques. Nanocapsules having a size less than 100 nm can be achieved using this technique, where the majority of the capsules exhibit a spherical shape with high embodiment effectiveness of around 75–96% with supported release and expanded assimilation.[53] However, this technique is dependent on other emulsification strategies such as homogenization, microfluidization and/or ultrasonication to produce nanoparticles of desired sizes. Factors such as disruption rate of homogenization, nature of the dispersing agent, temperature, viscosity of the two phases, etc. affect the size allocation and storage permanence of the molecules.

18.3.1.2 BOTTOM–UP APPROACH

In this, the development of nanomaterials is achieved through assembling and associating the molecules within themselves. This self-orientation of molecules depends on temperature, ionic strength, pH, and intensity of the particles used.[60] This approach includes coacervation, nanoprecipitation, inclusion-complexion, and supercritical fluid techniques.

18.3.1.2.1 Coacervation

This technique is based on phase separation of polyelectrolytes present in solution and accumulation of the formed coacervate phase across the dynamic component. A hydrocolloid shell can then be cross-linked to enhance the strength of the coacervate. This cross-linking is done by using enzymatic or chemical linkers like transglutaminase or glutaraldehyde, etc.[60] As it is carried out in a simultaneous agitation process for encapsulation, the technique is considered a colloidal phenomenon. Two types of polymers are used in this method; single type (called simple coacervation) and mixture type having two or more polymers (called complex coacervation). The interaction of these biopolymers and the complex structure formed thereafter depends on pH, molar mass, a charge of polymer, hydrogen bonding and hydrophobic interactions between the biopolymers, flexibility of the molecules, ionic strength and concentration of the biopolymers, and the ratio in which they are present.[48] This technique is ideal for the encapsulation of flavor oils. The only limitation of this method is the use of glutaraldehyde for cross-linking the polymers, which cannot be used commercially. Otherwise, it is a very promising and cost-effective nanoencapsulation technology that deals with organic solvents and requires moderate temperature conditions and also yields a high amount of payloads.[7]

18.3.1.2.2 Nanoprecipitation

In this method, biopolymer from organic solution is precipitated and the organic solvent is diffused in an aqueous medium, this displacement of solvent leads to the formation of nanospheres and nano-capsules. This is also known as the solvent displacement method.[14] Biopolymers such as polylactide (PLA), poly-alkyl cyanoacrylate (PACA), PCL and eudragit, etc. are used in nanoprecipitation.[13] This technique is proficient in producing nanodroplets having a size of 100 nm and less. The encapsulates so formed, demonstrate increased strength of encapsulation, stability, and bioavailability of the components in vivo contemplates. To achieve better results, this technique is used in combination with freeze-drying. However, this method is restricted to water-miscible solvents, where the dissemination rate is sufficient to deliver unconstrained emulsification.

18.3.1.2.3 Inclusion–Complexion

In this, the ligand associated with supra-molecules is encapsulated via H-bonding, van der Waals interactions, or hydrophobic impact into a shell substrate.[12] This technique of inclusion complexion is widely utilized to encapsulate volatile organic molecules such as vitamins and essential oils, and for aromatic preservation. This technique possesses high productivity and stability.[32]

18.3.1.2.4 Supercritical Fluid Technique

This technique uses liquids and gases over their critical thermodynamic pressure and temperature point that exhibit the properties such as low density and viscosity, enhanced solvating capacity, increased rate of mass transfer, and diffusion.[21] These critical gaseous liquids are applied for the encapsulation of temperature-sensitive compounds and are mixed with the biodynamic components and the biopolymers to form an expanded solution. From this solution, the solute particles are precipitated by evaporating the critical fluids using sprays. The compounds that are used in this technique are carbon dioxide, nitrogen, water, propane, etc. The size of particles, encapsulation effectivity, and distribution of nanocapsules are some of the factors that affect the efficiency of this method.[42]

18.3.1.3 PHYSICAL METHODS

This is the most available technology used in industries for micro/ nanoencapsulation.[15]

18.3.1.3.1 Extrusion

This method is used specifically to encapsulate unstable and highly volatile aromas in carbohydrate glass matrices. The encapsulated materials produced by this method possess longer shelf life as compared to any other method. The glass matrix acts as a protective wall, preventing the gases from diffusing out of its matrix and protecting the inside materials from oxidation reactions, thereby increasing its encapsulating efficiency for longer durations. The technique is widely used for nanoencapsulation of flavored oils such as citrus oils.[5,56]

18.3.1.3.2 Spray Drying

In this fluid, atomizers are used to spray liquids in a medium with hot drying gas. And the solvent present in the liquid droplets is collected by removing it from the air stream carrying solid particles of the solvent. Simply, spray drying includes the suspension of nanoparticles into dehydrating gas to safeguard the product, increasing the tolerance of the bioproduct to withstand harsh conditions such as acidic environment, increased temperature, extreme moisture, etc.[10] This is a rapid and cost-efficient approach that can be used commercially because of reproducibility. The most effective and often used method used in the food industry for the encapsulation of lipids and carotenoids.[18]

18.3.1.3.3 Spray–Chilling

It is a spray cooling method majorly used for the production of dynamic agents coated with lipids. In this lipid molecules are taken in the molten state, the droplets of which are sprayed in a chamber at freezing temperature. The lipid droplets are solidified in this chamber and are collected as fine particles.[18,45]

18.3.1.3.4 Fluid Bed Coating

In this, powdered particles are used which as suspended in an air stream, and are applied to a batch process as a coating material at definite temperature conditions. Different coating materials used in batch processes are proteins, dextrins, or sugar derivatives. The coating materials used in this method should have high viscosity and thermal stability so that a film is easily formed over the particle surface. Various factors such as humidity inside the chamber, air flow rate, temperature, nature of the material used, etc. should be maintained.[33,57]

18.4 BIOBASED MATERIAL FOR A NANOPARTICLE

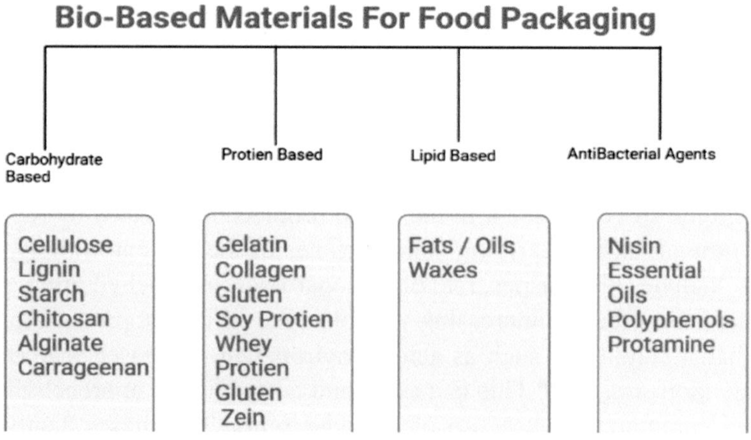

FIGURE 18.4 Different types of biobased material for food packaging.

18.4.1 CARBOHYDRATE-BASED BIOMATERIAL

Chitosan, starch, and cellulose are some of the polysaccharides. They are non-toxic and biodegradable. Due to some properties of these polysaccharides such as acid hydrolysis and semi-crystalline state, they can be considered as a valuable and potential source of nanosized reinforcement.[34] In food coating and packaging, chitosan is used as a biopolymer due to its excellent film-forming, biodegradable and antimicrobial properties.[59] The antibacterial activity is due to the protonated amine group and depends on the pH value, degree of deacetylation, and molecular weight. Modifications on chitosan such as gamma-aminobutyric acid-rich fermented soy protein and edible film based on (2–3% w/v) chitosan result in great antimicrobial properties.[13] Forms of chitosan like powdered, casted film, nanoparticles, and coating are mostly used and strongly exploited in recent years. Chitosan and its derivatives have various applications in different fields and are considered excellent materials in nanofiber production.[16]

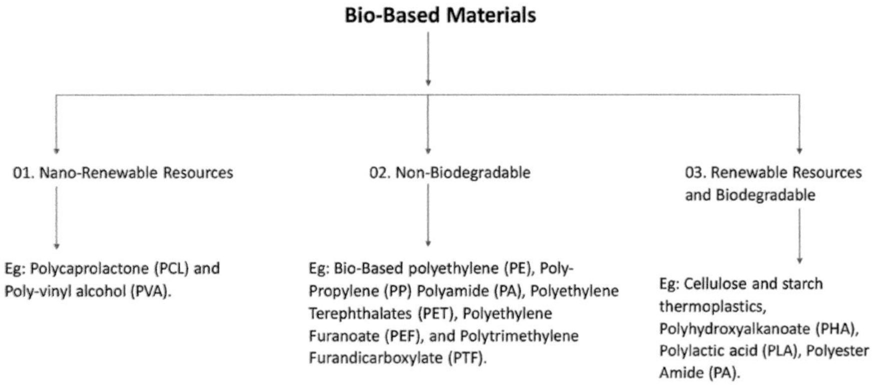

FIGURE 18.5 Types of biobased material for the formulation of nanoparticles.

Starch is one of the widely naturally explored carbohydrate polymers which is renewable and extracted from maize, corn, rice, barley, potato, wheat, soya, potato, vegetables, etc., it constitutes a linear molecule that links with each other through α-1,4-glycosidic bonds, that is, amylase and a branched chain molecule with branches forming α-1,6 bonds, that is, amylopectin.[34] Due to their nature of non-toxic, cost-effectiveness, and plasticity, these are widely used as raw materials in the production of bioplastics or biodegradable films.[13] Starch-based films usually have high properties of an oxygen barrier, and have poor properties of moisture barrier, alongside there is a change in sensitivity of water and enhancement in mechanical and barrier features of starch films with the addition of chitosan in starch.[30] In applications of food packaging, there is a preparation of bionanocomposite material with the use of starch and its derivatives, and with the use of biodegradable polymer material, there is a production of food spoilage detecting materials.[8] The current research is also going on starch which mainly focuses on certain properties like maintaining stiffness and mechanical storage during storage which is based on decrement in retrogradation and reducing water for absorption of material which leads to the packaging of food material in a cheaper, simpler and more active manner and the consumer can assess safety, quality, nutritional components and shelf life of packet.[4] Additionally, a potential source of nanosized reinforcements includes chitin and cellulose which are polysaccharides. A linear polysaccharide, that is, cellulose composes units of β-d-glucopyranose which link and form β-1–4 linkage, with the repeat in a unit of cellobiose and it includes amorphous and crystalline features.[13] The structural composition of cellulosic fibers which is reported by Siqueira et al.[43] are lignin, cellulose, and hemicelluloses, and based on origin there is the presence of extractives in different amounts.[59] For instance, cellulose acetate, carboxymethyl cellulose, and ethyl cellulose are some celluloses and their derivatives from which for packaging cellulose-based films prepare among them cellophane which is used in general.[13] Currently, nanocellulosic materials play bright attention in the improvement of properties like antimicrobial, thermal, mechanical, and barrier properties for packaging of materials, and these also retain characteristics like non-toxic and biodegradable.[51] For packaging of food like fresh produce, home personal products, dried foods, meat, dairy, etc., nanofibers play a major role in providing superior rigidity, thermal, tensile, and eco-friendly properties due to their recyclable and reusable nature.[17] With the use of cyclodextrin and oxidize cellulose nanocrystals, the preparation of materials for the packaging of food enables antibacterial molecules which release and load over a long period.[24] With the comparison of petroleum-based material, at the commercial level use of nanocellulosic materials that are limited.[1] Nowadays, innovative approaches are followed to design nanocellulosic materials for eco-friendly packaging, which provides an improved experience to manufacture efficiently. For coatings or edible films agar, alginate, and carrageenan are commonly used as hydrocolloids.[26] There is the use of alginate due to its non-toxic nature, and other properties such as biodegradability, stabilizing, biocompatibility, thickening, and gel production used for the preparation of the film.[16] In various fields of the food industry like frozen and fresh meat, fish, and poultry because of the properties such as good gas barrier use of carrageenan for edible films and coating.[46]

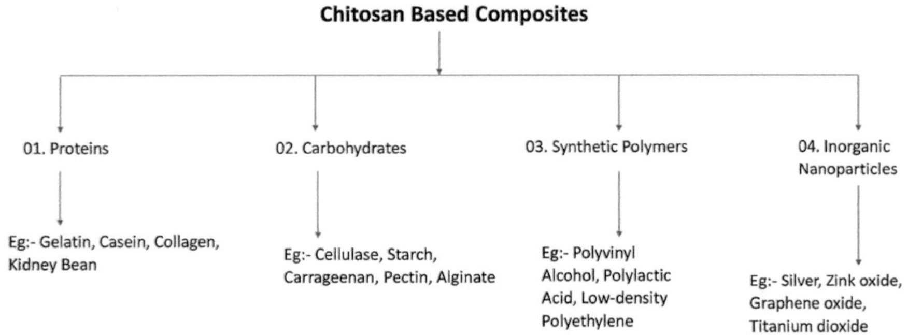

FIGURE 18.6 The chitosan-based material for nanoparticle formulation.

18.4.2 PROTEIN-BASED BIOMATERIAL

Packaging based on proteins (e.g., soy protein, casein, wheat gluten, whey protein, collagen, gelatin-based, and zein),[35] which have been widely investigated regarding their mechanical, thermal, and barrier properties comparatively more than any other biobased material used for packaging owing to their biodegradability, availability, extensive resources, and release regulation of additives and bioactive compounds /antimicrobial agents into the packaging system, just as in case of edible packagings such as films and coatings.[46]

For example, flexible, transparent films with low mechanical strength and high-water vapor permeability can be formed with Whey proteins isolates (WPI) because they contain a high amount of hydrophilic amino acids in their structure, as a result, they have better barrier properties when compared to lipid-based films or polysaccharides.

Gelatin is also a suitable protein-based polymer for the fabrication of biodegradable packaging because of its admirable properties such as oxygen barrier, biodegradable, low cost, functional agent carrier, and excellent film-forming activity, and it has been effectively and extensively deployed to create packaging films.[46] Researchers developed a bionanocomposite aiming for food packaging by embedding gelatin nanofibers into moringa oil or chitosan nanoparticles (CS–NPs) which are known to exhibit high antibacterial activity against *Listeria monocytogenes* and *Staphylococcus aureus* on cheese. Notably, the chief approach to obtaining protein films is via solution casting which includes protein solubilization in an appropriate solvent and further casting and vaporizing of solution on a non-stick surface.[24] However, protein coatings can be used mainly by dipping and spraying or by brushing a food product in a few cases.[30] Protein-based packaging could be carriers of antioxidants, antimicrobials, colorants, flavoring agents, nutrients, etc. generally used in cheese, meat products, fishery products, and fruits and vegetables to enhance their durability and acceptability.[50]

Protein-based films are generally considered as effective oxygen hinderers below intermediate humidity and have good mechanical properties, but their physical features such as tensile strength durability, etc. are still poor and thus their use in industries is tough.[51] Therefore, certain approaches to eliminate the limitations of protein-based films as well as cross-linking (physical, chemical, and enzymatic), reinforced with nanoparticles, and blending with other biopolymers have been explained.[37] Nanotechnology has proved to be a great approach for considerably improving protein-based films, and various investigations have also been carried out on this matter. Zubair et al.[59] pointed out that soy protein isolates, gelatin proteins, and wheat gluten proteins, are the most studied protein-based bionanocomposites suitable for food packaging. However, bionanocomposites based on proteins are still in their early stages and more research should be considered in the future.[3,20] Moreover, for the commercial production of protein films, future studies are advised which are recommended to focus on extrusion processes for the production of these films rather than that the solution casting at laboratory scales. It inspires scientists to develop innovative preservation strategies by incorporating bioactive agents such as bioactive reinforcements with antioxidant capacity and biobased reinforcements with antimicrobial capacity.

18.4.3 LIPID-BASED BIOMATERIAL

Lipids are the main components of animal and vegetable oils and fats for example butter, fatty acids, mono-, di- and triglycerides, etc., waxes such as paraffin, beeswax, jojoba, etc., edible films and coatings for food applications, natural resins, emulsifiers, and surface-active agents.[43]

Lipids can be incorporated into edible film coating by improving properties such as hydrophobicity, cohesiveness, flexibility, moisture barriers, etc. which help in the maintenance of quality for longer durations and inhibition of microorganisms in fresh and processed meats.[29]

Films made from lipids have a good water vapor barrier but are poor in mechanical strength and have increased oxygen permeability however this can be improved by lamination with hydrocolloid film lipid layer for example - study on the effects of saturated and unsaturated fatty acids on hydroxypropyl-methylcellulose based films it is observed that fatty acids can form stable layers in film matrices and improve the moisture barrier properties .also, reinforcing with nanoparticles is proved to be an effective method for considerably improving properties of the lipid-based film.[58]

18.4.4 ANTIBACTERIAL AGENTS

Natural antibacterial agents (e.g., bacteriocins, ε-polylysine, tea, polyphenols, enzymes, plant essential oils, etc.), inorganic antibacterial agents such as some of the oxidized nanoparticles and metals, and organic synthetic antibacterial agents have been used extensively in food packaging.[25]

Natural antibacterial agents are generally considered to be harmless, healthy, and eco-friendly antibacterial agents, and are extracted and purified from animal plants and microorganisms.[40] Nisin, natamycin, leucocin, enterocin, and pediocin or other natural antimicrobial peptides can be used for inhibiting or killing pathogens and other spoilage bacteria preventing spoilage of food Essential oils (EOs).[40]

Essential oils (EOs) are acknowledged to have broad-spectrum antibacterial activity making them a subject of numerous research studies and hence they are extensively used as a sustained release preparation in food packaging.[25]

EOs interact with the lipids of the microbial cell membrane and mitochondria with the help of their hydrophobicity properties along with their constituents, as a result, the structures get less organized becoming more permeable for ions and other cell contents to overflow are a complex mixture of compounds with main components such as alkaloids, monoterpenes, flavonoids, carotenoids, isoflavones, phenolic acids, and aldehydes.[52] Different chemical components can act through different mechanisms, and the same chemical composition when applied to different types of microorganisms, can have different effects. Their application may also be limited because EOs are almost insoluble in water, prone to oxidative decomposition, and are not photothermal resistant.[52]

With the advancement of nanotechnology antibacterial inorganic agents acquires improved and different biological characteristics. The nanoparticles of metal and their oxides have high thermal stability and they are easy to form these oxides of metal are advantageous for the transportation of essential minerals to the body.[38] Moreover, ZnO, TiO_2, MgO, and nanoparticles of metal oxides (such as ZnO–NPs, TiO_2–NPs, SiO_2–NPs, Al_2O_3–NPs, etc.) are some antibacterial agents that successfully showed antimicrobial activity against pathogens which are generally found in food by the generation of reactive oxygen species (ROS) which show biocompatibility with cells and have interaction with the cell wall.[51] On the basis of selectivity and specificity, these antibacterial agents are having an advantage over organic antibacterial agents.[47] Also, ZnO metal oxide is most commonly used in different fields due to its photocatalytic and antimicrobial activity.[36] Moreover, ZnO nanoparticles impact antibacterial activity by demolishing bacterial membranes due to the mordant surface as its texture, which inhibits microorganisms' growth. Recently, work has been done on those films of Al-doped ZnO that can change optical packaging properties.[6] Also, Moreover, Cu has the advantage of reduction of toxicity, and CuO–NPs plays a crucial role in active packaging as they have broad-spectrum effects of antimicrobial[28] and based on size and concentration of CuO–NPs, inhibition of growth of species.

Due to the low toxicity of TiO_2, it is widely used in cosmetics, healthcare, and food packaging materials and nanoparticles of TiO_2 (TiO_2–NPs) also offer antibacterial activity which invades bacteria's surface and also generates ROS.[41] Also, SiO_2–NPs mostly used as a contact surface for food packaging and to generate non-stick coatings for bottles, bags, and jars which prevents food items' spoilage.[11] Coating of Al_2O_3 also utilizes in food packaging as it provides an alternative to metallic films which have characteristics such as an effective barrier, microwavable, transparent, and biodegradable light.[26,55] Ultrathin films of around 10–100 nm in thickness or having nanocoating with barrier properties toward water and oxygen, and also has antimicrobial impacts which refer to Al_2O_3.[11] Also, Ag–NPs show characteristics like they have better bactericidal properties against different pathogenic microorganisms such as bacteria, viruses, and fungi. Bactericide effect of Ag–NPs by the increase in toxic influence to a bacterial cell on certain species such as *Salmonella choleraesuis*, *Escherichia coli*, *Listeria innocua*, *Pseudomonas aeruginosa*, *Bacillus cereus*, and *S. aureus* founded by researchers.[10] In addition to this, the synthetic organic polymer also plays a major role in the packaging of food, including fermentation-based biopolymers such as polyhydroxyalkanoates (PHA), polylactic acid (PLA), poly-b-hydroxyl butyrate (PHB), poly-3-hydroxybutyrate-co-3-hydroxyvalerate (PHBV), polyhydroxyvalerate (PHV), and exopolysaccharides (EPS), etc. and biobased polymers such as bio-polyethylene (Bio–PE), biopolypropylene (Bio–PP), biopolyethylene terephthalate(Bio-PET), etc.[27]

After certain modifications or with the combination of bioactive compounds, fermentation-based biopolymers show antimicrobial activity. Some of them such as PLA, PHBV, and PHA are under research and may show packaging applications with excellent gas barrier and biodegradable properties. On the other side, bio-PE, bio-PP, and bio-PET were also obtained from biological resources but they show the non-biodegradable property.[27] These materials show wide applications in the packaging of food and show a novel eco-friendly alternative to petrochemical-based plastics, majorly for single-use plastic goods, and the new Commission's European Green Deal of 2020 implement it.[27] For betterment in the future, it is recommendable to have a focus on the development of an appropriate combination of a mixture of materials, and proper additives to improve the functional properties of material packaging, thus from petroleum-based plastics, these show better properties and will have great achievement with this research work.

18.5 APPLICATIONS OF ENCAPSULATION

Encapsulation can be defined as a technique used to entrap the active agents within a carrier material which is useful to improve the delivery of bioactive molecules as well as living cells into foods. Encapsulation plays a very important role in the food industry, agricultural field, pharmaceuticals, etc. Following are the applications of encapsulation.[3]

18.5.1 FOOD PROCESSING—IMMOBILIZATION OF CELLS AND ENZYMES

Like in numerous industries food sector has also a large number of applications with nanoparticles and proved them a boon for every industry serving mankind. Encapsulation is the biggest advantage for the food industry to get entrapped food cells and enzymes for various purposes. Nowadays food sector is completely relying upon nanocaptured enzymes and items to enhance food quality and shelf life. At the time of fermentation or called bioprocessing various enzymes are used to reduce microbial contamination and augmented production. Organic carrier materials have turned into an expanding research area.[32] Hence, there is a number of techniques such as gel entrapment, covalent binding, and adsorption method for the entrapment of food enzymes and cells. Therefore, is it not a big deal to say immobilization can only be quite possible and time-reducing as well as stable with nanoparticles. Among a few unique methodologies described in the literature, the most valuable for the food processor is the entrapment of cells inside the matrix of natural polymers like alginate, agarose, carrageenan, chitosan, and pectin. Such natural gelling polysaccharides address an arising group due to their benefit of being non-toxic, biocompatible, and modest. [42] Immobilization of cells gives simplicity of biomass separation and recovery, lower risk of microbial contamination, better utilization of equipment and as an outcome of these and other benefits, higher productivity and proficiency. Immobilized cell techniques offer other various potential benefits contrasted with a free cell system, like higher cell densities, cell loads, shorter reaction times, smaller bioreactor size thereby lower capital expenses, reuse of similar biocatalysts for delayed timeframes, development of the continuous process which might be performed beyond the normal waste of time rate, better substrate use, decreased risk for microbial contamination, simplified process design, constant product quality and assurance of cells, and quicker fermentation rates. However, the two most significant disadvantages to be kept in mind are the complexity of the production process and cost constraints.[39]

Applications of immobilized yeast cells have been explored in wine production to reduce labor requirements, simplify time-consuming procedures, and reduce cost. To be convenient in wine production, this method is easy to perform in industrial conditions which avoids oxidation and contamination in wine. The main objective in wine and cider-making is to achieve the adequate quality of the product.[35] Microbial cell immobilization can improve the efficiency of malolactic fermentation. It has the ability for cell stability, cell recycling, cell viability, and improvement of product quality. The advantage of using immobilized cells in wine or cider is to simplify the systems for removing microbial cells from batch processes and generating greater tolerance of inhibitory substances, smaller scale fermentation facilities, and possible of using a variety of microbial strains including genetically modified organisms. Immobilization aims to increase the tolerance of malolactic bacteria and speed up the process. In cider production, researchers have focused on alcoholic and malolactic fermentation by co-immobilization of two different species /same species with the help of genetic modification.[11] There are also some disadvantages like cell overgrowth which increases the turbidity of fermented beverages and mechanical stability of the matrix which is used to immobilize microbial cells and loss of activity during prolonged operation.[24] The new challenge is to select a suitable carrier and bioreactor system. The factors should be taken such as product quality, safety investments, operating costs, stability, and legality.[26]

Immobilized cell/enzyme technology has been applied in dairy and meat fermentations and enzymatic processes. Immobilization may be beneficial in improving the stability of probiotics as well as protective cultures in fermented foods. This technology, immobilized, provides protection to cells during fermentation and drying process, protection against bacteriophage attack, inhibition or killing of undesirable flora, enhances the survival of cells to heating and freezing, improves the stability of cells during storage, accelerates flavor development.[4] Lactic acid bacteria are used in fermented products.[59] Basically, cells are microentrapped into natural gel particles and added to the growth medium. The most common techniques used for the immobilization of cultures in gels are extrusion and emulsion techniques. However, from the point of industrial manufacturers of probiotics, it becomes more desirable when system is used for continuous inoculation of milk such as in a dairy plant.[60]

18.5.2 PHARMACEUTICS

Pharmaceutical/biomedical applications for controlled/sustained drug administration are one of the major applications of encapsulation technology. This drug delivery technology could be a replacement for therapeutic agents, and gene therapy and can use for making vaccines for treating cancer, AIDS, tumors, and diabetes.[8] Proteins such as insulin, erythropoietin (used to treat anemia), and growth hormone are some of the best examples of those drugs that could be benefited compared to a new form of oral delivery. Based on this novel technique of drug delivery, Lupin has already developed the world's first cephalexin and cefadroxil antibiotic tablets in a market for the treatment of some bacterial infections. Moreover, ZORprin CR tablets which are controlled versions of aspirin are used for relieving the symptoms of arthritis. Other tablets are also launched in the market for the treatment of several diseases such as Niaspan CR which is used to improve cholesterol levels and therefore reduces the risk of a heart attack. An antidiabetic medicine, that is, Glucotrol (Glipizide SR) is used in order to control high blood pressure.

18.5.3 ENERGY GENERATION

Gaseous deuterium (a fusion fuel) in hollow plastic microspheres is used to harness nuclear fusion in order to produce electrical energy.[49, 60] their capsules are multilayered and the inner layer which helps to compress the fuel is made up of polystyrene shell which is about 3-mm thick. Another one is made up of a layer of polyvinyl alcohol which is about 3-mm thick which helps to retard the diffusion of deuterium from the capsule. Its outer layer, that is, the ablator is about 50-mm thick which is made up of highly cross-linked polymer (2-butene). During the process of fusion experiments, the energy released from high-powered laser beams is being absorbed by the surface of the microcapsule shell. As the outer part of the shell burns off, the reaction force steps up the rest of the shell inward. The deuterium present inside the shell gets compressed and heated. Due to this, the density and temperature in the center of the capsule increase and lead to the fusion of deuterium nuclei to give helium, tritium, and other particles to release a large amount of energy and this process of fusion is known as inertial confinement fusion (ICF).

18.6 SUMMARY

Undoubtedly, nanotechnology is proved to be an effective technology in the field of the food industry, offering a dynamic range of applications in the same industry. Currently, this technique is emerging in various domains of the food industry with and without some limitations. All the methods of nano-encapsulation have provided a quality-based packaging and prevented the food from microbial contamination. The approach of nanotechnology in food packaging is considered as active packaging and is the most recent way of preserving food in the modern era. This technology has proved its best by putting some high standards of food safety; however, the harmful effects of the nanomaterials (if any) used in this technique are yet unknown and a deep study concerning the same is required.

KEYWORDS

- **nanoencapsulation**
- **active food packaging**
- **nanotechnology**
- **nanocarriers**
- **nanomaterials**
- **food technology**
- **biobased materials**
- **antimicrobial agents**
- **biocatalytic**
- **emulsifications**

REFERENCES

1. Al-Tayyar, N. A.; Youssef, A. M.; Al-Hindi, R. Antimicrobial Food Packaging Based on Sustainable Bio-Based Materials for Reducing Foodborne Pathogens: A Review. *Food Chem.* **2020,** *310*, 125915.
2. Assadpour, E.; Jafari, S. M. Advances in Spray-Drying Encapsulation of Food Bioactive Ingredients: From Microcapsules to Nanocapsules. *Annu. Rev. Food Sci Technol.* **2019,** *10*, 103–131.
3. Autio, K. A.; Dreicer, R.; Anderson, J.; Garcia, J. A.; Alva, A.; Hart, L. L.; Scher, H. I. Safety and Efficacy of BIND-014, a Docetaxel Nanoparticle Targeting Prostate-Specific Membrane Antigen for Patients with Metastatic Castration-Resistant Prostate Cancer: A Phase 2 Clinical Trial. *JAMA Oncol.* **2018,** *4* (10), 1344–1351.
4. Beesley, L.; Moreno-Jiménez, E.; Gomez-Eyles, J. L. Effects of Biochar and Greenwaste Compost Amendments on Mobility, Bioavailability and Toxicity Of Inorganic and Organic Contaminants in a Multi-Element Polluted Soil. *Environ. Pollution* **2010,** *158* (6), 2282–2287.
5. Bun Ng, T.; Chi Fai Cheung, R.; Ho Wong, J.; Juan Y. X. Antimicrobial Activity of Defensins and Defensin-Like Peptides with Special Emphasis on Those from Fungi and Invertebrate Animals. *Curr. Protein Peptide Sci.* **2013,** *14* (6), 515–531.
6. Davis, M.; Etheridge, A. *Louis Bachelier's Theory of Speculation*. **2006,** p 23. https://f-origin.hypotheses.org/wp-content/blogs.dir/1596/files/2014/12/Mark-Davis-Talk.pdf
7. De Kruif, C. G.; Weinbreck, F.; de Vries, R. Complex Coacervation of Proteins and Anionic Polysaccharides. *Curr. Opin. Colloid Interf. Sci.* **2004,** *9* (5), 340–349.
8. Debeaufort, F.; Voilley, A. Lipid-Based Edible Films and Coatings. In *Edible Films and Coatings for Food Applications*; Springer International Publishing: New York, 2009; pp 135–168.

9. Devlieghere, F.; Vermeiren, L.; Debevere, J. New Preservation Technologies: Possibilities and Limitations. *Int. Dairy J.* **2004,** *14* (4), 273–285.
10. Đorđević, V.; Paraskevopoulou, A.; Mantzouridou, F.; Lalou, S.; Pantić, M.; Bugarski, B., Nedović, V. Encapsulation Technologies for Food Industry. In *Emerging and Traditional Technologies for Safe, Healthy and Quality Food*; Springer International Publishing: Cham, Switzerland, 2016; pp 329–382.
11. Ehyaei, M. A.; Ahmadi, A.; Assad, M. E. H.; Hachicha, A. A.; Said, Z. J. S. E. Energy, Energy and Economic Analyses for the Selection of Working Fluid and Metal Oxide Nanofluids in a Parabolic Trough Collector. *Solar Energy* **2019,** *187*, 175–184.
12. Galindo-Rodriguez, S.; Allemann, E.; Fessi, H.; Doelker, E. Physicochemical Parameters Associated with Nanoparticle Formation in the Salting-Out, Emulsification-Diffusion, and Nanoprecipitation Methods. *Pharm. Res.* **2004,** *21* (8), 1428–1439.
13. Gharsallaoui, A.; Roudaut, G., Chambin, O.; Voilley, A.; Saurel, R. Applications of Spray-Drying in Microencapsulation of Food Ingredients: An Overview. *Food Res. Int.*, **2007,** *40* (9), 1107–1121.
14. Gouin, S. Microencapsulation: Industrial Appraisal of Existing Technologies and Trends. *Trends Food Sci Technol.* **2004,** *15* (7–8), 330–347.
15. Hadaruga, N. G.; Hadaruga, D. I.; Paunescu, V.; Tatu, C.; Ordodi, V. L.; Bandur.; G Lupea, A. X. Thermal Stability of the Linoleic Acid/α-and β-Cyclodextrin Complexes. *Food Chem.* **2006,** *3*, 52–68.
16. Hadidi, M.; Pouramin, S.; Adinepour, F.; Haghani, S.; Jafari, S. M. Chitosan Nanoparticles Loaded with Clove Essential Oil: Characterization, Antioxidant and Antibacterial Activities. *Carbohyd. Polym.* **2020,** *236*, 116075.
17. Haj, Y.; Teng, D. Reaffirming the Importance of Managerial Discretion in Corporate Governance: A Comment on Andersen (2017). *Corporate Governance: Int. J. Busi. Soc.* **2019,** 33–49.
18. Heyang, J. I. N.; Fei, X. I. A.; Jiang, C.; Yaping, Z. H. A. O.; Lin, H. E. Nanoencapsulation of Lutein with Hydroxypropylmethyl Cellulose Phthalate by Supercritical Antisolvent. *Chinese J. Chem. Eng.* **2009,** *17* (4), 672–677.
19. Hoseinnejad, M.; Jafari, S. M.; Katouzian, I. Inorganic and Metal Nanoparticles and Their Antimicrobial Activity in Food Packaging Applications. *Crit. Rev. Microbiol.* **2018,** *44* (2), 161–181.
20. Ibrahim, I.; Hassan, M. A.; Abd-Aziz, S.; Shirai, Y.; Andou, Y.; Othman, M. R.; Zakaria, M. R. Reduction of Residual Pollutants from Biologically Treated Palm Oil Mill Effluent Final Discharge by Steam Activated Bioadsorbent from Oil Palm Biomass. *J. Clean. Prod.* **2017,** *141*, 122–127.
21. Ingrao, C.; Nikkhah, A.; Dewulf, J.; Siracusa, V.; Ghnimi, S.; Rosentrater, K. A. Introduction to the Special Issue "Sustainability Issues of Food Processing and Packaging: The Role of Life Cycle Assessment". *Int. J. Life Cycle Assess.* **2021,** *26*, 726–737.
22. Jafari, S. M.; Assadpoor, E.; He, Y.; Bhandari, B. Encapsulation Efficiency of Food Flavours and Oils During Spray Drying. *Dry. Technol.* **2008,** *26* (7), 816–835.
23. Jafari, S. M.; He, Y.; Bhandari, B. Production of sub-micron emulsions by ultrasound and microfluidization techniques. *Journal of Food Engineering*, **2007,** *82*(4), 478–488.
24. Jafarzadeh, S.; Jafari, S. M.; Salehabadi, A.; Nafchi, A. M.; Kumar, U. S. U.; Khalil, H. A. Biodegradable Green Packaging with Antimicrobial Functions Based on the

Bioactive Compounds from Tropical Plants and Their By-Products. *Trends Food Sci. Technol.* **2020,** *100*, 262–277.

25. Kalia, S.; Dufresne, A.; Cherian, B. M.; Kaith, B. S., Avérous, L.; Njuguna, J.; Nassiopoulos, E. Cellulose-Based Bio-and Nanocomposites: A Review. *Int. J. Polym. Sci.* **2011,** *2011*, 20–58.
26. Khaneghah, A. M.; Hashemi, S. M. B.; Limbo, S. Antimicrobial Agents and Packaging Systems in Antimicrobial Active Food Packaging: An Overview of Approaches and Interactions. *Food Bioprod. Process.* **2018,** *111*, 1–19.
27. Kosseva, M. R.; Panesar, P. S.; Kaur, G.; Kennedy, J. F. Use of Immobilised Biocatalysts in the Processing of Cheese Whey. *Int. J. Biol. Macromol.* **2009,** *45* (5), 437–447.
28. Liu, F.; Lin, L.; Zhang, Y.; Wang, Y.; Sheng, S., Xu, C.; Chen, X. A Tumor-Microenvironment-Activated Nanozyme-Mediated Theranostic Nanoreactor for Imaging-Guided Combined Tumor Therapy. *Adv. Mater.* **2019,** *31* (40), 1902885.
29. Luthans, F.; Youssef-Morgan, C. M. Psychological Capital: An Evidence-Based Positive Approach. *Annu. Rev. Org. Psychol. Org. Behav.* **2017,** *4*, 339–366.
30. Manojlović, V.; Nedović, V. A.; Kailasapathy, K.; Zuidam, N. J. Encapsulation of Probiotics for Use in Food Products. In *Encapsulation Technologies for Active Food Ingredients and Food Processing*; Springer International Publishing: New York, 2010; pp 269–302.
31. Mueller, F.; Procter, S.; Buchana, D. Teamworking in Its Context, Antecedents, Nature and Dimensions. *Human Relat.* **2000,** *53* (11), 1387–1424.
32. Nedovic, V.; Kalusevic, A.; Manojlovic, V.; Levic, S.; Bugarski, B. An Overview of Encapsulation Technologies for Food Applications. *Procedia Food Sci.* **2011,** *1*, 1806–1815.
33. Nile, S. H.; Baskar, V.; Selvaraj, D.; Nile, A.; Xiao, J.; Kai, G. Nanotechnologies in Food Science: Applications, Recent Trends, and Future Perspectives. *Nano-Micro Lett.* **2020,** *12*(1), 1–34.
34. Niranjana Prabhu, T.; Prashantha, K. A Review on Present Status and Future Challenges of Starch Based Polymer Films and Their Composites in Food Packaging Applications. *Polym. Compos.* **2018,** *39* (7), 2499–2522.
35. Pires, D. P.; Oliveira, H.; Melo, L. D.; Sillankorva, S.; Azeredo, J. Bacteriophage-Encoded Depolymerases: Their Diversity and Biotechnological Applications. *Appl. Microbiol. Biotechnol.* **2016,** *100* (5), 2141–2151.
36. Reis, H. T.; Crasta, D.; Rogge, R. D.; Maniaci, M. R.; Carmichael, C. L. Perceived Partner Responsiveness Scale (PPRS). *Sourcebook Listen. Res.: Methodol. Measures* **2017,** 516–521.
37. Ren, B.; Yan, F.; Kuang, Y.; Li, N.; Zhang, D.; Zhou, X.; Zhou, H. Improved Base Editor for Efficiently Inducing Genetic Variations in Rice with CRISPR/Cas9-Guided Hyperactive hAID Mutant. *Mol. Plant* **2018,** *11* (4), 623–626.
38. Sanguansri, P.; Augustin, M. A. Nanoscale Materials Development—A Food Industry Perspective. *Trends Food Sci. Technol.* **2006,** *17* (10), 547–556.
39. Sarwan, J.; Pandey, P.; Bose, J. C. Proteases Immobilization and Pharmacological Aspects of Matrix Metalloproteinases. *Ann. Roman. Soc. Cell Biol.* **2021,** 3604–3615.
40. Sekar, A.; Bialas, A. R.; De Rivera, H.; Davis, A.; Hammond, T. R.; Kamitaki, N.; McCarroll, S. A. Schizophrenia Risk from Complex Variation of Complement Component 4. *Nature* **2016,** *530* (7589), 177–183.

41. Shao, Y.; Gan, Z.; Epifanovsky, E.; Gilbert, A. T.; Wormit, M.; Kussmann, J.; Rassolov, V. A. Advances in Molecular Quantum Chemistry Contained in the Q-Chem 4 Program Package. *Mol. Phys.* **2015,** *113* (2), 184–215.
42. Sharma, H.; Bajwa, J. Potential Role and Mechanism of Probiotics. *Ann. Roman. Soc. Cell Biol.* **2021,** 3616–3624.
43. Siqueira, G.; Bras, J.; Dufresne, A. *Luffa cylindrica* as a Lignocellulosic Source of Fiber, Microfibrillated Cellulose and Cellulose Nanocrystals. *BioResources* **2010,** *5* (2), 727–740.
44. Solans, C.; Izquierdo, P.; Nolla, J.; Azemar, N.; Garcia-Celma, M. J. Nano-Emulsions. *Current Opinion Colloid Interf. Sci.* **2005,** *10*(3–4), 102–110.
45. Struller, C. F.; Kelly, P. J.; Copeland, N. J. Aluminum Oxide Barrier Coatings on Polymer Films for Food Packaging Applications. *Surface Coat. Technol.* **2014,** *241*, 130–137.
46. Tang, J.; Yu, J. X.; Hubbard-Lucey, V. M.; Neftelinov, S. T.; Hodge, J. P.; Lin, Y. Trial Watch: The Clinical Trial Landscape for PD1/PDL1 Immune Checkpoint Inhibitors. *Nat. Rev. Drug Discov.* **2018,** *17* (12), 854–856.
47. Tavassoli-Kafrani, E.; Shekarchizadeh, H.; Masoudpour-Behabadi, M. Development of Edible Films and Coatings from Alginates and Carrageenans. *Carbohydr. Polym.* **2016,** *137*, 360–374.
48. Tolstoguzov, V. Some Thermodynamic Considerations in Food Formulation. *Food Hydrocoll.* **2003,** *17* (1), 1–23.
49. Tsochatzis, E. D.; Mieth, A., Lopes, J. A.; Simoneau, C. A Salting-out Liquid-Liquid Extraction (SALLE) for the Analysis of Caprolactam and 2, 4-di-Tert Butyl Phenol in Water and Food Simulants. Study of the Salinity Effect to Specific Migration from Food Contact Materials. *J. Chromatogr. B*, **2020,** ***1156*,** 122301.
50. Venkatakrishnan, A. J.; Deupi, X., Lebon, G.; Tate, C. G., Schertler, G. F.; Babu, M. M. Molecular Signatures of G-Protein-Coupled Receptors. *Nature* **2013,** *494* (7436), 185–194.
51. Voilley, A.; Etiévant, P. (Eds.) *Flavour in food.* Elsevier BV: Amsterdam: The Netherlands, 2006; p 320.
52. Walkowiak, S.; Gao, L.; Monat, C.; Haberer, G.; Kassa, M. T.; Brinton, J.; Pozniak, C. J. Multiple Wheat Genomes Reveal Global Variation in Modern Breeding. *Nature* **2020,** *588* (7837), 277–283.
53. Wei, S.; Li, A.; Liu, J. C.; Li, Z.; Chen, W.; Gong, Y.; Li, Y. Direct Observation of Noble Metal Nanoparticles Transforming to Thermally Stable Single Atoms. *Nat. Nanotechnol.* **2018,** *13* (9), 856–861.
54. World Health Organization (WHO) *World Health Statistics*, **2015,** p 264. https://www.who.int/docs/default-source/gho-documents/world-health-statistic-reports/world-health-statistics-2015.pdf.
55. Wu, X.; Chen, Y.; Xing, Z.; Lam, C. W. K.; Pang, S. S.; Zhang, W.; Ju, Z. Advanced Carbon-Based Anodes for Potassium-Ion Batteries. *Adv. Energy Mater.* **2019,** *9* (21), 1900343.
56. Yang, L.; Huang, J. M.; Zu, Y. G., Ma, C. H.; Wang, H.; Sun, X. W., Sun, Z. Preparation and Radical Scavenging Activities of Polymeric Procyanidins Nanoparticles by a Supercritical Antisolvent (SAS) Process. *Food Chem.* **2011,** *128*(4), 1152–1159.
57. Zasypkin, D.; Porzio, M. Glass encapsulation of flavours with chemically modified starch blends. *Journal of microencapsulation*, **2004,** *21*(4), 385–397.

58. Zhao, Y.; Yin, X., Qin, H.; Zhu, F., Liu, H.; Yang, W.; Deng, H. Two Supporting Factors Greatly Improve the Efficiency of Human iPSC Generation. *Cell Stem Cell* **2008,** *3* (5), 475–479.
59. Zubair, M.; Wang, S., Zhang, P.; Ye, J.; Liang, J.; Nabi, M.; Cai, Y. Biological Nutrient Removal and Recovery from Solid and Liquid Livestock Manure: Recent Advance and Perspective. *Bioresour. Technol.* **2020,** *301*, 122823.
60. Zuidam, N. J.; Shimoni, E. Overview of Microencapsulates for Use in Food Products or Processes and Methods to Make Them. In *Encapsulation Technologies for Active Food Ingredients and Food Processing*; Springer International Publishing: New York, 2010; pp 3–29.

CHAPTER 19

ADVANCED BIONANOMATERIALS FOR CATALYSIS AND SURFACE PROCESSES

NOSHIN AFSHAN[1], SAJID MAHMOOD[2], ALINA BARI[3], and SOBIA NOREEN[4]

ABSTRACT

Owing to the programmable architecture and functions of the bionanomaterials, they have been considered the spine of nanoscience in this era showing potent applicability in various fields, including biotechnology, chemistry, material science, and specifically catalysis and surface chemistry. Introducing a variety of functional groups, for example, polymers, lipids, oligonucleotides (RNA/DNA), and peptides or proteins (viruses and enzymes), to the growing nanostructures impart modification in their architecture and improvement in their catalytic activity, thereby enabling them to mimic the biological activity with several advantages including accelerated reaction rate, enhanced catalytic activity, reusability, reduced reaction time, low energy consumption, minimum substrate concentration, and lack of secondary pollution. Moreover, they show excellent physical–chemical, biodegradable, and biocompatible properties, multi-active binding sites, high reactivity, tunable solubility, long-term stability, and shape accuracy, which is essential for the target-specific mechanism. For example, enzymes mimicking enzymatic activity have been potentially employed in immunoassays, biomedical, antimicrobial therapy, and biosensors. This chapter will deal with the types and recent advances in bionanomaterials, the main challenges of redesigning synthetic strategies for enhanced catalytic activity,

Enzyme Inactivation in Food Processing: Technologies, Materials, and Applications. Megh R. Goyal, PhD, Junaid Ahmad Malik, PhD & Ravi Pandiselvam, PhD (Eds.)

limitations, and emerging trends in catalysis and surface process. Moreover, the applications of catalysis and surface processes undergone by bionanomaterials would also be explained, along with the toxicity perspectives of bionanomaterials. Hence, it would present keen insight into the novel functionalities of bionanomaterials enduring catalysis.

19.1 INTRODUCTION

The field of biomaterials is related to the Greeks and the Chinese due to their work for the betterment of the human body affected by diseases. Using different methods and materials, many devices were introduced to treat diseases such as cardiovascular, coronary artery, orthopedic surgery, wound filling, rigidity to the skeleton, skin treatment, etc. Scientists are responsible for the development of soft and hard tissues replaced by human body organs. With the discovery of the nanoscale, scientists have worked too much on biomaterials and changed their properties. These nanoparticles (NPs) interact with organic ligands having a smaller size, therapeutic molecules, proteins, cell membranes, and DNA.[38] The interaction between biomolecules and NPs depends on colloidal forces as well as bio–physicochemical interactions. The interaction between nanomaterials and biomaterials imparts unique physical properties between them and gives us nanoconjugates. Later on, studies are performed to understand physical properties such as size, shape, physiochemical functionality, surface charge, and composition of bionanomaterials. The bionanomaterials help to give new nanomaterials like the last generation of a computer generates a new generation. The intrusion of nanomaterials can enhance enzyme catalysts in biochemical reactions and their activity. Various nanomaterials, such as dendrimers, polymers, cyclodextrins, biomolecules, metal complexes, fullerenes, and porphyrins, have been seen as mimics in many chemical reactions.

Because they are the same size and have solid catalytic activity, many NPs can behave like proteins. In catalysis, some NPs produce the same compounds as enzymes. Nanozymes have been employed as biosensing agents, cytoprotection against diverse biomolecular components of the cell, illness detection and management, environmental remediation, and so on. For the monitoring of biomolecules, inorganic substances, toxins, antioxidants, and analytes, nanostructures have been used as the horseradish peroxidase (HRP) mimicking enzyme.[8] Nanozymes such as magnetic nanocomposites and carbon nanotubes (CNTs) are employed in metal ion sensing to create this susceptible sensor for detecting minimal amounts of Cu^{+2}. By combining trinitrotoluene (TNT) with glucose oxidase, NPs are employed as glucose sensors. Bacterial DNA was used to identify nucleic acids using Fe_3O_4 MNPs in a sandwich test for evaluating bacteria in tap water and drinking water sources. A calorimetric technique for detecting DNA was developed by modulating the peroxide activity of Au NPs in graphene.[28] Nanozymes, such as cerium oxide NPs, are utilized in the identification of cancer cells. Nanoparticles have oxidase activity for a variety of organic substrates.

Cellulose nanofibrils are up to 520-nm in thickness and several millimeters in length, having greater surface area. Nanocellulose makes gel-like materials that are environmentally friendly. The cellulose-producing micro-organisms, including algae, microbes, and tunicates, participate in the synthesis of proteins causing the polymerization of glucan reactions. There is considerable evidence that nanomaterials have unique biological with molecules or bulk materials. Bionanomaterials can undergo surface modification by forming tree-like structures called dendrimers.[26] As a stepwise process forms the dendrimers, they can easily control molecular structure and size—a primary important function in surface modification of bionanomaterials attachment of functional groups. To get better efficient bionanomaterials, mixed functional groups are used.

Cerium oxides are used in green polymers to form for pharmacological treatment. Chitosan coated with iron oxide can be used as antibacterial activity against Pseudomonas aeruginosa. Since NPs exhibit bactericidal activity, chitosan-coated silver NPs and mesoporous silica have been successfully used for antibacterial activity. Bovine serum albumin (BSA) and chitosan show good activity against certain species of bacteria. Sida acuta plant of the Malvaceae family is used as a source of bionanomaterials by some modification. Cellulose can be modified into cellulose nanocrystals having a diameter (100–500 nm) and nanofibers. With the modification of cellulose crystals, different properties are studied, such as morphology, crystallinity, and thermal stability.[10]

This chapter presents keen insight into the recent advances in bionanomaterials, the main challenges to designing efficient synthetic strategies for enhanced catalytic activity, limitations, and emerging trends in catalysis and surface process. Moreover, the applications of catalysis and surface processes undergone by bionanomaterials, along with the toxicity perspectives of bionanomaterials, will also be explained. Hence, it would deal with the novel functionalities of bionanomaterials enduring catalysis.

19.2 BIONANOMATERIALS UNDERGOING CATALYSIS AND SURFACE PROCESS

Nanoparticles, specifically gold NPs, have been used for millennia by Chinese and Egyptian women for medicinal and entertainment purposes and for decorating different materials such as the famous Lycurgus cup. Though had already been documented research on nanoparticles since the early 17th century, the Faraday group came up with the idea of the impact of light in the middle of the nineteenth century, followed by Mie's ideology regarding Plasmon.[3,7,45] The metal-based NPs of palladium and platinum metals had been utilized as catalysts for hydrogenation which laid down the basis of catalysis for the first time in 1941.[22] It was followed by Paul Sabatier's discovery of catalytic hydrogenation employing nickel NPs. The reduction in the size of the conventional materials up to nanometer scale, for example, the gold NPs up to less than 5nm, rendered enhancement in their catalytic activity.[18] The nanomaterials exhibiting the characteristic of catalysis have found their applicability in such fields as optics, bionanotechnology, and bioengineering.[45] For example, plasmatic excitation of the gold NPs by visible light irradiation produces electrons that induce semiconducting materials including TiO_2 to modify the substrates. Catalytic nanoclusters have become more common, leading to special consideration in that the nanocluster framework is well-organized for theoretical models and discoveries supporting structure-function correlations. However, the most commonly studied catalytic systems are tiny metal-based nanostructures employed for homogeneous or heterogeneous catalysis. They have been found feebly stabilized by such ligands as inorganic molecules, polymers, solids, and ionic liquids. They retain simple preparation and the enhanced catalytic activity resulting from an easily accessible approach to the NPs surface.[34] Even though various successful high-temperature techniques have been developed throughout the twentieth century, a significant component of conventional heterogeneous catalysis has merged with nanotechnology.[46] The worth of catalytic nanostructures is equally unique for homogenous and heterogeneous catalysis.[34] Heterogeneous catalysis created templates using zeolites which were then expanded to microstructure silica, oxides, and alumina, allowing sub nanoclusters to be stabilized down to the nanoscale.[49] Novel 1-, 2- and 3D support has achieved remarkable nanotubes, metal-organic framework, and graphene derivatives.[43] In addition to stabilizing the nanomaterials, they work with the NP surface to stimulate the substrates such as the positive synergy found among the transition metal atoms present in alloys.[9] Because of these synergistic actions, N-doped carbon supports outperformed undoes analogs. Catalytic NP families include binary and ternary particles containing elements from the leading group and mono- and bimetallic nanomaterials.

Nanomaterials are NP-support ensembles that are complicated. The complexity of the nanocatalyst design is frequently reminiscent of Mother Nature's. These biomimetic features are essential in the construction of processes that require catalysis.[21] In this regard, light plays a vital function, particularly in activating tiny molecules, water splitting, and hydrogen generation.[37] Reducing the size of the nanocatalyst to a single atom has received a lot of interest recently since no atoms situated at the inside of the NPs are lost in this instance. However, single-atom supported catalysts exhibit greater specificity and improved selectivity as compared to the polydisperse nanocatalysts. Finally, an understanding of heterogeneous and homogeneous catalysis, bioinorganic chemistry, metallic organ chemistry, material science, surface science, and catalysis may enrich the concept of nanocatalysis.[33]

19.3 PROPERTIES OF BIONANOMATERIALS

Bionanomaterials show different properties compared to biomaterials such as physical, chemical, biological, mechanical, optical, electric, and catalytic properties. Different techniques such as microscopy and spectroscopy study the properties of bionanomaterials. It is understood as the size of the particle decreases, and the particle provides a maximum surface-to-volume ratio leading to the change in their properties. Owing to pharmacokinetics and other properties such as size, shape, and agglomeration state of bionanomaterials, compatibility of the bionanomaterials has been decided for different species such as binding with proteins and receptors drug delivery to target and transport across epithelial cells. The morphology of bionanomaterials is studied based on the surface area to volume ratio and particle size distribution. The smaller sized bionanomaterials have significant scope in our life as they interact easily with molecules of the human body. The primary benefit of bionanomaterials is that they decide the final fate of nanomaterials. The quantum effect, volume, and size of the bionanomaterials help describe the altered mechanical properties of biomaterials. As NPs are interlinked with another compatible material, they experience enhancement in their intergranular and intragranular structure leading to the change in their mechanical properties. For example, the addition of 3% palm oil as nanomaterial to epoxy composites increased their tensile strength.[19]

19.4 ADVANTAGES OF BIONANOMATERIALS

Bionanomaterials refer to biodegradable materials of nanometer size along any of its dimensions. They retain various shapes such as round, tubular, square, lozenge, hexagonal or irregular. They differ in their chemical composition, including organic or inorganic and amorphous or crystalline. The nanoscale size and peculiar structure predict their physicochemical characteristics dictating their role in the industry, including catalysis and surface process. Their widespread applications induce temptation and curiosity in their future perspectives. Moreover, owing to public awareness, reduction in fossil resources and increasing waste have attracted researchers' attention to the field of material chemistry; for example, waste from agriculture such as lignocellulose has been turned into bionanomaterials to strengthen bionanocomposite polymer, which retains multifunctional behavior and capability to give rise to naturally biodegradable materials.[20,25]

Although bionanomaterials are not abundant on our earth, several types of bionanomaterials have been developed with a unique set of characteristics and applications in various fields. Biomolecules such as vitamins, antioxidants, lipids, and proteins have been coupled with nanomaterials to rise to bionanomaterials. This technology has been bringing a revolution in catalysis and clean energy generation. For example, the photocatalytic activity of ZnO-based bionanomaterials has been employed to catalyze the biodegradation of phenol and chlorinated phenols.[30]

19.5 CATALYSIS AND SURFACE PROCESS

It deals with the kinetics of atomic-level surface reactions, their mechanisms, and the correlation between the composition and structural features of their surface. The surface composition of the bionanocatalyst has decided the fate of the catalytic reaction. Various techniques and specific equipment have been designed which serve as a bridge to connect the low-pressure domain required for these approaches with the high-pressure domain required for catalysis. The response generated from such a small discrete single crystal surface is equal to that of catalysts that retain a large surface area. The exemplary investigations mostly on surfaces show that surface imperfections and additives have a significant role in regulating surface reactions and the resulting product distribution.

In the context of heterogeneous catalysis, both dynamic and static dimensions on finely tuned surfaces have been considered. Zeolites have been reported as the first high-tech catalysts produced and developed utilizing these concepts. The alumina silicates which retain improved surface area have been synthesized in a range of compositions and pore sizes. The organic molecules integrated into the oxides during the synthetic procedure regulated their pore structure and size leaving behind channels and cavities with distinct shapes when treated under heat. Hence, owing to the capability of the pore structure to control the size and nature of the organic species passing through the catalyst, they have been employed in high amounts for shape-selective hydro-cracking.[3,7,45]

19.5.1 NEED FOR ENHANCED CATALYTIC ACTIVITY

Catalytic activity is the ability of the catalyst to how much speed up the rate of a chemical reaction. Some catalysts such as enzymes provide the surface to accelerate the chemical reaction rate. Similarly, bionanomaterials provide improved surface area and enhanced chemical reaction rate. Carbon nanotubes and graphene with modified biomolecules have been used in biosensing, diagnostics, and bioanalysis. The first biosensor of CNTs has advantages over the small size and large surface area, easy biomolecule mobilization, and excellent electron ability. Bionanomaterials modified by CNT show enhanced activity as a redox cofactor and redox enzymes. As fluorescent agents, bionanomaterials show enhanced activity in magnetic resonance imaging (MRI). Bionanomaterials targeted to specific disease provides anatomical data related to physiology and gives precise results about a diagnosis such as the detection of cancer at early stages.[47]

19.5.2 SURFACE PROCESS AT THE MOLECULAR LEVEL

The surface process of bionanomaterials depends on factors (1) the nanomaterial interface, (2) the material in which engineered nanomaterial is suspended, and (3) the properties and influence of biological entities. The primary factor that influences the properties of nanomaterials is the surface. The silica particle of 200 nm and larger size after adsorption conformational changes in myoglobin and BSA protein. Surface oxidation is an important parameter that controls the toxicity of cerium oxide NPs. From the results, it was determined only Ce^{3+} was the cause of toxicity. The second factor that causes a change in engineered bionanomaterials is the external medium. Like CuO NPs that dissolve entirely in the amino acid-rich environment, NaCl does not affect the solubility of these NPs.[35]

The factors that affect bionanomaterials' properties are their surface composition and their ability to affect the surrounding environment. It has been understood that natural organic matter can control NPs' aggregation by creating an adsorbed layer on the NPs' surface. It enhances the steric ability and colloidal stability and allows the engineered NPs to remain in the present media. The ability of engineered nanomaterials to interact with different cells and organisms depends on the substance they attach. The interaction between engineered NPs and biomolecules has been studied extensively in complex-shaped proteins. The attachment of NPs with proteins and their adsorption-desorption is studied by considering corona as standard. If the surface of engineered NPs is tightly adsorbing proteins, then it is the hard corona, the otherwise soft corona. It is considered, there is the presence of soft corona on the hard corona. Slow exchange time, significant conformational changes, solid binding ability, and long residence time are essential characteristics of this layer.[35]

19.5.3 CATEGORIES OF BIONANOMATERIALS CARRYING OUT CATALYSIS AND SURFACE PROCESS

There are different types of nanomaterials carrying out catalysis and surface process. Out of which are engineered, incidental, and natural bionanomaterials. Engineered bionanomaterials are engineered in such a way to give required properties for human beings. On the other hand, incidental bionanomaterials are produced unintentionally by side products. These may cause by side products due to mechanical or industrial processes through combustion or vaporization. For example, modification of cellulose or protein gives unexpected results as we want. The one fundamental and significant source of bionanomaterials is a natural source.[1]

Bio-NPs consist of rod-like and icosahedral plant viruses and ferritin. Bionanomaterials can exist as nanofibers such as lignocellulose. To produce nanofibers with minimal mechanical labor, high-intensity homogenizers, cryo-crushing, and ultrasonication are utilized. However, several chemical techniques such as oxidation/alkaline acid pretreatment and hydrolysis reduce the amount of energy needed to synthesize bionanomaterials. The oxidation-based homogenizing mechanics have been employed to manufacture ultrasonic nanofibers 3–5-nm wide and 2–3-m long from bleached wood pulp. Nanotubes are made up of graphite, and inorganic compounds such as zeolites, lipids, and cyclodextrins. They drew attention because of their excellent efficiency and control via design and self-organization. Helical nanotubes, hollow nanorods, nanowires, stacked rings, and helically juxtapositioned truncated wedges are examples of different shapes of bionanomaterials. The nanotube diameter is regulated by the quantitative cost of amino acid residues, and the amino acid side chains predict the characteristics of the nanotube boundary. Bionanomaterials, like piezoelectric materials, may exist as nanofilms. Piezoelectric bionanomaterials are those that can create electricity when subjected to external pressure. Due to the cytotoxic nature of lead-based materials, certain traditional piezoelectric materials, such as poly(vinylidene fluoride) (PVDF), are costly, non-renewable, non-biodegradable, and lack biocompatibility. Piezoelectricity may be conducted on a variety of biomaterials such as chitin and collagen. Peptides, amino acids, and cellulose are all examples of macromolecules. These piezoelectric bionanomaterials are utilized to gather energy as biosensors and as wound healers.[1,6,11]

19.5.3.1 LIPIDS

Biocompatible polymers such as chitosan and polyethylene glycol have been converted into a modified polymeric NP. Their applicability in various fields is evident as selegiline-coated NPs have been utilized in brain therapy and Nano-DOPA polymeric system to treat Parkinson's disease. Hollow gold nanoshells bonded to the surface of a robust liposome have been utilized for medication delivery. The lysosome can encapsulate dopamine and regulate drug release spatially. In Parkinson's disease, diacylglycerol is mixed with a linker and linked to a peptide. Lipids interact with various surfactants to generate solid-lipid NPs composed of various lipids and surfactants such as Tween, Span, and poloxamer. Curcumin and piperine-based NPs can enter brain tissues and improve the anti-Parkinson efficacy of the administered medication. Lipid binding proteins may bind to 125 amino acids to create antiparallel beta-strands that form shell-like structures. The primary function of bile acid proteins is to bind and transport bile acid throughout the cytoplasm. Although bile acid does not play a significant role in lipid adsorption, it can influence gene expression by acting as ligands.[11]

19.5.3.2 PEPTIDES, APTAMERS, AND PROTEINS

19.5.3.2.1 Peptides

Peptides are biological substances that exist naturally and are present in all living beings playing an essential part in catalytic activities. Peptides, including proteins and hormones, are naturally produced from the transcription procedure in which a specific gene sequence is transcribed into a messenger RNA followed by the subsequent decoding into respective proteins. They frequently operate as hormones, such as insulin, glucagon, and ghrelin, catalyzing biological functions and channeling information through the bloodstream. Being generated in various organs such as glands, intestines, stomach, and brain, they also regulate blood glucose. Their specific binding to the cell surface receptors induces activation, and catalytic function, leading to a profound biological effect. For example, the GLP-1 receptor present in the pancreatic beta cells is activated either by natural GLP-1 or its mimicking analog, lixisenatide, which catalyzes the release of insulin.[6,17]

19.5.3.2.2 Aptamers

Aptamers or artificial antibodies are small-sized single-stranded nucleic acids (ssRNA or ssDNA) isolated from chemically synthesized combined oligonucleotides through in vitro selectivity showing the profound ability of binding to the target site with enhanced selectivities such as drugs, proteins, DNA nanostructures, peptides, and complex tissues. It is their tertiary structure that determines the nature of aptamer binding. The 3D structure, base stacking, hydrophobic, and shape-dependent interactions are all involved in target recognition and binding. Various protocols, including next-generation sequencing, SELEX, and advanced bioinformatics for in vitro aptamers selection, have been designed. Unlike the conventional processes, which rely upon the complex sample pretreatment, expensive detection equipment, and experimental environment to exhibit catalysis and surface process, aptamers are exceptionally advantageous, rendering simple preparation, easy to label, highly stable, target-specific, exceedingly selective, sensitive, and efficient in their mode of action. Besides their widespread usage in the theranostic, detection of proteins, viruses, cells, antibodies, pesticides, heavy metals, and sugars, they have been showing potential applications in the fields of catalysis, surface process, targeted drug delivery, gene therapy, bionanosensors, as optical aptasensors, and electrochemical aptasensors and diagnostic reagents.[27,44]

19.5.3.2.3 Proteins

Engineering topology of protein as bionanomaterials has played a vital role in bionanotechnology, biomedical research, and biological sciences manifesting catalytic activity. It has been serving as an indispensable hazard-free technology tool for bionanotechnology for designing valuable motifs as building blocks to create various nano-size protein systems, proteins having immobilized peptides on NP, and those for application to bioimaging, gene therapy, biosensing, bioelectronic devices, and biocatalysis.[6,17] The amino acids sequence has determined the unique three-dimensional shape and function of the protein. Owing to their peculiar function, proteins have been classified as listed in Table 19.1.

TABLE 19.1 Brief outline of different functions carried out by the proteins.[6,17]

Function	Description
Antibodies	Foreign entities such as viruses bind to them maintaining normal body functions.
Enzymes	They catalyze various in vivo chemical reactions thereby facilitating to build novel materials through decoding the genetic information.
Information carrier	As messenger, they transport signals to coordinate biological activities across different domains.
Building blocks	They serve as shaping and sustenance of the cells and the movement of the body.
Transportation and store-house	They bind to and transport tiny structures across the body.

19.5.3.3 RIBOZYMES

Ribozymes, a subclass of bionanomaterials, are RNA molecules that catalyze phosphodiester bonds and specifically inhibit gene expression through

RNA substrate cleavage, as in cancer, genetic diseases, and viral therapy. The RNA catalytic motif's clamping sequence directs the ribozyme to its target site, thereby stabilizing the complex. They follow Watson-Crick base pairing to bind with the RNA substrate leading to the cleavage of the substrate sequence. In such cases, the divalent ions such as magnesium facilitate the fission reaction, which may occur after any NUH triplet (N being nucleotide and H being A, U, or C) in the target RNA. Kinetics of such reactions vary widely along with the varying combinations of triplet-flanking sequences.[27]

The ribozymes catalyze a series of chemical reactions such as formation, fission, and rearrangement, for example, RNA substrate breakdown, Diels-Alder reaction, N-glycoside bond formation, amide bond formation, acylation, and alkylation without producing the secondary pollutants. Instead, they serve as highly efficient, eco-friendly, and clean recovery tools for the environment. It may elevate their catalytic performance, nuclease resistance, and group diversity by introducing chemical modifications linked to the nucleotides. They may also be expressed from vectors offering the advantage of continuous intracellular production. They often require a carrier for efficient delivery into target cells. A range of nanomaterials such as polymer micelles, liposomes, and nucleic acids inhibit nuclease-based degradation and also improve intracellular transduction.[27,39]

19.5.3.4 ENZYMES

Enzymes are biological catalysts that are protein in nature and accelerate chemical processes in the cell. They are not destroyed during the process, instead, they are reused. There are millions of distinct enzymes, and every enzyme is specialized for a specific chemical reaction.[9,12,17] Some of their peculiar characteristics are as follows:

- They accelerate the biochemical reaction rate in either direction.
- Their activity is dependent on the pH of the medium.
- Enzymes can accelerate the reaction.
- They render active sites for binding to their respective substrates.

19.5.3.5 OLIGONUCLEOTIDES (DNA AND RNA)

Oligonucleotides are oligomers of DNA/RNA having several uses in genetic testing, investigation, and forensics. Any sequence of nucleic acids may be synthesized via the solid-phase chemical procedure, rendering them essential for a polymerase chain reaction, artificial gene synthesis, DNA sequencing, and molecular cloning/probes. They are often present in nature in small single-stranded molecules that tend to control gene expression.[13]

The nucleotide sequence predicts the nature of the whole RNA/DNA molecule, and the suffix indicates their length "-mer," for example, the "25-mer" strand consists of 25 nucleotides (nt). Oligonucleotides quickly attach to their complementary base pairs in a sequence-specific way to create duplexes or, less frequently, higher order hybrids. This fundamental feature underpins the employment of oligonucleotides as probes for identifying particular DNA or RNA sequences. DNA-based microarrays, Southern blots, PCR, fluorescence in-situ hybridization technique, and creation of gene mimics are all examples of such processes that employ oligonucleotides. They are made up of 2'-deoxyribonucleotides that may be modified either at the 2' sugar position or the backbone to undertake various medicinal applications.[13]

19.5.3.6 BIONANOCOMPOSITES

Bionanocomposite show unique properties as compared to individual materials from which they are prepared. They are formed by two materials matrix-like polymer and reinforcement. The structure formed after incorporation such as fibers, mesh whiskers, or lamellae. They have increased mechanical characteristics, with inorganic filler-reinforced polymers being the finest example. Bionanocomposite is also known as biocomposite, nanocomposite, green composite, biohybrids, and biobased polymers. They were first described in 1970 as an emergent domain of bionanohybrids with nanoscale dimensions arising from the conjugates of inorganic solids with biopolymers. Owing to their multidimensional characteristics of biodegradation, high mechanical/optical strength, biocompatible in nature, and antimicrobial activity, they have found applicability in various fields as compared to the composite materials at micro-and macro-level having at least one phase with nanoscale dimension.[2]

19.5.3.7 MICRO-ORGANISMS DRIVEN BIONANOMATERIALS

Several intense products have been developed in the recent decade by integrating nanomaterials and nanobiotechnology against theranostic modalities and clinical pathologies. Synthetic nanomaterials based on nanomedicine have significant hurdles in achieving particular targets and inadequate accumulations on diseased locations. Bacteria and viruses have evolved to the point that they can efficiently target specific types of cells using processes such as tropism, which acts as cargo to transport medicines in natural vesicles. It will thoroughly examine the views, prospective research frontiers, and remaining obstacles in this multidisciplinary field to give insights into the "swallowing the surgeon" concept.[41]

19.5.3.8 BIONANOHYBRIDS

Bionanomaterials in nanohybrids serve as layered doubly-hydroxy nanomaterials, which are stable on heat and biocompatible depending on their composition. In layered double hydroxy materials, the drug is intercalated between layers and hits a specific target. When layered double hydroxy material reaches its target, it starts releasing the drug to cure the disease. To treat influenza, A virus H1N1 strain is used. The vaccines contain aluminum oxide to improve immunogenicity.[24]

It is commonly recognized that solid interactions between solid adjuvants and antigens can cause structural changes in antigens, reducing their immunogenic potential. As a result, novel antigen carriers that can modulate these interfacial interactions are needed. Nanostructured biohybrids are intriguing candidates for addressing these difficulties in this context.[24] Previously, clay mineral sepiolite, microfibrous silicate in nature, had been treated with polysaccharides such as xanthan gum, causing stabilization of the influenza-A virus and retention of their immunogenicity. It has been serving as an intriguing component for bionanocomposites and bionanohybrids because of its low cost and reputation for biosafety and biocompatibility.

Furthermore, the surface properties of such material make it highly adaptable to alteration toward biomimetic interfaces. As a result, bioinspired lipid membranes made from organic solvents were produced on sepiolite and used as enzyme support. Sepiolite-lipid bionanohybrids produced from the conjugation of sepiolite with the phosphatidylcholine liposome were testified to the viral particle nanocarriers. Furthermore, the broadened range of acceptable nanocarriers included biocompatible Mg/Al-layered doubly-hydroxides, which were employed as medication delivery and gene vectors. Their inclusion has been promoted based on their similarity index with alum-based nanogels concerning their structure and composition. Moreover, they serve as the most widely used adjuvant in vaccines along with the modified surface exhibited by a biomimetic lipid membrane.[32]

19.6 APPLICATIONS OF CATALYSIS AND SURFACE PROCESS UNDERGONE BY BIONANOMATERIALS

19.6.1 CATALYTIC REDUCTION

The nitrogen oxides (NOx) are catalytically reduced by ammonia under a specific catalyst and transformed into their essential components in molecular form, oxygen, and nitrogen. Hence, under strict consideration, it serves as the best treatment technique to make the flue gas free from NOx while rendering its influence on emission-control devices, including submicronic particle formation. Since submicronic particle formation comes out to be a well-known problem associated with its applicability for NOx reduction, owing to ammonia slip (NH_3 slip), the submicronic particles formed in this technique can relate to ammonia and ammonium sulfate. Fine particles can enhance space charge load and alter an electrostatic precipitator (ESP) voltage–current properties. Particle cohesion is caused by styrene derivatives which can lead to air preheater fouling and clogging. Cohesion, on the other hand, improves ESP efficiency by increasing particle size and decreasing re-entrainment.[16]

19.6.1.1 HYDROGENATION

Since the turn of the century, large-scale hydrogenation of fats and oils has been employed to generate hard fat from various categories of fats and oils. In this process, the hydrogen atoms are introduced to an unsaturated species, such as triglycerides. It produces a more significant amount of saturated fat with a higher melting point value than its component species. Although the hydrogen gas interacts with liquid oil under an appropriate catalyst at high pressures and temperatures, it occurs in batches. Since it is an exothermic process that raises the batch temperature, it must regulate through cooling. The whole procedure goes on repetition unless the product of the desired characteristics is obtained. It is followed by the filtration used to recover the catalyst. In addition to saturation, double bond isomerization occurs during this process. There will be two categories of isomerization reactions: position and geometric isomerization. Because of the inclusion of various stages, the entire reaction procedure has been found highly intricate and still partially understood with gaps in their kinetic characterization. Due to the adverse effects of trans-fatty acids on the blood composition, the food industries have almost excluded trans-fatty acid-containing components from their fat phase composition, leading to a significant decline in solid fats. The complete hydrogenation of virtually all the unsaturated content decreases the percentage of trans-fatty acids to a satisfactory level producing an extraordinarily high melting product, which may utilize once it has been interesterified with liquid oil.[48]

19.6.1.2 CATALYTIC REDUCTION OF TRANSITION METALS

In the infield of green chemistry, TiO_2 is used as a source of photocatalytic degradation in solar cells. The bandgap between TiO_2 NPs lies between 3.0 and 3.4 eV, which will make it absorb mainly the UV light. This wide-band gap made it suitable to absorb UV light and conducting photochemical reactions. The primary benefit of TiO_2 NPs is that it is cost-effective and biocompatible over other semiconductor nanomaterials. These particles are used to generate hydrogen and oxygen from water as fuel and for CO_2 reduction. Later on, metal-doped and non-metal-doped, such as Nd, N, and C, TiO_2 nanomaterials were developed to enhance their photocatalytic performance.[16] TiO_2-doped photoactivity depends on many factors such as dopant concentration, the dopants' energy level, and the dopants' distribution, which have gained a lot of interest. It has been found that metal doping of TiO_2 shows mid-gap states by changing valence bonds and narrow bandgap. The Nd^{3+}-doped TiO_2 NPs band gap was decreased to 0.55 eV because it created new states. It can narrow the bandgap by doping with other nonmetals.[48]

Metal chalcogens such as CdS and CdSe can break down water into H_2 and O_2. However, due to the presence of S^{2-} and Se^{2-} anions, they are not stable in water because they are oxidized before water molecules. Oxynitrides and oxysulfides are used as photocatalysts. It is observed that $SrTiO_3$ NPs can be modified easily with Ru NPs as co-catalyst to enhance proton reduction ability. Pt and Ru can be used as co-catalyst for H_2 evolution, and Ru-loaded ZrO_2/TaON shows less activity in the presence of the IO^{3-}/I^- aqueous system. As for size decrease, it shows much activity. The enzyme can be used as a catalyst in photochemical reactions. Enzymes can only perform an only certain types of chemical reactions in whit it shows effective efficiency. For example, in CO deficiency, the reaction is highly competitive for the preparation of H_2, as they have similar redox oxidation potential, while most photocatalysts face product selection challenges. But enzymes can counteract this problem. Also, consider the exciting topic of carbon dioxide reduction, for example. Numerous studies have been done on enzymes, and people are trying to create a more efficient system using enzymes to reduce carbon dioxide, including linking photocatalysts with enzymes. However, in maximal reaction to renewable energy conversion, enzyme performance needs to be improved in addition to stabilization and moderate reaction, in which enzyme use is limited.[29]

19.6.1.3 CATALYTIC DEGRADATION OF DYES

Recently, synthetic dyes have been used in several fields such as printing cloth and dying, pulp treatment, paper, and leather. Soil and water bodies received a considerable amount of synthetic dyes from industrial effluents. So, contamination of water due to these chemical effluents from different industries has become a severe environmental protection concern. Hence, it is necessary to treat wastewater containing different dyes. Several methods have been used to treat the dyes contaminated wastewater. However, the catalytic reduction of synthetic dyes using metallic NPs has become the best approach.[5]

Metallic NPs such as Ag, Au, Pd, and Pt have been used for the catalytic degradation of synthetic dyes. The NPs can act as a center between the dyes and reducing agents that have electrons. The NPs get electrons from the reducing agent such as borohydride ions. The NPs have a negative charge due to stored electrons and act as nano-electrodes and transfer these electrons to dyes for degradation. The NPs coagulate very fast due to extraordinary surface energy, and as a result, the active sites for the adsorption of reactants have been decreased. Hence, the catalytic activity of naked NPs has been decreased. Different supporting material has been used to overcome this drawback.[23]

Supporting materials such as microgels, dendrimers, polymers, and micelles have been used to stabilize NPs. Microgels are the best stabilizing material for the NPs due to their sensitivity, sieve structure, and stability, which increase their catalytic activity.[23]

19.6.2 CATALYTIC OXIDATION

Metal oxide NPs are synthesized biological techniques using sources such as micro-organisms and plants. This method is an emerging method because it is clean, non-toxic, and environmentally friendly. Copper oxide (CuO) particles are synthesized by bacteria, fungi, actinomycetes, yeast, and plants. Streptomyces *Escherichia coli*, fungi Penicillium aurantiogriseum, Penicillium citrinum, Penicillium waksmanii, Morganella were used as a source of CuO NPs. The conversion of aldehydes to carboxylic acid is an essential chemical reaction. Many oxidation reactions were performed using varieties of oxidizing agents like chromic acid, potassium permanganate, nitric acid, bromine, and Jone's reagent. Hydrogen peroxide (H_2O_2) is considered environmentally friendly and used in industry and laboratories on a large scale. Oxidation of aldehydes to acids in the presence of CuO NPs is very simple, eco-friendly, and efficient.[42]

19.6.3 COUPLING REACTIONS

A combination of inorganics and biomolecules develops nanostructures. Such methods cause bionanomaterials for catalysis, biosensor, and batteries. Therefore, by understanding these interventions, it may be possible to design biomaterials with better performance. The PL_4 peptide was isolated using the palladium-compatible phase display technique. Using this sequence, palladium particles, approximately 1.9 nm in diameter, were active in the reaction of steel to water, and were prepared at room temperature.[4]

It was understood that histidine residues on different positions like 6 and 11 were responsible for bonding like structure kinked form and represents palladium surface. To get better results, histidine can be replaced by aniline. Couplings of different schemes formed different products. These peptides were synthesized by dissolving K_2PdCl_4 for 30 min by adding $NaBH_4$.[12]

19.6.4 ARTIFICIAL ENZYME CATALYSIS

Artificial enzymes for site-selective DNA scission are also valuable tools in bioengineering. They are made by fusing the Ce (IV)/iminodiacetate complex and DNA oligomers. The Ce (IV) complex hydrolyzes the target phosphodiester linkage when it is put near it. The length and sequence of the DNA oligomer may be arbitrarily changed, allowing the sequence specificity to be enhanced at a whim. Thus, these artificial enzymes can alter the massive DNA of higher plants and animals.[9,12]

19.6.5 ELECTROCATALYSIS

Electrocatalysis is the heterogeneous catalysis of electrochemical processes at the electrode-electrolyte interface, and the electrode serves as both an electron donor/acceptor and a catalyst. The study of the connection between the physicochemical characteristics of electrode materials and the process and pace of electrode reactions is known as electrocatalysis. Electrocatalysis is concerned with minimizing electrode overpotential, determined by the electrode material, products, reactants, and intermediates at the electrode-electrolyte interface. Electrocatalysis aims to increase the reaction rate, that is the generated electrical current, by lowering the activation energy of the target process. These can be reduced by forming advantageous bonds between the chemical and the solid electrode or by altering the reaction pathways.[3,46]

19.6.6 PHOTOCATALYSIS

Photocatalysis is described as the "acceleration of a photo-generated electron in the presence of a catalyst," when the catalyst neither changes nor is consumed in the process. Because the chemical process happens in more than one medium, it is known as heterogeneous photocatalysis. When a semiconductor surface is exposed to light, it produces an electron-hole pair. The energy of the infringing photon should be larger than or equal to the bandgap of the semiconductor material. The electron excites to the conduction band from the valence band, leaving a valence band.[16,34,48]

19.6.7 STEAM REFORMING REACTIONS

CLSR employs an oxygen transfer material, also defined as an oxygen carrier, to drive the reactions in a cyclic process, supplying the heat required by the oxidation reactions feeding air to the reactor while avoiding the use of external burners to heat the reformer and the expenses involved with the use of highly corrosion-resistant materials. Steam reforming is an established technique. Nowadays, steam reforming of natural gas is the primary method of producing hydrogen. Methanol, liquefied petroleum gas, naphtha, jet fuel, and diesel are other nonrenewable feedstocks for hydrogen generation by steam reforming. The optimum feedstock selection should consider the process's global economy; feedstock pricing is essential, and a sufficient operation is critical. Coking deactivates catalysts, and sulfur in feed can reduce hydrogen generation in steam reforming processes. Extensive research is being conducted in this sector to produce more stable and cost-effective catalysts. New methods are being created as a result of the high production of hydrogen by steam reforming and various disadvantages resulting from the complexity of the processes, including a large number of steps, thermodynamic limitations, and low energy efficiency, among others.[46]

19.7 ROLE OF BIONANOMATERIALS IN THE FIELD OF CATALYSIS

19.7.1 BIOMATERIALS AS NANOBIOSENSOR

Sensing is a fundamental property in bionanomaterials and has too much importance on Vivo and ex vivo level. In different areas have many applications such as screening food quality, pharmaceutical diagnostics environmental field. The significant component in biosensing is the transduction mechanism which converts responses of bio analyte interactions. It converts specific biochemical reaction energy to electrical energy by transduction mechanism. As nanomaterials have a high surface area to volume ratio, they perform a better role in the function of biomaterials. The field of nanotechnology has revolutionized molecular biology, which has provided the state for manipulating molecules and atoms and monitoring biological phenomena at the physiological level. The term "biosensor" can be defined as a sensing device designed to estimate materials by using biological interactions and assessing these interactions to readable form by electrochemical and transduction interpretations. The bioreceptor in the biosensor is a part that serves as a template for detection. The antibodies are detected using antigen, and the corresponding selective substrate detects proteins. The transducer covers one form of energy to another. The first form is biochemical nature, while the second is electrical. The first form is formed by the interaction between bio analyte and bio receptor, while the second is by electrical interaction. This biotechnology is affected by varying pH, temperature, contaminants, and other physicochemical variations.[13]

19.7.2 BIOMATERIALS AS BIONANOZYMES

Nowadays, scientists are working in the field of solid-supported catalysts to check whether matrices can speed up the catalytic behavior of nano-sized mimics. Materials filled with NPs show better performance and enhance catalytic activities, and improve stabilities. Iron (Fe^{+3}) porphyrin was combined with Zr^{+4} ions by Ling and Gao's group. Then further were combined with ultrasmall platinum Pt to get better results. In this way, nanocomposite shows property and stability such as peroxidase. The CeO_2 NPs were encapsulated on a porous carbonaceous framework which causes a high degree of very small, well dispersed, and stable cerium oxide NPs. The glucose oxidase selenium (GO–Se) nanocomposite could serve as mimic to remove dangerous H_2O_2 with the assistance of glutathione. Carbon nanotubes are used to support for enhancing the stability and activity of nanomaterial base artificial enzymes.[31]

19.7.3 BIOMATERIALS AS BIONANOCATALYSTS

With the rise in environmental consciousness and the depletion of petroleum supplies, biodiesel, classified as monoalkyl esters of fatty acids, has been the subject of extensive recent study as an alternative sustainable fuel. Many nations, including Germany, France, Italy, the United States, and Japan, are increasing their usage of biodiesel.[41]

19.7.4 MISCELLANEOUS

The bio or biomimetic catalysts on carbon nanomaterials can provide a pathway for catalytic enhancement having extra stable properties. In a nanostructured multi catalyst system, carbon on a mesoporous level can serve as support. Graphene-based and porous-based silica hybrid materials are used as a catalyst for artificial modulation. Gold-based NPs Au NPs are mixed in different places on graphene mesoporous hybrid materials based on silica. Bioinspired nanomaterials have worked in plants, biomass, and micro-organisms.[15] Bioinspired nanomaterials are divided into the following two types: (1) Water-soluble DNA proteins extracted from plants and micro-organisms and (2) biometrics with sizes ranging from nanometers to micrometers. Biometrics plays a role between metal NPs and bulk materials. Such materials can play a role in support. As for the concern with soluble metal NPs solubility, the metal NPs are transported from one place to another on support.[40]

19.8 TOXICITY PERSPECTIVES OF BIONANOMATERIALS

Nanobiomaterials are released in daily life in the environment. The risk of their release is their interaction with the environment. In nano-bio interactions, the nanomaterials interact closely and dynamically with biotic and abiotic factors. The ingredients of an abiotic environment such as solar radiation and dissolved organic materials have been studied. The generation of reactive oxygen species which enhances pro-inflammatory responses is responsible for toxicity. The effect of nanomaterials on cells of the human body is clarified in the form of quantum dots. Silica NPs enter the human body nucleoplasm, which forms thin dots in the cytoplasm. Gold nanoparticles interact with DNA and cause cell death.[14] The effect of toxicity in nanoparticles is studied based on the size of agglomerates and original nanoparticles size. Rats were exposed to nanoparticles with a size range of 20–250 nm TiO_2; the alveolar clearance of polystyrene halftime stands linearly with the surface area of TiO_2 NPs. A surface impurity is caused by water or air pollution such as bacterial endotoxin, especially in immunological responses. Residual materials such as surfactants and transition metals interacted with by bionanomaterials also affect the environment. Protein corona such as serum albumin or calcium formed by adsorption protein adsorption on metal NPs and CNTs. The albumin adsorbed on CNTs was involved in phagocytosis of the macrophage via a scavenger receptor. The nanomaterials are nondegradable and accumulate into cells and affect the degradation products.[36]

19.9 FUTURE ENDEAVORS

The biosphere provides us things in excess that fulfill the needs of life. These are proteins, sugars, glucose, peptides, and biopolymers. These have been studied for nearly 500 years, but their interaction with nanomaterials has made recent progress. These days bionanomaterials play a vital role in biomedical sciences such as tissue engineering, and gene–drug delivery with NPs derived from biotic and green sources as NPs are nontoxic and have wide applications. So, the gold NPs are found efficient and stabilizing agents for many biochemical reactions. Biopolymers and their composite are essential, but carbohydrates and derivatives have played a vital role in business. So, after all, the discussion, it is very clear that plant-based nanomaterials have too many applications for the future. However, some challenges include easy availability, purification, mass production, and utilization without any matrices. Chemists and biologists are working day and night for challenges. So, in one sentence, it can be said there is a need for technology and innovation in the field of bionanomaterials. We hope in future days, bionanomaterials would be available to all easily.[29,46]

19.10 SUMMARY

The field of catalysis dealing with the enhancement in the reaction output regardless of its consumption during the process has been receiving a lot of attention. Catalysis has emerged as a critical strategy for addressing many of today's issues. Nearly all manufacturing chemicals, from batteries to refineries, have used catalysts in various forms to enhance economic viability. Meanwhile, environmental concerns have prompted the creation of new catalysts capable of catalyzing desired biochemical reactions. Catalysis is widely used for effective energy conversion, fuel generation, chemical synthesis, and atmospheric mitigation. Nanotechnology has been serving as an essential engineering technique for catalysis-based purposes. Due to the significant characteristics such as regular shape and distinct structure with improved surface area, NPs serve as an optimum platform for catalyzing desired biochemical reactions with enhanced catalytic activity. The present limitation to the broad environmental and industrial impact is the control over the synthetic procedure of the nanocrystals demanding high energy surfaces and knowledge of the structure-performance interconnection. Through enhanced characterization techniques and modeling, multidisciplinary developments in chemistry, physics, and materials science have played a significant role in catalyst production, fabricated structure and composition, and mechanistic rationale with high-order nanoscale accuracy.

ACKNOWLEDGMENT

We pay special thanks to Mr. Kashif Javed, Department of Information Technology, Masood Textile Mills Faisalabad, Punjab, Pakistan, for his contribution and for providing all the technical assistance.

KEYWORDS

- **bionanocatalysts**
- **bionanomaterials**
- **bionanotechnology**
- **catalysis**
- **catalytic reduction**
- **surface process**

REFERENCES

1. Ahmad, S. Engineered Nanomaterials for Drug and Gene Delivery—A Review. *J. Nanopharma. Drug Deliv.* **2016,** *3* (1), 1–50.
2. Almeida-Marrero, V.; Van De Winckel, E.; Anaya-Plaza, E.; Torres, T.; De La Escosura, A. Porphyrinoid Biohybrid Materials As An Emerging Toolbox For Biomedical Light Management. *Chem. Soc. Rev.* **2018,** *47* (19), 7369–7400.
3. Astruc, D. Introduction: Nanoparticles in Catalysis. *Chem. Rev.* **2020,** *120* (2), 461–463.
4. Care, A.; Bergquist, P. L.; Sunna, A. Solid-Binding Peptides: Smart Tools for Nanobiotechnology. *Trends Biotechnol.* **2015,** *33* (5), 259–268.
5. Chowdhary, P.; Bharagava, R. N.; Mishra, S.; Khan, N. Role of Industries in Water Scarcity and Its Adverse Effects on Environment and Human Health. In *Environmental Concerns and Sustainable Development*; Shukla, V., Kumar, N., Eds.; Springer International Publishing: Basel, Singapore, 2020; pp 235–256.
6. Chueng, S. T. D.; Yang, L.; Zhang, Y.; Lee, K. B. Multidimensional Nanomaterials for the Control of Stem Cell Fate. *Nano Convergence* **2016,** *3* (1), 1–15.
7. Chugh, H.; Sood, D.; Chandra, I.; Tomar, V.; Dhawan, G.; Chandra, R. Role of Gold and Silver Nanoparticles in Cancer Nano-Medicine. *Artif. Cells Nanomed. Biotechnol.* **2018,** *46* (1), 1210–1220.
8. Eivazzadeh-Keihan, R.; Radinekiyan, F.; Aliabadi, H. A. M.; Sukhtezari, S.; Tahmasebi, B.; Maleki, A.; Madanchi, H. Chitosan Hydrogel/Silk Fibroin/$Mg(OH)_2$ Nanobiocomposite as a Novel Scaffold with Antimicrobial Activity and Improved Mechanical Properties. *Sci. Rep.* **2021,** *11* (1), 1–13.
9. Fiorenza, R. Bimetallic Catalysts for Volatile Organic Compound Oxidation. *Catalysts* **2020,** *10* (6), 661.
10. Gao, C.; Zhang, L.; Wang, J.; Jin, M.; Tang, Q.; Chen, Z.; Zhao, G. Electrospun Nanofibers Promote Wound Healing: Theories, Techniques, and Perspectives. *J. Mater. Chem. B* **2021,** *9* (14), 3106–3130.
11. Ghazy, E.; Rahdar, A.; Barani, M.; Kyzas, G. Z. Nanomaterials for Parkinson Disease: Recent Progress. *J. Mol. Struct.* **2021,** *1231*, 1–8.
12. Ghith, A.; Ismail, N. S.; Youssef, K.; Abouzid, K. A. Medicinal Attributes of Thienopyrimidine Based Scaffold Targeting Tyrosine Kinases and their Potential Anticancer Activities. *Archiv Der Pharmazie*, **2017,** *350* (11), 1700242.

13. Hashem, A.; Hossain, M. A. M.; Marlinda, A. R.; Almamun, M.; Simarani, K.; Johana, M. R. Nanomaterials Based Electrochemical Nucleic Acid Biosensors for Environmental Monitoring: A Review. *Appl. Surf. Sci. Adv.* **2021,** *4,* 1–19.
14. He, X.; Fu, P.; Aker, W. G.; Hwang, H. M. Toxicity of Engineered Nanomaterials Mediated by Nano–Bio–Eco Interactions. *J. Environ. Sci. Health C* **2018,** *36* (1), 21–42.
15. Huang, Y.; Yu, D.; Qiu, Y.; Chu, L.; Lin, Y. The Role of Nanomaterials in Modulating the Structure and Function of Biomimetic Catalysts. *Front. Chem.* **2020,** *8* (764) 1–6.
16. Ismael, M. A Review and Recent Advances in Solar-to-Hydrogen Energy Conversion Based on Photocatalytic Water Splitting Over Doped-TiO_2 Nanoparticles. *Solar Energy* **2020,** *211*, 522–546.
17. Kafi, M. A.; Cho, H. Y.; Choi, J. W. Engineered Peptide-Based Nanobiomaterials for Electrochemical Cell Chip. *Nano Convergence* **2016,** *3* (1), 1–12.
18. Kamiuchi, N.; Sun, K.; Aso, R.; Tane, M.; Tamaoka, T.; Yoshida, H.; Takeda, S. Self-Activated Surface Dynamics in Gold Catalysts Under Reaction Environments. *Nat. Commun.* **2018,** *9* (1), 1–6.
19. Karak, N. Fundamentals of Nanomaterials and Polymer Nanocomposites. In *Nanomaterials and Polymer Nanocomposites*; Elsevier Inc.: Amsterdam, Netherlands, 2019; pp 1–45.
20. Kaur, M.; Mehta, A.; Bhardwaj, K. K.; Gupta, R. Bionanomaterials from Agricultural Wastes. In *Green Nanomaterials*; Series: Advanced Structured Materials; Ahmed, S.; Ali, W., Eds., Vol. 126; Springer International Publishing: Basel, Singapore, 2020; pp 243–260.
21. Kornienko, N.; Zhang, J. Z.; Sakimoto, K. K.; Yang, P.; Reisner, E. Interfacing Nature's Catalytic Machinery with Synthetic Materials for Semi-Artificial Photosynthesis. *Nat. Nanotechnol.* **2018,** *13* (10), 890–899.
22. Li, J.; Wang, Y.; Jin, X.; Wang, Y.; Li, H. Poly (Amic Acid) Salt-Mediated Palladium and Platinum Nanoparticles as Highly Active and Recyclable Catalysts for Hydrogenation of Nitroarenes in Water Under Ambient Conditions. *Appl. Organometall. Chem.* **2019,** *33* (2), e4717.
23. Messih, M. A.; Ahmed, M. A.; Soltan, A.; Anis, S. S. Synthesis and Characterization of Novel Ag/ZnO Nanoparticles for Photocatalytic Degradation of Methylene Blue Under UV and Solar Irradiation. *J. Phys. Chem. Solids* **2019,** *135*, 1–13.
24. Mishra, G.; Dash, B.; Pandey, S. Layered Double Hydroxides: A Brief Review from Fundamentals to Application as Evolving Biomaterials. *Appl. Clay Sci.* **2018,** *153*, 172–186.
25. Moon, R. J.; Martini, A.; Nairn, J.; Simonsen, J.; Youngblood, J. Cellulose Nanomaterials Review: Structure, Properties and Nanocomposites. *Chem. Soc. Rev.* **2011,** *40* (7), 3941–3994.
26. Munir, S.; Shah, A. A.; Rahman, H.; Bilal, M.; Rajoka, M. S. R.; Khan, A. A.; Khurshid, M. Nanozymes for Medical Biotechnology and its Potential Applications in Biosensing and Nanotherapeutics. *Biotechnol. Lett.* **2020,** *42* (3), 357–373.
27. Nagamune, T. Biomolecular Engineering for Nanobio/Bionanotechnology. *Nano Convergence,* **2017,** *4* (1), 1–56.
28. Nasir, M.; Nawaz, M. H.; Latif, U.; Yaqub, M.; Hayat, A.; Rahim, A. An Overview on Enzyme-Mimicking Nanomaterials for Use in Electrochemical and Optical Assays. *Microchim. Acta* **2017,** *184* (2), 323–342.

29. Nelson, A. W. *Chemical Transformations of Nanocrystals: Structure, Surface Chemistry, and Diffusion*. MSc Thesis; Cornell University, Ithaca, NY, 2019; p 186.
30. Noman, M. T.; Ashraf, M. A.; Ali, A. Synthesis and Applications of Nano-TiO_2: A Review. *Environ. Sci. Pollution Res.* **2019,** *26* (4), 3262–3291.
31. *Nyoka, M.; Choonara, Y. E.; Kumar, P.; Kondiah, P. P. D.; Pillay, V. Synthesis of Cerium Oxide Nanoparticles using various Methods: Implications for Biomedical Applications. Nanomaterials* **2020,** *10* (2), 242–263.
32. Orta, M. M.; Martin, J.; Santos, J. L.; Aparicio, I.; Medina-Carrasco, S.; Alonso, E. Biopolymer-Clay Nanocomposites as Novel and Ecofriendly Adsorbents for Environmental Remediation. *Appl. Clay Sci.* **2020,** *198,* 1–19.
33. Palys, M. J.; Wang, H.; Zhang, Q.; Daoutidis, P. Renewable Ammonia for Sustainable Energy and Agriculture: Vision and Systems Engineering Opportunities. *Curr. Opin. Chem. Eng.* **2021,** *31*, 100667.
34. Pei, X.; Deng, Y.; Li, Y.; Huang, Y.; Yuan, K.; Lee, J. F. Zhang, L. Size-Controllable Ultrafine Palladium Nanoparticles Immobilized on Calcined Chitin Microspheres as Efficient and Recyclable Catalysts for Hydrogenation. *Nanoscale* **2018,** *10* (30), 14719–14725.
35. Peijnenburg, W. J.; Baalousha, M.; Chen, J.; Chaudry, Q.; Von Der Kammer, F.; Kuhlbusch, T. A.; Koelmans, A. A. A Review of the Properties and Processes Determining the Fate of Engineered Nanomaterials in the Aquatic Environment. *Crit. Rev. Environ. Sci. Technol.* **2015,** *45* (19), 2084–2134.
36. Prajitha, N.; Athira, S. S.; Mohanan, P. V. Bio-Interactions and Risks of Engineered Nanoparticles. *Environ. Res.* **2019,** *172*, 98–108.
37. Prasad, C.; Yang, X.; Liu, Q.; Tang, H.; Rammohan, A.; Zulfiqar, S.; Shah, S. Recent Advances MXenes in Supported Semiconductors Based Photocatalysts: Properties, Synthesis and Photocatalytic Applications. *J. Ind. Eng. Chem.* **2020,** *85*, 1–33.
38. Rezaie, H. R.; Bakhtiari, L.; Ochsner, A., Eds. *Biomaterials and Their Applications*; Springer International Publishing: Berlin, Germany, 2015; p 49.
39. Rouge, J. L.; Sita, T. L.; Hao, L.; Kouri, F. M.; Briley, W. E.; Stegh, A. H.; Mirkin, C. A. Ribozyme–Spherical Nucleic Acids. *Journal of the American Chemical Society,* **2015,** *137* (33), 10528–10531.
40. Sangeetha, J.; Thangadurai, D.; Hospet, R.; Purushotham, P.; Manowade, K. R.; Mujeeb, M. A.; Harish, E. R. Production of Bionanomaterials from Agricultural Wastes. In *Nanotechnology*; Springer International Publishing: Berlin, Germany, 2017; pp 33–58.
41. Sason, H.; Shamay, Y. Nanoinformatics in Drug Delivery. *Israel J. Chem.* **2020,** *60* (12), 1108–1117.
42. Shah, M.; Fawcett, D.; Sharma, S.; Tripathy, S. K.; Poinern, G. E. J. Green Synthesis of Metallic Nanoparticles via Biological Entities. *Materials* **2015,** *8* (11), 7278–7308.
43. Shen, M.; Forghani, F.; Kong, X.; Liu, D.; Ye, X.; Chen, S.; Ding, T. Antibacterial Applications of Metal-Organic Frameworks and their Composites. *Compreh. Rev. Food Sci. Food Saf.* **2020,** *19* (4), 1397–1419.
44. Sun, H.; Zu, Y. A Highlight of Recent Advances in Aptamer Technology and its Application. *Molecules* **2015,** *20* (7), 11959–11980.
45. Ten Hove, J. B.; Schijven, L. M.; Wang, J.; Velders, A. H. Size-Controlled and Water-Soluble Gold Nanoparticles using UV-Induced Ligand Exchange and Phase Transfer. *Chem. Commun.* **2018,** *54* (95), 13355–13358.

46. Trunschke, A.; Bellini, G.; Boniface, M.; Carey, S. J.; Dong, J.; Erdem, E.; Wrabetz, S. Towards Experimental Handbooks in Catalysis. *Topics Cataly.* **2020,** *63* (19), 1683–1699.
47. Varma, R. S. Biomass-Derived Renewable Carbonaceous Materials for Sustainable Chemical and Environmental Applications. *ACS Sustain. Chem. Eng.* **2019,** *7* (7), 6458–6470.
48. Wang, S.; Ding, Z.; Chang, X.; Xu, J.; Wang, D. H. Modified Nano-TiO_2 Based Composites for Environmental Photocatalytic Applications. *Catalysts* **2020,** *10* (7), 759–797.
49. Zhang, Z.; Huang, Z.; Yuan, H. Direct Conversion of Cellulose to Ethyl Levulinate Catalysed by Modified Fibrous Mesoporous Silica Nanospheres in a Co-Solvent System. *New J. Chem.* **2021,** *45* (12), 5526–5539.

INDEX

E

P

R

S

T

U

V

X